	IB	IIB	IIIA	IVA	VA	VIA	VIIA	VIIIA
								2 He 4.00260
			5 B 10.81	6 C 12.011	7 N 14.0067	8 O 15.9994	9 F 18.998403	10 Ne 20.179
			13 Al 26.98154	14 Si 28.0855	15 P 30.97376	16 S 32.06	17 Cl 35.453	18 Ar 39.948
28 Ni 58.69	29 Cu 63.546	30 Zn 65.38	31 Ga 69.72	32 Ge 72.59	33 As 74.9216	34 Se 78.96	35 Br 79.904	36 Kr 83.80
46 Pd 106.42	47 Ag 107.868	48 Cd 112.41	49 In 114.82	50 Sn 118.69	51 Sb 121.75	52 Te 127.60	53 I 126.9045	54 Xe 131.29
78 Pt 195.08	79 Au 196.9665	80 Hg 200.59	81 Tl 204.383	82 Pb 207.2	83 Bi 208.9804	84 Po (209)	85 At (210)	86 Rn (222)

Metals ← → Nonmetals

63 Eu 151.96	64 Gd 157.25	65 Tb 158.9254	66 Dy 162.50	67 Ho 164.9304	68 Er 167.26	69 Tm 168.9342	70 Yb 173.04	71 Lu 174.967
95 Am (243)	96 Cm (247)	97 Bk (247)	98 Cf (251)	99 Es (252)	100 Fm (257)	101 Md (258)	102 No (259)	103 Lr (260)

Introduction to Chemical Principles

Introduction to Chemical Principles

H. Stephen Stoker
Weber State College, Ogden, Utah

Macmillan Publishing Co., Inc.
NEW YORK
Collier Macmillan Publishers
LONDON

Front cover (clockwise from top): copper sulfate ($CuSO_4$), cobalt nitrate ($Co(NO_3)_2$), cobalt chloride ($CoCl_2$).

Back cover (clockwise from top right): nickel chloride ($NiCl_2$), manganese sulfate ($MnSO_4$), sodium carbonate (Na_2CO_3), potassium dichromate (KCr_2O_7), chromium sulfate ($CrSO_4$).

Printed in the United States of America.

Macmillan Publishing Co., Inc.
866 Third Avenue, New York, New York 10022

Collier Macmillan Canada, Inc.

Library of Congress Cataloging in Publication Data

Stoker, H. Stephen (Howard Stephen), Date
Introduction to chemical principles.

Includes index.
1. Chemistry. I. Title.
QD33.S86 540 82-6537
ISBN 0-02-417620-6 AACR2

Printing: 2 3 4 5 6 7 8 Year: 3 4 5 6 7 8 9 0

ISBN 0-02-417620-6

Introduction to Chemical Principles is a text intended for students who have had little or no previous instruction in chemistry or who have had such instruction long enough ago that a thorough review is needed. The text's purpose is to give students the background (and confidence) needed for a subsequent successful encounter with a main sequence college level general chemistry course.

Many texts written for preparatory chemistry courses are simply "watered-down" versions of general chemistry texts. Such texts treat almost all topics found in the general chemistry course but only at a superficial level. *Introduction to Chemical Principles* does not fit this mold. The author's philosophy is that it is better to treat extensively fewer topics and have the student understand those topics in greater depth. Topics treated in this text are those most necessary for a solid foundation upon which further chemistry courses can build.

Because of the varied degrees of understanding of chemical principles possessed by students taking a preparatory course, development of each topic in this text starts at "ground level" and continues step by step until the level of sophistication required for further courses is attained.

Problem solving is a major emphasis of this text. Years of teaching experience indicate to the author that student "troubles" in general chemistry courses are almost always centered in the student's inability to set up and solve problems.

Whenever possible, dimensional analysis is used in problem solving. This method, which requires no mathematics beyond arithmetic and elementary algebra, is a most powerful problem-solving tool with wide applicability. Most important, it is a method that an average student can master with an average amount of diligence. Mastering dimensional analysis also helps build the confidence that is so valuable for success in further chemistry courses.

Numerous Examples — worked-out problems — are found in the text. In each case the solution is given step by step, and detailed commentary accompanies each step in the worked-out solution. In addition, many types of problems are explored through more than one Example.

Working practice exercises is a necessary activity for students if they are to become proficient at problem solving. Consequently, an abundant selection of end-of-chapter Questions and Problems is provided. Most chapters offer at least forty exercises, many of which contain subparts. The Questions and Problems are grouped according to topic and arranged in order of chapter coverage. Thus, instructors can assign some problems before students have completed the chapter. This section-by-section arrangement of exercises also enables the student to test his or her understanding of each section before moving to another section. Answers to many of the Questions and Problems (all those whose numbers or letters appear in color) are given in the back of the text.

The availability of inexpensive electronic calculators has introduced into general chemistry courses a "new problem" that is treated in depth in this text. This "problem" relates to significant figures. Routinely, electronic calculators display answers that contain more digits than are needed. It is a mistake to record these extra digits, since they are meaningless — that is, they are not significant figures. In every Example students are reminded of this potential for error by the appearance of two answers after the sample problem has been solved: the calculator answer (which does not take into account significant figures) and (in color) the correct answer (which is the calculator answer adjusted to the correct number of significant figures).

Many learning aids are found in the text. The important skills and ideas that students need to absorb are summarized as Learning Objectives at the end of each chapter. There follows a list of new terms and concepts defined in the chapter. Throughout the text important terms and statements are highlighted. The key words are printed in boldface when defined and repeated in the margin (in color) for easy reference. Related topics are extensively cross-referenced with particular attention to informing students of previous sections needed as a foundation for understanding the topic currently under discussion.

More material than can be conveniently covered in one semester or quarter is present in the text. Thus, instructors will have a choice of chapters or sections to cover in class. The later chapters particularly lend themselves to selective assignment.

Three topics not traditionally included in preparatory chemistry texts are included in this one: (1) chemical calculations involving simultaneous reactions and series of consecutive reactions (Sec. 8.9), (2) quantum numbers (Secs. 9.6–9.8), and (3) molecular geometry using VSEPR theory (Sec. 10.12). Instructors should consider these subjects as optional topics that can provide additional insights for some students or that can be omitted without affecting overall topic coverage. All three optional topics are completely self-contained and do not serve as prerequisites for any material that follows.

Supporting materials, to assist both the student and the instructor, have been especially written for use with this text. For the student both a study guide and a laboratory manual (for courses that include a laboratory period) are available. Both of these items reinforce many of the concepts covered in the text. They also employ the same notation and methodology for problem solving as are used in the text. For the instructor a solutions manual is available.

Writing a text such as this is never accomplished without cooperation and contributions from many sources. The help of the following people who reviewed the text prior to its publication is acknowledged: Elliott L. Blinn, Bowling Green State University; Owen C. Gayley, San Antonio College; Ethelreda Laughlin, Cuyahoga Community College; Ruth Sherman, Los Angeles City College; and Linda N. Sweeting, Towson State University. The reviewers offered many valuable comments and suggestions that have been incorporated into the text. The help and prodding of Elisabeth Belfer of Macmillan Publishing Co. during the production stages of this project are also appreciated.

H. S. S.

Contents

11 States of Matter 274

12 Gas Laws 302

1 The Science of Chemistry

1.1 Chemistry — A Scientific Discipline

During the entire time of their existence on earth, human beings have been concerned with and fascinated by their surroundings. This desire to understand their surroundings — an attribute that distinguishes them from all other living organisms — has led them to accumulate vast amounts of information concerning themselves, their world, and the universe. **Science** *is the study in which humans attempt to organize and explain in a systematic and logical manner knowledge about themselves and their surroundings.*

science

The enormous range of types of information covered by science, the sheer amount of accumulated knowledge, and the limitations of human mental capacity relative to mastering such a large and diverse body of knowledge have led to the division of the whole of science into smaller subdivisions called scientific disciplines. **Scientific disciplines** *are branches of scientific knowledge limited in their size and scope to make them more manageable.* Chemistry is one of these disciplines. Astronomy, botany, geology, physics, and zoology are some of the other disciplines that have resulted from this substructuring of science.

scientific disciplines

In a sense, the boundaries between scientific disciplines are artificial, because all scientific disciplines borrow information and methods from each other. No scientific discipline is totally independent. This overlap requires that scientists, in addition to having in-depth knowledge of a selected discipline, also have limited knowledge of other disciplines. Problems scientists have encountered in the last decade have particularly pointed out the interdependence of disciplines. For example, chemists attempting to solve the problem of chemical contamination of the environment find that they need some knowledge of geology, zoology, and botany. Because of this overlap, it is now common to talk not only of chemists, but also of geochemists, biochemists, chemical physicists, etc. Figure 1-1 shows how chemistry merges with selected other scientific disciplines.

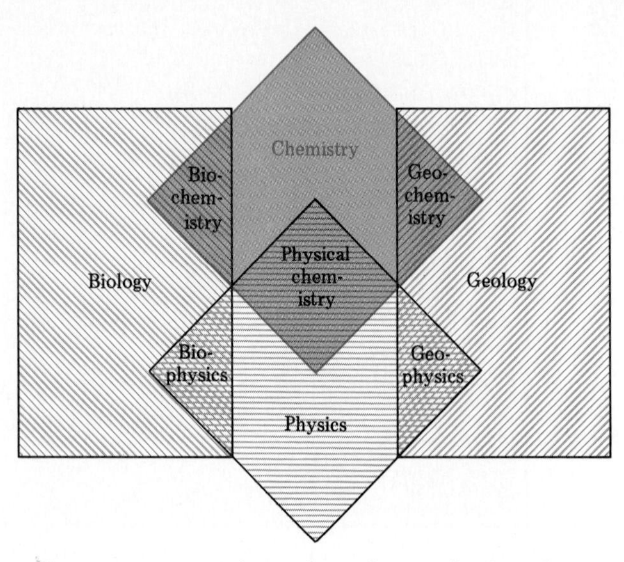

Figure 1-1. Interrelationship of scientific disciplines.

The overlap of the various scientific disciplines affects not only practicing scientists but also today's college students. It means that students must necessarily study in disciplines other than those of primary interest to them. The knowledge they gain from such studies is often useful and many times essential if they are to be competent in their chosen field. Many students in courses for which this textbook is written are studying chemistry because of its applicability to other disciplines in which they have a more specific interest.

1.2 The Scope of Chemistry

Although chemistry is concerned with only a part of the scientific knowledge that has been accumulated, it in itself is an enormous and broad field. Chemistry touches all parts of our lives because of its broadness in scope. There is no getting away from chemistry, nor should anyone try to avoid it.

Most of the clothes we wear are made from synthetic fibers produced by chemical processes. Even natural fibers, such as cotton or wool, are the products of naturally occurring chemical reactions within living systems. Our transportation usually involves vehicles powered with energy obtained by burning chemical mixtures, such as gasoline. Drugs used to cure many of our illnesses are the result of much chemical research. The paper on which this textbook is printed was created through a chemical process and the ink used in printing the words on each page is a mixture of many chemicals. The movies we watch are possible because of synthetic materials called film. The images on film are produced through interaction of selected chemicals. Almost all of our recreational pursuits involve objects containing materials produced by chemical industries. Skis, boats, basketballs, bowling balls, musical instruments, and television sets all contain materials that are not naturally occurring.

Our bodies are a complex mixture of chemicals. Principles of chemistry are fundamental to an understanding of all processes of the living state. Chemical secretions (hormones) produced within our bodies help determine our outward physical characteristics such as height, weight, and appearance. Digestion of food involves a complex series of chemical reactions. Food itself is an extremely complicated array of chemical substances. Chemical reactions govern our thought processes and how knowledge is stored in and retrieved from our brains. In short, chemistry runs our lives.

A formal course in chemistry can be a fascinating experience, because it helps us understand ourselves and our surroundings. One cannot truly understand or even know very much about the world we live in or about our own bodies without being conversant with the fundamental ideas of chemistry.

1.3 How Chemists Discover Things — The Scientific Method

Chemistry is an experimental science; that is, chemical discoveries are made as the result of experimentation. This feature, experimentation, is what distinguishes chemistry (and other sciences) from other types of intellectual activity.

A majority of the scientific and technological advances of the twentieth century are the result of systematic experimentation using a method of problem solving known as the scientific method. The **scientific method** *is a set of specific procedures for acquiring knowledge and explaining phenomena.* Procedural steps in this method are

scientific method

1. Collecting data through observation and experimentation.
2. Analyzing and organizing the data in terms of general statements (generalizations) that summarize the experimental observations.
3. Suggesting probable explanations for the generalizations.
4. Experimenting further to prove or disprove the proposed explanations.

Occasionally a great discovery is made by accident, but the majority of scientific discoveries are the result of the application of the above steps over long periods of time. There are no instantaneous steps in the scientific method; applying them requires considerable amounts of time.

There is a vocabulary associated with the scientific method and its use. This vocabulary includes the terms fact, law, hypothesis, and theory. Understanding the relationship between these terms is the key to a real understanding of how chemical knowledge has been and still is obtained.

The beginning step in the search for chemical knowledge is to identify a problem concerning some chemical system which needs study. After determining what other chemists have already learned concerning the selected problem, one may begin experimentation. New firsthand information is collected about the system via observation; that is, new facts about the system are obtained. A **fact** *is a valid observation about some natural phenomenon.* Facts are reproducible pieces of information. If a given experiment is repeated, under exactly the same conditions, the same facts should be obtained. All facts, to be acceptable, must be verifiable by anyone who has the time, means, and knowledge needed to repeat the experiments that led to their discovery.

fact

Next an effort is made to determine ways in which the facts about a given chemical system relate to each other and to those known for similar chemical systems. Quite often repeating patterns become apparent among the collected facts. These patterns lead to generalizations, called laws, about how the chemical systems of concern behave under specific conditions. A **law** *is a generalization that summarizes in a concise way facts about natural phenomena.*

law

It should not be assumed that laws are easy to discover. Often, years and years of work and thousands upon thousands of facts are needed before the true relationships between variables in the area under study become apparent.

A law is a description of what happens in a given type of experiment. There is no mention in a law as to why what happens does happen. It simply summarizes experimental observations without attempting to clarify why.

A law may be expressed as a verbal statement or as a mathematical equation. An example of a verbally stated law is "If hot and cold pieces of metal are placed in contact with each other, the hot piece always cools off and the cold piece always warms up."

It is important to distinguish between the use of the word law in science and its use in a societal context. Scientific laws are *discovered* by research. Researchers have *no control* over what the laws turn out to be. Societal laws, which are designed to control aspects of human behavior, are *arbitrary conventions* agreed upon (in a democracy) by the majority of those to whom the laws apply. Such laws *can be* and are *changed* when necessary. For example, the speed limit on a particular highway (a societal law) may be decreased or increased due to various safety and/or political conditions.

Chemists, and scientists in general, are not content with just knowing about natural laws. They want to know why a certain type of observation is always made. Thus, after a law is discovered, plausible tentative explanations of the behavior encompassed by the law are worked out by scientists. Such explanations are called hypotheses. A **hypothesis** *is a tentative model or picture that offers an explanation for a law.*

hypothesis

Once a hypothesis has been proposed, experimentation begins again. Many, many more experiments, under varied conditions, are run to test the reliability of the proposed explanation. The hypothesis must be able to predict the outcome of as yet untried experiments. The validity of the hypothesis is dependent upon its predictions being true.

It is much easier to disprove a hypothesis than it is to prove it. A negative result from an experiment indicates that the hypothesis is not valid as formulated and that it must be modified. Obtaining positive results supports the hypothesis, but doesn't definitely prove it. There is always the chance that someone will carry out a new type of experiment, never before thought of, which disproves the hypothesis.

As further experimentation continues to validate the concepts of a hypothesis, its acceptance in scientific circles increases. If after extensive testing the reliability of a hypothesis is still very high, confidence in it increases to the extent of its acceptance by the scientific community at large. After further lapse of time and additional accumulation of positive support, the hypothesis assumes the status of a theory. A **theory** *is a hypothesis that has been tested and validated over long periods of time.* The dividing line between a hypothesis and a theory is arbitrary and cannot be precisely defined. There is not a set number of supportive experiments that must be performed to give a hypothesis theory status.

theory

Theories serve two important purposes: (1) They allow scientists to predict what will happen in experiments that have not yet been run, and (2) they simplify the very real problem of being able to remember all the scientific facts that have already been discovered.

Even theories often must undergo modification. As scientific tools, particularly instrumentation, become more sophisticated, there is an increasing probability that

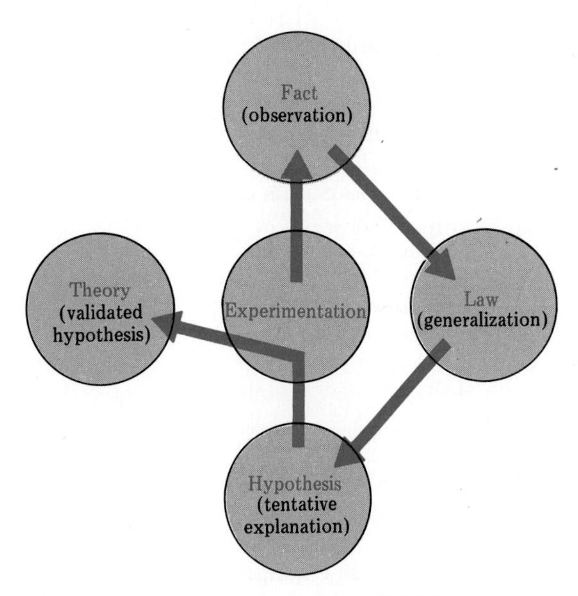

Figure 1-2. The steps in the scientific method.

some experimental observation will not be consistent with all aspects of a given theory. A theory affected in this manner must undergo modification to accommodate the new results or be restated in such a way that scientists know where it is useful and where it is not. All "truths" (theories) in science are provisional and subject to change in light of new experimental observations. Any science is like a living organism. It continues to grow and change. Science develops through a constant interplay between theory and experimentation.

The term theory is often misused by nonscientists in everyday contexts. "I have a theory that such and such is the case" is a comment often heard. The word theory as used here is meant to mean a "speculative guess," which is not what a theory is. A more appropriate term to describe speculation is hypothesis.

Figure 1-2 summarizes the sequence of steps scientists use when applying the scientific method to a given research problem.

Learning Objectives

After completing this chapter you should be able to

- Understand the relationship between science as a whole and various scientific disciplines (Sec. 1.1).
- Give examples of how chemistry directly affects your life (Sec. 1.2).
- Outline the major steps in applying the scientific method to a problem (Sec. 1.3).
- Explain the differences among a fact, a law, a hypothesis, and a theory (Sec. 1.3).

Terms and Concepts for Review

The new terms or concepts defined in this chapter are

science (Sec. 1.1)
scientific discipline (Sec. 1.1)
scientific method (Sec. 1.3)
fact (Sec. 1.3)
law (Sec. 1.3)
hypothesis (Sec. 1.3)
theory (Sec. 1.3)

Questions and Problems

Much of this chapter is narrative material that is better reviewed by rereading than by working exercises.

Scientific Method

1-1 Describe the four steps of the scientific method.

1-2 What is the difference between a hypothesis and a theory?

1-3 What are two important purposes that theories serve?

1-4 What would you have to do in order to show that a theory was incorrect?

1-5 Indicate whether each of the following statements is true or false.

- **a.** The word hypothesis describes ideas proposed to explain a law which has little or no supporting evidence.
- **b.** A theory is a hypothesis that has not yet been subjected to experimental testing.
- **c.** A law is a general statement that summarizes a number of experimental facts.
- **d.** A hypothesis is a tentative law.
- **e.** A theory is a summary of experimental observations.
- **f.** A law is an explanation of why a particular natural phenomenon occurs.

1-6 Classify each of the following statements as a fact, a law, or a hypothesis. Pay no attention to the truthfulness of the statements.

- **a.** Mr. Pferdhoof's typewriter, when dropped down the elevator shaft, makes a loud noise.
- **b.** All typewriters, except yellow ones, burn brightly when thrown out the window.
- **c.** Any two typewriters, no matter how far apart, attract each other.
- **d.** Typewriter keys become activated and hit the typewriter ribbon as the result of minute changes in the humidity of the air around the typewriter.

1-7 If, after applying the scientific method to a particular problem, it is impossible to reach any conclusions, which of the following is the most likely next step?

- **a.** Discontinue the work.
- **b.** Carry out additional experimentation.
- **c.** Formulate a new hypothesis.
- **d.** Repeat each step of the previous experiment.

2 Numbers from Measurements

2.1 The Importance of Measurement

Most of the concepts now considered to be the basic principles of chemistry had their origin in extensive tabulations of experimental data obtained by making measurements. The concepts were "discovered" as these data tabulations (measurements) were subjected to the procedures of the scientific method (Sec. 1.3). Thus, measurements are the "backbone" of chemistry.

It would be extremely difficult for a carpenter to build cabinets without being able to use tools such as hammers, saws, and drills. They are a carpenter's "tools of the trade." Chemists also have "tools of the trade." Their most used tool is the tool called measurement. Questions such as "how much . . . ?", "how long . . . ?", and "how many . . . ?" simply cannot be answered without resorting to measurements. Understanding measurement is indispensable in the study of chemistry.

Measurements made by scientists need to be precise and accurate. Although the terms *precise* and *accurate* are used somewhat interchangeably in nonscientific discussion, they have distinctly different meanings in science. **Precision** *refers to how closely multiple measurements of the same quantity come to each other.* **Accuracy** *refers to how close a measurement (or the average of multiple measurements) comes to the true or accepted value.* A simple analogy not directly involving measurement—shooting at a target—illustrates nicely the difference between these two terms (see Fig. 2-1). Accuracy depends on how close the shots are to the center (bull's-eye) of the target. Precision depends on how close the shots are to each other.

precision

accuracy

The preciseness of a measurement is directly related to the actual physical measuring device used; that is, precision is an inherent part of any measuring device. You would expect, and it is the case, that the reproducibility (preciseness) of temperature readings obtained from a thermometer with a scale marked in tenths of a degree would be greater than readings obtained from a thermometer whose scale has only degree

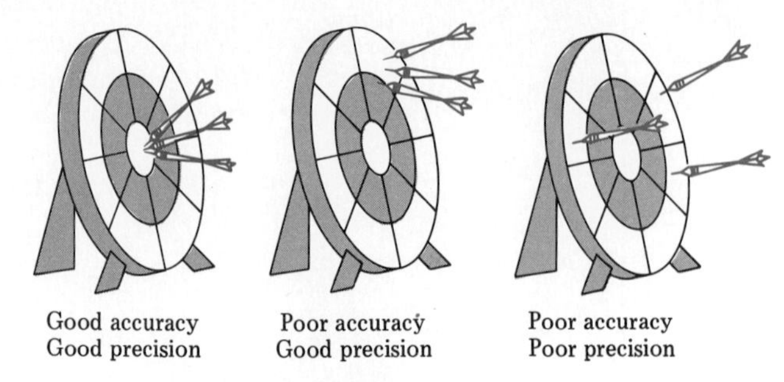

Figure 2-1. The difference between precision and accuracy.

marks. A stopwatch whose dial shows tenths of a second is preferred over one that shows only seconds if a precise time measurement is to be made.

In contrast to precision, accuracy depends not only on the measuring device used but also on the technical skill of the person making the measurement. How well can that person read the numerical scale of the instrument? How well can that person calibrate the instrument prior to its use?

Normally, high precision also results in high accuracy. However, high precision and low accuracy are also possible. Results obtained using a high-precision, poorly calibrated instrument would give precision but not accuracy. All measurements would be off by a constant amount as a result of the improper calibration.

All measurements consist of two parts: a *number* that tells the amount of the quantity measured and a *unit,* which tells the nature of the quantity that is being measured. Both are absolutely necessary. If you were to ask someone to loan you four, the immediate response would be "four what?" You would then have to indicate that you wished to borrow four dollars, four eggs, four bottles of pickles, or whatever it was that you wanted. An understanding of both numerical notation and unit systems, as used by scientists, is essential to an understanding of the significance of scientific observations.

Scientists are concerned with handling numbers associated with measurements *sensibly* (significant figures, Secs. 2.2 and 2.3) and *conveniently* (scientific notation, Secs. 2.4 and 2.5). Concerns about measurement units include *convenience of use* (metric system, Secs. 2.6 through 2.9) and their *relationship to problem-solving processes* (dimensional analysis, Sec. 2.10).

2.2 Uncertainty in Measurement — Significant Figures

The ability to interpret and manipulate numbers is a necessary skill for the effective use of the tool of measurement. It is the purpose of this section and others that follow to help you get "into shape" mathematically for your encounter with chemistry. The math involved is not awesome, but it cannot be ignored.

Physical exertion in sports can be fun, relaxing, and challenging if one is in good physical shape. If one is not in good physical condition, such exertion is unsatisfying and may be downright painful (especially the day after). By analogy, being in good shape mathematically has the same effect on the study of chemistry. It can cause that study to be a very satisfying and enjoyable experience. On the other hand, lack of necessary mathematical skills can cause "chemical exercise" to be somewhat painful.

In science there are two kinds of numbers — those that are *counted* or *defined* and those that are *measured.* The difference between counted or defined numbers and measured numbers is that we may know the exact values of the former but can never know the exact values of the latter.

You can count the number of peaches in a bushel of peaches or the number of toes on your left foot with absolute certainty. Counting does not involve reading the scale of a measuring device, and thus, counted numbers are not subject to the uncertainties inherent in a measurement.

An example of a defined number is the number of objects in a dozen — twelve. By definition 12 and exactly 12 (not 12.01 or 12.02) objects make a dozen. There are exactly 24 hours in a day, never 24.07 hours. A square has 4 sides, never 3.75 or 3.83 sides. Thus, a defined number always has one exact value.

In contrast to counted or defined numbers, every measured number carries with it a degree of uncertainty or error. Even when very elaborate and expensive measuring devices are used, some degree of uncertainty in measurement will always be present. The size of the error in a measurement depends upon the precision of the measuring device used and on the skill with which the person uses the device.

Uncertainty in a measurement can be illustrated by Figure 2-2, which shows how two different thermometer scales measure a given temperature. Determining the temperature involves determining the height of the mercury column in the thermometer.

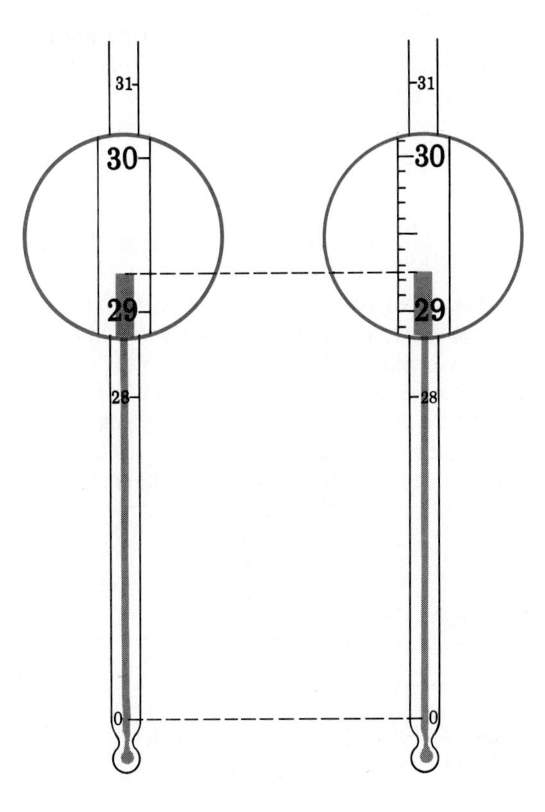

Figure 2-2. Measuring a temperature using a thermometer. A portion of the degree scale on each of two different scales has been magnified.

The scale on the left in Figure 2-2 is marked off in one degree intervals. Using this scale we can say with certainty that the temperature is between 29 and 30 degrees. We can further say that the actual temperature is closer to 29 degrees than to 30 and estimate it to be 29.2 degrees. The scale on the right has more subdivisions, being marked off in tenths of a degree rather than in degrees. Using this scale we can definitely say that the temperature is between 29.2 and 29.3 degrees and can estimate it to be 29.25 degrees. Note how both temperature readings contain some digits (all those except the last one) that are exactly known and one digit (the last one) that is estimated. Note also that the uncertainty in the second temperature reading is less than that in the first reading — an uncertainty in the hundredths place compared to an uncertainty in the tenths place. We say that the scale on the right is more precise than the one on the left.

Because measurements are never exact, any time a scientist writes down a numerical value for a measurement two kinds of information must be conveyed: (1) the magnitude of the measurement and (2) the precision or reliability of the measurement. The digit values give the magnitude. Precision is indicated by the number of significant figures recorded.

significant figures

Significant figures *are the digits in any measurement that are known with certainty plus one digit that is uncertain.* Only one estimated digit is ever recorded as part of a measurement. It would be incorrect for a scientist to report that the height of the mercury column in Figure 2-2, as read on the scale on the right, corresponds to a temperature of 29.247 degrees. The value 29.247 contains two estimated digits (the 4 and the 7) and would indicate a measurement of precision greater than that actually obtainable with that particular measuring device.

The magnitude of the uncertainty in the last significant digit in a measurement (the estimated digit) may be indicated using a "plus-minus" notation. The following three time measurements illustrate this notation.

15 ± 1 second

15.3 ± 0.1 second

15.34 ± 0.03 second

Most often the uncertainty in the last significant digit is one unit (as in the first two time measurements), but it may be larger (as in the last of the three time measurements). In this text we will follow the almost universal practice of dropping the "plus-minus" notation if the magnitude of the uncertainty is one unit. Thus, in the absence of "plus-minus" notation you will be expected to assume that there is an uncertainty of one unit in the last significant digit. A measurement reported simply as 27.3 inches means 27.3 ± 0.1 inches. Only in the situation where the uncertainty is greater than one unit in the last significant digit will the amount of the uncertainty be explicitly shown.

EXAMPLE 2.1

How many significant figures should be reported in each of the following measurements?

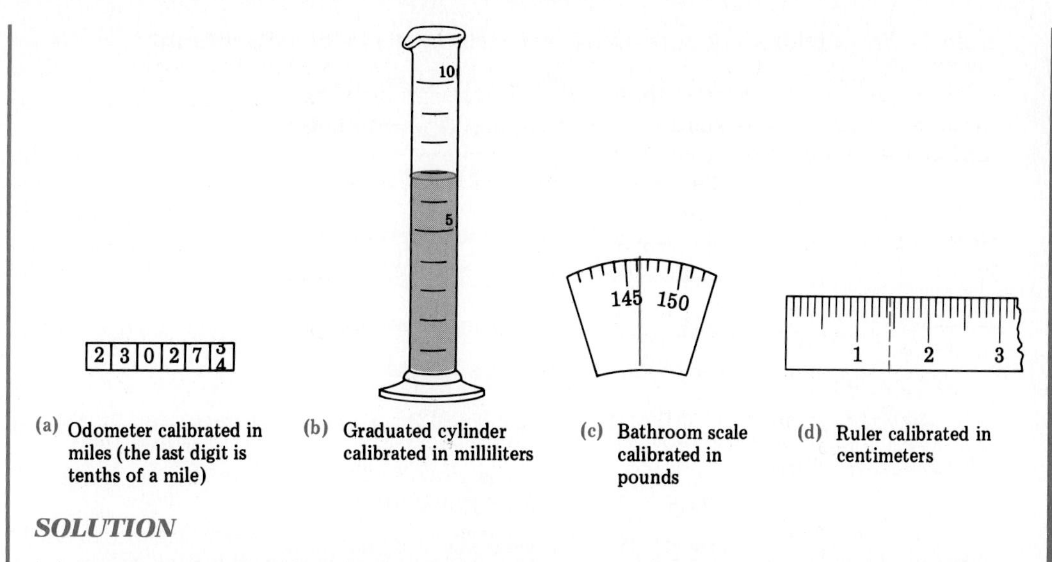

(a) Odometer calibrated in miles (the last digit is tenths of a mile)

(b) Graduated cylinder calibrated in milliliters

(c) Bathroom scale calibrated in pounds

(d) Ruler calibrated in centimeters

SOLUTION

(a) We know definitely that the mileage shown on the odometer is between 23027.3 and 23027.4 miles. We estimate the final digit (hundredths of a mile) to be a 4, giving a reading of 23027.34 miles. Thus, *seven* significant figures are reportable.
(b) The level of the liquid is between the 6 and 7 milliliter marks. We estimate the level to be at 6.8 milliliters. *Two* significant figures are to be recorded.
(c) The weight shown on the scale is definitely between 146 and 147 pounds. Estimating the last digit (tenths of a pound) we get 146.2 pounds, which is a measurement involving *four* significant figures.
(d) The ruler is calibrated in tenths of a centimeter. The broken line is definitely between 1.4 and 1.5 centimeters, being closer to 1.5. Estimating the next digit to be 7 we get a measurement of 1.47 centimeters. Thus, a measurement involving *three* significant figures is obtained.

Determining the number of significant figures in a measurement is not always as straightforward as Example 2.1 would lead one to believe. In this example you knew the type of instrument used for each measurement and its limitations because you made the measurement. Quite often when someone else makes a measurement such information is not available. All that is known is the reported final result — the numerical value of the measured quantity. In this situation questions do arise about the "significance" of various digits in the measurement. For example, consider the published value of the distance from the earth to the sun, which is 93,000,000 miles. Intuition tells you that it is highly improbable that this distance is known to the closest mile. You suspect that this is an estimated distance. To what digit has this number been estimated? Is it to the nearest million miles, the closest hundred thousand miles, the nearest ten thousand miles, or what?

A set of standard rules has been developed to aid scientists in interpreting the "significance" of reported measurements or values calculated from measurements. These rules deal mostly with how to treat zeros. A zero in a measurement may or may not be significant.

Rule 1. In numbers not containing any zeros all digits are significant.

13.753	five significant figures
3,123	four significant figures
1.31	three significant figures

Rule 2. All zeros between significant digits are significant.

1.032	four significant figures
3033	four significant figures
3,000,111	seven significant figures

Rule 3. In a number less than one, zeros used to fix the position of the decimal point are *not* significant.

$0.00\underline{23}$	two significant figures
$0.0\underline{103}$	three significant figures
$0.000000\underline{5}$	one significant figure

Another way of stating rule 3 is: In numbers less than one, zeros to the left of the first nonzero digit are *not* significant.

Rule 4. When a number has a decimal point, zeros to the right of the last nonzero digit are significant.

147.000	six significant figures
16.30	four significant figures
120.0	four significant figures
$0.000\underline{230}$	three significant figures
$0.\underline{300040}$	six significant figures

Rule 5. When a *number without a decimal point explicitly shown* ends in one or more zeros, such as 4300, the zeros that end the number may or may not be significant. That is, the number is ambiguous in terms of significant figures. If the zeros are there simply to indicate the magnitude of the number, they are not significant. One needs further information about how the number was obtained in order to know whether such is the case. In absence of such information we will assume, in this text, that all "ambiguous" zeros are *not* significant. Applying this rule to the number 93,000,000 — the distance from the earth to the sun in miles — results in the conclusion that this distance is stated only to the closest million miles since none of the zeros is considered significant.

One method available for avoiding the ambiguity problem in numbers that fit rule 5 is to write a small bar (—) above the terminal zeros that are significant. Thus, the notation $43\overline{00}$ indicates both zeros in 4300 are significant, and $43\bar{0}0$ indicates only the first of the two zeros is significant. The number $43\overline{00}$ may also be written as 4300., the decimal point after the number indicating that all digits preceding it are significant (rule 4).

Another method, more convenient than the bar method, for dealing with the confusion inherent in numbers like 4300 is to express the number in scientific notation. In this notation, to be presented in Section 2.4, only significant figures are shown.

EXAMPLE 2.2

Determine the number of significant figures in the following measured values.

(a) 10.423 **(b)** 1230.0 **(c)** 0.0032 **(d)** 2.527 **(e)** 3,3$\bar{0}$0
(f) 3,004,000 **(g)** 100.

SOLUTION

(a) There are *five* significant figures. The zero is significant since it is between two digits (1 and 4) that are significant (rule 2).
(b) There are *five* significant figures. Since the number contains a decimal point, zeros to the right of the last nonzero digit are significant (rule 4).
(c) There are *two* significant figures. The zeros are not significant as they are used to fix the position of the decimal (rule 3).
(d) There are *four* significant figures. All nonzero digits are significant (rule 1).
(e) There are *three* significant figures. The zero with the bar over it is significant and the other zero is not significant (rule 5).
(f) There are *four* significant figures. The zeros between the 3 and 4 are significant (rule 2). The end zeros are not significant (rule 5).
(g) There are *three* significant figures. Explicitly showing the decimal point after the zeros makes them significant. Without the decimal point present they would be assumed to be not significant (rule 5).

EXAMPLE 2.3

What meaning, in terms of significant figures and magnitude of uncertainty, is conveyed by each of the following notations for a measurement involving the number twenty thousand.

(a) 20,000 **(b)** 2$\bar{0}$,000 **(c)** 2$\overline{0,0}$00 **(d)** 2$\overline{0,00}$0 **(e)** 2$\overline{0,000}$
(f) 20,000.

SOLUTION

(a) When only *one* significant figure is present, which is the case here, it is an estimated digit. Since the significant digit, the 2, is located in the fifth place to the left of the decimal (the ten-thousands place), the uncertainty is ±10,000. This uncertainty means that the actual value of the number could be anywhere between 10,000 and 30,000.
(b) This number has *two* significant figures since the first of the four zeros is significant, as indicated by the bar over it. The last significant digit, the zero, occupies the thousands place in the number. Thus, the uncertainty is ±1,000. This uncertainty

implies that the actual value of the number is definitely known to be between 19,000 and 21,000.
(c) Here we have *three* significant figures since two of the four zeros are significant. The uncertainty now lies with the hundreds digit (±100), and the number's actual value is definitely known to be between 19,900 and 20,100.
(d) With *four* significant figures present (the 2 and the three zeros) the uncertainty is reduced to the tens digit (±10), and the number is known to be between 19,990 and 20,010.
(e) *All* digits are now significant. The uncertainty now lies in the last digit, which occupies the ones place (±1), and the number is known to be between 19,999 and 20,001.
(f) Explicitly placing a decimal point at the end of the number makes *all* of the zeros significant. The meaning is thus the same as in part (e).

2.3 Significant Figures and Calculated Quantities

Most experimental measurements are not end results in themselves. Instead they function as intermediates used to calculate other quantities. For example, we might measure in a laboratory the height, width, depth, and weight of a rectangular-shaped solid object and from this information calculate its volume and density.

In doing a calculation using experimental data, a scientist must give major consideration to the number of significant figures in the computed result. Correct calculations never improve (or decrease) the precision of experimental measurements.

Concern about the number of significant figures in a calculated number is particularly critical when an electronic calculator is used to do the arithmetic of the calculation. Hand calculators now in common use are not programmed to take into account significant figures. Consequently, the digital read-outs on them more often than not display more digits than are needed. It is a mistake to record these extra digits since they are meaningless; that is, they are not significant figures.

In order to record correctly numbers obtained through calculations students must be able to (1) adjust (usually decrease) the number of digits in a number to give it the correct number of significant figures and (2) determine the allowable number of significant figures in the result of any mathematical operation. We will consider these skills in the order listed.

rounding off

Rounding off is *the process of deleting unwanted (nonsignificant) digits from a calculated number.* Three simple rules govern this process.

Rule 1. If the first digit to be dropped is less than 5, that digit and all digits that follow it are simply dropped. Thus, 62.312 rounded off to three significant figures becomes 62.3.

Rule 2. If the first digit to be dropped is a digit greater than 5, or a 5 followed by digits other than zero, the excess digits are all dropped and the last retained digit is increased in value by one unit. Thus, 62.36 rounded off to three significant figures becomes 62.4.

Rule 3. If the first digit to be dropped is a 5 not followed by any other digit or a 5 followed only by zeros, an odd–even rule applies. If the last retained digit is odd, that digit is increased in value by one unit after dropping the 5 and any zeros that follow it. If the last retained digit is even, its value is not changed, and the 5 and any zeros that follow are simply dropped. Thus, 62.150 and 62.450 rounded to three significant figures become, respectively, 62.2 (odd rule) and 62.4 (even rule).

These rounding rules must be modified slightly when digits to the left of the decimal point are to be dropped. In order to maintain the position of the decimal point in such situations, zeros must replace all of the dropped digits that are to the left of the decimal point. Parts (c) and (f) of Example 2.4 illustrate this point.

EXAMPLE 2.4

Round off each of the following numbers to three significant figures.

(a) 384.53 **(b)** 0.180623 **(c)** 1,021,001 **(d)** 0.3335 **(e)** 5.5050
(f) 102,500

SOLUTION

(a) Rule 2 applies, which means that the last retained digit (the 4) is increased in value by one unit.

384.53 becomes 385

(b) Rule 2 applies again. The last retained digit (a zero) is increased in value by one unit.

0.180623 becomes 0.181

(c) Since the first digit to be dropped is a 1, rule 1 applies.

1,021,001 becomes 1,020,000

Note that to maintain the position of the decimal, zeros must replace all of the dropped digits. This will always be the case when digits to the left of the decimal place are dropped.

(d) Rule 3 applies. The first (and only) digit to be dropped is a 5. The last retained digit (the 3) is an odd number, so using the odd–even rule its value is increased by one unit.

0.3335 becomes 0.334

(e) Rule 3 applies again. This time the last digit retained is even (a zero), so its value is not changed.

5.5050 becomes 5.50

(f) This is again a rule 3 situation. Since an even digit (a 2) occupies the third significant figure place, its value is not changed.

102,500 becomes 102,000

Note again that zeros must take the place of all digits to the left of the decimal place that are dropped.

The number of significant figures *allowable* in a calculated result depends on the number of significant figures in the data used to obtain the result and on the type of mathematical operation(s) used in obtaining the result. There are separate rules for multiplication and division operations and for additions and subtractions.

For *multiplication* and *division* an *answer should not have any more significant figures than the least number present in any of the numbers that were multiplied or divided.* A number of applications of this rule are shown in Example 2.5.

EXAMPLE 2.5

Perform the following computations, all of which involve multiplication and/or division. Express your answers to the proper number of significant figures. Assume that all numbers are measured numbers.

(a) 3.751×0.42 (b) $\dfrac{3420}{174.3}$ (c) $\dfrac{1{,}810{,}000}{3.1453}$ (d) $\dfrac{8}{3.0}$

(e) $\dfrac{1800.0}{6.0000}$ (f) $\dfrac{3.130 \times 3.140}{3.15}$

SOLUTION

(a) The calculator answer to this problem is

$$3.751 \times 0.42 = 1.57542$$

The input number with the least number of significant figures is 0.42.

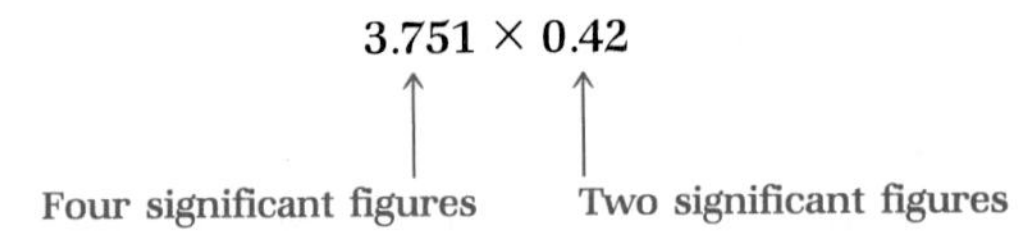

Thus the calculator answer must be rounded off to two significant figures.

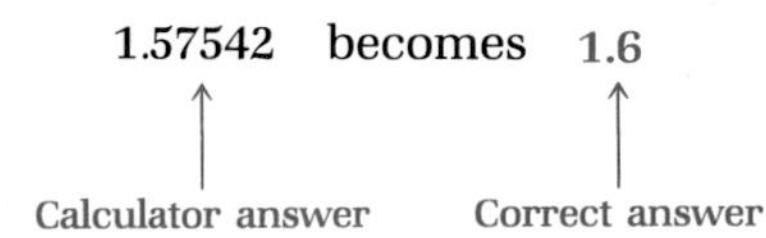

(b) The calculator answer to this problem is

$$\frac{3420}{174.3} = 19.621343$$

The input number with the least number of significant figures is 3420.

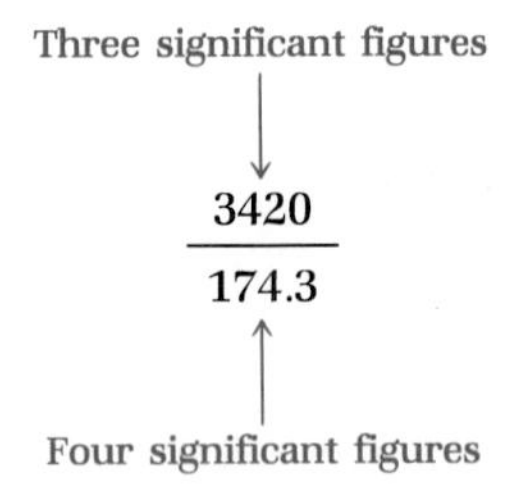

Thus the calculator answer must be rounded to three significant figures.

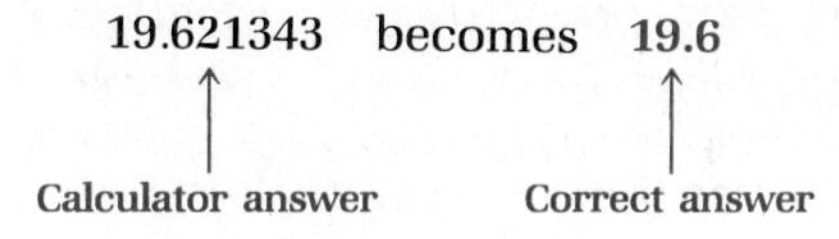

(c) The calculator answer to this problem is

$$\frac{1{,}810{,}000}{3.1453} = 575{,}461.8$$

The input number with the least number of significant figures is 1,810,000.

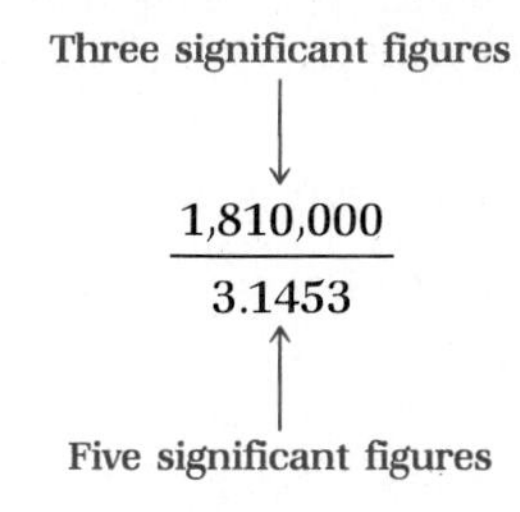

Thus the calculator answer must be rounded to three significant figures.

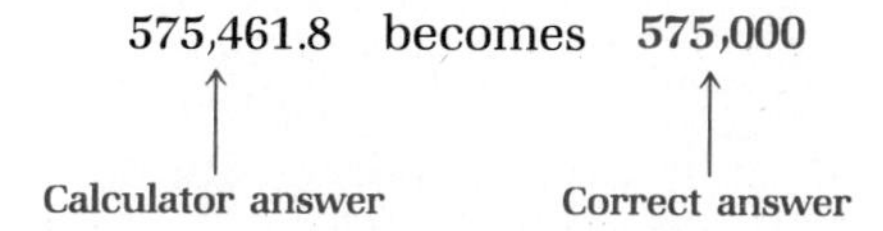

(d) The calculator answer to this problem is

$$\frac{8}{3.0} = 2.6666667$$

The input number with the least number of significant figures is 8.

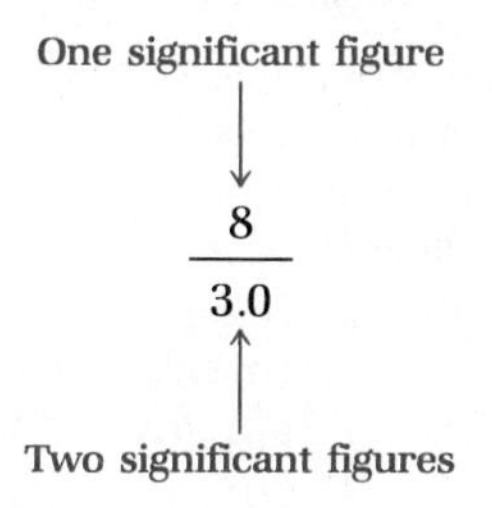

Thus the calculator answer must be rounded to one significant figure.

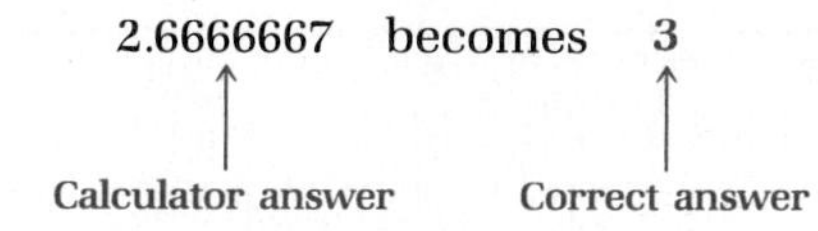

(e) The calculator answer to this problem is

$$\frac{1800.0}{6.0000} = 300$$

Both input numbers contain five significant figures. Thus the correct answer must also contain five significant figures.

300 becomes 300.00

Calculator answer → 300; Correct answer → 300.00

Note here how the calculator gave too few significant figures. Most calculators cut off zeros after the decimal point even if they are significant. Using too few significant figures is just as wrong as using too many.

(f) The calculator answer to this problem is

$$\frac{3.130 \times 3.140}{3.15} = 3.1200635$$

The input number with the least number of significant figures is 3.15.

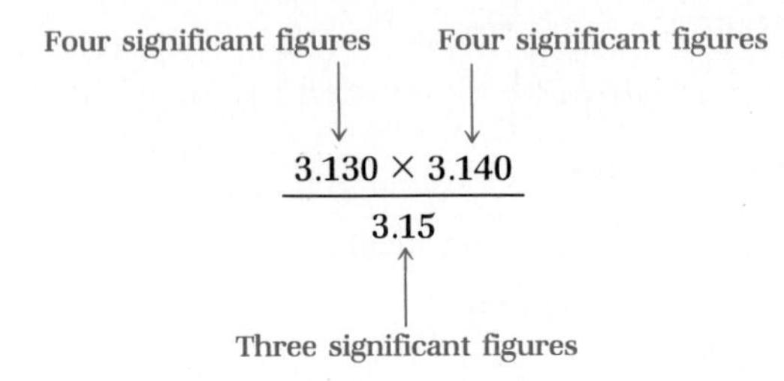

Thus the calculator answer must be rounded off to three significant figures.

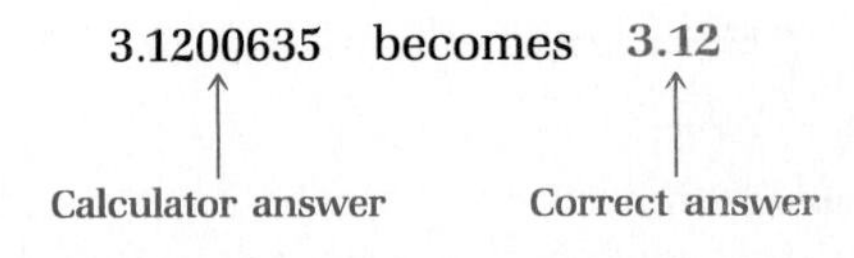

In addition and subtraction operations it is not the fewest number of significant figures present that determines the allowable number of significant figures in the answer, as was the case with multiplication and division. Instead, in *addition* and *subtraction* the answer *should not have digits beyond the last digit position common to all numbers being added or subtracted.* As will be seen in some of the sample additions and subtractions in Example 2.6, the answers from either addition or subtraction can have more or less significant figures than any of the numbers that have been added or subtracted.

EXAMPLE 2.6

Perform the following computations, all of which involve addition or subtraction. Express your answers to the proper number of significant figures. Assume that all numbers are measured numbers (each reported with the correct number of significant figures).

(a) 26.8 + 13.7 + 22 (b) 8.3 + 1.2 + 2.4 (c) 13.01 + 13.001 + 13.0001
(d) 0.4378 − 0.4370 (e) 34.7 + 0.00007

SOLUTION

(a) The last digit position common to all numbers is the units place.

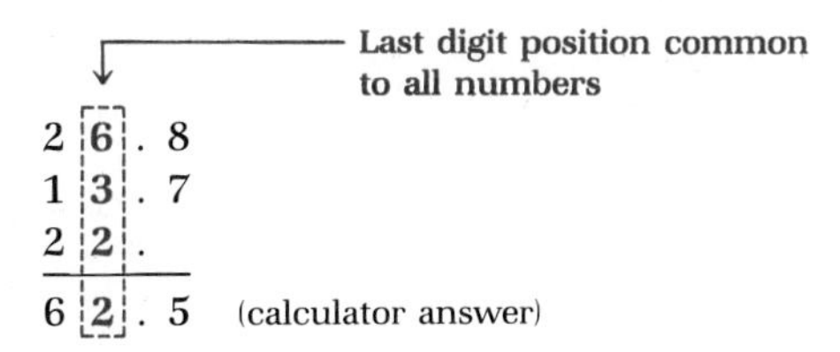

Hence, the calculator answer of 62.5 must be rounded off to the units place.

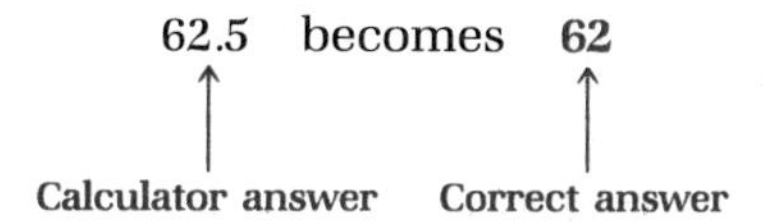

(b) All three numbers are known to the tenths place, which becomes the last digit position common to all numbers.

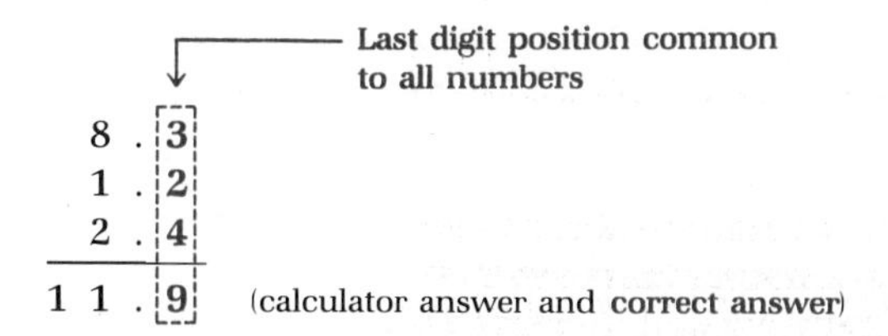

Since the calculator answer is to the tenths place, this answer is acceptable as is. (Once in a while the calculator answer and the correct answer are one and the same.) Note in this problem how each of the numbers in the sum contains two significant figures and yet the correct answer has three significant figures. Again, the number of significant figures is not the determining factor in addition and subtraction as it is in multiplication and division.

(c) The last digit position common to all numbers is the hundredths place.

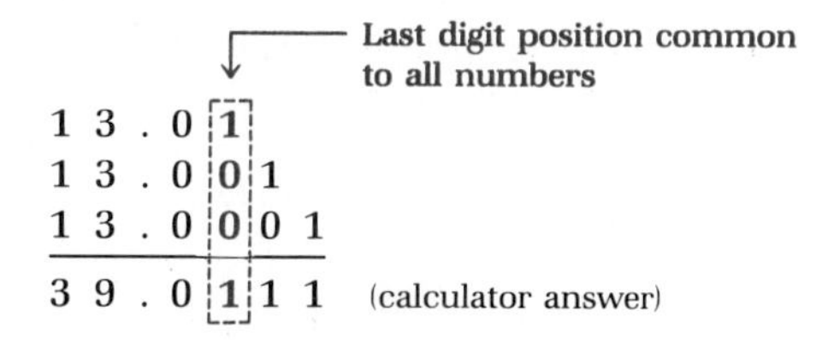

Hence, the calculator answer of 39.0111 must be rounded off to the hundredths place.

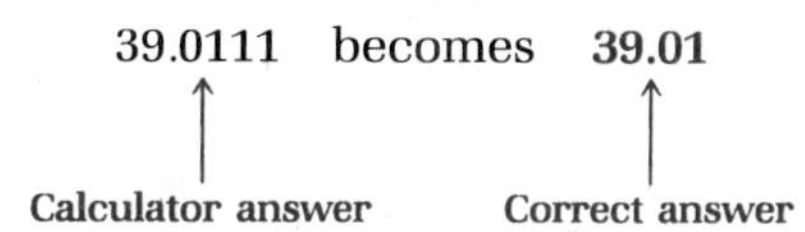

(d) Both numbers are known to the ten-thousandths place. Thus the answer is reportable to the ten-thousandths place.

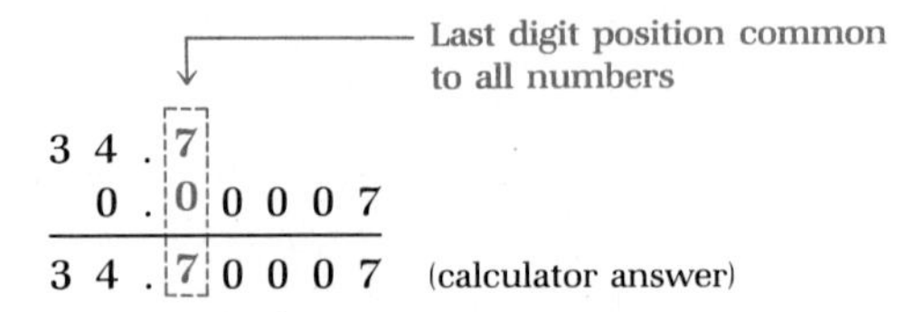

Again, the calculator and correct answer are one and the same. Note that three significant figures were "lost" in the subtraction. The answer has one significant figure. The two input numbers each have four significant figures.

(e) The last digit position common to both numbers is the tenths digit.

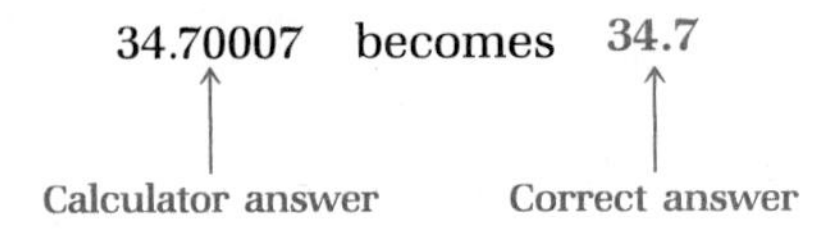

Hence, the calculator answer of 34.70007 must be rounded to the tenths place.

34.70007 becomes 34.7

Calculator answer — Correct answer

The answer of 34.7 is the same as one of the input numbers. From a significant figure standpoint, the number 0.00007 is negligible compared to 34.7.

Not all numbers in a calculation need originate from a measurement. Sometimes *exact numbers* (counted or defined numbers) are part of a calculation. Since there is no uncertainty in an exact number, it is considered to possess an infinite number of significant figures. Consequently, exact numbers, when they appear in a calculation, will never limit the number of significant figures allowable in the answer. We do not consider them when determining the number of allowable significant figures. The allowable number is determined in the usual way taking into account only those numbers that were experimentally determined via measurement.

EXAMPLE 2.7

A chemical reaction is allowed to continue for 6323 seconds. How many minutes is this time equivalent to?

SOLUTION

If we divide the number of seconds by 60 we will get the number of minutes. The calculator answer to this division is

$$\frac{6323}{60} = 105.38333$$

The input number with the least number of significant figures is 6323 (4 significant figures) since 60 is an exact number with an unlimited number of significant figures. Thus, the answer may contain four significant figures.

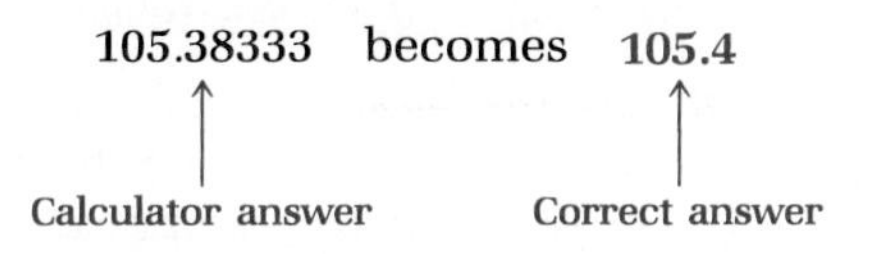

2.4 Scientific Notation

In scientific work, very large and very small numbers are frequently encountered. For example, in one cup of water — an "ordinary" amount of water — there are approximately

7,910,000,000,000,000,000,000,000 molecules.

(A molecule of water is the smallest possible unit of water — Sec. 5.2). A single water molecule has a weight of

0.000000000000000000000000066 pound.

Such large and small numbers as these are difficult to use. Recording them is not only time consuming but also very prone to errors — too many or too few zeros recorded. Also such numbers are awkward to work with in calculations. Consider the problem of multiplying the above two numbers together. Handling all of those zeros is "mind boggling."

A method exists for expressing cumbersome many-digit numbers, such as the previous examples, in compact form. Called scientific notation, this method eliminates the need to write all the zeros. **Scientific notation** *is a system in which an ordinary decimal number is expressed as a product of a number between 1 and 10 times 10 raised to a power.* The two previously cited numbers, dealing with molecules of water, expressed in scientific notation are, respectively,

scientific notation

7.91×10^{24} molecules and 6.6×10^{-26} pound

Note that in scientific notation the first number involves seven digits (compared to 25 previously), and the second number six digits (compared to 28 previously).

Exponents

Understanding exponents is the key to understanding scientific notation. A brief review of exponents and their use is in order prior to considering the rules for converting numbers from ordinary decimal notation to scientific notation and vice versa.

In mathematics a convenient method exists for showing that a number has been multiplied by itself two or more times. It involves the use of an exponent. An **exponent** *is a number written as a superscript following another number and indicates how many times the first number is to be multiplied by itself.* The following examples illustrate the use of exponents.

exponent

$$6^2 = 6 \times 6 = 36$$

$$3^5 = 3 \times 3 \times 3 \times 3 \times 3 = 243$$

$$10^3 = 10 \times 10 \times 10 = 1000$$

Exponents are also frequently referred to as *powers* of numbers. Thus, 6^2 may be verbally read as "six to the second power," and 3^5 as "three to the fifth power." Raising a number to the second power is often called "squaring," and to the third power "cubing."

In scientific work, powers of the number 10 are almost always used. This is because multiplying or dividing by 10 coincides with moving the decimal point in a number one place to the right or left. Scientific notation exclusively uses powers of ten.

When ten is raised to a power, its decimal equivalent is the number 1 followed by as many zeros as the power. This one-to-one correlation between power magnitude and number of zeros is shown in color in the following examples.

$10^2 = 100$ (two zeros and a power of 2)

$10^4 = 10{,}000$ (four zeros and a power of 4)

$10^6 = 1{,}000{,}000$ (six zeros and a power of 6)

The notation 10^0 is a defined quantity.

$$10^0 = 1$$

The above generalization easily explains why. Ten to the zero power is the number 1 followed by no zeros, which is simply 1.

All of the examples of exponential notation presented so far have had positive exponents. This is because each example represented a number of magnitude greater than one. Negative exponents are also possible. They are associated with numbers of magnitude less than one.

A negative sign in front of an exponent is interpreted to mean that the number and the power to which it is raised is in the denominator of a fraction in which 1 is the numerator. The following examples illustrate this interpretation.

$$10^{-1} = \frac{1}{10^1} = \frac{1}{10} = 0.1$$

$$10^{-2} = \frac{1}{10^2} = \frac{1}{10 \times 10} = \frac{1}{100} = 0.01$$

$$10^{-3} = \frac{1}{10^3} = \frac{1}{10 \times 10 \times 10} = \frac{1}{1000} = 0.001$$

When the number 10 is raised to a negative power, the absolute value of the power (the value ignoring the minus sign) is always one more than the number of zeros between the decimal point and the one. This correlation between power magnitude and number of zeros is shown in color in the following examples.

$10^{-2} = 0.01$ (one zero and a power of 2)

$10^{-4} = 0.0001$ (three zeros and a power of 4)

$10^{-6} = 0.000001$ (five zeros and a power of 6)

order of magnitude

Differences in magnitude between numbers that are powers of 10 are often described with the phrase "orders of magnitude." An **order of magnitude** *is a single exponential value of the number 10.* Thus, 10^6 is four orders of magnitude larger than 10^2, and 10^7 is three orders of magnitude larger than 10^4.

Writing Numbers in Scientific Notation

A number written in scientific notation has two parts: (1) a *coefficient,* written first, which is a number between 1 and 10, and (2) an *exponential term,* which is 10 raised to a power. The coefficient part is always multiplied by the exponential term. Using the scientific notation form of the number 703 as an example, we have

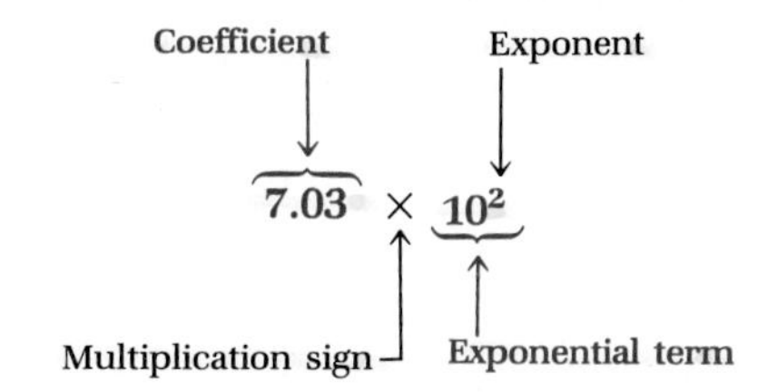

Rules for converting numbers from decimal to scientific notation are very simple.

Rule 1. The value of the exponent is determined by counting the number of places the original decimal point must be moved to give the coefficient.

Rule 2. If the decimal point is moved to the *left* to get the coefficient, the exponent is a *positive number.* If the decimal point is moved to the *right* to get the coefficient, the exponent is a *negative number.*

Numerous examples of the use of these two rules are found in Examples 2.8 and 2.9.

When a number is expressed in scientific notation, *only significant digits become part of the coefficient.* Because of this there is never any confusion (ambiguity) in determining the number of significant figures in a number expressed in scientific notation. There are five possible precision interpretations for the number 10,000 (1, 2, 3, 4, or 5 significant figures). In scientific notation each of these interpretations assumes a different form.

1×10^4 (10,000 with one significant figure)

1.0×10^4 (10,000 with two significant figures)

1.00×10^4 (10,000 with three significant figures)

1.000×10^4 (10,000 with four significant figures)

1.0000×10^4 (10,000 with five significant figures)

EXAMPLE 2.8

Express the following numbers in scientific notation.

(a) 654 **(b)** 11.013 **(c)** 101.0 **(d)** 0.000781 **(e)** 0.3210 **(f)** 0.043

SOLUTION

(a) The coefficient, which must be a number between 1 and 10, is obtained by moving the decimal point in 654 two places to the left.

6.54.

This gives us 6.54 as the coefficient. The value of the exponent in the exponential term is +2, since the decimal point was shifted two places to the left. A shift of the decimal point to the left will always result in a positive exponent. Multiplying the coefficient by the exponential term, 10^2, gives us the scientific notation form of the number.

$$6.54 \times 10^2$$

(b) The coefficient, 1.1013, is obtained by moving the decimal point one place to the left.

1.1.013

The power of 10 is +1, indicating the decimal point had to be moved one place to the left to obtain the coefficient. Multiplying the coefficient and exponential terms together gives

$$1.1013 \times 10^1$$

as the scientific notation form of the number 11.013.

(c) The coefficient, obtained by moving the decimal point two places to the left, is 1.010.

1.01.0

A movement of the decimal point two places to the left sets the power of ten at +2. Thus, the scientific notation form of the number is

$$1.010 \times 10^2$$

Note that the zero after the decimal point in the original number must be shown in the coefficient. Dropping that zero gives the number 1.01×10^2, which contains only three significant figures. The original number had four significant figures. You can never lose or gain significant figures in converting a number from decimal notation to scientific notation.

(d) Since this number has a value less than one, the decimal point must be moved to the right in order to obtain a coefficient with a value between one and ten.

0.00007.81

Movement of the decimal point to the right means the exponent will have a negative value, in this case −4. Thus, in scientific notation we have

$$7.81 \times 10^{-4}$$

The zeros to the left of the 7 in the original number do not appear in the coefficient as they are nonsignificant zeros.

(e) Movement of the decimal point one place to the right gives a coefficient of 3.210.

The power of ten will be -1. Multiplying the coefficient and exponential terms gives us an answer of

$$3.210 \times 10^{-1}$$

The zero following the 1 in the coefficient is necessary because four significant figures are required in the coefficient. The original number had four significant figures.
(f) This time the decimal point is moved two places to the right, giving a coefficient of 4.3 and an exponential term of 10^{-2}. Multiplying the two we get the scientific notation, which is

$$4.3 \times 10^{-2}$$

EXAMPLE 2.9

Express the number 223,400,000 in scientific notation to

(a) five significant figures **(b)** seven significant figures
(c) three significant figures

SOLUTION

To obtain a coefficient whose magnitude is between 1 and 10, the decimal point needs to be moved eight places to the left.

2,23,400,000.

The exponential term thus becomes 10^8. The answers to the various parts of this example will differ only in the number of significant figures in the coefficient.
(a) Expressed to five significant figures the coefficient is 2.2340, and the answer is 2.2340×10^8
(b) Expressed to seven significant figures the coefficient is 2.234000, and the answer is 2.234000×10^8
(c) Expressed to three significant figures the coefficient is 2.23 and the answer is 2.23×10^8

Converting from Scientific Notation to Decimal Notation

To convert a number in scientific notation, such as 6.02×10^{23}, into a regular decimal number, we start by examining the exponent. The value of the exponent tells how many places the decimal point must be moved. If the exponent is positive, movement is to the right to give a number greater than one; if it is negative, movement is to the left to give a number less than one. Zeros may have to be added to the number as the decimal point is moved.

EXAMPLE 2.10

Write the following exponential numbers in ordinary decimal notation.

(a) 6.24×10^{-4} **(b)** 3.4×10^5 **(c)** 3.000×10^2 **(d)** 7.5×10^{-3}

SOLUTION

(a) The exponent −4 tells us the decimal is to be located four places to the left of where it is in 6.24. Three zeros will have to be added to accommodate the decimal point change. Remember, when the exponent is negative, the number of added zeros is one less than the exponent.

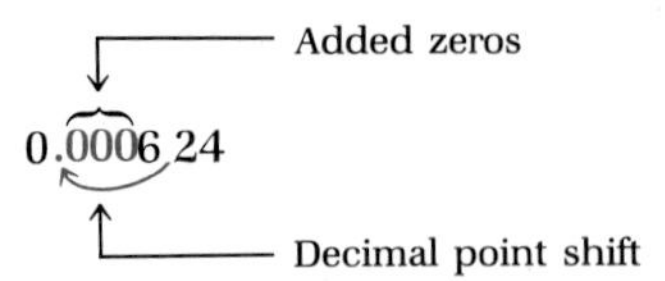

(b) The exponent +5 tells us the decimal is to be located five places to the right of where it is in 3.4. Again, zeros must be added to mark the new decimal place correctly.

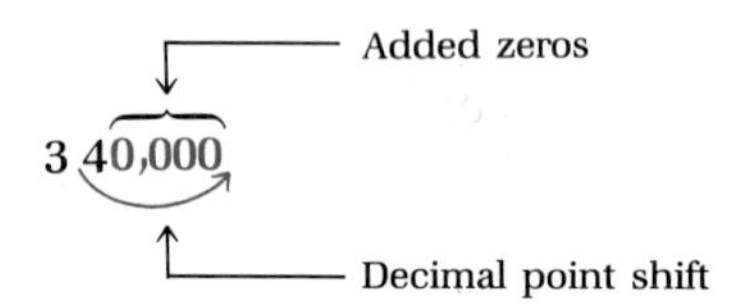

(c) The decimal point is to be moved two places to the right of where it is in 3.000. This gives the number 300.0 as the answer. The zero to the right of the decimal point in 300.0 must be retained. Otherwise the number of significant digits would drop from four to three as a result of the notation change. The number of significant figures must always remain the same unless specific instructions relative to rounding off are given.
(d) The decimal point is to be moved three places to the left. Zeros must be added to mark the new decimal place correctly.

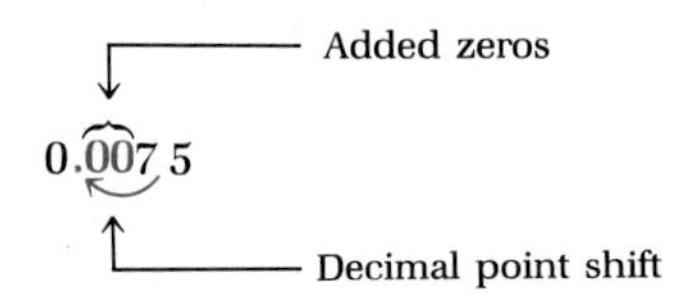

2.5 Scientific Notation and Mathematical Operations

A major advantage of writing numbers in scientific notation is that it greatly simplifies the mathematical operations of multiplication and division.

Multiplication in Scientific Notation

Multiplication of two or more numbers expressed in scientific notation involves two separate steps. *First, we multiply together the coefficients* (*the decimal numbers between* 1 *and* 10) *in the usual manner. Second, we* ADD *algebraically the exponents of the powers of* 10.

EXAMPLE 2.11

Carry out the following multiplications in scientific notation. Be sure to take into account significant figures in obtaining your final answer.

(a) $(1.113 \times 10^3) \times (7.200 \times 10^5)$ **(b)** $(2.05 \times 10^{-3}) \times (1.19 \times 10^{-7})$

(c) $(4.21 \times 10^{-9}) \times (2.107 \times 10^6)$ **(d)** $(3.92 \times 10^{10}) \times (2.3 \times 10^{-4})$

SOLUTION

(a) Multiplying the two coefficients together gives

$$1.113 \times 7.200 = 8.0136 \quad \text{(calculator answer)}$$

Since each of the input numbers for the multiplication has four significant figures, the answer should also have four significant figures, not five as given by the calculator.

Multiplication of the two powers of ten to give the exponential part of the answer requires adding the exponents to give a new exponent.

$$10^3 \times 10^5 = 10^{(3+5)} = 10^8$$

Combining the new coefficient with the new exponential term gives the answer.

$$8.014 \times 10^8$$

(b) Multiplying the two coefficients together gives

$$2.05 \times 1.19 = 2.4395 \quad \text{(calculator answer)}$$

Since both input numbers for the multiplication contain three significant figures, the calculator answer must be rounded down to three significant figures.

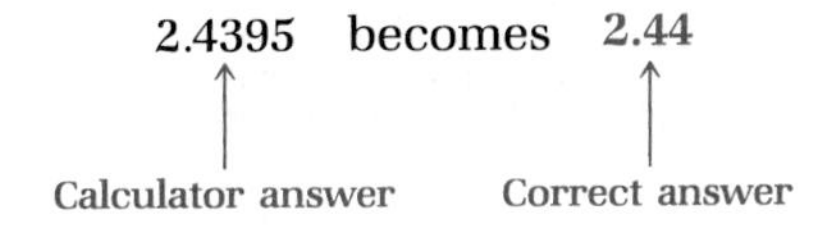

Next, the exponents of the powers of ten are added to generate the new power of ten.

$$10^{-3} \times 10^{-7} = 10^{(-3)+(-7)} = 10^{-10}$$

(To add two numbers of the same sign — positive or negative — just add the numbers and place the common sign in front of the sum.) Combining the coefficient and the exponential term gives an answer of

$$2.44 \times 10^{-10}$$

(c) Multiplying the two coefficients together gives

$$4.21 \times 2.107 = 8.87047 \quad \text{(calculator answer)}$$

Because the input number 4.21 contains only three significant figures, the answer is also limited to three significant figures. Thus,

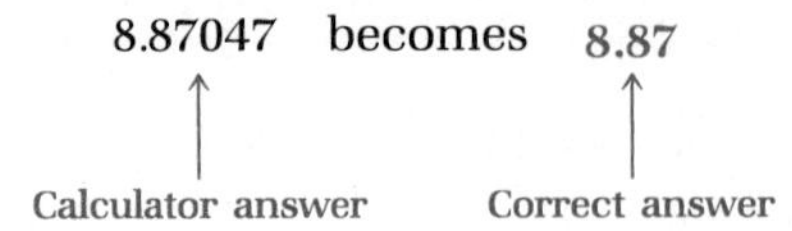

In multiplying the exponential terms we will have to add exponents with different signs, a −9 and a +6. To do this, we first determine the larger number (9 is larger than 6) and then subtract the smaller number from it (9 − 6 = 3). The sign is always the sign of the larger number (minus in this case). Thus,

$$(-9) + (+6) = -3$$

$$10^{-9} \times 10^{6} = 10^{(-9)+(+6)} = 10^{-3}$$

Combining the coefficient and the exponential parts gives

$$8.87 \times 10^{-3}$$

(d) Multiplying coefficients gives

$$3.92 \times 2.3 = 9.016 \quad \text{(calculator answer)}$$

This answer must be rounded to two significant figures since the input number 2.3 has only two significant figures.

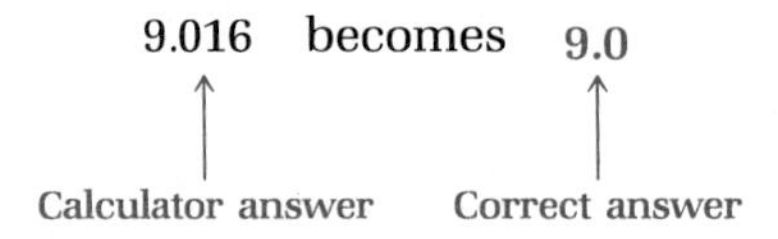

Again, in combining the exponential terms we have exponents of different signs. The smaller exponent (4) is subtracted from the larger exponent (10), and the sign of the larger number (+) is used.

$$(+10) + (-4) = (+6)$$

$$10^{10} \times 10^{-4} = 10^{6}$$

Combining the coefficient and the exponential term gives

$$9.0 \times 10^{6}$$

Division in Scientific Notation

To divide two numbers written in scientific notation, *first perform the indicated division of the coefficients in the usual manner and then subtract algebraically the exponent in the denominator (bottom) from the exponent in the numerator (top) to give the exponent of the new power of* 10.

EXAMPLE 2.12

Carry out the following divisions in scientific notation. Be sure to take into account significant figures in obtaining your final answer.

(a) $\dfrac{2.05 \times 10^{5}}{1.19 \times 10^{3}}$ (b) $\dfrac{7.200 \times 10^{-3}}{1.113 \times 10^{-7}}$ (c) $\dfrac{4.21 \times 10^{-9}}{2.107 \times 10^{6}}$ (d) $\dfrac{3.92 \times 10^{10}}{2.3 \times 10^{-4}}$

SOLUTION

(a) Performing the indicated division involving the coefficients gives

$$\frac{2.05}{1.19} = 1.7226891 \quad \text{(calculator answer)}$$

Since both input numbers for the division are three-significant-figure numbers, the calculator answer must be rounded off to three significant figures.

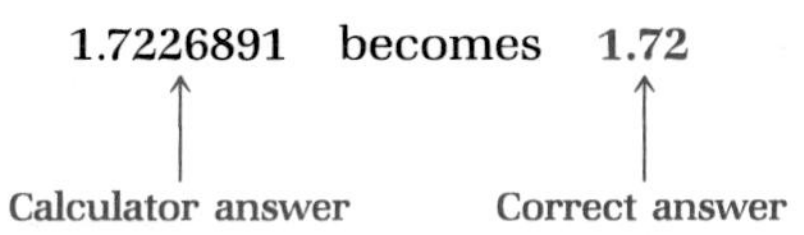

The division of exponential terms involves the algebraic subtraction of exponents.

$$\frac{10^5}{10^3} = 10^{(+5)-(+3)} = 10^2$$

Algebraic subtraction involves changing the sign of the number to be subtracted and then following the rules for addition (as outlined in Ex. 2.11). In this problem the number to be subtracted (+3) becomes, upon changing the sign, a (−3). Then we add (+5) and (−3). The answer is (+2), as shown above. Combining the coefficient and the exponential term gives

$$1.72 \times 10^2$$

(b) The new coefficient is obtained by dividing 1.113 into 7.200.

$$\frac{7.200}{1.113} = 6.4690027 \quad \text{(calculator answer)}$$

The correct answer will contain only four significant figures, the same number as in both of the input numbers for the division.

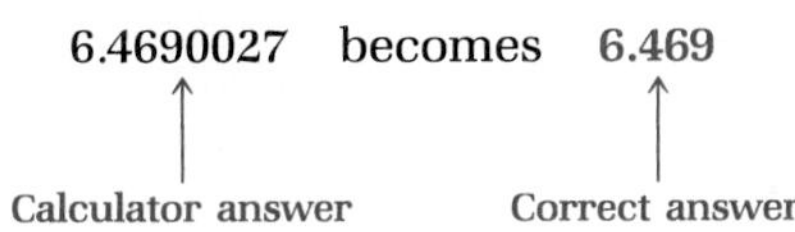

The exponential part of the answer is obtained by subtracting (−7) from (−3). Changing the sign of the number to be subtracted gives (+7). Adding (+7) and (−3) gives (+4). Therefore,

$$\frac{10^{-3}}{10^{-7}} = 10^{(-3)-(-7)} = 10^4$$

Combining the coefficient and the exponential term gives

$$6.469 \times 10^4$$

(c) One input number for the coefficient division has three significant figures, and the other four significant figures. Therefore, the answer should contain three significant figures.

$$\frac{4.21}{2.107} = 1.9981016 \quad \text{(calculator answer)}$$

$$= 2.00 \quad \text{(correct answer)}$$

The exponential-term division involves subtracting (+6) from (−9). Changing the sign of the number to be subtracted gives a (−6). Adding a (−6) and a (−9) gives a (−15).

$$\frac{10^{-9}}{10^{6}} = 10^{(-9)-(+6)} = 10^{-15}$$

Combining the two parts of the problem yields an answer of

$$2.00 \times 10^{-15}$$

(d) The new coefficient, obtained by dividing 3.92 by 2.3, should contain only two significant figures, the same number as in 2.3.

$$\frac{3.92}{2.3} = 1.7043478 \quad \text{(calculator answer)}$$

$$= 1.7 \quad \text{(correct answer)}$$

Performing the exponential-term division by subtracting the powers of the exponential terms gives

$$\frac{10^{10}}{10^{-4}} = 10^{(+10)-(-4)} = 10^{(+10)+(+4)} = 10^{14}$$

The final answer is thus

$$1.7 \times 10^{14}$$

Justification of Exponential Numbers

justified number

A **justified number** *is a number, expressed in exponential form, whose coefficient is a number between one and ten.* When we convert decimal numbers directly to scientific notation, they will be justified automatically. Quite often, however, numbers in exponential form obtained as the result of a calculation (multiplication, division, etc.) are not justified. For example,

$$(4.00 \times 10^2) \times (5.00 \times 10^2) = 20.0 \times 10^4$$

The answer, 20.0×10^4, is not justified because the coefficient is not a number between 1 and 10. When unjustified numbers are obtained through calculation, they must be justified as the last step in the calculation.

Justification of a number is accomplished by writing the coefficient itself as an exponential number and then multiplying it and the original power of ten together. For example, to justify 354×10^3, we write 354 as 3.54×10^2 and then multiply it by 10^3.

$$354 \times 10^3 = (3.54 \times 10^2) \times 10^3 = 3.54 \times 10^5$$

EXAMPLE 2.13

Justify the following exponential numbers.

(a) 63.2×10^2 **(b)** 0.343×10^{-3} **(c)** 234×10^{-4} **(d)** 0.0111×10^6

SOLUTION

(a) First we rewrite 63.2 as 6.32×10^1. We then multiply it and the original power of 10.

$$63.2 \times 10^2 = (6.32 \times 10^1) \times 10^2 = 6.32 \times 10^3$$

(b) We rewrite 0.343 as 3.43×10^{-1}. Multiplying it and the original power of ten gives

$$0.343 \times 10^{-3} = (3.43 \times 10^{-1}) \times 10^{-3} = 3.43 \times 10^{-4}$$

(c) Following the same procedures as in parts (a) and (b), we get

$$234 \times 10^{-4} = (2.34 \times 10^2) \times 10^{-4} = 2.34 \times 10^{-2}$$

(d) Following the same procedures again, we get

$$0.0111 \times 10^6 = (1.11 \times 10^{-2}) \times 10^6 = 1.11 \times 10^4$$

Changing the Sign of an Exponent

In performing calculations it is frequently advantageous to shift an exponential term from the denominator to the numerator of a fraction (or vice versa). Mathematically this is allowable if the sign of the exponent is changed. The following examples illustrate this operation.

$$\frac{5}{2 \times 10^{-2}} = \frac{5 \times 10^2}{2}$$

(Negative exponent 10^{-2} → Positive exponent 10^2)

$$\frac{1}{10^6} = 1 \times 10^{-6}$$

(Positive exponent 10^6 → Negative exponent 10^{-6})

$$4 \times 10^{-3} = \frac{4}{10^3}$$

(Negative exponent 10^{-3} → Positive exponent 10^3)

Note that *only exponential terms* may be shifted from numerator to denominator (or vice versa) in the preceding manner. This sign change method is not mathematically valid for coefficients.

Addition and Subtraction in Scientific Notation

To add or subtract numbers written in scientific notation, *the power of ten for all numbers must be the same.* More often than not, one or more exponents must be adjusted. Adjusting the exponent requires rewriting the number in an unjustified form. With exponents all the same, *the coefficients are then added or subtracted and the exponent is maintained at its now common value.* Although any of the exponents may be changed, changing the smaller numbers to larger numbers will usually eliminate the need to justify the resulting answer.

Changing an exponent (unjustifying a number) involves the reverse of the steps previously discussed for justifying a number. This process is illustrated several times in Example 2.14.

EXAMPLE 2.14

Perform the following additions or subtractions with all numbers expressed in scientific notation. Answers will need to be checked for justification and also for the correct number of significant figures.

(a) $(2.661 \times 10^3) + (3.011 \times 10^3)$ **(b)** $(2.66 \times 10^3) - (2.66 \times 10^2)$
(c) $(6.48 \times 10^{-3}) + (2.04 \times 10^{-5})$ **(d)** $(2.1 \times 10^5) + (3.05 \times 10^2)$

SOLUTION

(a) The exponents are the same to begin with. Therefore, we can proceed with the addition immediately.

$$\begin{array}{r} 2.661 \times 10^3 \\ +3.011 \times 10^3 \\ \hline \mathbf{5.672 \times 10^3} \end{array} \quad \text{(calculator and correct answer)}$$

Note that the exponent values are not added.

(b) The exponents are different, so before subtracting we must change one of the exponents. Let us change 10^2 to 10^3.

$$10^2 \text{ can be rewritten as } 10^3 \times 10^{-1}$$

Then,

$$2.66 \times 10^2 = \underbrace{2.66 \times 10^{-1}}_{} \times 10^3$$
$$= \quad 0.266 \quad \times 10^3$$

Subtracting, we have

Common exponent

$$\begin{array}{rr} 2.66 \times 10^3 = & 2.66 \times 10^3 \\ -2.66 \times 10^2 = & -0.266 \times 10^3 \\ \hline & 2.394 \times 10^3 \end{array} \quad \text{(calculator answer)}$$

The calculator answer must be adjusted for significant figures. The last digit common to both input numbers is in the hundredths place, so the correct answer cannot have any digits to the right of the hundredths place (recall Ex. 2.6 on adding and significant figures).

$$2.394 \times 10^3 \quad \text{becomes} \quad \mathbf{2.39 \times 10^3}$$

Calculator answer Correct answer

Note a general principle here concerning adjusting positive exponents upward. To adjust a positive exponent upward requires the introduction of a 10^{-n} factor, where the value of n is the magnitude of the increase in exponent value.

$$10^2 = 10^3 \times 10^{-1} \qquad 3 - 2 = 1$$
$$10^2 = 10^4 \times 10^{-2} \qquad 4 - 2 = 2$$
$$10^2 = 10^5 \times 10^{-3} \qquad 5 - 2 = 3$$
$$10^2 = 10^6 \times 10^{-4} \qquad 6 - 2 = 4$$

(c) Again, the exponents are different. Let us change the 10^{-5} exponential term to a 10^{-3} term, an increase of two powers of ten.

$$10^{-5} = 10^{-3} \times 10^{-2}$$

Then,

$$2.04 \times 10^{-5} = \underbrace{2.04 \times 10^{-2}} \times 10^{-3}$$
$$= 0.0204 \times 10^{-3}$$

Now, upon adding, we have

Common exponent

$$\begin{array}{rcl} 6.48 \times 10^{-3} & = & 6.48 \times 10^{-3} \\ 2.04 \times 10^{-5} & = & 0.0204 \times 10^{-3} \\ \hline & & 6.5004 \times 10^{-3} \end{array}$$ (calculator answer)

Adjusting the calculator answer for significant figures gives us a final answer of 6.50×10^{-3}. The last digit position common to both input numbers in the addition is the hundredths place.

(d) The exponents are 10^5 and 10^2. Let us have 10^5 as the common exponent.

10^2 can be rewritten as $10^5 \times 10^{-3}$

Then, we have

$$3.05 \times 10^2 = \underbrace{3.05 \times 10^{-3}} \times 10^5$$
$$= 0.00305 \times 10^5$$

With both numbers now containing a 10^5 term we can proceed to add.

$$\begin{array}{rcl} 2.1 \times 10^5 & = & 2.1 \times 10^5 \\ 3.05 \times 10^2 & = & 0.00305 \times 10^5 \\ \hline & & 2.10305 \times 10^5 \end{array}$$ (calculator answer)

The last common digit place for both coefficients is the tenths place. Therefore, the answer must be rounded to the tenths place.

2.10305×10^5 becomes 2.1×10^5

Calculator answer — Correct answer

> Note that the answer, 2.1×10^5, is the same as one of the original numbers added. The message of this problem is: Adding a small number to a large number causes little or no change in the larger number if the orders of magnitude of the two numbers are far apart relative to the number of significant figures involved. An everyday example of this concept would be the effect on your weight, as shown by a bathroom scale, of putting a Band-Aid on your big toe.

Learning Objectives

After completing this chapter you should be able to

- Illustrate the difference between the precision and accuracy of a measurement (Sec. 2.1).
- Determine the number of significant figures in a number resulting from a measurement (Sec. 2.2).
- Identify numbers that have an unlimited number of significant figures and explain why they do (Sec. 2.2).
- Distinguish between significant and nonsignificant zeros in numbers containing zeros (Sec. 2.2).
- Round off measured values to a specified number of significant figures (Sec. 2.3).
- Add, subtract, multiply, and divide numbers, giving your answer to the correct number of significant figures (Sec. 2.3).
- Convert numbers from decimal notation to scientific notation and vice versa (Sec. 2.4).
- Carry out mathematical operations with all numbers expressed in scientific notation (Sec. 2.5).

Terms and Concepts for Review

The new terms or concepts defined in this chapter are

precision (Sec. 2.1)
accuracy (Sec. 2.1)
significant figures (Sec. 2.2)
rounding off (Sec. 2.3)
scientific notation (Sec. 2.4)
exponent (Sec. 2.4)
order of magnitude (Sec. 2.4)
justified number (Sec. 2.5)

Questions and Problems

Accuracy and Precision

2-1 With a very precise measuring device, the length of an object is determined to be 6.32141 centimeters. Three students, I, II, and III, are asked to determine the length of this same object using a less-precise measuring device. Each student measures the object's length, in centimeters, six times with the following results.

I: 6.54, 6.53, 6.55, 6.55, 6.56, and 6.52 (average = 6.54)
II: 7.31, 5.55, 5.82, 6.32, 6.67, and 7.92 (average = 6.60)
III: 6.31, 6.31, 6.33, 6.32, 6.34, and 6.31 (average = 6.32)

Characterize each student's performance in terms of accuracy (good or poor) and precision (good or poor).

Significant Figures

2-2 Determine the number of significant figures in each of the following measured values.

a. 13.27 **b.** 0.00347 **c.** 2.1010
d. 0.00300 **e.** $2\overline{70,0}00$ **f.** 10.0
g. 3,140,000 **h.** 0.10034

2-3 Which of the following numbers contain three significant figures?

a. 3,330 **b.** 3.330 **c.** 3.0030
d. 300 **e.** 0.3330 **f.** $33\overline{0},000$
g. 33.0 **h.** 0.033

2-4 The following laboratory measurements are all properly recorded relative to the uncertainty of the measurement. Circle the uncertain digit(s) in each number.

a. 0.000033 lb **b.** 3.3 lb **c.** $33\overline{0}$ lb
d. $3\overline{0}00$ lb **e.** 0.0303 lb **f.** 3 lb
g. 300 ± 10 lb **h.** 333.1131 lb

2-5 What is the magnitude of the uncertainty conveyed by each of the following notations for a measurement involving the number one thousand?

a. 1000 **b.** $1\overline{000}$ **c.** 1000 ± 100
d. 1000. **e.** 1000.000 **f.** $1\overline{00}0$

2-6 How many significant digits are there in each underlined number? If there is an infinite number, explain why.

a. 1 ream = $\underline{500}$ sheets
b. U.S. population = $\underline{223,000,000}$
c. object's length = $\underline{37.2}$ inches
d. outside temperature = $\underline{23}$°F

e. 7 oranges
f. 36 inches = 1 yard
g. chapter length = 37 pages
h. 1 bottle = 4.05 lb

Rounding Off

2-7 Round off each of the following numbers to three significant figures.

a. 3883 b. 0.277623 c. 1,030,001
d. 0.55555 e. 4.4050 f. 2.335
g. 12,343 h. 1000.

2-8 Drop the last two digits in each of the following numbers, using correct rounding techniques. Be careful not to decrease the number of significant figures by more than two.

a. 0.123 b. 1,231 c. 0.0001255
d. 210 e. 0.10012 f. 3.350
g. 1.2245 h. 1,000,012

Significant Figures in Calculations

2-9 Carry out the following multiplications, expressing your answer to the correct number of significant figures. Assume that all numbers are measured numbers.

a. 4.76×3.0
b. 1.001×2.0
c. 3.7666×4.7666
d. 1.745763×0.000023
e. 4×4
f. 8×3.576384
g. $1234 \times 2341 \times 3.003$
h. $3.00 \times 3.00 \times 3.000 \times 3.0000$

2-10 Carry out the following divisions, expressing your answer to the correct number of significant figures. Assume that all numbers are measured numbers.

a. $\frac{2750}{3.3333}$ b. $\frac{1000}{2.3}$ c. $\frac{8}{3.1}$
d. $\frac{3.00}{3.01}$ e. $\frac{2700.0}{3.0000}$ f. $\frac{4}{16}$
g. $\frac{3.12376}{2.0}$ h. $\frac{1,000,000}{2.73}$

2-11 Carry out the following mathematical operations, expressing your answers to the correct number of significant figures. Assume that all numbers arise from measurement.

a. $\frac{2.0 \times 8.0}{1.1}$ b. $\frac{3.333 \times 4.444}{5.50}$
c. $\frac{2.475}{3.03 \times 4.03}$ d. $\frac{6.671 \times 3.001}{0.03134 \times 0.0012}$

2-12 Perform the following addition or subtraction computations. Report your results to the proper number of significant figures. Assume that all numbers are measured numbers.

a. $4 + 8 + 22$ b. $4.0 + 8.00 + 22.00$
c. $3.17 - 3.1$ d. $3.233 - 2.5709$
e. $999.0 + 0.7 + 0.37$ f. $1200 + 53 + 3.137$
g. $0.2317 - 0.00317$ h. $50.0 + 0.00000550$

Exponents and Orders of Magnitude

2-13 Indicate the meaning of the following exponential terms.

a. 2^3 b. 3^4 c. 10^6 d. 10^0
e. 10^{10} f. 10^{-3} g. 10^{-5} h. 10^{-2}

2-14 Perform the indicated changes on the following exponential terms.

a. 10^4; increase by 4 orders of magnitude
b. 10^6; decrease by 2 orders of magnitude
c. 10^2; decrease by 4 orders of magnitude
d. 10^{-3}; increase by 2 orders of magnitude

Scientific Notation

2-15 Express the following numbers in scientific notation.

a. 234 b. 0.00317 c. 30.3002
d. 1,000,000 e. 0.230 f. 0.0000100
g. 3235 h. 23.75

2-16 Express the following numbers in decimal notation.

a. 2.34×10^6 b. 2.707×10^{-3}
c. 1.20×10^6 d. 3.000×10^{-4}
e. 3×10^{-2} f. 6.75×10^5
g. 3.0×10^1 h. 3×10^1

2-17 Express the number one million in scientific notation to

a. one significant figure
b. three significant figures
c. five significant figures
d. ten significant figures

2-18 Express the numerical portion of the following important facts in scientific notation.

a. The alligator population of the United States is an estimated 734,400.
b. A sheet of paper is 0.0042 inch thick.
c. The world's oldest captive baboon is 45 years old.
d. The highest price ever paid for a stuffed bird was 23,400 dollars for a Great Auk taken in Iceland.
e. More than 2,500,000 nontournament pool balls are manufactured in the U.S. each year.
f. The Asiatic elephant has a period of gestation of 609 days.
g. The maximum speed of a three-toed sloth is 0.15 mile per hour.
h. The loudest noise ever created in a laboratory came from a 576 inch steel and concrete horn in Huntsville, Alabama.

2-19 Express the following numbers in scientific notation to three significant digits.

a. 234,123 b. 0.003300 c. 1000.000
d. 0.30003 e. 1,115,000 f. 6,666
g. 0.000333 h. 3.000000

Mathematical Operations in Scientific Notation

2-20 Carry out the following multiplications, expressing each answer in scientific notation. Be sure to consider significant figures in obtaining your final answer.

a. $(3 \times 10^2) \times (2 \times 10^4)$
b. $(2 \times 10^{-4}) \times (4 \times 10^{-5})$
c. $(1 \times 10^{-10}) \times (9 \times 10^6)$
d. $(3 \times 10^7) \times (3 \times 10^{-4})$
e. $(2.63 \times 10^2) \times (1.783 \times 10^2)$
f. $(3.00 \times 10^1) \times (2.0 \times 10^{-4})$
g. $(2.7 \times 10^{-3}) \times (2.63 \times 10^{-4}) \times (1.011 \times 10^8)$
h. $(6 \times 10^8) \times (1.023 \times 10^{-2}) \times (1.111 \times 10^{-3})$

2-21 Carry out the following divisions, expressing each answer in scientific notation. Be sure to consider significant figures in obtaining your final answer.

a. $\dfrac{4.0 \times 10^4}{2.0 \times 10^2}$ b. $\dfrac{4.0 \times 10^4}{2.0 \times 10^{-2}}$

c. $\dfrac{4.0 \times 10^{-2}}{2.0 \times 10^4}$ d. $\dfrac{4.0 \times 10^{-2}}{2.0 \times 10^{-4}}$

e. $\dfrac{8.73 \times 10^9}{2.63 \times 10^8}$ f. $\dfrac{3.0 \times 10^{-9}}{2.73 \times 10^4}$

g. $\dfrac{7 \times 10^6}{3 \times 10^{-4}}$ h. $\dfrac{6.237 \times 10^{-5}}{3.0 \times 10^{-4}}$

2-22 Justify the following exponential-notation numbers.

a. 63.2×10^4 b. 602×10^2
c. 0.32×10^2 d. 0.003×10^5
e. 63.2×10^{-4} f. 602×10^{-2}
g. 0.32×10^{-2} h. 0.003×10^{-5}

2-23 Perform the following mathematical operations. Be sure your answer contains the correct number of significant figures and that it is justified.

a. $(6 \times 10^2) \times (3 \times 10^3)$
b. $(9.2 \times 10^{-4}) \times (1.3 \times 10^7)$
c. $\dfrac{2.3 \times 10^{-7}}{9.6 \times 10^4}$
d. $\dfrac{1.1 \times 10^4}{8.79 \times 10^{-2}}$
e. $\dfrac{(6 \times 10^2) \times (8 \times 10^6)}{3 \times 10^7}$
f. $\dfrac{(6.32 \times 10^{-2}) \times (5.43 \times 10^{-3})}{2.0 \times 10^{-10}}$
g. $\dfrac{(8 \times 10^3) \times (8 \times 10^3)}{4 \times 10^6}$
h. $\dfrac{6.37384 \times 10^8}{(9.123 \times 10^6) \times (9.030 \times 10^{-6})}$

2-24 Perform the following mathematical operations. Be sure your answer contains the correct number of significant figures and that it is justified.

a. $(2.661 \times 10^3) + (3.133 \times 10^3)$
b. $(3.03 \times 10^3) - (1.2 \times 10^3)$
c. $(4.0 \times 10^6) - (4.0 \times 10^2)$
d. $(3.67 \times 10^4) + (3.67 \times 10^5)$
e. $(2.000 \times 10^6) - (2.00 \times 10^4)$
f. $(6.00 \times 10^6) + (5.0 \times 10^5) + (5 \times 10^4)$
g. $(3.1 \times 10^5) - (2.67 \times 10^1)$
h. $(3.1111 \times 10^5) + (3.1 \times 10^2)$

3 Units Systems and Dimensional Analysis

3.1 The Metric System of Units

All measurements consist of two parts: a number that tells the amount of the quantity measured, and a unit that tells the nature of the quantity measured. Chapter 2 dealt with the interpretation and manipulation of the number part of a measurement. We now turn our attention to units.

A unit is a "label" that describes what is being measured (or counted). It can be almost anything: 4 quarts, 4 dimes, 4 dozen frogs, 4 bushels, 4 inches, or 4 pages. Two formal systems of units of measure are in use in the United States today. Common measurements in commerce, such as those done in a grocery store, are made in the **English system.** The units of this system include the familiar inch, foot, pound, quart, and gallon. A second system, the **metric system,** is used in scientific work. Units in this system include the gram, meter, and liter. The United States is one of only a very few countries which use differing unit systems in commerce and scientific work. On a worldwide basis, almost universally, the metric system is used for both commercial and scientific work.

The United States is in the process of a voluntary conversion to the metric system. Many metric system units are now appearing on consumer products. Soft drinks may now be bought in 2 liter containers. Road signs in some states display distances in both miles and kilometers. Canned and packaged goods (cereals, mixes, fruits, etc.) on grocery store shelves now have their content weights listed in grams as well as in ounces or pounds.

Why should the United States convert to the metric system? The answer is simple. The metric system is superior to the English system. Its superiority lies in the area of interrelationships between units of the same type (volume, length, etc.). Metric unit interrelationships are less complicated than English unit interrelationships, because the metric system is a decimal unit system. In the metric system conversion from one unit size to another can be accomplished simply by moving the decimal point to the right or left an appropriate number of places. The metric system is no more precise than the English system, only more convenient.

A recent modification of the metric system has been developed. It is called the International System of Units and is abbreviated SI (after the French name Systeme International). Since SI units are still not in universal use, we shall use the more traditional metric system in this text.

Because the SI and metric systems are very similar, switching to the SI system sometime in the future, when its use is more extensive, will present no major difficulties to the student who properly understands the traditional metric system.

Metric System Prefixes

In the metric system there is one base unit for each type of measurement — length, volume, mass, etc. These base units are then multiplied by appropriate powers of ten to form smaller or larger units. The names of the larger and smaller units are constructed from the base unit name by attaching to it a prefix that tells which power of ten is involved. These prefixes are given in Table 3-1, along with their symbols or abbreviations and mathematical meanings. The prefixes in color are those most frequently used.

The use of numerical prefixes should not be new to you. Consider the use of the prefix tri- in the following words: triangle, tricycle, trio, trinity, triple. Every one of these words conveys the idea of three of something. In the same way we will use the metric system prefixes.

The meaning of a prefix always remains constant; it is independent of the base it modifies. For example, a kilosecond is a thousand seconds; a kilowatt a thousand watts; and a kilocalorie a thousand calories. The prefix kilo- will always mean a thousand.

Table 3-1. Metric System Prefixes and Their Mathematical Meanings

		Mathematical Meaning			
Prefix	**Symbol**	**Exponential Number**	**Common Number**	**Origin**	**Original Meaning**
Tera-	T	10^{12}	1,000,000,000,000	Greek	monstrous
Giga-	G	10^{9}	1,000,000,000	Greek	gigantic
Mega-	M	10^{6}	1,000,000	Greek	great
Kilo-	k	10^{3}	1,000	Greek	thousand
Hecto-	h	10^{2}	100	Greek	hundred
Deca-	da	10^{1}	10	Greek	ten
Deci-	d	10^{-1}	1/10	Latin	tenth
Centi-	c	10^{-2}	1/100	Latin	hundredth
Milli-	m	10^{-3}	1/1000	Latin	thousandth
Micro-	μ	10^{-6}	1/1,000,000	Greek	small
Nano-	n	10^{-9}	1/1,000,000,000	Greek	very small
Pico-	p	10^{-12}	1/1,000,000,000,000	Spanish	extremely small

Metric Units of Length

meter

The **meter** *is the basic unit of length in the metric system.* (*Metre* has been adopted as the preferred international spelling for this unit, but *meter* is the spelling used in the United States and in this book.)

Other units of length in the metric system are derived from the meter by using the prefixes listed in Table 3-1. The kilometer (km) is 1000 times larger than the meter, whereas the centimeter (cm) and millimeter (mm) are, respectively, 100 and 1000 times smaller than the meter. Note that the abbreviation for meter is a lower case m.

A comparison of metric lengths with the commonly used English system lengths of mile, yard, and inch gives us some size perspective. The meter is slightly larger than a yard, a kilometer is approximately three fifths of a mile, and a centimeter is about two fifths of an inch (see Fig. 3-1).

A better "feel" for lengths equivalent to a meter, centimeter, and millimeter can, perhaps, be obtained by relating these metric units to objects and situations we encounter in everyday life. This is done in Figure 3-2 in terms of paper clips, nickels, index fingers, basketball players, and preschool children.

Metric Units of Mass

gram

Mass is a measurement of the total quantity of matter present in an object. A **gram** *is the basic unit of mass in the metric system.* The gram is relatively small compared to the commonly used English mass units of ounce and pound. It takes approximately 28 grams to equal an ounce and nearly 454 grams to equal a pound. Because of the small size of the gram, the kilogram (kg) is a very commonly used unit. (The abbreviation for gram is a lower case g.) A kilogram is equal to approximately 2.2 pounds, as shown in Figure 3-3.

To give you a "feel" for metric mass unit magnitude, Figure 3-4 relates the mass units of milligram (mg), gram, and kilogram to everyday objects. Comparisons involve a single staple, two thumb tacks, a coin, a quart of milk in a paper carton, and a "good-sized" football player.

The terms mass and weight are frequently used interchangeably in discussions. Although in most cases this practice does no harm, technically it is incorrect to

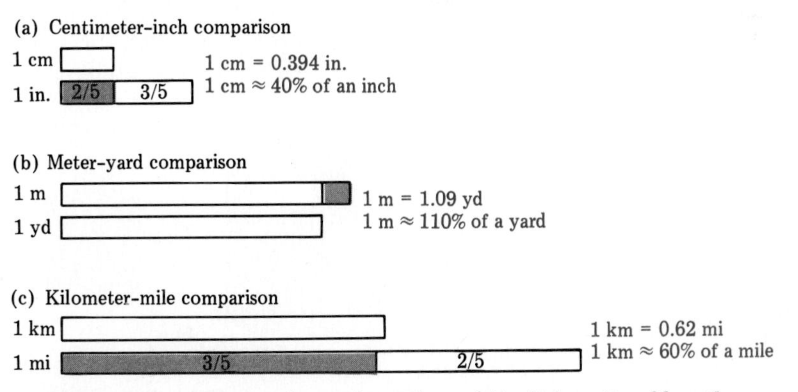

Figure 3-1. A comparison of metric and English units of length.

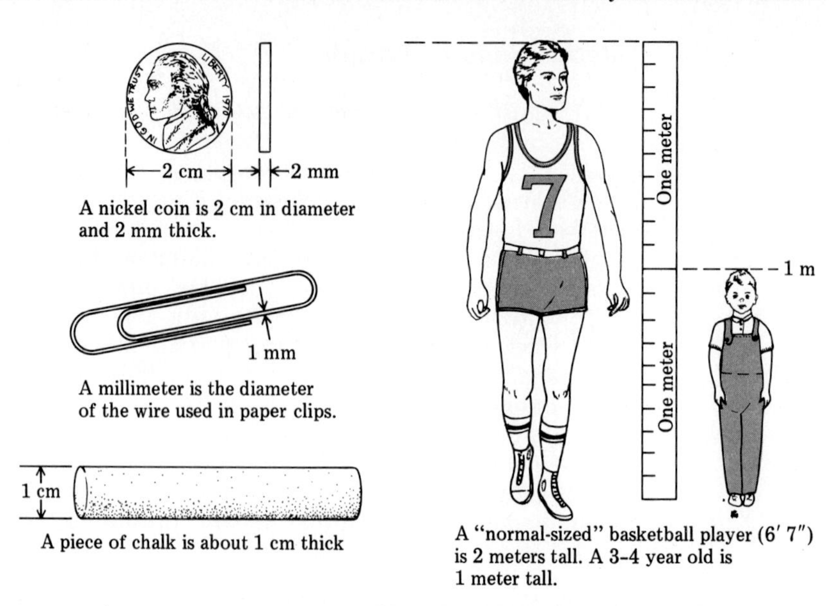

Figure 3-2. Metric units of length and the "everyday realm."

interchange the terms. Mass and weight refer to different properties of matter and their difference in meaning should be understood.

mass

weight

Mass *is a measure of the total quantity of matter in an object.* **Weight** *is a measure of the force exerted on an object by the pull of gravity.* The mass of a substance is a constant; the weight of an object is a variable dependent upon the geographical location of that object.

Matter at the equator weighs less than it would at the North Pole because the earth is not a perfect sphere, but bulges at the equator. As a result, an object at the equator is farther from the center of the earth. It therefore weighs less, because the magnitude of gravitational attraction (the measure of weight) is inversely proportional to the distance between the centers of the attracting objects; that is, the gravitational attraction is larger when the objects are closer together and smaller when the objects are farther apart. Gravitational attraction also depends on the masses of the attracting bodies; the greater the masses, the greater the attraction. For this reason, an object would weigh much less on the moon than on earth due to the smaller size of the moon and the correspondingly lower gravitational attraction. Quantitatively, a 22.0 pound mass weighing 22.0 pounds at the earth's North Pole would weigh 21.9 pounds at the earth's

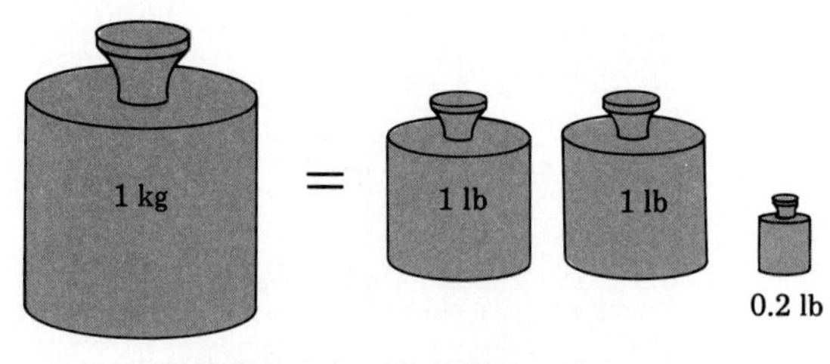

Figure 3-3. A comparison of the metric kilogram with the English pound.

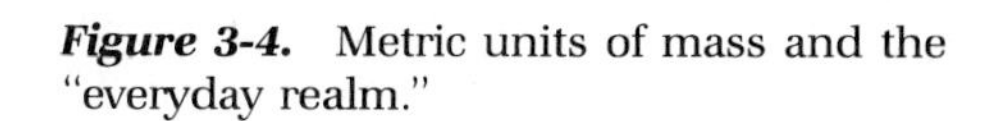

Figure 3-4. Metric units of mass and the "everyday realm."

A single staple has a mass of about 3 mg.

2 thumbtacks have a mass of 1 g.

A nickel has a mass of about 5 g.

1 quart of milk in a cardboard container has a mass of 1 kg.

A 220-lb football player is equivalent to a mass of 100 kg.

equator and only 3.7 pounds on the moon. In outer space an astronaut may be weightless, but he can never be massless. In fact, he has the same mass in space as on earth.

Different types of equipment are used to determine mass and weight. A *balance* is used to determine the mass of an object. The left side of Figure 3-5 shows three common types of balances. With balances such as these, the actual mass of the object can be obtained by balancing the object against objects whose masses are known. The object whose mass is to be determined is placed on the left pan of balance (a) and (b), and objects ("weights") of known mass are placed on the right-hand pan and/or sliding scale to counterbalance the object. Balance (c) appears to have only one pan. The counterbalancing masses for this balance are located within the balance case and are put in place with "knobs and levers." Gravitational attraction will have the same effect on the object whose mass is to be determined and the counterbalancing masses, thus making the measurement independent of geographical location.

Spring *scales* are used to determine weight. Bathroom scales and grocer's scales are spring scales. The object to be weighed stretches the spring (Fig. 3-5d) with the extent of stretching dependent on gravitational attraction.

The processes of determining mass and weight are both referred to as "weighing" since the word "massing" is not an accepted part of the English language (although it should be). This dual usage of the term weighing leads to the practice (somewhat common) of using mass and weight interchangeably. Many textbooks follow this practice, although it will not be followed in this one. A mass is not a weight. A mass will be called a mass. Chemists always use balances rather than scales. Hence they always measure the mass of an object.

Metric Units of Volume

volume

An object's **volume** *is the amount of space it occupies.* Frequently it is convenient and often more rapid to measure the volume rather than the mass of a substance. This is particularly true for liquids and gases. For example, volume is used instead of mass to determine the amount of gasoline put into an automobile fuel tank. A **liter** *is the basic unit of volume in the metric system.* (As with meter, we will use the United States

liter

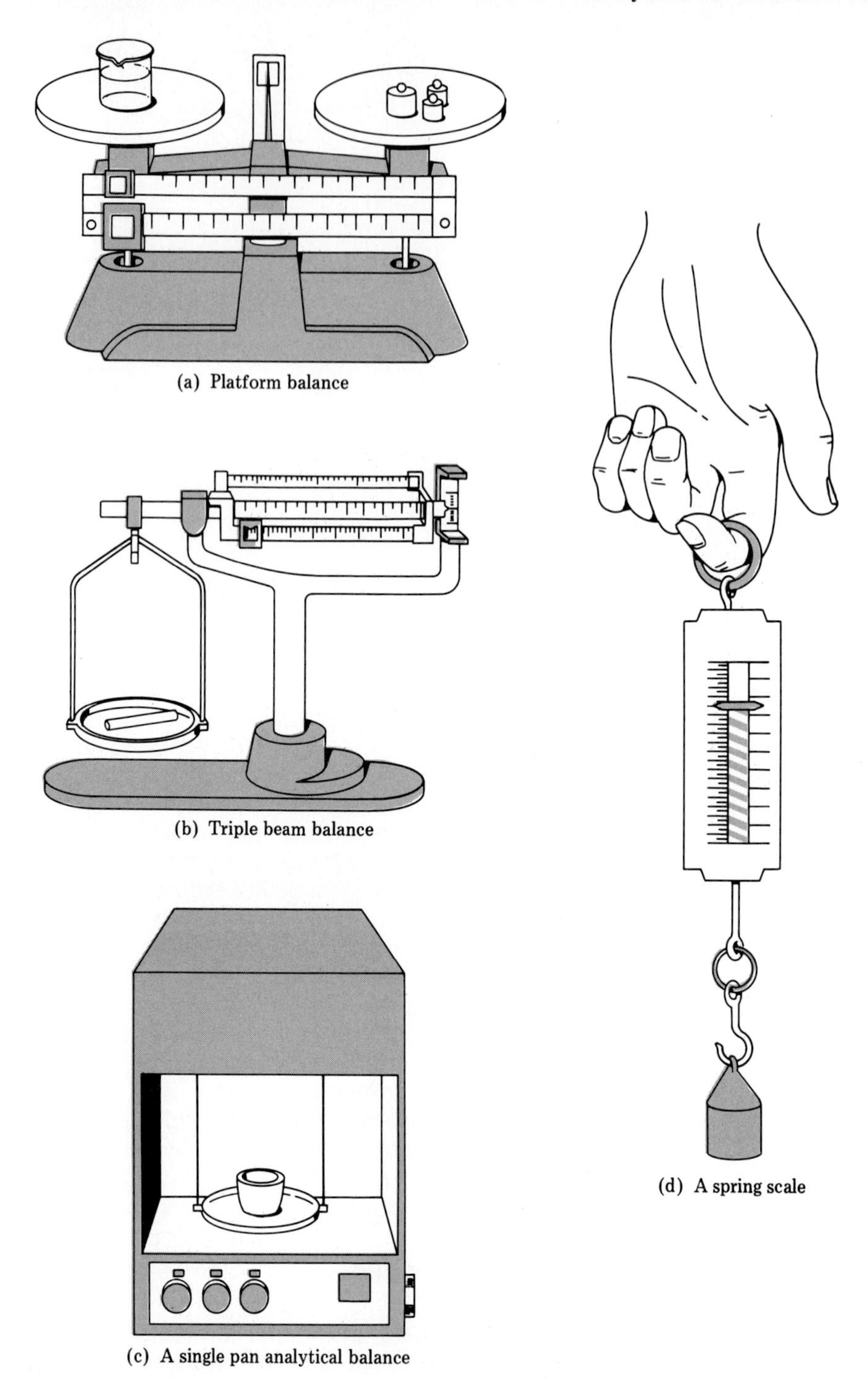

(a) Platform balance

(b) Triple beam balance

(c) A single pan analytical balance

(d) A spring scale

Figure 3-5. Equipment used to determine mass and weight. Balances to measure mass are on the left (a, b and c) and a scale to measure weight (d) is on the right.

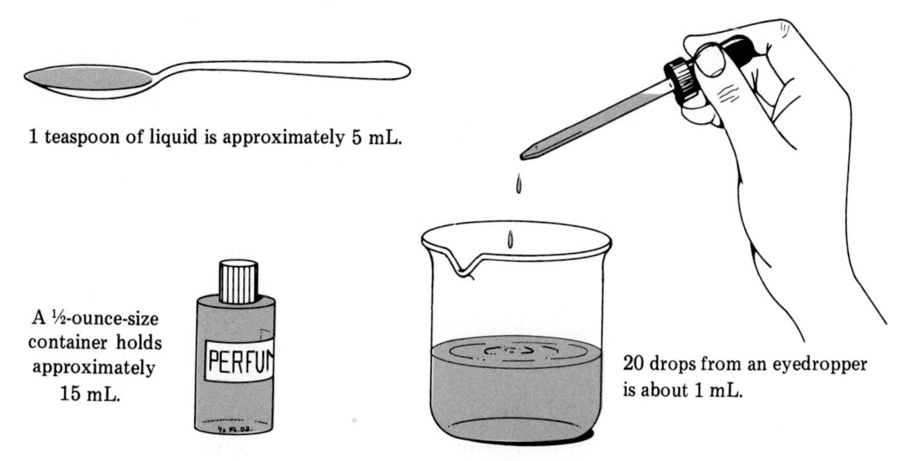

Figure 3-6. The metric volume unit of milliliter and the "everyday realm."

spelling rather than the international spelling which is litre.) A liter is abbreviated by a capital L. If a lower case l was used for the abbreviation, the problem of confusing the abbreviation for liter with the number one would exist.

As with the units of length and mass, the basic unit of volume is modified with prefixes to represent smaller or larger units. The most commonly used prefixed volume unit is the milliliter (mL), which is 1/1000 L. A milliliter is much smaller than a fluid ounce; it takes approximately 30 milliliters to equal 1 fluid ounce. A quart contains 946 milliliters. Figure 3-6 contains comparisons to give you a better "feel" for the size of a milliliter.

A liter is defined as the volume occupied by a perfect cube, 10 centimeters on each edge. The volume of a cube is calculated by multiplying length × width × height. Since all three dimensions are the same for a cube, we have

$$\begin{aligned} 1\text{ L} &= \text{volume of a 10-cm-edged cube} \\ &= 10\text{ cm} \times 10\text{ cm} \times 10\text{ cm} \\ &= 1000\text{ cm}^3 \text{ (centimeters cubed or cubic centimeters)} \end{aligned}$$

An alternate abbreviation for cubic centimeter is cc. The terminology cc is most frequently used in medically oriented situations.

A liter is equal to 1000 milliliters. We have also just seen that it is equal to 1000 cubic centimeters. Therefore,

$$1000\text{ mL} = 1000\text{ cm}^3$$

Dividing through by a thousand, we find

$$1\text{ mL} = 1\text{ cm}^3$$

Consequently, the units mL and cm^3 are interchangeable. In practice, mL is usually used for volumes of liquids and gases and cm^3 for volumes of solids. Figure 3-7 shows relationships between English and metric units of volumes in terms of cubic volumes.

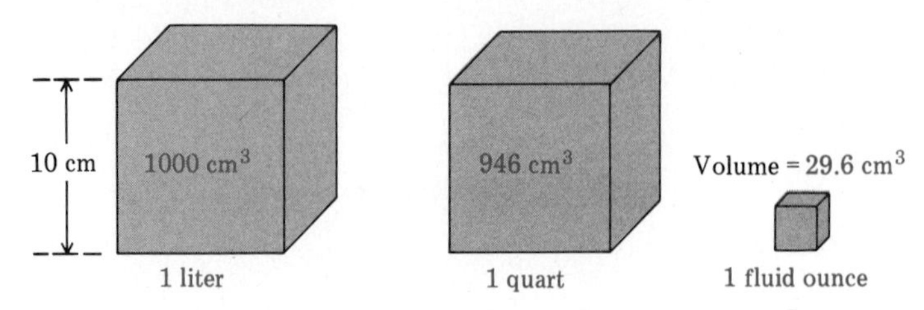

Figure 3-7. A comparison of metric and English units of volume.

3.2 Conversion Factors and Dimensional Analysis

Many times a need arises to change the units of a quantity or measurement to different units. In this section we deal with this topic. The new units needed may be in the same measurement system as the old ones or in a different one. With two unit systems in common use in the United States, the need to change measurements in one system to their equivalent in the other frequently occurs.

The mathematical tool we will use to accomplish the task of changing units is a general method of problem solving called *dimensional analysis.* By this method, unit conversion is accomplished by multiplying a given quantity or measurement by one or more conversion factors to obtain the desired quantity or measurement.

$$\text{Given quantity} \times \text{conversion factor(s)} = \text{desired quantity}$$

Prior to discussing the mechanics involved in using dimensional analysis to solve "unit-conversion problems" some comments concerning conversion factors are in order. A proper understanding of conversion factors is the key to being able to solve problems "comfortably" with dimensional analysis.

conversion factor

Formally defined, a **conversion factor** *is a fraction obtained from a fixed relationship between two quantities which when multiplied by a measurement or quantity does not change the size of the measurement or quantity even though it changes the number and units.*

Multiplication by one or more conversion factors does not change the value of a measurement or quantity because *all conversion factors have a numerical value equal to unity (one).* Multiplication by unity does not change the value of an expression.

A fixed relationship between two quantities gives sufficient information to construct a conversion factor. Let us construct some conversion factors to see (1) how they originate and (2) why they always have values of unity.

The quantities "1 minute" and "60 seconds" both describe the same amount of time. We may write an equation describing this fact.

$$1 \text{ min} = 60 \text{ sec}$$

This fixed relationship may be used to construct a pair of conversion factors that relate seconds and minutes. (Conversion factors always occur in pairs, as will become obvious shortly.)

Dividing both sides of our minute–second equation by the quantity "1 minute" gives

$$\frac{1 \text{ min}}{1 \text{ min}} = \frac{60 \text{ sec}}{1 \text{ min}}$$

Since the numerator and denominator of the fraction on the left are identical, this fraction has a value of unity.

$$1 = \frac{60 \text{ sec}}{1 \text{ min}}$$

The fraction on the right side of the equation is our conversion factor. Its value is one. Note that the numerator and denominator of the conversion factor describe the same "amount" of time.

Two conversion factors are always obtainable from any given equality. For the equality we are considering (1 min = 60 sec) the second conversion factor is

$$\frac{1 \text{ min}}{60 \text{ sec}}$$

It is obtained by dividing both sides of the equality by "60 seconds" instead of "1 minute." The relationship between the two conversion factors is that of reciprocals.

$$\frac{1 \text{ min}}{60 \text{ sec}} \quad \text{and} \quad \frac{60 \text{ sec}}{1 \text{ min}}$$

In general, we will always be able to construct a set of two conversion factors, each with a value of unity, from any two terms which describe the same "amount" of whatever we are considering. The two conversion factors will always be reciprocals of each other. (Conversion factors can also be constructed from two quantities that are equivalent rather than equal. We will consider this situation in Sec. 3.3.)

For convenience in discussing additional information about conversion factors we will classify them into three general categories: (1) metric–metric, (2) English–English, and (3) metric–English or English–metric.

Metric–Metric Conversion Factors

Metric–metric conversion factors are used to change one metric unit into another metric unit. Both the numerator and the denominator of such conversion factors involve metric system units.

Conversion factors relating a prefixed metric unit to the base unit for that type of measurement are derived from the meaning of the prefix of concern. For example, the set of conversion factors involving kilogram and gram is derived from the meaning of the prefix kilo-, which is 10^3. The two conversion factors are

$$\frac{10^3 \text{ m}}{1 \text{ km}} \quad \text{and} \quad \frac{1 \text{ km}}{10^3 \text{ m}}$$

Note again the reciprocal relationship between the two conversion factors of the set. The conversion factors relating centimeter and meter involve 10^{-2}, the mathematical equivalent of centi-, and are

$$\frac{1 \text{ cm}}{10^{-2} \text{ m}} \quad \text{and} \quad \frac{10^{-2} \text{ m}}{1 \text{ cm}}$$

For conversion factors of the type now under discussion, the numerical equivalent of the prefix always goes with the base unit.

All metric–metric conversion factors are derived from exact definitions. Thus, they all contain an unlimited number of significant figures. This observation about significant figures will be of major importance when these conversion factors become part of an actual calculation.

English–English Conversion Factors

English–English conversion factors also contain an unlimited number of significant figures, since they also all result from defined equalities. Twelve inches equals 1 foot — exactly. Three feet equal 1 yard — exactly. The majority of English–English conversion factors should be "second nature" to most students. You have used these relationships all your life, even though you may not have formally thought of them as conversion factors. Representative of these "second-nature" conversion factors are the following.

$$\frac{2\ \text{pt}}{1\ \text{qt}} \qquad \frac{36\ \text{in.}}{1\ \text{yd}} \qquad \frac{1\ \text{mi}}{5280\ \text{ft}} \qquad \frac{16\ \text{oz}}{1\ \text{lb}}$$

Metric–English and English–Metric Conversion Factors

Conversion factors that relate metric units to English units or vice versa are not exact defined quantities, since they involve two different systems of measurement. The numbers associated with such conversion factors must be determined experimentally. Since they are not "exact," concern about the number of significant figures present will have to be a consideration every time they are used. In most instances we will quote these conversion factors to *four* significant figures. It should be noted, however, that such factors are known to greater precision than that. For example,

1 in. = 2.540 cm to *four* significant figures

and

1 in. = 2.540005 cm to *seven* significant figures

Table 3-2 lists some commonly encountered relationships between metric and English system units. These few factors are sufficient to solve most of the problems that we will encounter. In fact, later in this section we will see that only one factor of a given type (volume, mass, length) is sufficient to solve most problems.

Dimensional Analysis

dimensional analysis

Formally defined, **dimensional analysis** *is a general problem-solving method that uses the units associated with numbers as a guide in setting up the calculation.* This method treats units in the same way as numbers, that is, they may be multiplied, divided, canceled, etc. For example, just as

$$3 \times 3 = 3^2 \text{ (3 squared)}$$

we have

$$\text{km} \times \text{km} = \text{km}^2 \text{ (km squared)}$$

Table 3-2. Equalities and Conversion Factors That Relate the English and Metric Systems of Measurement to Each Other

	Metric to English	English to Metric
Length		
1 in. = 2.540 cm	$\frac{1 \text{ in.}}{2.540 \text{ cm}}$	$\frac{2.540 \text{ cm}}{1 \text{ in.}}$
1 m = 39.37 in.	$\frac{39.37 \text{ in.}}{1 \text{ m}}$	$\frac{1 \text{ m}}{39.37 \text{ in.}}$
1 km = 0.6214 mi	$\frac{0.6214 \text{ mi}}{1 \text{ km}}$	$\frac{1 \text{ km}}{0.6214 \text{ mi}}$
Mass		
1 lb = 453.6 g	$\frac{1 \text{ lb}}{453.6 \text{ g}}$	$\frac{453.6 \text{ g}}{1 \text{ lb}}$
1 kg = 2.205 lb	$\frac{2.205 \text{ lb}}{1 \text{ kg}}$	$\frac{1 \text{ kg}}{2.205 \text{ lb}}$
1 oz = 28.34 g	$\frac{1 \text{ oz}}{28.34 \text{ g}}$	$\frac{28.34 \text{ g}}{1 \text{ oz}}$
Volume		
1 qt = 0.9463 L	$\frac{1 \text{ qt}}{0.9463 \text{ L}}$	$\frac{0.9463 \text{ L}}{1 \text{ qt}}$
1 L = 2.113 pt	$\frac{2.113 \text{ pt}}{1 \text{ L}}$	$\frac{1 \text{ L}}{2.113 \text{ pt}}$
1 fl oz = 29.57 mL	$\frac{1 \text{ fl oz}}{29.57 \text{ ml}}$	$\frac{29.57 \text{ ml}}{1 \text{ fl oz}}$

Also, just as the two's cancel in the expression

$$\frac{\cancel{2} \times 3 \times 6}{\cancel{2} \times 5}$$

the inches cancel in the expression

$$\frac{\cancel{(\text{inch})} \times (\text{cm})}{\cancel{(\text{inch})}}$$

Like units found in the numerator and denominator of a fraction will always cancel, just as like numbers do.

The steps followed in setting up a problem by dimensional analysis are as follows.

Step 1. *Identify the known or given quantity (both a numerical value and units) and the units of the new quantity to be determined.*

This information, which serves as the starting point for setting up the problem, will always be found in the statement of the problem. Write an equation with the given quantity on the left and the units of the desired quantity on the right.

Step 2. *Multiply the given quantity by one or more conversion factors in a manner such that the unwanted (original) units are canceled out leaving only the new desired units.*

The general format for the multiplications is

$$\frac{\text{Information}}{\text{given}} \times \frac{\text{conversion}}{\text{factor(s)}} = \frac{\text{information}}{\text{sought}}$$

The number of conversion factors used depends on the individual problem. Except in the simplest of problems, it is a good idea to predetermine formally the sequence of unit changes to be used. This sequence will be called the unit "pathway."

Step 3. *Perform the mathematical operations indicated by the conversion factor "setup."*

In performing the calculation one needs to double check that all units except the desired set have canceled out. Also the numerical answer needs to be checked to see that it contains the proper number of significant figures.

Now let us work a number of sample problems using dimensional analysis and the steps just outlined. Our first two examples involve only metric system units and thus will involve only metric–metric conversion factors. The first problem (Ex. 3.1) is very simple. Despite being able to do this problem "in your head," let us formally set it up using the steps of dimensional analysis. It is better to "cut our teeth" on something soft than hard. Much can be learned about dimensional analysis from this simple problem.

EXAMPLE 3.1

A basketball player is 192 cm tall. What is the player's height in meters?

SOLUTION

Step 1. The given quantity is 192 cm, the height of the basketball player. The unit of the desired quantity is meters.

$$192 \text{ cm} = ?\text{ m}$$

Step 2. Only one conversion factor will be needed to convert from centimeters to meters, one that relates centimeters and meters. Two forms of this factor exist.

$$\frac{1 \text{ cm}}{10^{-2} \text{ m}} \quad \text{and} \quad \frac{10^{-2} \text{ m}}{1 \text{ cm}}$$

The second factor is used because it will allow for the cancellation of the centimeter units, leaving us with meters as the new units.

$$192 \cancel{\text{cm}} \times \left(\frac{10^{-2} \text{ m}}{1 \cancel{\text{cm}}}\right) = ?\text{ m}$$

For cancellation, a unit must appear in both numerator and denominator. Since the given quantity (192 cm), has centimeters in the numerator, the conversion factor must be the one with centimeters in the denominator.

If the other conversion factor had been used, we would have

$$192 \text{ cm} \times \left(\frac{1 \text{ cm}}{10^{-2} \text{ m}}\right) = ? \frac{\text{cm}^2}{\text{m}}$$

No unit cancellation is possible in this setup. Multiplication thus gives cm^2/m as the final units, which is certainly not what we want. In all cases only one of the two conversion factors of a reciprocal pair will correctly "fit" into a dimensional analysis setup.

Step 3. Step 2 takes care of the problem of units. All that is left to do in step 3 is to combine numerical terms to get a final answer, that is, we still have to do the "arithmetic." Collecting the numerical terms gives

$$\frac{192 \times 10^{-2}}{1} \text{ m} = 192 \times 10^{-2} \text{ m}$$

$$= 1.92 \text{ m} \quad \text{(calculator answer and correct answer)}$$

Since the conversion factor used in this problem is derived from a definition, it contains an unlimited number of significant figures and will not limit in any way the allowable number of significant figures in the answer. Therefore, the answer should have three significant figures, the same number as in the given quantity. Since the calculator answer does contain three significant figures it is also the correct answer.

EXAMPLE 3.2

The moon is 3.9×10^5 km from the earth. Express this distance in millimeters.

SOLUTION

Step 1. The given quantity is 3.9×10^5 km and the units of the desired quantity are millimeters.

$$3.9 \times 10^5 \text{ km} = ? \text{ mm}$$

Step 2. In dealing with metric–metric unit changes where both the original and desired units carry prefixes (which is the case in this problem), it is recommended that you always "channel" units through the basic unit (the unprefixed unit). If this is done, one does not need to deal with any conversion factors other than those resulting from prefix definitions. Following this recommendation, the unit "pathway" for this problem is

$$\text{km} \longrightarrow \text{m} \longrightarrow \text{mm}$$

(↑ m: Basic unit)

In the setup for this problem we will need two conversion factors, one for the km → m change and one for the m → mm change.

$$(3.9 \times 10^5 \cancel{\text{km}}) \times \left(\frac{10^3 \cancel{\text{m}}}{1 \cancel{\text{km}}}\right) \times \left(\frac{1 \text{ mm}}{10^{-3} \cancel{\text{m}}}\right) = ?\ \text{mm}$$

This conversion factor converts km to m

This conversion factor converts m to mm

Note how all the units cancel except for millimeters.

Step 3. Carrying out the indicated numerical calculation gives

Number from the first factor

$$\frac{(3.9 \times 10^5) \times 10^3 \times 1}{1 \times 10^{-3}} \text{ mm} = 3.9 \times 10^{11} \text{ mm}$$ (calculator answer and correct answer)

Numbers from the second factor

Numbers from the third factor

The correct answer is the same as the calculator answer for this problem. The given quantity has two significant figures and both conversion factors are "exact." Thus, the correct answer should contain two significant figures.

Our next two worked-out sample problems involve the use of English–English conversion factors. The conversion factors themselves should pose no problems for you. The examples are intended to give you further insights into dimensional analysis, particularly in the area of setting up the pathway for unit change. In addition, Example 3.4 exposes you to the complication of having to deal with "compound" units.

EXAMPLE 3.3

It is determined that a large container can hold 7.63 gal of apple cider. What is the container capacity stated in terms of fluid ounces?

SOLUTION

Step 1. The given quantity is 7.63 gal, and the units of the desired quantity are fluid ounces.

$$7.63 \text{ gal} = ?\ \text{fl oz}$$

Step 2. The logical pathway to follow to accomplish the desired unit change is

$$\text{gal} \longrightarrow \text{qt} \longrightarrow \text{fl oz}$$

Always use logical steps in setting up the pathway for a unit change. It does not have to be done in one "big jump." Use smaller steps for which conversion factors are known. Big steps usually get you involved with unfamiliar conversion factors. Most people do not carry around in their head the number of ounces in a gallon, but they do know that there are 32 fl oz in a quart.

The setup for this problem will require two conversion factors — "gallons to quarts" and "quarts to fluid ounces."

$$7.63\ \cancel{\text{gal}} \times \left(\frac{4\ \cancel{\text{qt}}}{1\ \cancel{\text{gal}}}\right) \times \left(\frac{32\ \text{fl oz}}{1\ \cancel{\text{qt}}}\right) = ?\ \text{fl oz}$$

The units all cancel except for the desired fluid ounces.

Step 3. Performing the indicated multiplications, we get

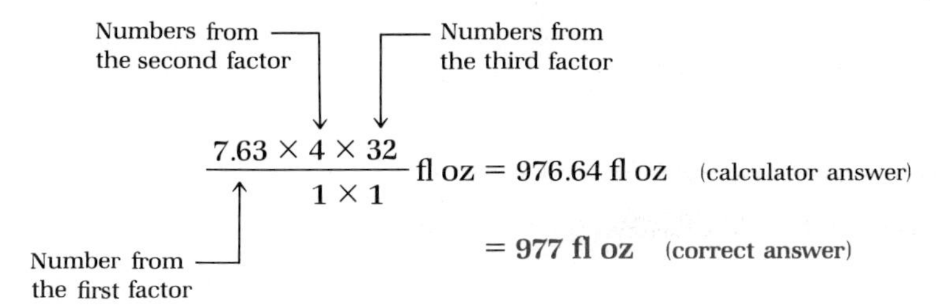

$$\frac{7.63 \times 4 \times 32}{1 \times 1}\ \text{fl oz} = 976.64\ \text{fl oz} \quad \text{(calculator answer)}$$

$$= 977\ \text{fl oz} \quad \text{(correct answer)}$$

The calculator answer contains too many significant figures. The correct answer should contain only *three* significant figures, the number in the given quantity. Again, both conversion factors involve exact definitions and will not enter into significant figure considerations.

EXAMPLE 3.4

A bumblebee is flying at a speed of 96 ft/sec. What is its speed in in./hr?

SOLUTION

Step 1. The given quantity is 96 ft/sec, and the units of this quantity are to be changed to in./hr.

$$96\ \frac{\text{ft}}{\text{sec}} = ?\ \frac{\text{in.}}{\text{hr}}$$

Step 2. This problem is a little more complex than previous examples because two different types of units are involved: length and time. The approach we use is, however, the same, other than that we must change two units instead of one. The feet must be converted to inches and the seconds to hours.

The logical pathway for the length change is the direct one-step path.

$$\text{ft} \longrightarrow \text{in.}$$

For time, it is logical to make the change in two steps.

$$\text{sec} \longrightarrow \text{min} \longrightarrow \text{hr}$$

It does not matter whether time or length is handled first in the conversion factor setup. We will arbitrarily choose to handle time first. The setup becomes

$$96\,\frac{\text{ft}}{\cancel{\text{sec}}} \times \left(\frac{60\,\cancel{\text{sec}}}{1\,\cancel{\text{min}}}\right) \times \left(\frac{60\,\cancel{\text{min}}}{1\text{ hr}}\right)$$

The units at this point are ft/min (after the first conversion factor). The units at this point are ft/hr (after the second conversion factor).

Note that in the first conversion factor, seconds, had to be in the numerator in order to cancel the seconds in the denominator of the given quantity.

We are not done yet. The time conversion from seconds to hours has been accomplished, but nothing has been done with length. To take care of length we do not start a new conversion factor setup. Rather, an additional conversion factor is tacked onto those we already have in place.

$$96\,\frac{\cancel{\text{ft}}}{\cancel{\text{sec}}} \times \left(\frac{60\,\cancel{\text{sec}}}{1\,\cancel{\text{min}}}\right) \times \left(\frac{60\,\cancel{\text{min}}}{1\text{ hr}}\right) \times \left(\frac{12\text{ in.}}{1\,\cancel{\text{ft}}}\right) = ?\,\frac{\text{in.}}{\text{hr}}$$

The feet in the denominator of the last factor cancel the feet in the numerator of the given quantity. With this cancellation the units now become in./hr.

Step 3. Collecting the numerical factors and performing the indicated math gives

$$\frac{96 \times 60 \times 60 \times 12}{1 \times 1 \times 1}\,\frac{\text{in.}}{\text{hr}} = 4{,}147{,}200\,\frac{\text{in.}}{\text{hr}} \quad \text{(calculator answer)}$$

$$= 4{,}100{,}000\text{ in./hr.} \quad \text{(correct answer)}$$

The original speed, 96 ft/sec, involves only two significant figures. Rounding the calculator answer to *two* significant figures gives 4,100,000 in./hr (or 4.1×10^6 in./hr). Again, none of the conversion factors plays a role in significant figure considerations since they all originated from definitions.

Incidentally, 4.1×10^6 in./hr is equivalent to a speed of 65 mi/hr.

We will now consider some sample problems that involve both the English and the metric systems of units. As mentioned previously, conversion factors between these two systems do not arise from definitions, but rather are determined experimentally. Hence they are not exact. Some of these experimentally determined factors were given in Table 3-2.

Instead of trying to remember all of the conversion factors listed in Table 3-2, memorize only one factor for each type of measurement (mass, volume, length). Knowing only one factor of each type is sufficient information to work metric–English or English–metric conversion problems. The relationships it is suggested that you memorize are the following.

Length:	1 in. = 2.540 cm
Mass:	1 lb = 453.6 g
Volume:	1 qt = 0.9463 L

These three equalities may be considered "bridge relationships" connecting English and metric system measurement units of various types. These bridge relationships are depicted in Figure 3-8.

These "bridge relationships" are always applicable in problem solving. For example, no matter what mass units are given or asked for in a problem, we can convert to pounds or grams (bridge units), cross the bridge with our memorized conversion factor, and then convert to the desired final unit. The only advantage that would be gained by memorizing all the factors in Table 3-2 would be that you could work some problems with fewer conversion factors, usually only one factor less. The reduction in the number of conversion factors used is usually not worth the added complication of keeping track of the additional conversion factors.

EXAMPLE 3.5

Your overweight Aunt Zelda Zucclakeley weighs 244 lb. What is her mass in kilograms?

SOLUTION

Step 1. The given quantity is 244 lb and the units of the desired quantity are kilograms.

$$244 \text{ lb} = ? \text{ kg}$$

Step 2. This is an English–metric mass-unit conversion problem. The mass-unit "measurement bridge" (Fig. 3-8b) involves pounds and grams. Since pounds are the given units, we are at the bridge to start with. Pounds are converted to grams (crossing the bridge), and then the grams are converted to kilograms.

$$\text{lb} \longrightarrow \text{g} \longrightarrow \text{kg}$$

The setup for this problem is

$$244 \cancel{\text{lb}} \times \left(\frac{453.6 \cancel{\text{g}}}{1 \cancel{\text{lb}}}\right) \times \left(\frac{1 \text{ kg}}{10^3 \cancel{\text{g}}}\right) = ? \text{ kg}$$

The units all cancel except for kilograms.

Step 3. Performing the indicated arithmetic gives

$$\frac{244 \times 453.6 \times 1}{1 \times 10^3} \text{ kg} = 110.6784 \text{ kg} \quad \text{(calculator answer)}$$

$$= \mathbf{111 \text{ kg}} \quad \text{(correct answer)}$$

The calculator answer must be rounded off to *three* significant figures, since the given quantity (244 lb) has only three significant figures. The first conversion factor has four significant figures and the second one is exact.

An alternate pathway for working this problem makes use of the conversion factor in Table 3-2 involving pounds and kilograms. If we use this conversion factor, we can go

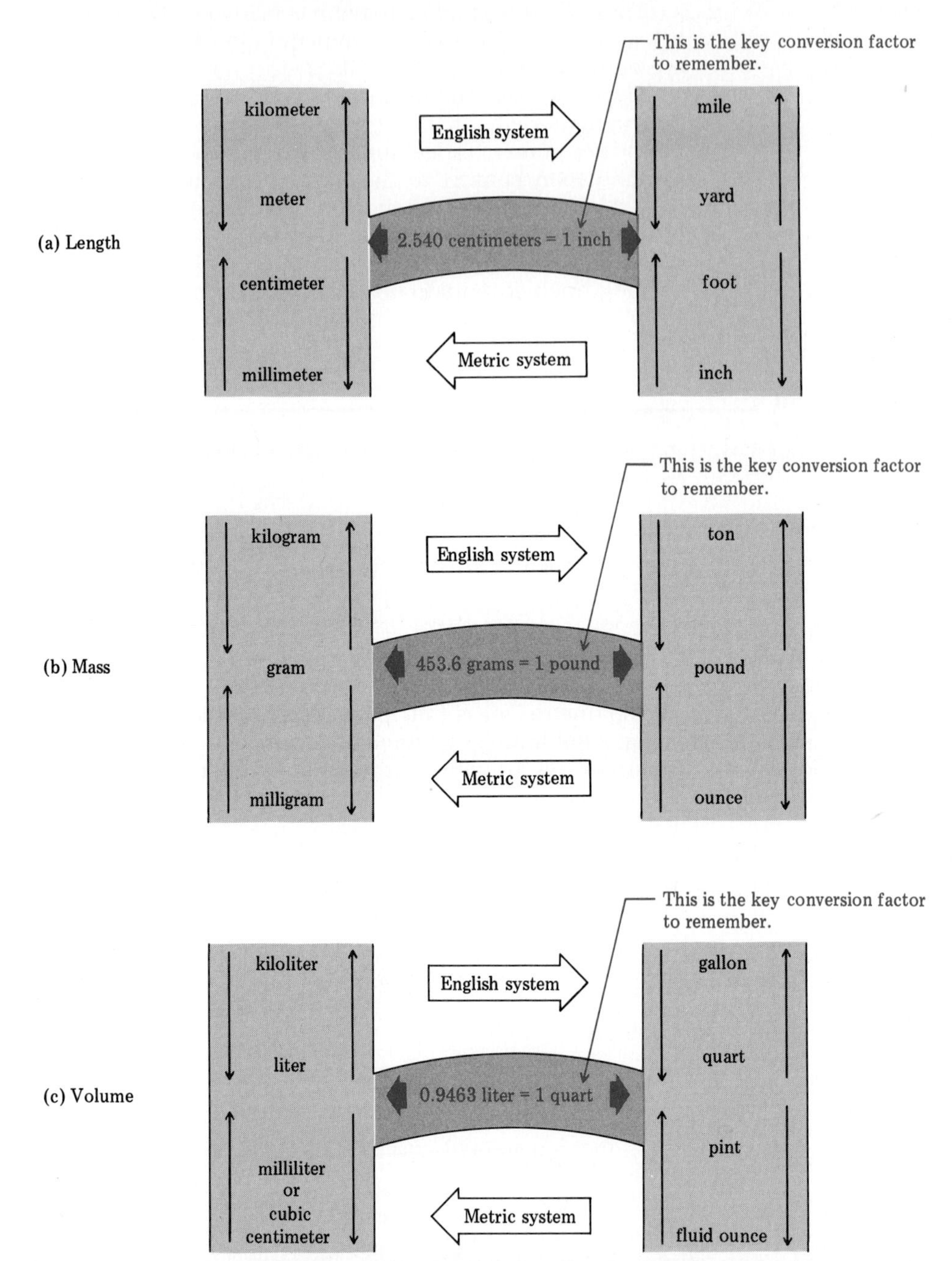

Figure 3-8. Measurement bridges connecting the metric and English systems of measurement for (a) length, (b) mass, and (c) volume. [From *Essential Chemistry* by Edward C. Stark. New York: Glencoe Publishing Co., Inc., 1979.]

directly from the given to the desired unit in one step. The setup is

$$244\ \cancel{\text{lb}} \times \left(\frac{1\ \text{kg}}{2.205\ \cancel{\text{lb}}}\right) = 110.6576\ \text{kg} \quad \text{(calculator answer)}$$

Rounding the calculator answer to *three* significant figures gives 111 kg.

This alternate solution was specifically shown to point out that for many problems there is more than one pathway to the answer. When alternate pathways exist, it cannot be said that one way is more correct than another. It is important that you select a pathway, choose a correct set of conversion factors consistent with that pathway, and get the answer.

Although both pathways used in this problem are "correct," we prefer the first solution because it uses our mass "measurement bridge." Use of the "measurement bridge" system cuts down on the number of conversion factors you are required to know (or look up in a table).

A last comment concerning the alternate setups for the problem. The calculator answers obtained for the two setups are slightly different — 110.6784 and 110.6576. The difference, which involves digits that are not significant digits, arises from metric–English and English–metric conversion factors being measured rather than defined quantities. The two conversion factors used have different "rounding-off" errors in them. When the calculator answers are corrected for significant figures, note that both answers become the same.

EXAMPLE 3.6

In the laboratory it is experimentally determined that a particular soda pop container has in it 2031 mL of root beer. How many gallons of root beer is this equivalent to?

SOLUTION

Step 1. The given quantity is 2031 mL and the units of the desired quantity are gallons.

$$2031\ \text{mL} = ?\ \text{gal}$$

Step 2. The "bridge relationship" for volume (Fig. 3-8c) involves liters and quarts. Thus, we need to convert the milliliters to liters, cross the bridge to quarts, and then convert quarts to gallons.

$$\text{mL} \longrightarrow \text{L} \longrightarrow \text{qt} \longrightarrow \text{gal}$$

Following this pathway, the setup becomes

$$2031\ \cancel{\text{mL}} \times \left(\frac{10^{-3}\ \cancel{\text{L}}}{1\ \cancel{\text{mL}}}\right) \times \left(\frac{1\ \cancel{\text{qt}}}{0.9463\ \cancel{\text{L}}}\right) \times \left(\frac{1\ \text{gal}}{4\ \cancel{\text{qt}}}\right) = ?\ \text{gal}$$

Step 3. The numerical calculation involves the following setup.

$$\frac{2031 \times 10^{-3} \times 1 \times 1}{1 \times 0.9463 \times 4}\ \text{gal} = 536.56346 \times 10^{-3}\ \text{gal} \quad \text{(calculator answer)}$$

$$= \mathbf{5.366 \times 10^{-1}\ gal} \quad \text{(correct answer)}$$

The calculator answer must be rounded to four significant figures and must also be justified.

Again, to illustrate that there is more than one way to set up almost any unit conversion problem, let us work this problem two alternate ways using the other two volume conversion factors in Table 3-2.

To use the conversion factor based on $1 \text{ L} = 2.113 \text{ pt}$ requires a pathway of

$$\text{mL} \longrightarrow \text{L} \longrightarrow \text{pt} \longrightarrow \text{qt} \longrightarrow \text{gal}$$

The setup is

$$2031 \text{ mL} \times \left(\frac{10^{-3} \text{ L}}{1 \text{ mL}}\right) \times \left(\frac{2.113 \text{ pt}}{1 \text{ L}}\right) \times \left(\frac{1 \text{ qt}}{2 \text{ pt}}\right) \times \left(\frac{1 \text{ gal}}{4 \text{ qt}}\right) = 536.43788 \times 10^{-3} \text{ gal}$$

(calculator answer)

$$= 5.364 \times 10^{-1} \text{ gal}$$

(correct answer)

Rounding the calculator answer to four significant figures and justifying gives the correct answer.

The conversion factor $1 \text{ fl oz} = 29.57 \text{ mL}$ requires a pathway of

$$\text{mL} \longrightarrow \text{fl oz} \longrightarrow \text{qt} \longrightarrow \text{gal}$$

The setup is

$$2031 \text{ mL} \times \left(\frac{1 \text{ fl oz}}{29.57 \text{ mL}}\right) \times \left(\frac{1 \text{ qt}}{32 \text{ fl oz}}\right) \times \left(\frac{1 \text{ gal}}{4 \text{ qt}}\right) = 536.59748 \times 10^{-3} \text{ gal}$$

(calculator answer)

$$= 5.366 \times 10^{-1} \text{ gal}$$

(correct answer)

Rounding the calculator answer to four significant figures and justifying again gives the correct answer.

Again, although all of these setups are correct, we still prefer the method of having a "bridge relationship" for volume, mass, and length, which is always used in crossing the metric–English bridge.

Although the emphasis in this section has been on using conversion factors to change units within the English or metric systems or from one to the other, the applications of conversion factors go far beyond this type of activity. We will resort to using conversion factors time and time again throughout this textbook in solving problems. What has been covered in this section is only the "tip of the iceberg" relative to dimensional analysis and conversion factors.

3.3 Density and Specific Gravity

density

Density *is the ratio of the mass of an object to the volume occupied by that object,* that is

$$\text{Density} = \frac{\text{mass}}{\text{volume}} \tag{3-1}$$

Table 3-3. ***Densities of Selected Solids, Liquids, and Gases***

Solids	Density (g/cm^3 at 25°C)[a]	Liquids	Density (g/mL at 25°C)[a]	Gases	Density (g/L at 25°C, 1 atm)[a]
Gold	19.3	Mercury	13.55	Chlorine	3.17
Lead	11.3	Milk	1.028–1.035	Carbon dioxide	1.96
Copper	8.93	Blood plasma	1.027	Oxygen	1.42
Aluminum	2.70	Urine	1.003–1.030	Air (dry)	1.29
Table salt	2.16	Water	0.997	Nitrogen	1.25
Bone	1.7–2.0	Olive oil	0.92	Methane	0.66
Table sugar	1.59	Ethyl alcohol	0.79	Hydrogen	0.08
Wood, pine	0.30–0.50	Gasoline	0.56		

[a] Density changes with temperature. (In most cases it decreases with increasing temperature, since almost all substances expand when heated.) Consequently the temperature must be recorded along with a density value. In addition, the pressure of gases must be specified.

The most frequently encountered density units in chemistry are grams per cubic centimeter (g/cm^3) for solids, grams per milliliter (g/mL) for liquids, and grams per liter (g/L) for gases. Use of these units avoids the problem of having density values that are extremely small or extremely large numbers. Table 3-3 gives density values for a number of substances.

People often speak of one substance being "heavier" or "lighter" than another. For example, it is said that "lead is a heavier metal than aluminum." What is actually meant by this statement is that lead has a higher density than aluminum; that is, there is more mass in a specific volume of lead than there is in the same volume of aluminum. The density of an object is a measure of how closely the object's mass is packed into a given volume. Even though the density of lead (11.3 g/cm^3) is greater than that of aluminum (2.70 g/cm^3), one pound of lead weighs exactly the same as one pound of aluminum. One pound is one pound. Because the aluminum is less dense than the lead, the mass in the one pound of aluminum will occupy a larger volume than the mass in the one pound of lead. Said another way, if equal volume samples of lead and aluminum are weighed the lead will have the greater mass. So we say lead is "heavier" than aluminum; although we actually mean lead is more dense than aluminum.

The densities of solids and liquids are often compared to the density of water. Anything less dense ("lighter") than water floats on it, and anything more dense ("heavier") sinks. In a similar vein, densities of gases are compared to that of air. Any gas less dense ("lighter") will rise in air, and anything more dense ("heavier") will sink in air.

To calculate an object's density, we must make two measurements: one involves determining the object's mass and the other its volume.

EXAMPLE 3.7

In the laboratory it is determined that the mass of a piece of copper rod is 21.09 g. The volume of this same piece of copper rod is found to be 2.361 cm^3. What is the density of copper?

SOLUTION

Using the formula density = mass/volume, we have

$$\text{Density} = \frac{21.09\ \text{g}}{2.361\ \text{cm}^3} = 8.9326557\frac{\text{g}}{\text{cm}^3} \quad \text{(calculator answer)}$$

$$= 8.933\ \text{g/cm}^3 \quad \text{(correct answer)}$$

Since both input numbers for the calculation contain four significant figures, the calculator answer must be rounded to *four* significant figures to give the correct answer.

In a mathematical sense, density can be thought of as a conversion factor that relates the volume and mass of an object. Interpreting density in this manner enables one to calculate a substance's volume from its mass and density or its mass from its volume and density. Density is thus a "bridge" connecting weight and volume (see Fig. 3-9).

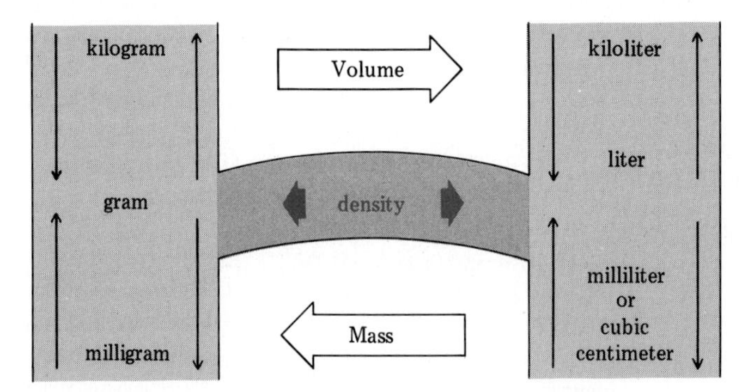

Figure 3-9. Density, a measurement bridge connecting mass and volume.

EXAMPLE 3.8

The metal osmium, with a density of 22.57 g/cm^3, is the most dense of all metals. The least dense metal is lithium with a density of 0.534 g/cm^3. What would be the volumes of "bricks" of osmium and lithium, each weighing 400.0 g? How many times greater is the volume of the lithium brick than the osmium brick?

SOLUTION

We will use density as a conversion factor in solving this problem by dimensional analysis.

Step 1. The given quantity is 400.0 g of metal (osmium or lithium). The units of the desired quantity (volume) are cubic centimeters. Thus,

$$400.0\ \text{g} = ?\ \text{cm}^3$$

Step 2. The pathway going from grams to cubic centimeters involves a single step, since density, used as a conversion factor, directly relates grams to cubic centimeters.

For osmium: $$400.0\,\cancel{g} \times \left(\frac{1\ cm^3}{22.57\,\cancel{g}}\right) = ?\ cm^3$$

For lithium: $$400.0\,\cancel{g} \times \left(\frac{1\ cm^3}{0.534\,\cancel{g}}\right) = ?\ cm^3$$

Step 3. Doing the indicated math gives the following answers.

For osmium: $$\frac{400.0 \times 1}{22.57}\ cm^3 = 17.722641\ cm^3 \quad \text{(calculator answer)}$$

$$= 17.72\ cm^3 \quad \text{(correct answer)}$$

The correct answer is the calculator answer rounded to *four* significant figures. Each of the input numbers contained four significant figures.

For lithium: $$\frac{400.0 \times 1}{0.534}\ cm^3 = 749.06367\ cm^3 \quad \text{(calculator answer)}$$

$$= 749\ cm^3 \quad \text{(correct answer)}$$

This time the correct answer contains only *three* significant figures, since the density value is quoted to only three significant figures.

Dividing the larger lithium volume by the smaller osmium volume shows that the lithium "brick" has a volume 42.3 times that of the osmium "brick".

$$\frac{749\ \cancel{cm^3}}{17.72\ \cancel{cm^3}} = 42.268623 \quad \text{(calculator answer)}$$

$$= 42.3 \quad \text{(correct answer)}$$

Three significant figures

EXAMPLE 3.9

A hospital patient's urine sample was found to have a density of 1.023 g/mL. If this individual normally excretes an average of 1275 mL of urine daily, how many grams of urine are excreted?

SOLUTION

Step 1. This is a milliliters-to-grams problem.

$$1275\ mL = ?\ g$$

Step 2. We will use density as a conversion factor to take us from milliliters to grams.

$$1275\ \cancel{mL} \times \left(\frac{1.023\ g}{1\ \cancel{mL}}\right) = ?\ g$$

Step 3. Performing the indicated mathematical operations gives

$$\frac{1275 \times 1.023}{1} \text{ g} = 1304.325 \text{ g} \quad \text{(calculator answer)}$$

$$= 1304 \text{ g} \quad \text{(correct answer)}$$

Since both the input volume and the density have *four* significant figures, the correct answer must also be limited to four significant figures. Rounding the calculator answer gives 1304 g.

Density is a conversion factor of a different type than those previously used in this chapter. It is a conversion factor whose numerator and denominator are *equivalent* rather than *equal*. All previously used conversion factors have been fractions where the numerator and denominator have been the same quantity under two different names. Twelve inches and one foot are different names for the same distance. The mass and volume involved in density are not different names for the same thing, but related equivalent quantities.

A major difference between equivalence conversion factors and equality conversion factors is that the former have applicability only in the particular problem setting for which they were derived, whereas the latter are applicable in all problem-solving situations. Many different gram to cubic centimeter mass–volume relationships (densities) exist, but only one foot–inch relationship exists. Mathematically, equivalence conversion factors can be used the same way equality conversion factors are. Use of equivalence conversion factors will be a common occurrence in later chapters of the text.

Specific Gravity

specific gravity

A quantity closely related to density is specific gravity. The **specific gravity** *of a solid or liquid is the ratio of the density of that substance to the density of water at 4°C.*

$$\text{Specific gravity of solid or liquid} = \frac{\text{density of substance}}{\text{density of water at 4°C}} \qquad (3\text{-}2)$$

At 4°C the density of water is 1.000 g/mL. For gases, specific gravity involves a density comparison with air rather than water. At 25°C the density of dry air is 1.29 g/L.

In calculating the specific gravity of a substance, both densities must be expressed in the same units. Specific gravity, itself, is a *unitless* quantity, since the identical sets of density units will always cancel.

EXAMPLE 3.10

The density of lead is 11.3 g/cm^3 at 25°C. What is the specific gravity of lead?

SOLUTION

The specific gravity of lead is equal to the density of lead divided by the density of water at 4°C.

$$\text{Specific gravity of lead at } 25^{\circ}\text{C} = \frac{11.3 \cancel{g/cm^3}}{1.000 \cancel{g/cm^3}} = 11.3$$ (calculator answer and correct answer)

Note how all the units cancel to make specific gravity a unitless quantity. Lead is thus 11.3 times as dense as water.

The specific gravity of a solid or liquid will always be equal numerically to the density of the substance (dropping the units) provided the densities are expressed in grams per cubic centimeter. With this set of units the density of water is 1.000 g/cm^3 and dividing by one will never change the numerical value of the numerator. When units other than g/cm^3 are used, the density and specific gravity of solids and liquids are not numerically equal as shown in Example 3.11.

EXAMPLE 3.11

Expressed in the units of pounds per cubic foot the density of lead is 705 at 25°C and the density of water is 62.4 at 4°C. Calculate the specific gravity of lead at 25°C.

SOLUTION

Dividing the density of lead by the density of water will give us the specific gravity of lead.

$$\text{Specific gravity of lead at } 25^{\circ}\text{C} = \frac{705 \cancel{lb/ft^3}}{62.4 \cancel{lb/ft^3}} = 11.298077$$ (calculator answer)

$$= 11.3$$ (correct answer)

Rounding the calculator answer to *three* significant figures, which is all that is allowed, gives the same answer as that obtained in Example 3.10.

Note that the density of lead (705 lb/ft^3) and its specific gravity (11.3) are not numerically equal.

Although the determination of the specific gravity of a solid or a gas usually requires both a mass and a volume measurement, it is possible to measure the specific gravity of a liquid directly by using a hydrometer. A hydrometer is a glass float with a weighted bottom and calibrations on its stem (see Fig. 3-10). The higher the density of the liquid the higher the hydrometer tube will float.

Specific gravity is used in many ways in many occupations. The attendant at a service station can tell the extent to which a car battery is charged by measuring the specific gravity of the battery acid (sulfuric acid). The strength of the antifreeze in a car radiator may also be checked with a hydrometer. In clinical and hospital laboratories, the specific gravity of urine samples is often determined; such information is helpful in diagnosing certain diseases.

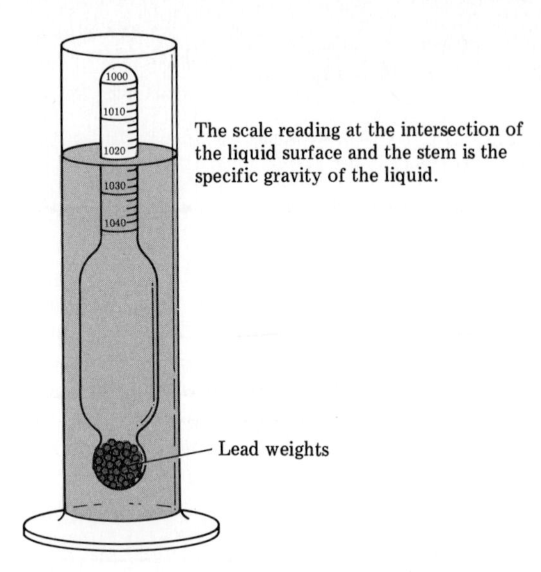

Figure 3-10. Specific gravity measurement using a hydrometer.

3.4 Percent as a Conversion Factor

percent

The word percent is actually a way of expressing a conversion factor. **Percent** *means "parts per 100 parts"; that is, it is the number of specific items in a group of 100 items.* The quantity 45% means 45 items per 100 total items.

Let us look at the mechanics of writing a percentage as a conversion factor, paying particular attention to the units involved. The percent value in the statement "A gold alloy is found upon analysis to contain 77% gold by mass" will be our focus point for the discussion. What are the mass units associated with the value 77%? The answer is that many mass units could be appropriate. The following statements, written as fractions (conversion factors) are all consistent with our 77% gold analysis.

$$\frac{77 \text{ oz of gold}}{100 \text{ oz of gold alloy}} \qquad \frac{77 \text{ lb of gold}}{100 \text{ lb of gold alloy}}$$

$$\frac{77 \text{ g of gold}}{100 \text{ g of gold alloy}} \qquad \frac{77 \text{ kg of gold}}{100 \text{ kg of gold alloy}}$$

Thus, in writing a percent as a conversion factor the unit choice is arbitrary. In practice, the complete context of the problem in which the percentage is found usually makes obvious the appropriate unit choice.

In conversion factors derived from percentages, confusion about cancellation of units frequently arises. Both numerator and denominator can contain the same units, for example, ounces. Yet, the ounces do not cancel. Not considering the *complete unit* is what causes confusion in the minds of students. In the conversion factor

$$\frac{77 \text{ oz of gold}}{100 \text{ oz of gold alloy}}$$

the numerator and denominator dimensions are not simply "ounces." The dimensions are, respectively, "ounces of gold" and "ounces of gold alloy." Ounces cannot be canceled, because they are only a part of the complete dimension. Cancellation is possible only when *complete* units are identical.

To avoid the problem of mistakenly canceling dimensions that are not the same, complete dimensions must always be written. Percent will always involve the same dimensional units (pounds, grams, meters) of different things. The identity of the different things, gold and gold alloy in our example, must always be included as part of the units.

EXAMPLE 3.12

A particular brand of canned chicken soup is found to contain 3.7% chicken by weight. How many pounds of chicken will be found in a soup serving weighing 0.875 lb?

SOLUTION

Step 1. The given quantity is 0.875 lb of chicken soup and the desired quantity is pounds of chicken.

$$0.875 \text{ lb chicken soup} = ? \text{ lb chicken}$$

Step 2. This is a one-conversion-factor problem with the conversion factor, obtained from the given percentage, being

$$\frac{3.7 \text{ lb of chicken}}{100 \text{ lb of chicken soup}}$$

The setup for the problem is

$$0.875 \cancel{\text{lb chicken soup}} \times \left(\frac{3.7 \text{ lb chicken}}{100 \cancel{\text{lb chicken soup}}}\right) = ? \text{ lb chicken}$$

Step 3. The numerical calculation involves the following arrangement of numbers.

$$\frac{0.875 \times 3.7}{100} \text{ lb chicken} = 0.032375 \text{ lb chicken} \quad \text{(calculator answer)}$$

$$= 0.032 \text{ lb chicken} \quad \text{(correct answer)}$$

The original percentage, 3.7%, has only two significant figures. This limits the correct answer to *two* significant figures. The number 100 in the setup is an exact number, since it originates from the definition of percent.

EXAMPLE 3.13

A study involving the inhabitants of a small town (population 1821) in southwestern Weber County in the state of Utah includes the following facts: (1) 21.3% of the inhabitants are adult males, (2) 32.3% of the adult males are bald-headed, and (3) 98.3% of the bald-headed adult males are handsome. How many handsome bald-headed adult males live in the town?

SOLUTION

Step 1. The given quantity is 1821 inhabitants, with the units of the desired quantity being handsome bald-headed adult males.

$$1821 \text{ inhabitants} = ? \text{ handsome bald-headed adult males}$$

Step 2. When conversion factors, particularly unusual ones, are given in word form in the statement of a problem, it is suggested that you first extract them from the problem statement and write them down before starting to solve the problem. You then have them available to use as you solve the problem. The given conversion factors in this problem are

$$\frac{21.3 \text{ adult males}}{100 \text{ inhabitants}}$$

$$\frac{32.3 \text{ bald-headed adult males}}{100 \text{ adult males}}$$

$$\frac{98.3 \text{ handsome bald-headed adult males}}{100 \text{ bald-headed adult males}}$$

The unit-conversion pathway for this problem will be

Inhabitants ⟶ adult males ⟶ bald-headed adult males ⟶ handsome bald-headed adult males

The setup for the problem is

$$1821 \cancel{\text{inhabitants}} \times \left(\frac{21.3 \cancel{\text{adult males}}}{100 \cancel{\text{inhabitants}}}\right) \times \left(\frac{32.3 \cancel{\text{bald-headed}}}{100 \cancel{\text{adult males}}}\right) \times \left(\frac{98.3 \text{ handsome}}{100 \cancel{\text{bald-headed}}}\right)$$

$$= ? \text{ handsome bald-headed adult males}$$

All of the units cancel except those in the numerator of the last conversion factor.

Step 3. Performing the indicated arithmetic gives

$$\frac{1821 \times 21.3 \times 32.3 \times 98.3}{100 \times 100 \times 100} \text{ handsome bald-headed adult males}$$

$$= 128.84365 \text{ handsome bald-headed adult males} \quad \text{(calculator answer)}$$

$$= 129 \text{ handsome bald-headed adult males} \quad \text{(correct answer)}$$

The calculator answer must be rounded off to *three* significant figures since all three of the given percentages contain only three significant figures.

3.5 Temperature Scales

The most common instrument for measuring temperature is the mercury-in-glass thermometer, which consists of a glass bulb containing mercury sealed to a slender glass capillary tube. The higher the temperature the farther the mercury will rise in the capillary tube. Graduations on the capillary tube indicate the height of the mercury column in terms of defined units called *degrees.* A small superscript zero is used as the symbol for a degree.

Three different degree scales (temperature scales) are in common use — Celsius, Kelvin, and Fahrenheit. Both the Celsius and Kelvin scales are part of the metric measurement system, and the Fahrenheit scale belongs to the English measurement system. Different size degrees and different reference points are what produce the various temperature scales.

The **Celsius** scale, named after Anders Celsius (1701–1744), a Swedish astronomer, is the scale most commonly encountered in scientific work. On this scale the boiling and freezing points of water serve as reference points, with the former having a value of 100° and the latter 0°. Thus, there are 100-degree intervals between the two reference points. The Celsius scale was formerly called the centigrade scale.

The **Kelvin** scale is a close relative of the Celsius scale. The size of the degree is the same on both scales as is the identity of the reference points. The two scales differ only in the numerical degree values assigned to the reference points. On the Kelvin scale the boiling point of water is at 373° and the freezing point of water at 273°. This scale, proposed by the British mathematician and physicist William Kelvin (1824–1907) in 1848, is particularly useful when working with relationships between temperature and pressure-volume behavior of gases (see Sec. 12.3).

A unique feature of the Kelvin scale is that negative temperature readings never occur. The lowest possible temperature thought to be obtainable occurs at 0° on the Kelvin scale. This temperature, known as *absolute zero,* has never been produced experimentally, although scientists have come within a fraction of a degree of reaching it.

The **Fahrenheit** scale was designed by the German physicist Gabriel Fahrenheit (1686–1736) in the early 1700's. After proposing several scales he finally adopted a system that used a salt/water mixture and boiling mercury as the reference points. These were the two extremes in temperature available to him at that time. A reading of 0° was assigned to the salt–ice mixture and 600° to the boiling mercury. The distance between these two points was divided into 600 equal parts or degrees. On this scale, water freezes at 32° and boils at 212°. Thus, there are 180 degrees between the freezing and boiling points of water on this scale as contrasted to 100 degrees on the Celsius and Kelvin scales. Figure 3-11 shows a comparison of the three temperature scales.

When changing a temperature reading on one scale to its equivalent on another scale, we must take two factors into consideration. (1) The size of the degree unit on the two scales may differ and (2) the zero points on the two scales will not coincide.

Difference in degree size will be a factor any time the Fahrenheit scale is involved in a conversion process. The conversion factors necessary to relate Fahrenheit degree size to metric degree size (Celsius or Kelvin) are obtainable from the information in Figure 3-11. From that figure we see that 180 Fahrenheit degrees are equivalent to 100 Celsius or Kelvin degrees. Using this relationship and the fact that 180/100 = 9/5, we obtain the following equalities.

5 Celsius degrees = 9 Fahrenheit degrees

5 Kelvin degrees = 9 Fahrenheit degrees

Conversion factors derived from these equalities will contain an infinite number of significant figures; that is, they are exact conversion factors.

Adjustment for differing zero-point locations is carried out by considering how

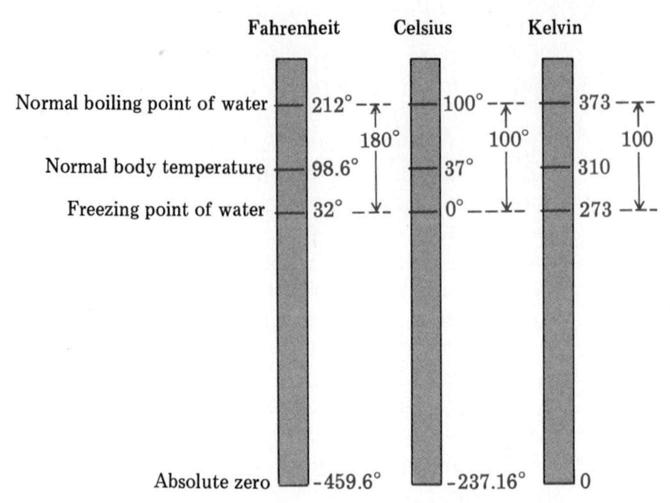

Figure 3-11. The relationships among the Fahrenheit, Celsius, and Kelvin temperature scales.

many degrees above or below the freezing point of water (the ice point) the original temperature is. Examples 3.14 and 3.15 show how this zero-point adjustment is carried out in addition to illustrating the use of temperature-scale conversion factors.

EXAMPLE 3.14

Normal body temperature on the Fahrenheit scale is 98.6°. What temperature is this equivalent to on the Celsius scale?

SOLUTION

First, we determine the number of degrees between the ice point and the temperature on the original scale.

$$98.6°\text{F} - 32.0°\text{F} = 66.6°\text{F}$$

Second, we convert from Fahrenheit units to Celsius units.

$$66.6 \text{ \sout{Fahrenheit degrees}} \times \left(\frac{5 \text{ Celsius degrees}}{9 \text{ \sout{Fahrenheit degrees}}}\right) = 37.0 \text{ Celsius degrees above the ice point}$$

Third, taking into account the ice point on the new scale, we determine the new temperature. On the Celsius scale, the temperature will be 37.0 degrees above the ice point. Since the ice point is 0°C, the new temperature reading is 37.0°C.

EXAMPLE 3.15

An oven for baking pizza operates at approximately 275°C. What is this temperature on the Fahrenheit scale?

SOLUTION

First, we determine how many degrees there are between the original temperature and the ice point. On the Celsius scale this will always be equal numerically to the temperature since the ice point on this scale is at zero degrees.

Second, we change this number of degrees — 275 in this case — from Celsius to Fahrenheit.

$$275 \text{ ~~Celsius degrees~~} \times \left(\frac{9 \text{ Fahrenheit degrees}}{5 \text{ ~~Celsius degrees~~}}\right) = 495 \text{ Fahrenheit degrees above the ice point}$$

Third, taking into account the ice point on the new scale, we determine the new temperature. The new temperature will be 495° above the ice point, which is 32°F.

$$495°\text{F} + 32°\text{F} = 527°\text{F}$$

Examples 3.14 and 3.15 point out that temperature-scale conversions are a little more complicated than the unit conversions of the last three sections. Not only is multiplication by a conversion factor required, but also addition and subtraction.

Since the size of the degree is the same, the relationship between the Kelvin and Celsius scales is very simple. No conversion factors are needed. All that is required is an adjustment for the differing zero points. The adjustment factor is 273, the number of degrees by which the two scales are offset from each other. (The difference between Celsius and Kelvin temperatures is actually 273.16 degrees, which we round off to 273, the nearest degree.)

$$\text{K} = °\text{C} + 273 \tag{3-3}$$

$$°\text{C} = \text{K} - 273 \tag{3-4}$$

Note that the symbol for degrees Kelvin is K, not °K.

The relationship between the Fahrenheit scale and the Celsius scale can also be stated in an equation format.

$$°\text{F} = \frac{9}{5}(°\text{C}) + 32 \tag{3-5}$$

$$°\text{C} = \frac{5}{9}(°\text{F} - 32) \tag{3-6}$$

Some students prefer to use these equations rather than the dimensional-analysis approach used in Examples 3.14 and 3.15. Use of these equations is illustrated in Example 3.16.

EXAMPLE 3.16

Assume that the temperature on a hot summer day in July reaches 105°F out of doors. What is this temperature on **(a)** the Celsius and **(b)** the Kelvin scale?

SOLUTION

(a) Substituting into Equation 3-6, we get

$$°C = \frac{5}{9}(105 - 32) = \frac{5}{9}(73) = 40.555556 \quad \text{(calculator answer)}$$

The calculator answer must be rounded off to *two* significant figures since 73 (the degrees above the ice point) contains two significant figures. The correct answer is thus 41°C.

(b) Using the answer from part (a) and substituting into Equation 3-3 gives

$$K = 41 + 273 = 314 \quad \text{(calculator answer)}$$

Using the rule for significant figures in addition (Sec. 2.3), we find that the correct answer cannot have any digits past the "units" place. The correct answer is thus 314 K.

Learning Objectives

After completing this chapter you should be able to

- List the basic metric system units and the prefixes used to indicate multiple and subunits of the basic unit (Sec. 3.1).
- Explain the difference between the mass and weight of an object (Sec. 3.1).
- Set up and work unit system conversion problems using dimensional analysis (Sec. 3.2).
- Construct two conversion factors from a given equivalence between two quantities (Sec. 3.2).
- Distinguish between density and specific gravity, including the units of each (Sec. 3.3).
- Calculate a substance's density given its mass and volume and use density as a conversion factor between mass and volume or vice versa (Sec. 3.3).
- Set up and solve, using dimensional analysis, problems involving percentages, treating percent as a conversion factor (Sec. 3.4).
- Know and be able to use the interrelationships between Fahrenheit, Celsius, and Kelvin temperature scales (Sec. 3.5).

Terms and Concepts for Review

The new terms or concepts defined in this chapter are

meter (Sec. 3.1)
gram (Sec. 3.1)
mass (Sec. 3.1)
weight (Sec. 3.1)
volume (Sec. 3.1)
liter (Sec. 3.1)
conversion factor (Sec. 3.2)
dimensional analysis (Sec. 3.2)
density (Sec. 3.3)
specific gravity (Sec. 3.3)
percent (Sec. 3.4)

Questions and Problems

Metric System Units

3-1 Write out the names of the metric system units having the following abbreviations.
a. cg **b.** μm **c.** mg **d.** kL
e. ng **f.** Gm **g.** mL **h.** Mm

3-2 Arrange each of the following in an increasing sequence, from smallest to largest.
a. milligram, centigram, decigram
b. microliter, nanoliter, decaliter
c. kilometer, millimeter, centimeter
d. microliter, megaliter, milliliter
e. centigram, kilogram, nanogram
f. hectometer, decameter, decimeter
g. gigagram, teragram, picogram
h. kiloliter, megaliter, gigaliter

3-3 What are the prefixes for the following multipliers of a gram?
a. 10^{-6} **b.** 1/100 **c.** 10^{-3}
d. 1000 **e.** 10^{6} **f.** one-millionth
g. 10^{2} **h.** 10^{-9}

3-4 For each of the pairs listed, indicate whether the first unit is larger or smaller than the second unit and then tell how many times larger or smaller it is.
a. centigram, gram
b. liter, milliliter
c. nanogram, microgram
d. picogram, gigagram

e. kilogram, microgram
f. megameter, kilometer
g. hectoliter, deciliter
h. milligram, megagram

Conversions Within the Metric System

3-5 Using the dimensional analysis method of problem solving, carry out the following metric–metric conversions.

a. 25 mg = ? g b. 323 km = ? m
c. 1×10^{-4} g = ? mg d. 0.003 kL = ? L
e. 753 mm = ? cm f. 0.00131 μg = ? kg
g. 300,000 cL = ? nL h. 3.67×10^{10} pm = ? Tm

3-6 Five blueberries of differing sizes have the following masses, respectively: 375 mg, 0.500 g, 0.000200 kg, 25.0 cg, and 1.00 dg. What is the combined mass, in grams, of the five blueberries?

3-7 Which of the following has (a) the smallest mass and (b) the largest mass? (1) a 1400 mg cockroach, (2) a 0.002 kg bumble bee, (3) a 1,100,000 microgram grasshopper, or (4) a 0.000003 Mg giant ant.

Metric–English and English–Metric System Conversions

3-8 Successful participants in a traditional marathon race traverse a distance of 26.00 miles. How many centimeters is this distance equivalent to?

3-9 A standard unsharpened pencil with eraser is 7.5 in. long. How many pencils, laid out end-to-end, are needed to form a "pencil column" a kilometer in length?

3-10 The smallest bone in the human body, which is in the ear, has a mass of approximately 3 mg. What is the approximate mass of this bone in pounds?

3-11 The earth has an estimated mass of 6.6×10^{21} tons. What is the mass of the earth in kilograms?

3-12 A standard basketball has a volume of 7.47 L. What is the volume of the basketball in (a) cubic centimeters and (b) quarts?

3-13 Assume a milliliter of water contains 20 drops. How many drops of water are there in a gallon of water?

3-14 How many hours would it take to count the drops of water in a gallon of water (Problem 3-13) assuming you can count them at the rate of 10 per second?

3-15 Your motorcycle averages 195 mi/gal of gasoline consumed. How many gallons of gasoline are required to travel 64333 ft on your motorcycle?

3-16 How many seconds would it take to send radio signals from the earth to the moon—a distance of 2.4×10^5 mi—if the signals travel at the speed of light which is 3.0×10^{10} cm/sec?

3-17 Seedless grapes are on sale in a grocery store for 69¢/lb. It is found that there are an average of 194 grapes per pound. How much money, in dollars, will you have to borrow from the bank in order to purchase 1 million (1.000000×10^6) grapes?

Density and Specific Gravity

3-18 A sample of sand is found to have a mass of 51.3 g and a volume of 20.2 cm^3. What is its density in g/cm^3?

3-19 Use Table 3-3 to calculate the mass in grams of each of the following samples.

a. 20.0 mL of olive oil b. 7.6 cm^3 of gold
c. 3.74 L of oxygen d. 2.0 qt of gasoline

3-20 Use Table 3-3 to calculate the volume in milliliters of each of the following samples.

a. 27 g of mercury
b. 1.75 g of nitrogen
c. 100.0 g of table salt
d. 100.0 g of table sugar

3-21 An automobile gasoline tank holds 24.0 gal when full. How many pounds of gasoline will it hold, if the gasoline has a density of 0.56 g/mL?

3-22 What weight of copper (density = 8.93 g/cm^3) occupies the same volume as 100.0 g of aluminum (density = 2.70 g/cm^3).

3-23 The density of silver is 10.40 g/cm^3. What is the specific gravity of silver?

3-24 What is the specific gravity of lithium, the least dense of all metals, if 5.34 g of lithium occupies a volume of 10.0 cm^3?

3-25 The specific gravities of corn oil and steel are, respectively, 0.90 and 7.9. For each of these materials calculate the volume in cm^3 which will weigh 50.0 g.

Percentage as a Conversion Factor

3-26 A 75.0 g solution containing table sugar and water is 7.0% sugar by weight. Using dimensional analysis, find the mass of the sugar in the solution.

3-27 A sample of copper ore was found to be 38.7% by weight copper. How many pounds of ore are needed to produce 500.0 lb of copper?

3-28 A 5 lb box of chocolates contains 150 chocolates. Dark chocolates are more prevalent than light chocolates—64% versus 36%. Exactly 25% of the dark chocolates are cream filled. How many cream-filled dark chocolates are there in the 5 lb box?

Temperature-Scale Conversions

3-29 Do the following temperature-scale conversions.

a. 500°C to degrees Kelvin
b. 362°C to degrees Fahrenheit
c. −200°F to degrees Kelvin
d. −40°F to degrees Celsius
e. 400 K to degrees Fahrenheit
f. 200 K to degrees Celsius

3-30 On a hot summer day the temperature outside may reach 105°F. What is this temperature in degrees Celsius?

3-31 Tungsten light bulb filaments have a melting point of approximately 3370°C. What temperature does this correspond to on the Fahrenheit scale?

3-32 A comfortable temperature for bathtub water is 95°F. What temperature is this in degrees Kelvin?

3-33 The temperature difference between the boiling and melting points of aluminum is 1807 Celsius degrees. Find this temperature difference in degrees Fahrenheit and in degrees Kelvin.

3-34 Which is the higher temperature, −10°C or 10°F?

3-35 Which is colder, −50°C or −60°F?

4 Basic Concepts About Matter

4.1 Chemistry — The Study of Matter

chemistry

Chemistry *is the branch of science that is concerned with matter and its properties.* What is matter? What is it that chemists study?

Intuitively, most people have a general feeling for the meaning of the word matter. They consider matter to be the materials of the physical universe — that is, the "stuff" from which the universe is made. Such an interpretation is a correct one.

matter

Formally defined, **matter** *is anything that has mass and occupies space.* A substance need not be visible to the naked eye to be labeled as matter as long as it meets the two qualifications of having mass and occupying space. Wood, paper, stone, the food we eat, the air we breathe, the fluids we drink, our bodies themselves, our clothing, and our shelter are all examples of matter.

The question may be asked, "What, then, does matter not include?" Various forms of energy, such as heat, light, and electricity, are not included. Neither are wisdom, friendship, ideas, thoughts, and emotions (such as anger and love) included.

Consider the extraordinary breadth of our definition of matter. This concept encompasses all objects we know about from the largest objects in outer space to the most minute objects seen under a microscope to objects so small they cannot be seen with any known type of instrumentation. Individuals first encountering this breadth automatically suppose that the study of matter will be a most complicated subject because of the literally millions of different types of matter that exist. One purpose of this chapter is to show that all matter can be classified into a surprisingly small number of categories. The naturally occurring materials of the universe and the synthetic materials humans have fashioned from them are, indeed, much simpler in makeup than they outwardly appear.

4.2 Physical States of Matter

solid state

liquid state

gaseous state

All matter may be conveniently classified into three physical states: solid, liquid, and gas. This classification scheme, familiar to everyone, relates to the definiteness of the shape and volume of the sample of matter. Matter in the **solid state** *is found to always have a definite shape and a definite volume.* The **liquid state** *is recognized by its properties of an indefinite shape and a definite volume.* A liquid always takes the shape of its container to the extent it fills it. The **gaseous state** *is characterized by both an indefinite shape and an indefinite volume.* A gas always completely fills its container, adopting both its volume and shape. Figure 4-1 summarizes the shape and volume characteristics of the three physical states of matter.

In nature we find matter in all three physical states. Rocks and minerals are in the solid state, water and petroleum are usually in the liquid state, and air, a mixture of many gases, represents the gaseous state.

The state of matter observed for a particular substance is always dependent on the temperature and pressure under which the observation is made. Because we live on a planet characterized by relatively narrow temperature extremes, we tend to fall into the error of believing that the commonly observed states of substances are the only states in which they occur. Under laboratory conditions, states other than the "natural" ones may be obtained for almost all substances. Oxygen, which is nearly always thought of as a gas, may be obtained in the liquid and solid states at very low temperatures. People seldom think of the metal gold as being a gas, its state at very high temperatures. At intermediate temperatures gold is in the liquid state. Water is one of the very few substances familiar to everyone in all three of its physical states: solid ice, liquid water, and gaseous steam.

Not all matter may exist in all three physical states as does oxygen, gold, and water. Some types of matter decompose when an attempt is made to change their physical states. For example, paper does not exist in the liquid state. Attempts to change paper to a liquid, through heating, always result in charring and decomposition.

Chapter 11 will consider in detail further properties of the different physical states of matter, changes from one state to another, and the question of why some substances decompose. Suffice it to say at present that physical state is one of the ways by which matter may be classified.

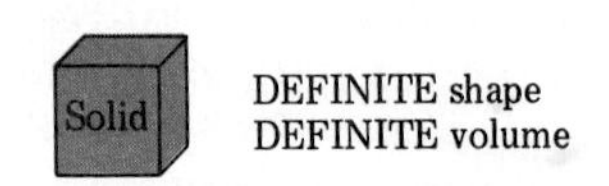

Figure 4-1. Shape and volume characteristics of the three physical states of matter.

4.3 Properties of Matter

properties

How are the various kinds of matter differentiated from each other? The answer is simple — by their properties. **Properties** *are the distinguishing characteristics of a substance used in its identification and description.* Each substance has a unique set of properties that distinguishes it from all other substances. If two substances had an identical set of properties, they would necessarily have to be one and the same substance.

The use of properties to identify an object is a technique not restricted to chemistry. It is an everyday practice in all of our lives. We recognize other people, including members of our own families, by properties. Height, weight, color of hair and eyes, walk, shape of nose and chin, and sound of voice are some of the properties we use in identifying another person.

physical properties

There are two general categories of properties of matter: physical and chemical. **Physical properties** *are properties observable without changing a substance into another substance.* Color, odor, taste, size, physical state, boiling point, melting point, and density are all examples of physical properties.

The physical appearance of a substance may change while determining a physical property but not the substance's identity. For example, measuring the melting point of a solid cannot be accomplished without melting it, which changes the solid to a liquid. Although the liquid's appearance is much different from that of the solid, the substance is still the same. Its chemical identity has not changed. Hence, melting point is a physical property.

chemical properties

Chemical properties *are properties that matter exhibits as it undergoes changes in chemical composition.* Most often such composition changes result from the interaction (reaction) of the matter with other substances. The fact that iron rusts when it is exposed to moist air is a chemical property of iron. The reddish brown substance formed (the rust) is a new substance with different properties from those of iron.

Sometimes, under proper conditions, a single substance will undergo chemical change in the absence of any other substance, a process called *decomposition.* For example, when water is heated to an extremely high temperature, it breaks down into the substances hydrogen and oxygen.

The *failure* of a substance to undergo change in the presence of another substance is also considered a chemical property even though no change has occurred. Flammability is a chemical property as is nonflammability.

Most often, in describing chemical properties, conditions such as temperature and pressure are specified since they can and do influence interactions between two or more substances. For example, two substances may interact explosively at an elevated temperature yet not interact at all at room temperature.

Table 4-1 contrasts the physical and chemical properties of the substance water. Note how chemical properties cannot be described without reference to other substances. It does not make sense to just say that a substance reacts spontaneously. The substance with which it combines must be specified since there may be many substances with which interaction will occur.

The properties of substances can be used in a number of practical ways, such as

1. *Identifying an unknown substance.* Identifying a confiscated drug as marijuana involves comparing the properties of the drug to those of known marijuana samples.

Table 4-1. ***Selected Physical and Chemical Properties of Water***

Physical Properties	Chemical Properties
1. Colorless	1. Reacts with bromine to form a mixture of two acids.
2. Odorless	2. Can be decomposed by means of electricity to form hydrogen and oxygen.
3. Boiling point = 100°C	3. Reacts vigorously with the metal sodium to produce hydrogen.
4. Freezing point = 0°C	4. Does not react with gold even at high temperatures.
5. Density = 1.000 g/mL at 4°C	5. Reacts with carbon monoxide at elevated temperatures to produce carbon dioxide and hydrogen.

2. *Distinguishing between different substances.* A dentist can quickly tell the difference between a real tooth and a false tooth because of property differences.

3. *Characterizing a newly discovered substance.* Any new substance must have a unique set of properties different from those of any previously characterized substance.

4. *Predicting the usefulness of a substance for specific applications.* Water-soluble substances obviously should not be used in the manufacture of bathing suits.

4.4 Changes in Matter

Changes in matter are common and familiar occurrences. Many occur spontaneously, independent of human influence. Others must be forced to occur. Representative of the wide variety of known change processes are melting of snow, digestion of food, burning of wood, detonation of dynamite, rusting of iron, and sharpening of a pencil.

physical change

chemical change

Like properties (Sec. 4.3), changes in matter may be classified as physical or chemical. A **physical change** *is a process that does not alter the basic nature (chemical composition) of the substance under consideration.* No new substances are ever formed as a result of a physical change. A **chemical change** *is a process that involves a change in the basic nature (chemical composition) of the substance.* Such changes always involve conversion of the material or materials under consideration into one or more new substances with distinctly different properties and composition from those of the original materials.

A change in physical state is the most common type of physical change. Melting, freezing, evaporation, condensation, and sublimation all represent changes of state (see Fig. 4-2). With the exception of sublimation these terms should be familiar to most everyone. Although the process of sublimation — going from a solid directly to the gaseous state — is not common, it is encountered in "everyday" life. Dry ice sublimes, as do mothballs placed in a clothing storage area. Wet wash hung out to dry in subfreezing weather dries as a result of sublimation. Water in the clothing first freezes and then sublimes.

In any change of state, the composition of the substance undergoing change re-

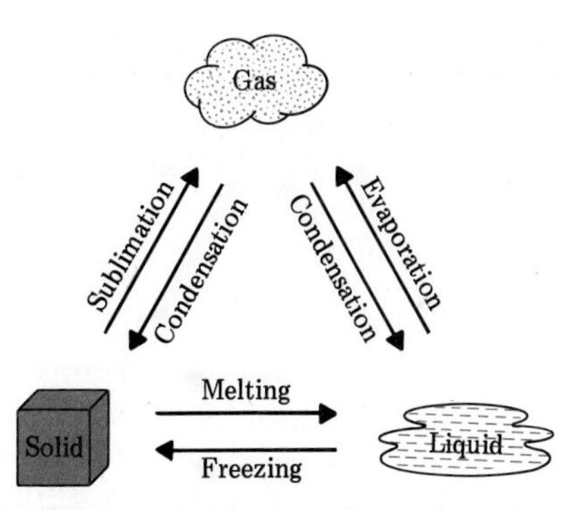

Figure 4-2. Terminology associated with changes of state.

mains the same even though its physical state is changed. The melting of ice does not produce a new substance. The substance is water before and after the change. Similarly, the steam produced from boiling water is still water.

Changes in size, shape, and state of subdivision are examples of physical changes which are not changes of state. Pulverizing a lump of coal into a fine powder and tearing a piece of aluminum foil into small pieces are examples of physical changes that involve only the solid state.

The appearance of one or more new substances is always a characteristic of a chemical change. Carbon dioxide and water are two new substances produced when the chemical change associated with burning of gasoline occurs. Ashes are among the new products produced when wood is burned. Chemical changes are often called chemical reactions. A **chemical reaction** *is a process in which at least one new substance is produced as a result of chemical change.*

chemical reaction

Table 4-2 classifies a number of changes as physical or chemical.

Most changes in matter can easily be classified as either physical or chemical. However, not all changes are "black" or "white." There are some "gray" areas. For example, the formation of certain solutions falls in the "gray" area. Common salt dissolves easily in water to form a solution of salt water. The salt can easily be recovered

Table 4-2. Classification of Changes as Physical or Chemical

Change	Classification
Rusting of iron	chemical
Melting of snow	physical
Sharpening a pencil	physical
Digesting food	chemical
Taking a bite of food	physical
Burning gasoline	chemical
Slicing an onion	physical
Detonation of dynamite	chemical
Souring of milk	chemical
Breaking of glass	physical

by the physical process of evaporating the water. When gaseous hydrogen chloride is dissolved in water, again a solution results; but in this case the starting materials cannot be recovered by evaporation. The formation of salt water is considered a physical change because the original components can be recovered in an unchanged form using physical methods. The second solution presents classification problems because of the possibility that a chemical reaction took place.

The changes involved in the cooking of an egg also present classification problems. The cooked egg contains the same structural units as the uncooked egg. However, some changes in structural arrangement have taken place — so, is the change physical or chemical? Despite the existence of "gray" areas, we shall continue to use the concepts of physical and chemical change because their usefulness far outweighs the problems created by a few exceptions.

The difference between the word change (the topic of this section) and the word property (Sec. 4.3) must be kept clear. A change always involves a transformation from one form to another. The new form may or may not be a new substance depending on whether the change was chemical or physical. Properties distinguish one substance from other substances.

Note that the term physical, when used to modify another term, as in physical property or physical change, always conveys the idea that the composition of the substances involved does not change. Similarly, the term chemical is always associated with the concept of change in composition.

The terms physical and chemical are commonly used to qualify the meaning of general scientific terms. For example, techniques used to accomplish physical change are called physical methods or physical means. Chemical methods and chemical means are used to bring about chemical change. A physical separation would be a separation process where none of the components experience composition changes. Composition changes would be part of a chemical separation process. The message of the "modifiers" physical and chemical is constant: Physical denotes no change in composition and chemical denotes change in composition.

4.5 Pure Substances and Mixtures

All specimens of matter can be divided into two categories — pure substances and mixtures. What are the distinctions between these two classifications of matter?

pure substance

A **pure substance** *is a form of matter that always has a definite and constant composition.* This constancy of composition is reflected in the properties of the pure substance. They never vary, always being the same under a given set of conditions. All samples of a pure substance, no matter what their source, must have the same properties under the same conditions. Collectively, these definite and constant physical and chemical properties of a pure substance form a set not duplicated by any other pure substance. This unique set of properties provides the identification for the pure substance.

A pure substance is exactly what its name implies — a single uncontaminated type of matter. All samples of a pure substance contain only that pure substance and nothing else. Hence, the constancy of properties. Pure water is water and nothing else. Pure table salt (sodium chloride) contains only that substance and nothing else.

It is important to note that there is a significant difference between the terms substance and pure substance. Substance is a general term used to denote any variety of matter. Pure substance is a specific term that applies only to matter with those characteristics we have just noted.

mixture

A **mixture** *is a physical combination or collection, with a variable composition, of two or more pure substances.* Because a mixture is a physical rather than a chemical combination of components, each component substance retains its own identity and properties. In a mixture of sand and gravel the properties of both sand and gravel are observable.

Once a mixture is made up, its composition will be constant. However, mixtures of the same components with other compositions can also be made up. Hence, mixtures are characterized as having variable compositions. Consider the large number of salt–water mixtures, each with a different composition, that can be produced by varying the ratio of the amounts of salt and water present.

Sometimes the fact that a sample of matter is a mixture can be determined by just looking at it. The individual components can be seen as separate entities. For example, small grains of different colored materials (pink, gray, tan, etc.) are readily discernible in a piece of granite rock. On the other hand, sometimes the separate ingredients of a mixture cannot be seen, even with the aid of a microscope. A mixture prepared by dissolving some sugar in water falls in this category. Outwardly, such a mixture has the same appearance as pure water.

heterogeneous mixture

On the basis of recognizability of the components present in a mixture, mixtures may be classified as heterogeneous or homogeneous. A **heterogeneous mixture** *contains visibly different parts, or phases, each with different properties.* A sausage-and-cheese pizza is obviously a heterogeneous mixture containing a number of identifiable components. Concrete is another example of such a mixture. The phases in a heterogeneous mixture may or may not be in the same physical state. Rocky soil is made up of a number of phases, all of which are in the solid state. A mixture of sand and water contains two phases, each in a different state (solid and liquid). It is possible to have heterogeneous mixtures in which all components are liquids. For such mixtures to occur, the mixed liquids must have limited or no solubility in each other. When this is the case, the mixed liquids form separate layers with the least dense liquid on top (see Sec. 13.2).

homogeneous mixture

A **homogeneous mixture** *contains only one phase with uniform properties throughout it.* Clean air, a mixture of numerous gases, is a homogeneous mixture as is water containing dissolved salt or sugar. A thorough intermingling of components in a homogeneous mixture must exist for only a single phase to be present. Sometimes this occurs almost instantaneously during the preparation of the mixture, for example, when alcohol is added to water. At other times an extended period of mixing or stirring is required. For example, a hard sugar cube added to a glass of water does not immediately form a homogeneous mixture. Only after the sugar completely dissolves is the mixture homogeneous.

An additional characteristic of any mixture is that its components can be retrieved intact from the mixture by physical means, that is, without a chemical reaction occurring. In many cases the differences in properties of the various components make the separation very easy. For example, in a sand–water mixture the water may be evaporated off to leave the sand behind. A mixture of salt dissolved in water may be separated in the same way, even though the salt is not visible in the mixture. Theoretically, physical separation is possible for all mixtures. In practice, however, separation is sometimes very difficult or nearly impossible. Consider the logistics involved in trying to separate out the components of a pizza once it has been cooked.

Figure 4-3 summarizes the major concepts developed in this section.

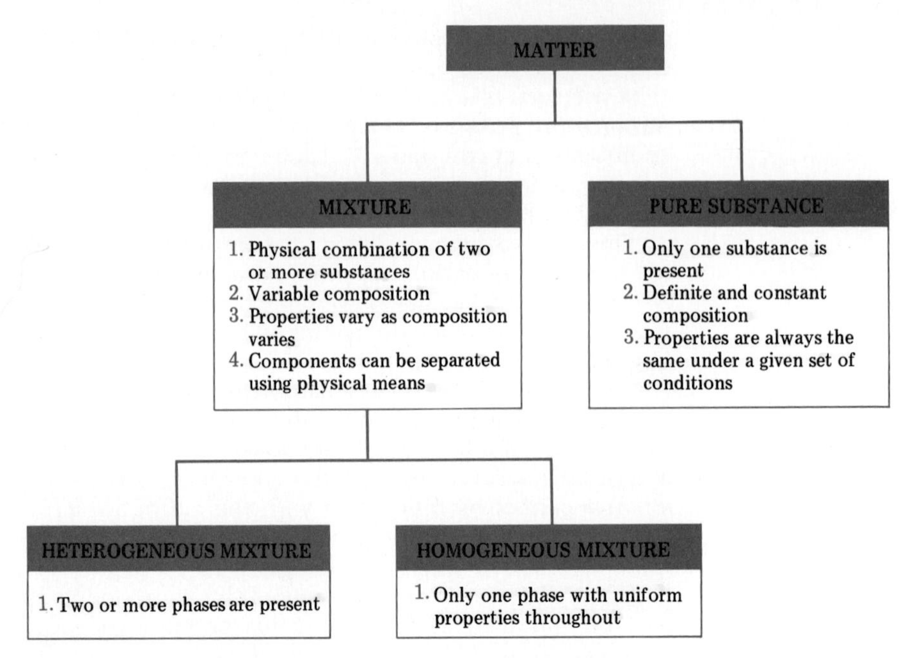

Figure 4-3. A comparison of the characteristics of pure substances and mixtures.

4.6 Types of Pure Substances: Elements and Compounds

element

compound

There are two kinds of pure substances: elements and compounds. An **element** *is a pure substance that cannot be broken down into simpler substances by ordinary chemical means.* Elements resist all attempts to fragment them into simpler pure substances. The metals gold, silver, and copper are all elements. A **compound** *is a pure substance that can be broken down into two or more simpler substances by chemical means.* Water is a compound. By means of electric current water can be broken down into the gases hydrogen and oxygen, both of which are elements. The properties of the simpler substances obtained from compound breakdown are always distinctly different from those of the parent compound.

Ultimately the products from the breakdown of any compound are elements. In practice the breakdown often occurs in steps, with simpler compounds resulting from the intermediate steps, as illustrated in Figure 4-4.

Presently 106 pure substances are classified as elements. These elements, the simplest known substances, are considered the building blocks of all other types of matter. Every object, regardless of its complexity, is a collection of substances made up of these elements.

Compared to the total number of compounds characterized by chemists, the number of known elements is extremely small. Over four million different compounds are known, each a definite combination of two or more of the known elements.

Before a substance can be classified as an element, all possible attempts must be made chemically to subdivide it into simpler substances. If a sample of pure substance, S, is subjected to a chemical process and two new substances, X and Y, are produced, S

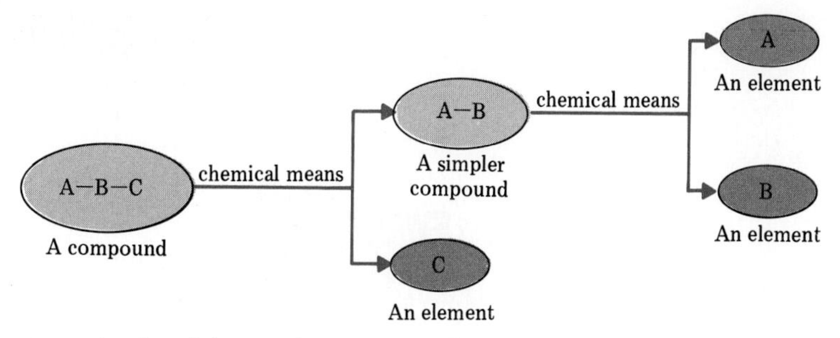

Figure 4-4. Stepwise breakdown of a compound containing three elements (A, B, and C) to yield the constituent elements.

would be classified as a compound. If, on the other hand, a number of attempts made chemically to subdivide S proved unsuccessful, we might correctly call it an element, but until all possible reactions have proved unsuccessful, such a classification could be in error. Figure 4-5 contrasts the differences between elements and compounds.

Compounds are not mixtures, even though two or more simpler substances may be obtained from them. It is very important that the distinction between a mixture and a compound be understood. A compound is the result of the *chemical* combination of two or more elements, whereas a mixture forms from the *physical* combination of two or more substances (elements or compounds). Compounds always have properties distinctly different from those of the elements used in producing the compound. Mixtures retain the properties of their individual components.

Other differences between compounds and mixtures exist. Compounds have a definite composition, a property of all pure substances, whereas mixtures have variable compositions. Also, mixtures, since they are only physical combinations of substances, can be separated by physical means. The separation of a compound into its constituent elements always requires chemical means. Compounds will never yield their constituent elements with only physical separation techniques.

Samples of pure substances are homogeneous, except if two physical states (solid,

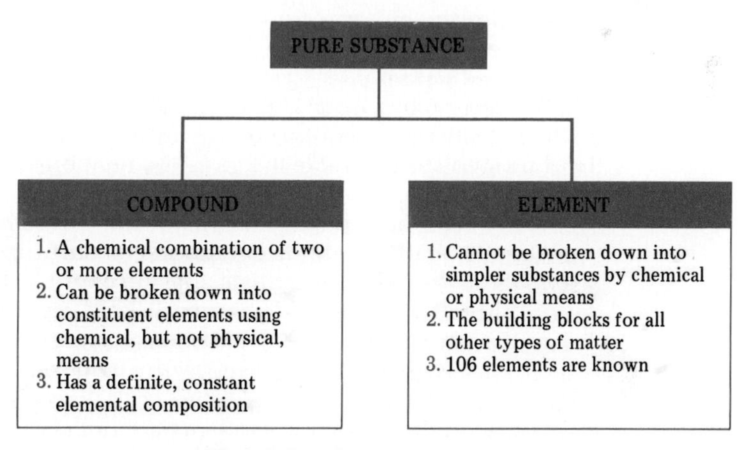

Figure 4-5. A comparison of the characteristics of elements and compounds.

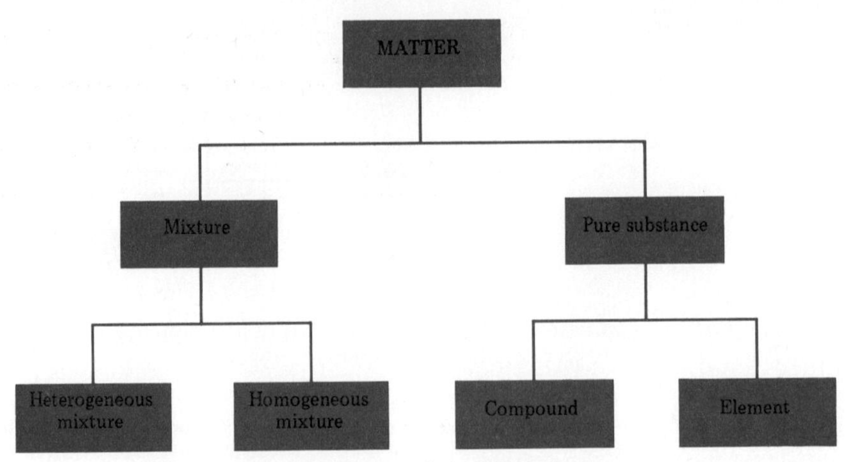

Figure 4-6. Categories of classification for matter.

liquid, gas) are present. Elements are homogeneous, since they consist of just one component throughout. Compounds are also homogeneous, even though two or more elements are present, because they have a definite composition, which assures a fixed proportion of all components throughout a sample.

A heterogeneous sample of a pure substance is possible. For this situation to occur the pure substance must be present in two or more states. An ice cube floating in a container of water is such a system. The solid phase (ice) obviously has some physical properties different from those of the liquid phase (liquid water). An ice–water system, although heterogeneous, is not a mixture. Only one substance is present—water. Mixtures require the presence of two substances.

Figure 4-6 summarizes the overall scheme developed in Sections 4.5 and 4.6 for classifying matter.

4.7 Discovery and Abundance of the Elements

The discovery and isolation of the 106 elements have taken place over a period of several centuries. Discovery, for the most part, has occurred since 1700, and in particular during the 1800s. Table 4-3 gives the relationship between time and the number of known elements.

Eighty-eight of the 106 elements occur naturally, and eighteen have been synthesized by bombarding atoms of naturally occurring elements with small particles. It is generally accepted by scientists that no more naturally occurring elements will be found, although it is possible that additional ones may be prepared synthetically. The last of the naturally occurring elements was discovered in 1925. The elements synthesized by scientists are all radioactive (unstable) and change (usually rapidly) into stable elements as the result of radioactive emissions (see Sec. 17.10).

The naturally occurring elements are not evenly distributed in our world and uni-

Table 4-3. Number of Elements Discovered During Various Fifty-Year Periods

Time Period	Number of Elements Discovered During Time Period	Total Number of Elements Known at End of Time Period
Ancient–1700	13	13
1701–1750	3	16
1751–1800	18	34
1801–1850	25	59
1851–1900	23	82
1901–1950	16	98
1951–present	8	106

verse. What is startling is the degree of unequalness of the distribution. A very few elements account for the majority of atoms. In answering the question of which elements are most common one must define the area to be considered. The abundances of the elements in the earth's crust are considerably different from those of the entire earth; they are even more different from abundances for the universe. When living organisms such as vegetation or the human body are considered, an all together different perspective emerges. Table 4-4 gives information on element abundances from a

Table 4-4. Abundance of the Elements from Various Viewpoints

Element	Abundance (atom %)	Element	Abundance (atom %)
Universe		*Atmosphere*	
Hydrogen	91	Nitrogen	78.3
Helium	9	Oxygen	21.0
Earth (including core)		*Hydrosphere*	
Oxygen	49.3	Hydrogen	66.4
Iron	16.5	Oxygen	33
Silicon	14.5	*Human body*	
Magnesium	14.2	Hydrogen	63
Earth's crust		Oxygen	25.5
Oxygen	55.1	Carbon	9.5
Silicon	16.3	Nitrogen	1.4
Hydrogen	16.0	*Vegetation*	
Aluminum	4.9	Hydrogen	49.8
Sodium	2.0	Oxygen	24.9
Iron	1.5	Carbon	24.9
Calcium	1.5		
Magnesium	1.4		
Potassium	1.1		

number of viewpoints. All numbers in the table are the percentage of total atoms that are those of the given element. Only element abundances greater than one percent are listed in any given analysis.

Note from Table 4-4 how the most abundant elements in the universe are not the same as the most abundant ones on earth. The cosmic figures reflect the composition of stars that are almost entirely hydrogen and helium. Significant differences exist between the composition of the earth as a whole and its crust. These differences reflect the fact that the figures for the earth's crust — taken to mean the earth's waters, atmosphere, and outer covering (to a depth of 10 miles) — do not take into account the composition of the earth's core, which is mostly iron. Note also how only two elements occur in large amounts (greater than 1% of atoms) in both the atmosphere and hydrosphere and how the relative numbers of carbon atoms increase and hydrogen atoms decrease in vegetation compared to human body composition.

4.8 Names and Symbols of the Elements

Each element has a unique name, which in most cases has been selected by its discoverer. A wide variety in rationale for choosing a name is found when name origins are studied. Some elements bear geographical names. Germanium was named after the native country of its German discoverer. The elements francium and polonium acquired names in a similar manner. The elements mercury, uranium, neptunium, and plutonium all are named for planets. Helium gets its name from the Greek word *helios* for sun, since it was first observed spectroscopically in the sun's corona during an eclipse. Some elements carry names that relate to specific properties of the element or compounds containing it. Chlorine's name is derived from the Greek *chloros* denoting greenish-yellow, the color of chlorine gas. Iridium gets its name from the Greek *iris* meaning rainbow, alluding to the varying color of the compounds from which it was isolated.

chemical symbols

In the early 1800s chemists adopted the practice of assigning chemical symbols to the elements. **Chemical symbols** *are abbreviations for the names of the elements.* These chemical symbols are used more frequently in referring to the elements than are the names themselves. A complete list of all the known elements and their symbols is given in Table 4-5. The symbols and names of the more frequently encountered elements are shown in color in this table. You would do well to learn the symbols of these more common elements. Learning them is a key to having a successful experience in studying chemistry.

Fourteen elements have one-letter symbols. All other symbols consist of two letters. If a symbol consists of a single letter it is capitalized. In all double-letter symbols, the first letter is capitalized, but the second letter is not. Double-letter symbols usually involve the first letter of the element's name. The second letter of the symbol is frequently, but not always, the second letter of the name. Consider the three elements, terbium, technetium, and tellurium, whose symbols are respectively Tb, Tc, and Te. Obviously, a variety of choices of second letters is necessary since the first two letters of each element's name are the same.

Eleven elements have "weird" symbols, that is, symbols that bear no relationship to the element's English language name. In ten of these cases the symbol is derived from the Latin name of the element; in the eleventh case a German name supplies the

Table 4-5. *The Elements and Their Chemical Symbols*[a]

Ac	actinium	Ge	germanium	Po	polonium
Ag	silver	H	hydrogen	Pr	praseodymium
Al	aluminum	Ha	hahnium[b]	Pt	platinum
Am	americium	He	helium	Pu	plutonium
Ar	argon	Hf	hafnium	Ra	radium
As	arsenic	Hg	mercury	Rb	rubidium
At	astatine	Ho	holmium	Re	rhenium
Au	gold	I	iodine	Rf	rutherfordium[b]
B	boron	In	indium	Rh	rhodium
Ba	barium	Ir	iridium	Rn	radon
Be	beryllium	K	potassium	Ru	ruthenium
Bi	bismuth	Kr	krypton	S	sulfur
Bk	berkelium	La	lanthanum	Sb	antimony
Br	bromine	Li	lithium	Sc	scandium
C	carbon	Lu	lutetium	Se	selenium
Ca	calcium	Lr	lawrencium	Si	silicon
Cd	cadmium	Md	mendelevium	Sm	samarium
Ce	cerium	Mg	magnesium	Sn	tin
Cf	californium	Mn	manganese	Sr	strontium
Cl	chlorine	Mo	molybdenum	Ta	tantalum
Cm	curium	N	nitrogen	Tb	terbium
Co	cobalt	Na	sodium	Tc	technetium
Cr	chromium	Nb	niobium	Te	tellurium
Cs	cesium	Nd	neodymium	Th	thorium
Cu	copper	Ne	neon	Ti	titanium
Dy	dysprosium	Ni	nickel	Tl	thallium
Er	erbium	No	nobelium	Tm	thulium
Es	einsteinium	Np	neptunium	U	uranium
Eu	europium	O	oxygen	V	vanadium
F	fluorine	Os	osmium	W	tungsten
Fe	iron	P	phosphorus	Xe	xenon
Fm	fermium	Pa	protactinium	Y	yttrium
Fr	francium	Pb	lead	Yb	ytterbium
Ga	gallium	Pd	palladium	Zn	zinc
Gd	gadolinium	Pm	promethium	Zr	zirconium

[a] Only 105 elements are listed in Table 4-5. The most recently discovered element (No. 106) does not have a name yet.

[b] Elements whose name and symbol have not yet been officially accepted.

symbol. Most of these elements have been known for hundreds of years and date back to the time when Latin was the language of scientists. Table 4-6 shows the relationship between the symbol and the non-English name of these eleven elements.

In addition to the listings in this section, the names and symbols of the elements are also found on the inside front and back covers of this book. The chart of elements on the inside front cover is called a periodic table. More will be said about it in later chapters. Both of these cover listings give additional information about the elements besides name and symbol. This additional information will be discussed in Chapter 5.

Table 4-6. Elements Whose Symbols Are Derived from a Non-English Name of the Element

English Name of Element	Non-English Name of Element	Symbol
	Symbols from Latin	
Antimony	stibium	Sb
Copper	cuprum	Cu
Gold	aurum	Au
Iron	ferrum	Fe
Lead	plumbum	Pb
Mercury	hydrargyrum	Hg
Potassium	kalium	K
Silver	argentum	Ag
Sodium	natrium	Na
Tin	stannum	Sn
	Symbol from German	
Tungsten	wolfram	W

Learning Objectives

After completing this chapter you should be able to

- Understand what is meant by the term matter (Sec. 4.1).
- Know the shape and volume characteristics of the three states of matter (Sec. 4.2).
- Given a property of a substance, classify it as a physical or chemical property (Sec. 4.3).
- Classify the changes that occur in matter as physical or chemical (Sec. 4.4).
- Know the terminology used to describe the various changes of state (Sec. 4.4).
- Distinguish between the characteristics of pure substances and mixtures (Sec. 4.5).
- Explain the major differences between heterogeneous and homogeneous mixtures (Sec. 4.5).
- Understand the differences between an element and a compound (Sec. 4.6).
- Know general trends concerning the discovery and abundance of the elements (Sec. 4.7).
- Write the names when given the symbols, or the symbols when given the names of the more common elements (Sec. 4.8).

Terms and Concepts for Review

The new terms or concepts defined in this chapter are

chemistry (Sec. 4.1)
matter (Sec. 4.1)
solid state (Sec. 4.2)
liquid state (Sec. 4.2)
gaseous state (Sec. 4.2)
properties (Sec. 4.3)
physical properties (Sec. 4.3)
chemical properties (Sec. 4.3)
physical change (Sec. 4.4)
chemical change (Sec. 4.4)
chemical reaction (Sec. 4.4)
pure substance (Sec. 4.5)
mixture (Sec. 4.5)
heterogeneous mixture (Sec. 4.5)
homogeneous mixture (Sec. 4.5)
element (Sec. 4.6)
compound (Sec. 4.6)
chemical symbol (Sec. 4.8)

Questions and Problems

States of Matter

4-1 Use the word *definite* or *indefinite* to describe the shape and the volume of
a. solids **b.** liquids **c.** gases

4-2 Gold has a melting point of 1063°C and a boiling point of 2966°C. Gallium has a melting point of 30°C and a boiling point of 2403°C. Oxygen has a melting point of −218°C and a boiling point of −183°C. Specify the

physical state of each of these three substances at
a. −200°C b. 0°C c. 200°C
d. 1000°C e. 2000°C f. 3000°C

4-3 Criticize the statement: "All solids will melt if heated to a high enough temperature."

Properties of Matter

4-4 The following are properties of the element lithium. Classify them as physical or chemical properties.
- **a.** light enough to float on water
- **b.** silvery gray in color
- **c.** changes from silvery gray to black when placed in moist air
- **d.** in the liquid state boils at 1317°C
- **e.** least dense of all elements that are solids at room temperature
- **f.** can be cut with a sharp knife
- **g.** reacts violently with chlorine to form a white solid
- **h.** in the liquid state reacts spontaneously with its glass container producing a hole in the container
- **i.** burns in oxygen with a bright red flame
- **j.** a solid at room temperature

Changes in Matter

4-5 Classify each of the following changes as physical or chemical.
- **a.** A block of ice melts.
- **b.** Your chemistry book is burned.
- **c.** A dry leaf is crushed.
- **d.** Water evaporates from a lake.
- **e.** Grass in your front yard is cut.
- **f.** A table leg is fashioned from a piece of wood.
- **g.** A bar of copper is formed into wire.
- **h.** Silverware tarnishes.

4-6 Give the name of the change of state associated with each of the following processes.
- **a.** Water is made into ice cubes.
- **b.** The inside of your car window "fogs" up.
- **c.** Dry Ice (solid carbon dioxide) changes directly to a gas.
- **d.** Copper metal becomes "molten" at a high temperature.
- **e.** Mothballs in the clothes closet "disappear" with time.
- **f.** Perspiration "dries."
- **g.** Dew on the lawn "disappears" when the sun comes out.

Pure Substances and Mixtures

4-7 Consider the following classes of matter: heterogeneous mixture, homogeneous mixture, and pure substance.
- **a.** In which of these classes must two or more substances be present?
- **b.** Which of these classes could not possibly have a variable composition?
- **c.** Which of these classes could not possibly be separated into simpler substances using physical means?
- **d.** In which of these classes must the composition be uniform throughout?

4-8 Assign each of the following descriptions of matter to one of the following categories: heterogeneous mixture, homogeneous mixture, pure substance.
- **a.** two substances present, two phases present
- **b.** two substances present, one phase present
- **c.** one substance present, one phase present
- **d.** one substance present, two phases present

4-9 Classify each of the following as a heterogeneous mixture, a homogeneous mixture, or a pure substance. Also indicate how many phases are present. (All substances are assumed to be present in the same container.)
- **a.** water
- **b.** water and dissolved salt
- **c.** water and sand
- **d.** water and oil
- **e.** liquid water, oil, and ice
- **f.** carbonated water (soda water) and ice
- **g.** oil, ice, salt water solution, sugar water solution, and pieces of copper metal

Elements and Compounds

4-10 Based on the information given, classify each of the pure substances A through K as elements or compounds, or indicate that no such classification is possible because of insufficient information.
- **a.** Analysis with some elaborate instrument indicates that *Substance A* contains two elements.
- **b.** *Substance B* and *Substance C* react to give a new *Substance D* (three answers — one for B, one for C, and one for D).
- **c.** *Substance E* decomposes upon heating to give *Substance F* and *Substance G* (three answers).
- **d.** Heating *Substance H* to 1000°C causes no change in it.
- **e.** By chemical means *Substance I* cannot be broken down into simpler substances.
- **f.** By physical means *Substance J* cannot be broken down into simpler substances.
- **g.** Heating *Substance K* to 500°C causes it to change from a solid to a liquid.

4-11 Indicate whether each of the following statements is true or false.
- **a.** Both elements and compounds are pure substances.
- **b.** A compound results from the physical combination of two or more elements.

c. In order for matter to be heterogeneous at least two pure substances must be present.
d. Pure substances cannot have a variable composition.
e. Compounds, but not elements, can have a variable composition.
f. Compounds may be separated into their constituent elements using chemical means.
g. A compound must contain at least two elements.

Discovery and Abundance of the Elements

4-12 What percentage of the 106 known elements were known prior to
a. 1700 **b.** 1800 **c.** 1900 **d.** 1950

4-13 In which of the seven categories of Table 4-4 is each of the following elements most abundant?
a. oxygen **b.** iron **c.** hydrogen
d. nitrogen **e.** carbon **f.** silicon

Names and Symbols of the Elements

4-14 What chemical elements are represented by the following symbols?
a. Cu **b.** Be **c.** F **d.** Na
e. Fe **f.** Mg **g.** N **h.** He
i. Si

4-15 What are the chemical symbols of the following elements?
a. chlorine **b.** gold **c.** aluminum
d. potassium **e.** iodine **f.** carbon
g. tin **h.** calcium **i.** lead

4-16 Make a list of the elements which have symbols that
a. begin with a letter other than the first letter of the English name of the element.
b. are the first two letters of the element's English name.
c. contain only one letter.

5 The Atom and Its Structure

5.1 The Atom

atom

If one takes a sample of the element gold and starts breaking it into smaller and smaller and smaller pieces, it seems reasonable that one will eventually reach a "smallest possible piece" of gold that could not be divided further and still be called gold. This smallest possible unit of gold would be a gold atom. An **atom** *is the smallest particle of an element that can exist and still have the properties of the element.* Thus, the atom is the limit of chemical subdivision for an element.

The concept of an atom is an old one, dating back to ancient Greece. Records indicate that around 460 B.C., Democritus, a Greek philosopher, suggested that continued subdivision of matter ultimately would yield small indivisible particles which he called atoms (from the Greek word *atomos* meaning "uncut or indivisible"). Democritus's ideas about matter were, however, lost (forgotten) during the Middle Ages, as were the ideas of many other people.

It was not until the beginning of the nineteenth century that the concept of the atom was "rediscovered." John Dalton (1776–1844), an English school teacher, proposed in a series of papers published in the period 1803–1807 that the fundamental building block for all kinds of matter was an atom. Dalton's proposal had as its basis experimentation that he and other scientists had conducted. This is in marked contrast to the early Greek concept of atoms, which was based solely on philosophical speculation. Because of its experimental basis, Dalton's idea got wide attention and stimulated new work and thought concerning the ultimate building blocks of matter.

Additional research, carried out by many scientists, has now validated Dalton's basic conclusion that the building blocks for all types of matter are atomic in nature. Some of the details of Dalton's original proposals have had to be modified in the light of recent more sophisticated experiments, but the basic concept of atoms remains.

Today, among scientists, the concept that atoms are the building blocks for matter is a foregone conclusion. The large accumulated amount of supporting evidence for atoms is most impressive. The following five statements, collectively referred to as the **atomic theory of matter,** summarizes modern-day scientific thought about atoms.

atomic theory of matter

1. All matter is made up of small particles called atoms, of which 106 different "types" are known, with each "type" corresponding to atoms of a different element.
2. All atoms of a given type are similar to one another and significantly different from all other types.
3. The relative number and arrangement of different types of atoms contained in a pure substance (its composition and structure) determine its identity.
4. Chemical change is a union, separation, or rearrangement of atoms to give new substances.
5. Only whole atoms can participate in or result from any chemical change, since atoms are considered indestructible during such changes.

Atoms are incredibly small particles. No one has seen or ever will see an atom with the naked eye. The question may thus be asked: "How can you be absolutely sure that something as minute as an atom really exists?" The achievements of twentieth-century scientific instrumentation have gone a long way toward removing any doubt about the existence of atoms. Electron microscopes, capable of producing magnification factors in the millions, have made it possible to photograph "images" of individual atoms. In 1976 physicists at the University of Chicago were successful in obtaining motion pictures of the movement of single atoms. One of these pictures is shown in Figure 5-1.

Just how small is an atom? Atomic dimensions and masses, although not directly measurable, are known quantities obtained by calculation. The data used for the calculations come from measurements made on macroscopic amounts of pure substances.

The diameter of an atom is on the order of 10^{-8} centimeter. If one were to arrange atoms of diameter 1×10^{-8} centimeter in a straight line, it would take 10 million of them to extend a length of 1 millimeter and 254 million of them to reach 1 inch (see Fig. 5-2). Indeed, atoms are very small.

The mass of an atom is also a very small quantity. For example, the mass of a uranium atom, one of the heaviest of known kinds of atoms, is 4×10^{-22} gram or 9×10^{-25} pound. It would require 1×10^{24} atoms of uranium to give a mass of 1 pound. This number, 1×10^{24}, is so large that it is difficult to visualize fully. The following comparison perhaps gives some idea of its magnitude. Assume that each of the 1×10^{24} atoms was represented by a dollar. Also assume that the 1×10^{24} dollars were divided equally among the world's inhabitants (4 billion people). Each person would receive 3×10^{14} dollars and become a multitrillionaire. Recall that 10^9 is a trillion and each person would have over 10^{14} dollars.

5.2 The Molecule

Free isolated atoms are rarely encountered in nature. Instead, under normal conditions of temperature and pressure, atoms are almost always found associated together in aggregates or clusters ranging in size from two atoms to numbers too large to count. When the group or cluster of atoms is relatively small and bound together tightly the

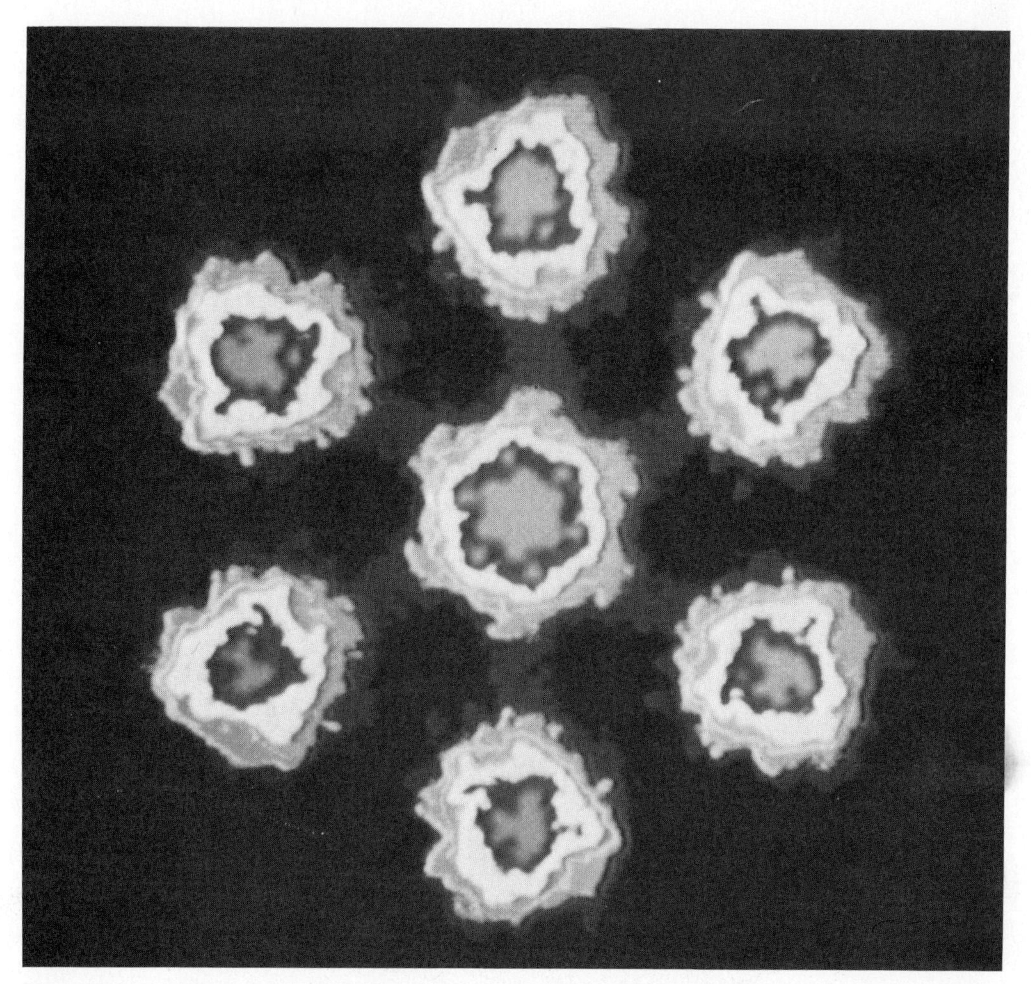

Figure 5-1. A uranyl acetate cluster on a very thin carbon substrate. The individual uranium atoms are the roundish spots with darker gray centers. [Courtesy of M. Isaacson, Cornell University, and M. Ohtsuki, The University of Chicago.]

molecule

resulting entity is called a molecule. Thus a **molecule** *is a group of two or more atoms that functions as a unit because the atoms are tightly bound together.* Reasons why atoms tend to collect together into molecules and information on the binding forces involved will be presented in Chapter 10. The important point at this time is that the collection of atoms called a molecule functions as a single composite particle.

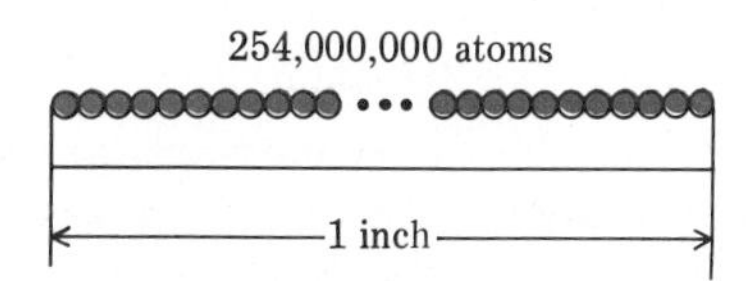

Figure 5-2. Comparison of atomic diameters and the common measuring unit 1 inch.

A *diatomic molecule,* a molecule containing just two atoms, is the simplest type of molecule. Next in complexity are *triatomic* (three atoms) and *tetraatomic* (four atoms) molecules. Numerous examples of all three of these types of molecules are known. The molecular unit present in water, the most common of all compounds, is triatomic, containing two hydrogen atoms and an oxygen atom. The molecules present in the poisonous gas carbon monoxide are diatomic — one carbon atom and one oxygen atom; tetraatomic molecules are found in the common household cleaning agent called ammonia — one nitrogen atom and three hydrogen atoms. Substances with a much larger molecular unit than two, three, or four atoms can and do exist. Tetrahydrocannabinol (THC), the active ingredient in marijuana, has a molecular unit containing 44 atoms: 14 carbon, 25 hydrogen, and 5 oxygen atoms. Figure 5-3 gives space filling models for selected types of di-, tri-, and tetraatomic molecules.

The atoms contained in a molecule may be all of the same kind or two or more kinds may be present. On the basis of this observation molecules are classified into two categories — homoatomic and heteroatomic. **Homoatomic molecules** *are molecules in which all atoms present are the same kind.* A substance containing homoatomic molecules is, thus, an element. **Heteroatomic molecules** *are molecules in which two or more different kinds of atoms are present.* Substances containing heteroatomic molecules must be compounds. All of the molecules in Figure 5-3 are heteroatomic as are the previously mentioned molecules for water, carbon monoxide, ammonia, and THC.

homoatomic molecules

heteroatomic molecules

The fact that homoatomic molecules exist indicates that individual atoms are not always the preferred structural unit for an element. Oxygen is an element that exists in molecular form. Almost all of the oxygen present in air is in the form of diatomic molecules. The elements hydrogen, nitrogen, and chlorine are most commonly encountered with their atoms in groups of two, that is, as diatomic molecules. Under

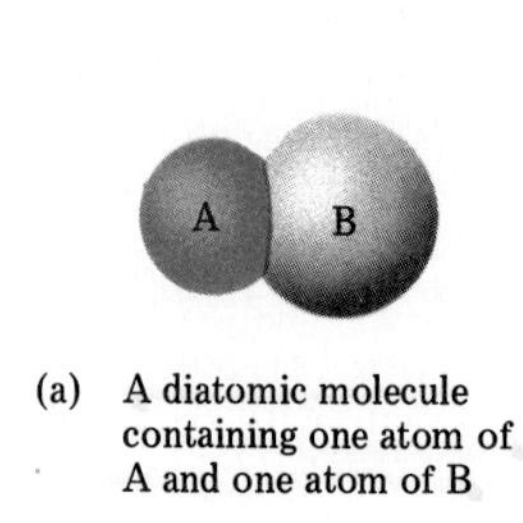

(a) A diatomic molecule containing one atom of A and one atom of B

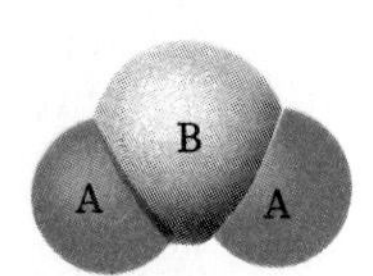

(b) A triatomic molecule containing two atoms of A and one atom of B

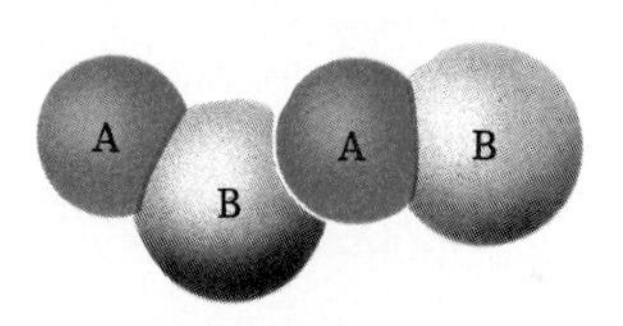

(c) A tetraatomic molecule containing two atoms of A and two atoms of B

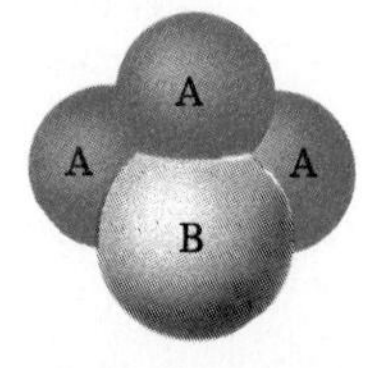

(d) A tetraatomic molecule containing three atoms of A and one atom of B

Figure 5-3. Space-filling models for simple molecules. Spheres of different colors are used to indicate different kinds of atoms.

most conditions sulfur atoms collect together in groups of eight; phosphorus atoms readily form tetraatomic molecules. Some guidelines for determining which elements have individual atoms as their basic unit and which exist in a molecular form will be given in Section 8.2.

Many, but not all, compounds have molecules as their basic structural unit. Such molecules must necessarily be heteroatomic since the presence of at least two elements is required for a compound to exist (Sec. 4.6). Compounds containing a molecular structural unit are called *molecular compounds.*

Some compounds, in the solid and liquid states, are not molecular, that is, the atoms present are not collected together into discrete heteroatomic molecules. These nonmolecular compounds still contain atoms of at least two kinds, but the form of aggregation is different, involving an extended three-dimensional assembly of positively and negatively charged particles called ions (see Sec. 5.9). Compounds that contain ions are called *ionic compounds.* The familiar substances sodium chloride (table salt) and calcium carbonate (limestone) are ionic compounds. The reasons why some compounds have an ionic rather than molecular structure will be considered in Section 10.4.

For molecular compounds, the molecule is the smallest particle of the compound capable of a stable independent existence. It is the limit of physical subdivision for the compound. Consider the molecular compound sucrose (table sugar). Continued subdivision of a quantity of table sugar to yield smaller and smaller amounts would ultimately lead to the isolation of one single particle of table sugar — a molecule of table sugar. This particle of sugar could not be broken down any further and still maintain the physical and chemical properties of table sugar. The sugar molecule could be broken down further by chemical (not physical) means to give atoms, but if that occurred we would no longer have sugar. The *molecule* is the limit of *physical* subdivision. The *atom* is the limit of *chemical* subdivision.

Every molecular compound has as its smallest, characteristic unit a *unique* molecule. If two compounds had the same molecule as a basic unit, both would have the same properties; thus, they would be one and the same compound. An alternate way of stating the same conclusion is: There is only one kind of molecule for any given molecular substance.

Since every molecule in a sample of a molecular compound is the same as every other molecule in the sample, it is commonly stated that molecular compounds are made up of a single kind of particle. Such terminology is correct as long as it is remembered that the particle referred to is the molecule. In a sample of a molecular compound there are at least two kinds of atoms present but only one kind of molecule.

The properties of molecules are much different from the properties of the atoms that make up the molecules. Molecules do not maintain the properties of their constituent elements. Table sugar is a white crystalline molecular compound with a sweet taste. None of the three elements present in table sugar (carbon, hydrogen, and oxygen) are white solids or have a sweet taste. Carbon is a black solid and hydrogen and oxygen are colorless gases.

Figure 5-4 summarizes the relationships between hetero- and homoatomic molecules and elements, compounds, and pure substances.

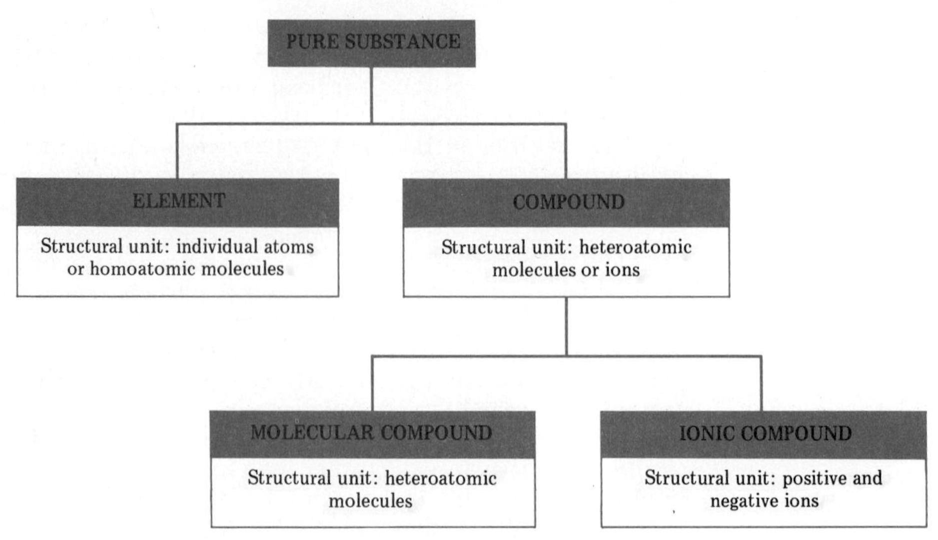

Figure 5-4. Structural units for various types of pure substances.

5.3 Chemical Formulas

chemical formula

A most important piece of information about a compound is its composition. Chemical formulas represent a concise manner for specifying compound compositions. A **chemical formula** *is a notation that contains the symbols of the elements present in a compound and numerical subscripts (located to the right of each symbol) that indicate the relative number of atoms of each element present.*

The chemical formula for the molecular compound sucrose (table sugar) is $C_{12}H_{22}O_{11}$. This formula tells us that table sugar contains the elements carbon (C), hydrogen (H), and oxygen (O) and that a molecule of table sugar contains 45 atoms — 12 of carbon, 22 of hydrogen, and 11 of oxygen.

When only one atom of a particular element is present in a molecule of a compound, that element's symbol is written without a numerical subscript in the formula for the compound. The formulas for water (H_2O) and ammonia (NH_3) both reflect the practice of not writing the subscript 1.

Always following capitalization rules for elemental symbols (Sec. 4.8) is a necessity if correct formulas are to be written. For example, making the error of capitalizing the second letter of an element's symbol can dramatically alter the meaning of a chemical formula. The formulas $CoCl_2$ and $COCl_2$ illustrate this point. The symbol Co stands for the element cobalt, whereas CO stands for one atom of carbon and one atom of oxygen.

For molecular compounds chemical formulas give the composition of the molecules making up the compounds. For ionic compounds, which have no molecules, a chemical formula gives the ion ratio found in the compound. For example, the ionic compound sodium oxide contains sodium ions and oxygen ions in a two to one ratio — twice as many sodium ions as oxygen ions. The formula of this compound is Na_2O — the ratio between the two types of ions present. The term *formula unit* is used to describe this smallest ratio between ions. The distinction between the formula unit

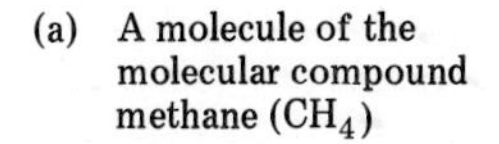
(a) A molecule of the molecular compound methane (CH_4)

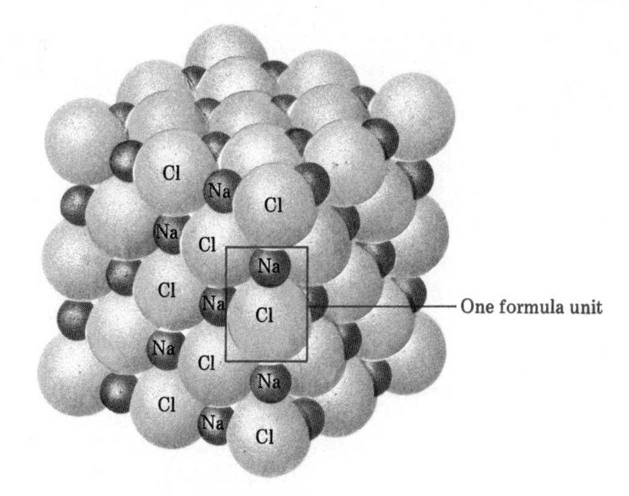

(b) A formula unit of the ionic compound sodium chloride (NaCl)

Figure 5-5. A comparison of a formula unit of an ionic compound and a molecule of a molecular compound. A molecule can exist as a separate unit, whereas a formula unit is simply two or more ions "plucked" from a much larger array of ions.

of an ionic compound and the molecule of a molecular compound is graphically portrayed in Figure 5-5.

Sometimes chemical formulas are written which contain parentheses, an example being $Ca(NO_3)_2$. The interpretation of this formula is straightforward—the atoms within the parentheses, NO_3, are repeated twice. The subscript following the parentheses always indicates the number of units of the multi-atom entity inside the parentheses. As another example of parentheses use, consider the compound $Pb(C_2H_5)_4$. Four units of C_2H_5 are present. In terms of atoms present, the formula $Pb(C_2H_5)_4$ represents 29 atoms—1 lead (Pb) atom, 8 (4×2) carbon (C) atoms, and 20 (4×5) hydrogen (H) atoms. The formula $Pb(C_2H_5)_4$ could be (but is not) written as PbC_8H_{20}. Both versions of the formula convey the same information in terms of atoms present; however, the former gives some information about structure (C_2H_5 units are present) that the latter does not. For this reason the former is the preferred way of writing the formula. Further information concerning the use of parentheses (when and why) will be presented in Section 6.8. The important concern now is being able to interpret formulas that contain parentheses in terms of total atoms present. Example 5.1 deals with this skill in greater detail.

EXAMPLE 5.1

Interpret each of the following formulas in terms of how many atoms of each element are present in one structural unit of the substance.

(a) H_2SO_4 **(b)** $Ca(HCO_3)_2$ **(c)** $(NH_4)_2CO_3$

SOLUTION

(a) We simply look at the subscripts following the symbols for the elements, remembering that the subscript 1 is implied when no subscript is written. This formula contains seven atoms — two hydrogen atoms, one sulfur atom, and four oxygen atoms.
(b) The subscript following the parentheses, a 2, indicates that two HCO_3 units are present. We therefore have two hydrogen, two carbon, and six oxygen atoms present in addition to the one calcium atom.
(c) The parentheses tell us that the subscript 2 applies to everything inside the parentheses. Doubling the amount of nitrogen and hydrogen gives us two nitrogen atoms and eight hydrogen atoms. We also have one carbon atom and three oxygen atoms present.

The phrases *atoms of an element* and *molecules of an element* are concepts often confused by students. An atom of an element is the smallest particle of that element that can combine to form a compound. A molecule of an element, if the element exists in a molecular form, is the preferred structural unit for the element as it is found in nature. It is not the preferred structural unit, however, for the element in compounds. Thus, the symbols and numbers found in chemical formulas deal with the number of *atoms* of various elements present; they have nothing to do with molecules of the elements. The formula for a molecule of oxygen is O_2. A molecule of carbon dioxide, CO_2, does not contain an oxygen molecule; it contains two oxygen atoms.

5.4 Subatomic Particles: Protons, Neutrons, and Electrons

subatomic particles

Until the closing decades of the nineteenth century, scientists believed that atoms were solid indivisible spheres without substructure. Today this concept is known to be incorrect. Evidence from a variety of sources indicates that atoms themselves are made up of smaller more fundamental particles called subatomic particles. **Subatomic particles,** *particles smaller than atoms, are the building blocks from which all atoms are made.*

protons

electrons

neutrons

Three major types of subatomic particles exist: the proton, the neutron, and the electron. The properties of these subatomic particles that are of most concern to us are mass and electrical charge. **Protons** *are subatomic particles that possess a positive charge* (+) *and have a mass of* 1.673×10^{-24} *gram.* Protons were discovered in 1886 by the German physicist Eugen Goldstein (1850–1930). **Electrons** *are subatomic particles that possess a negative charge* (−) *and have a mass of* 9.109×10^{-28} *gram.* Electrons were discovered in 1897 by the English physicist J. J. Thomson (1856–1930). **Neutrons** *are subatomic particles that are neutral* (*i.e., have no charge*) *and have a mass of* 1.675×10^{-24} *gram.* The neutron was the last of the three subatomic particle types to be identified in 1932 by the English physicist James Chadwick (1891–1974). Table 5-1 summarizes the mass and electrical charge characteristics for subatomic particles. In this table relative masses are given in addition to actual masses.

How subatomic particle mass values compare to each other, that is, relative mass values, serve just as well as actual mass values in most instances. Using the relative

Table 5-1. ***Charges and Masses of the Major Subatomic Particles***

	Electron	Proton	Neutron
Charge	−1	+1	0
Actual mass (g)	9.11×10^{-28}	1.672×10^{-24}	1.675×10^{-24}
Relative mass (based on the electron being one unit)	1	1836	1839
Relative mass (based on the neutron being one unit)	0 (1/1839)	1	1

mass values from Table 5-1, we see that the neutron has a mass value only slightly greater than that of the proton. For most purposes their masses may be considered equal. Both neutrons and protons are very massive particles when compared to an electron, being nearly 2000 times heavier. The mass of an electron is almost negligible when compared to that of the other two types of subatomic particles.

Electrons and protons, the two types of *charged* subatomic particles, possess the same amount of electrical charge; the character of the charge is, however, opposite (negative versus positive). The fact that these subatomic particles are charged is most important because of the way in which charged particles interact. *Particles of opposite or unlike charge attract each other; particles of like charge repel each other.* This behavior of charged particles will be of major concern in many of the discussions in later portions of the text.

The arrangement of subatomic particles within an atom is not haphazard. As is shown in Figure 5-6, the atom may be considered to be composed of two regions: (1) a nuclear region and (2) an extranuclear region, which is everything except the nuclear region.

nucleus

At the center of every atom is a nucleus. The **nucleus** *is the center region (core) of an atom and contains within it all protons and neutrons present in the atom.* Because of their presence in the nucleus, neutrons and protons are often collectively called *nucleons.* Almost all (over 99.9%) of the mass of an atom is concentrated in its nucleus; all of

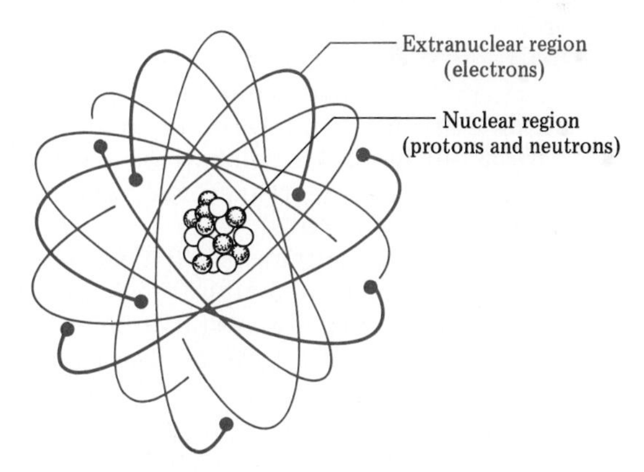

Figure 5-6. Arrangement of subatomic particles in an atom.

the heavy subatomic particles (protons and neutrons) are there. A nucleus always carries a positive charge because of the presence of the positively charged protons.

The extranuclear region contains all of the electrons. It is an extremely large region, mostly empty space, in which the electrons move rapidly about the nucleus. The motion of the electrons in this extranuclear region determines the volume (size) of the atom in the same way as the propeller of an airplane determines an area by its motion. The volume occupied by the electrons is sometimes referred to as the *electron cloud.* Since electrons are negatively charged, the electron cloud is also negatively charged.

An atom as a whole is neutral. How can it be that an entity possessing positive charge (the nuclear region) and negative charge (the extranuclear region or electron cloud) can end up neutral overall? For this to occur the same amount of positive and negative charge must be present in the atom. Equal amounts of positive and negative charge would cancel each other. Atom neutrality thus requires that there be the same number of electrons and protons present in an atom, which is always the case.

It is important that the size relationships between the parts of an atom be correctly visualized. The nucleus is extremely small (supersmall) compared to the total atom size. A nonmathematical conceptual model to help visualize this relationship involves clouds. Visualize a large spherical or near-spherical fluffy cloud in the sky. Let the cloud represent the negatively charged extranuclear region of the atom. Buried deep within the cloud would be the positively charged nucleus, the size of a small pebble. As another example, consider the electron cloud to be the size of a major league baseball park. The nucleus would be no larger than a small fly located somewhere in the region behind second base. A more quantitative example involves the enlargement (magnification) of the nucleus until it would be the size of a baseball (7.4 cm in diameter). If the nucleus were this large, the whole atom would have a diameter of approximately 2.5 miles. The nuclear volume is only one one-hundred thousandth that of the atom's total volume. Again, almost all (over 99.9%) of the volume of an atom involves the electron cloud.

Our just-completed discussion of the makeup of atoms in terms of subatomic particles is based on the existence of three types of subatomic particles — protons, neutrons, and electrons. Actually this model of the atom is an oversimplification. In recent years, as the result of research carried out by nuclear physicists, our picture of the atom has lost its simplicity. Experimental evidence now available indicates that protons and neutrons themselves are made up of even smaller particles. Numerous other particles, with names such as leptons, mesons, baryons, and quarks, have been discovered. No theory is yet available that can explain all of these new discoveries relating to the complex nature of the nucleus.

Despite the existence of other nuclear particles we will continue to use the three-subatomic-particle model of the atom. It readily explains almost all chemical observations about atoms. We will have no occasion to deal with any of the recently discovered types of subatomic (or sub-subatomic) particles. Protons, neutrons, and electrons will meet all of our needs.

We will also continue to use the concept of atoms being the fundamental building blocks for all types of matter (Sec. 5.1) despite the existence of protons, neutrons, and electrons. This is because subatomic particles do not lead an independent existence for any appreciable length of time. The only way they gain stability is by joining together to form an atom.

5.5 Atomic Number and Mass Number

atomic number

The characteristic of an atom that identifies what kind of atom it is is the atomic number. The **atomic number** *is equal to the number of protons in the nucleus of an atom.* Atomic numbers are always integers (whole numbers). All atoms of a given element must contain the same number of protons. If two atoms differ in the number of protons present, they must be atoms of two different elements.

Since an atom has the same number of electrons as protons, the atomic number also specifies the number of electrons present.

Atomic number = number of protons = number of electrons

mass number

A second necessary quantity in specifying atomic identities is the mass number. The **mass number** *is equal to the number of protons plus neutrons in the nucleus of the atom, that is, the total number of nucleons present.* The mass of an atom is almost totally accounted for by the protons and neutrons (Sec. 5.4); hence the term mass number. Like the atomic number, the mass number is always an integer.

Mass number = number of protons + number of neutrons

Knowing the atomic number and mass number of an atom uniquely specifies the atom's makeup in terms of subatomic particles. The following equations show the relationship between subatomic particle makeup and the two numbers.

Number of protons = atomic number

Number of electrons = atomic number

Number of neutrons = mass number − atomic number

EXAMPLE 5.2

For an atom having an atomic number of 6 and a mass number of 14 determine

(a) the number of protons present (b) the number of neutrons present
(c) the number of electrons present

SOLUTION

(a) The number of protons is six. The atomic number is always equal to the number of protons present.
(b) The number of neutrons is eight. The number of neutrons is always obtained by subtracting the atomic number from the mass number.

(Protons + neutrons) − protons = neutrons

Mass number Atomic number

(c) The number of electrons present is six. In a neutral atom the number of protons and the number of electrons are always the same.

Printed on the inside back cover of this text is an alphabetical listing of the 106 known elements along with selected information about each element. One of the pieces of information given for each element is its atomic number. If you were carefully to check the atomic number data you would find that an element exists for each atomic number in the numerical sequence 1 through 106. There are no missing numbers in this sequence. The existence of an element with each of these numbers is an indication of the order existing in nature. Thus, elements exist with any number of protons (or electrons) up to and including 106.

Mass numbers are not tabulated in a manner similar to atomic numbers because, as we shall see in the next section, most elements lack a unique mass number.

5.6 Isotopes

The identity of an atom is determined by its atomic number, that is, by the equal number of protons and electrons that it contains. All atoms of an element must contain the same number of these two types of subatomic particles.

The chemical properties of an atom, which are the basis for its identification, are determined by the number and arrangement of the electrons about the nucleus. All atoms with the same atomic number will have the same number of electrons and hence the same chemical properties.

Atoms of an element need not, however, all be identical. They may differ from each other in the number of neutrons present. The presence of one or more extra neutrons in the tiny nucleus of an atom has essentially no effect on the way it behaves chemically. For example, all oxygen atoms have eight protons and eight electrons. Most oxygen atoms also contain eight neutrons. Some oxygen atoms exist, however, that contain nine neutrons, and a few exist that contain ten neutrons. Thus, three different kinds of oxygen atoms exist — each with the same chemical properties. Three oxygen *isotopes* are said to exist. **Isotopes** *are atoms that have the same number of protons and electrons (thus being atoms of the same element) but different numbers of neutrons.* Isotopes of an element will always have the same atomic number and differing mass numbers.

isotopes

When it is necessary to distinguish between isotopes, the following notation is used.

$$^{A}_{Z}\text{symbol}$$

The atomic number, whose general symbol is Z, is written as a subscript to the left of the elemental symbol for the atom. The mass number, whose general symbol is A, is also written to the left of the elemental symbol, but as a superscript. By this symbolism, the three previously mentioned oxygen isotopes would be designated, respectively,

$$^{16}_{8}\text{O}, \quad ^{17}_{8}\text{O}, \quad \text{and} \quad ^{18}_{8}\text{O}$$

The existence of isotopes clarifies some of the wording used earlier (Sec. 5.1) in stating the postulates of the atomic theory. Postulate 1 states: "All matter is made up of small particles called atoms, of which 106 different 'types' are known." It should now be apparent why the word types was put in quotation marks. Because of the existence of isotopes, types means "similar, but not identical." Atoms of a given element are similar in that they have the same atomic number, but not identical since the atoms may have different mass numbers.

Postulate 2 states: "All atoms of a given type are similar to one another and significantly different from all other types." All atoms of an element are similar in chemical properties and differ significantly from atoms of other elements with different chemical properties.

Most elements occurring naturally are mixtures of isotopes. The various isotopes of a given element are of varying abundance; usually one isotope is predominate. Typical of this situation is the element magnesium, which exists in nature in three isotopic forms: $^{24}_{12}Mg$, $^{25}_{12}Mg$, and $^{26}_{12}Mg$. The percentage abundances for these three isotopes are, respectively, 78.70%, 10.13%, and 11.17%. Percentage abundances are number percentages (number of atoms) rather than mass percentages. A sample of 10,000 magnesium atoms would contain 7870 $^{24}_{12}Mg$ atoms, 1013 $^{25}_{12}Mg$ atoms, and 1117 $^{26}_{12}Mg$ atoms. Table 5-2 gives natural isotopic abundances and isotopic masses for selected elements. The units used for specifying the mass of the various isotopes (last column of Table 5-2) will be discussed in Section 5.8. A few elements, twenty in number, exist in nature in only one form. They are listed at the bottom of Table 5-2.

The percentage abundances of the isotopes of an element may vary slightly in samples obtained from different locations, but such variations are ordinarily extremely small. We will assume in this text that the isotopic composition of an element is a constant.

EXAMPLE 5.3

Indicate whether the members of each of the following pairs of atoms are isotopes.

(a) $^{65}_{30}X$ and $^{65}_{31}Y$ **(b)** $^{25}_{12}X$ and $^{26}_{12}Y$
(c) (an atom with 17 protons and 18 neutrons) and (an atom with 16 protons and 18 neutrons)

SOLUTION

(a) These atoms are not isotopes. Isotopes must have the same atomic number. The first atom of this pair has an atomic number of 30 and the second atom 31. The mass numbers of the two atoms are the same, but that does not make them isotopes.
(b) These atoms are isotopes. Both atoms have the same atomic number of 12. Isotopes differ from each other only in neutron content, which is the case here. The first atom has 13 neutrons and the second atom 14 neutrons.
(c) These atoms are not isotopes. Since they differ in the number of protons present, they differ in atomic number. The atomic number is the number of protons.

It is because of the existence of isotopes that mass numbers are not unique for elements as are atomic numbers (Sec. 5.5). It is common for isotopes of two different elements to have the same mass number. For example, the element iron (atomic number = 26) exists in nature in four isotopic forms, one of which is $^{58}_{26}Fe$. The element nickel, with an atomic number two units greater than that of iron, exists in nature in five isotopic forms, one of which is $^{58}_{28}Ni$. Thus, atoms of both iron and nickel exist with a mass number of 58. Note again, that even though atoms of two different elements can

Table 5-2. ***Naturally Occurring Isotopic Abundances for Some Common Elements***

Element	Isotope	Percent Natural Abundance	Isotopic Mass (amu)
Hydrogen	$^{1}_{1}H$	99.985	1.0078
	$^{2}_{1}H$	0.015	2.0141
Carbon	$^{12}_{6}C$	98.89	12.0000
	$^{13}_{6}C$	1.11	13.0033
Nitrogen	$^{14}_{7}N$	99.63	14.0031
	$^{15}_{7}N$	0.37	15.0001
Oxygen	$^{16}_{8}O$	99.759	15.9949
	$^{17}_{8}O$	0.037	16.9991
	$^{18}_{8}O$	0.204	17.9992
Sulfur	$^{32}_{16}S$	95.0	31.9721
	$^{33}_{16}S$	0.76	32.9715
	$^{34}_{16}S$	4.22	33.9679
	$^{36}_{16}S$	0.014	35.9671
Chlorine	$^{35}_{17}Cl$	75.53	34.9689
	$^{37}_{17}Cl$	24.47	36.9659
Copper	$^{63}_{29}Cu$	69.09	62.9298
	$^{65}_{29}Cu$	30.91	64.9278
Titanium	$^{46}_{22}Ti$	7.93	45.95263
	$^{47}_{22}Ti$	7.28	46.9518
	$^{48}_{22}Ti$	73.94	47.94795
	$^{49}_{22}Ti$	5.51	48.94787
	$^{50}_{22}Ti$	5.34	49.9448
Uranium	$^{234}_{92}U$	0.0057	234.0409
	$^{235}_{92}U$	0.72	235.0439
	$^{238}_{92}U$	99.27	238.0508

Twenty elements have only one naturally occurring form: $^{9}_{4}Be$, $^{19}_{9}F$, $^{23}_{11}Na$, $^{27}_{13}Al$, $^{31}_{15}P$, $^{45}_{21}Sc$, $^{55}_{25}Mn$, $^{59}_{27}Co$, $^{75}_{33}As$, $^{89}_{39}Y$, $^{93}_{41}Nb$, $^{103}_{45}Rh$, $^{127}_{53}I$, $^{133}_{55}Cs$, $^{141}_{59}Pr$, $^{159}_{65}Tb$, $^{165}_{67}Ho$, $^{169}_{69}Tm$, $^{197}_{79}Au$, $^{209}_{83}Bi$.

have the same mass number, atoms of different elements cannot have the same atomic number. All atoms of a given atomic number must necessarily be atoms of the same element.

5.7 Atomic Weights

The mass of an atom of a specific element can have one of several values if the element exists in isotopic forms. For example, oxygen atoms can have any one of three masses, since three isotopes of oxygen exist. Because of the existence of isotopes, it might seem necessary to specify isotopic identity every time atomic masses must be dealt with; however, this is not the case. In practice, isotopes are seldom mentioned in discussions and calculations involving atomic masses; instead, the atoms of an element are treated as if they all had a single mass. The mass used is an average relative mass that takes into account the existence of isotopes. The use of this average relative mass concept reduces the number of masses needed for calculations from many hundreds to 106 — one for each element. These average masses are called *atomic weights* and are determined from the following information: the number of isotopes that exist for an element, the isotopic masses (on a relative scale), and the percentage abundance of each isotope. The term atomic weight is actually a misnomer; the quantity we are talking about should be called atomic mass. However, the term atomic weight has become so firmly entrenched in the working vocabulary of scientists that it is futile to attempt to change the terminology. Use of the term atomic weight is tolerable as long as it is remembered that atomic weights are really mass values.

atomic weight

Formally defined, an **atomic weight** *is the relative mass of an average atom of an element on a scale using atoms of* $^{12}_{6}C$ *as the reference.* The key to understanding this formal definition involves understanding clearly what is meant by two phrases used in the definition: *relative mass* and *average atom.* Let us consider in detail their meanings.

Relative Mass

The usual standards of mass, such as grams or pounds, are not convenient for use with atoms, because very small numbers are always encountered. For example, the mass in grams of a $^{238}_{92}U$ atom, one of the heaviest atoms known, is 3.95×10^{-22}. To avoid repeatedly encountering such small numbers scientists have chosen to work with relative rather than actual mass values.

A relative mass value of an atom is the mass of that atom relative to some standard rather than the actual mass value of the atom in grams. The term relative means "as compared to." The choice of the standard is arbitrary, giving scientists control over the magnitude of the numbers on the relative scale; thus, very small numbers may be avoided.

For most purposes in chemistry relative mass values serve just as well as actual mass values. Knowing how many times heavier one atom is than another, information obtainable from a relative mass scale, is just as useful a piece of information as knowing the actual mass values of the atoms involved. Example 5.4 illustrates the procedures involved in constructing a relative mass scale and also points out some of the characteristics of such a scale.

EXAMPLE 5.4

Construct a relative mass scale for the hypothetical atoms A, B, and C given the following information about them.

1. Atoms of A are four times heavier than those of B.
2. Atoms of B are three times heavier than those of C.

SOLUTION

Atoms of C are the lightest of the three types of atoms. We will arbitrarily assign atoms of C a mass value of one unit. The unit name can be anything we wish, and we shall choose "snurk." On this basis, one atom of C has a mass value of 1 snurk. Atoms of C will be our scale reference point. Atoms of B will have a mass value of 3 snurks (three times as heavy as C) and A atoms a value of 12 snurks (four times as heavy as B).

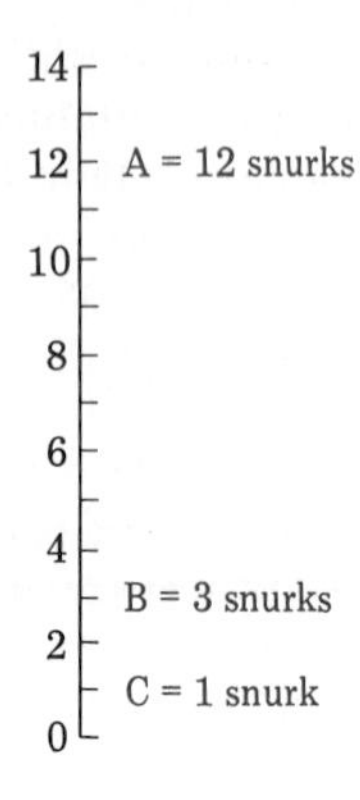

The name chosen for the mass unit was arbitrary. The assignment of the value 1 for the mass of C, the reference point on the scale, was also arbitrary. What if we have chosen to call the unit a "bleep" and had chosen a value of 3 for the mass of an atom of C? If this had been the case, the resulting relative scale would have appeared as

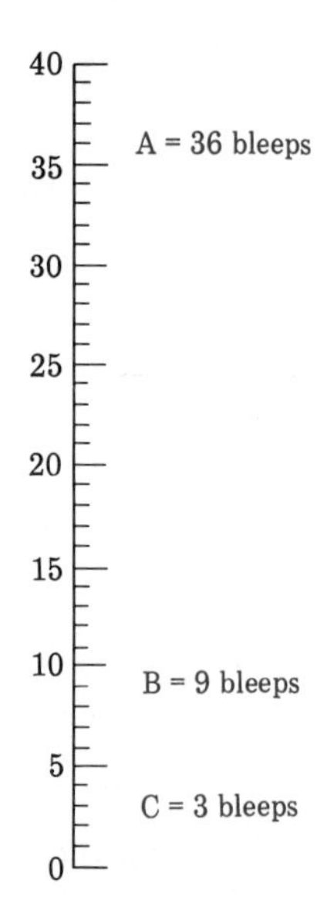

Which of the above relative scales is the "best" scale? The answer is that the scales are equivalent. The relationships between the masses of A, B, and C are the same on the

two scales, even though the reference points and unit names differ. On the "snurk" scale, for example, A is four times heavier than B (12/3); on the "bleep" scale, A is also four times heavier than B (36/9).

Notice that we did not need to know the actual masses of A, B, and C to set up either the "snurk" or "bleep" scale. All that is needed to set up a relative scale is a set of interrelationships between quantities. One value is arbitrarily assigned, the reference point, and all other values are determined by using the known interrelationships between quantities.

The information given at the start of this example is sufficient to set up an infinite number of relative mass scales. Each scale would differ from the others in choice of reference point and unit name. All the scales would, however, be equivalent to each other, and each scale would provide all of the mass relationships obtainable from an actual mass scale, except for actual mass values.

A relative scale of atomic masses has been set up in a manner similar to that used in Example 5.4. The unit is called the *atomic mass unit,* abbreviated amu. The arbitrary reference point involves a particular isotope of carbon, ${}^{12}_{6}C$. The mass of this isotope is set at 12.00000 amu. The masses of all other atoms are then determined relative to that of ${}^{12}_{6}C$. For example, if an atom is twice as heavy as a ${}^{12}_{6}C$ atom, its mass is 24.00000 amu on the scale, and if an atom weighs half as much as a ${}^{12}_{6}C$ atom, its scale mass is 6.00000 amu.

The masses of all atoms have been determined relative to each other experimentally. Actual values for the masses of selected isotopes on the ${}^{12}_{6}C$ scale are given in Figure 5-7.

Reread the formal definition of atomic weight given at the start of this section. Note how ${}^{12}_{6}C$ is mentioned explicitly in the definition because of the central role it plays in the setting up of the relative atomic mass scale.

On the basis of the values given in Figure 5-7, it is possible to make statements such as: ${}^{238}_{92}U$ is 4.256 times as heavy as ${}^{56}_{26}Fe$ (238.05 amu/55.93 amu = 4.256), and ${}^{56}_{26}Fe$ is 2.798 times as heavy as ${}^{20}_{10}Ne$ (55.93 amu/19.99 amu = 2.798). In order to make statements such as these we do not need to know the actual masses of the atoms involved; relative masses are sufficient to calculate the information.

Average Atom

If isotopes exist, the mass of an atom of a specific element may have one of several values. For example, oxygen atoms can have any one of three masses, since three isotopes exist: ${}^{16}_{8}O$, ${}^{17}_{8}O$, and ${}^{18}_{8}O$. Despite mass variances because of isotopes the atoms of an element are treated as if they all had a single common mass. The common mass value used is a *weighted average mass,* which takes into account the natural abundances and atomic masses of the isotopes of an element. The use of weighted average masses reduces the number of masses used in calculations from the hundreds characteristic of all isotopes to 106, one for each known element.

The validity of the weighted average mass concept rests on two points. First, extensive studies of naturally occurring elements have shown that the percentage abundance of the isotopes of a given element is generally constant. No matter where the element sample is obtained on earth, it generally contains the same percentage of each

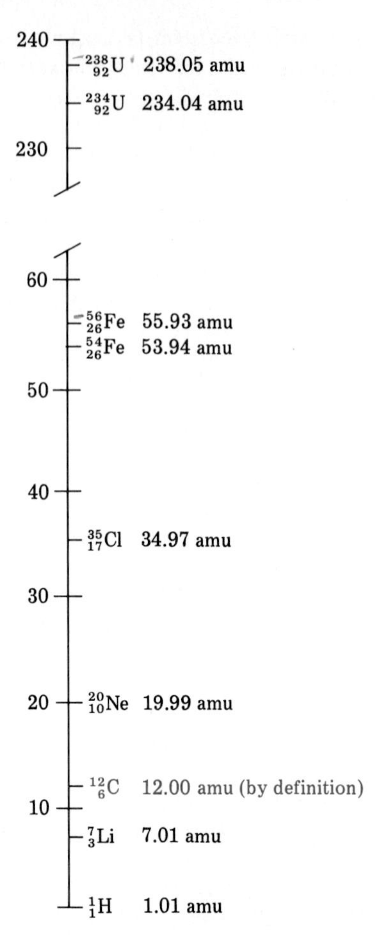

Figure 5-7. Relative masses of selected isotopes on the $^{12}_{6}C$ atomic mass scale.

isotope. Because of these constant isotopic ratios, the mass of an "average atom" does not vary. Second, chemical operations are always carried out with very large numbers of atoms. The tiniest piece of matter visible to the eye contains more atoms than can be counted by a person in a lifetime. The numbers are so great that any collection of atoms that a chemist works with will be representative of naturally occurring isotopic ratios.

Weighted Averages

Atomic weights are weighted averages calculated using the following three pieces of information.

1. The *number* of isotopes that exist for the element.
2. The *isotopic mass* for each isotope, that is, the relative mass of each isotope on the $^{12}_{6}C$ scale.
3. The *percentage abundance* of each isotope.

This type of information is found in Table 5-2 for selected elements.

Examples 5.5 and 5.6 illustrate the operations needed to calculate weighted averages. Example 5.5 is a general exercise concerning weighted averages, and Example 5.6 illustrates the calculation of an atomic weight by the method of weighted averages.

EXAMPLE 5.5

Five cups are found to contain three, three, seven, seven, and ten glass marbles, respectively. What is the average number of glass marbles in a cup?

SOLUTION

Let us solve this problem using two different methods. The first method involves procedures familiar to you — the "normal" way of taking an average. By this method, the average is found by dividing the sum of the numbers by the number of values summed.

$$3 + 3 + 7 + 7 + 10 = 30$$

$$\frac{30}{5} = 6 = \text{average value}$$

Now let us solve this same problem again, this time treating it as a "weighted-average" problem. To do this, we organize the given information in a different way. In our list of given information we have three kinds of numbers — threes, sevens, and a ten.

40% of the numbers are 3 (two out of five numbers = 40%)

40% of the numbers are 7 (two out of five numbers = 40%)

20% of the numbers are 10 (one out of five numbers = 20%)

We will use the data in this "percentage form" for our weighted average calculation.

To find the weighted average we multiply each number (3, 7, or 10) by its fractional abundance (i.e., the percentage expressed in fractional or decimal form), and then we sum the products from the multiplications.

$$\frac{40}{100} \times 3 = (0.40) \times 3 = 1.2$$

$$\frac{40}{100} \times 7 = (0.40) \times 7 = 2.8$$

$$\frac{20}{100} \times 10 = (0.20) \times 10 = \underline{2.0}$$

$$6.0 \quad \text{(same average value as before)}$$

This averaging method, although it appears somewhat more involved than the "normal" method, is the one that must be used in calculating atomic weights because of the form in which data about isotopes is obtained. The percentage abundances of isotopes are experimentally determinable quantities. The total number of atoms present in nature of various isotopes, a prerequisite for using the "normal" average method, is not easily determined. Hence we use the "percentage" method.

EXAMPLE 5.6

Magnesium occurs in nature in three isotopic forms: $^{24}_{12}Mg$, $^{25}_{12}Mg$, and $^{26}_{12}Mg$. The isotopic mass of $^{24}_{12}Mg$ is 23.985 amu, and its percentage abundance is 78.70%. For $^{25}_{12}Mg$ and $^{26}_{12}Mg$, the corresponding figures are 24.986 (10.13%) and 25.983 (11.17%), respectively. Calculate the atomic weight of magnesium.

SOLUTION

The atomic weight is calculated using the "weighted-average" method illustrated in the previous example. Each of the isotopic masses is multiplied by the fractional abundance and then the products are summed.

$$(0.7870)(23.985 \text{ amu}) = 18.876195 = 18.88$$

$$(0.1013)(24.986 \text{ amu}) = 2.5310818 = 2.531$$

$$(0.1117)(25.983 \text{ amu}) = 2.9023011 = 2.902$$

↑ Percentage abundance converted to a decimal; ↑ Isotope mass; ↑ Calculator answers; ↑ Correct answers (limited to four significant figures)

Remember significant figures! Significant figures must be taken into account in any calculation that involves experimentally determined numbers. Both isotopic masses and percentage abundances are experimentally determined.

$$18.88 + 2.531 + 3.902 = 24.313 \quad \text{(calculator answer)}$$

$$= 24.31 \quad \text{(correct answer)}$$

Since the number 18.88 is known only to the hundredths place, the answer can be expressed only to the hundredths place.

The above calculation involved an element which exists in three isotopic forms. An atomic weight calculation for an element having four isotopes would be carried out in an almost identical fashion. The only difference would be four products to calculate (instead of three) and four terms in the resulting sum.

In Example 5.6 the atomic weight of magnesium was calculated to be 24.31 amu. How many magnesium atoms have a mass of 24.31 amu? The answer is none. Magnesium atoms have a mass of 23.985 amu, 24.986 amu, or 25.983 amu, depending on which isotope they are. The mass 24.31 is the mass of an "average" magnesium atom. It is this average value that is used in calculations. Again, no magnesium atoms have masses exactly equal to this average value. Only in the case where there are no isotopes of an element (all atoms are the same kind) will the isotopic mass and the atomic weight be the same.

5.8 The Periodic Law and the Periodic Table

During the early part of the nineteenth century, many chemical facts became available from detailed studies of the elements then known. In the hope of providing a systematic approach to the study of chemistry, scientists began to look for order in the increasing amount of chemical information. They were encouraged in their search by the unexplained but well-known fact that certain elements had properties very similar to those of other elements. Numerous attempts were made to find reasons for these similarities and to use them as a way to provide some method of arranging or classifying the elements.

periodic law

In 1869 these efforts culminated in the discovery of what is now called the *periodic law.* Proposed independently by both the Russian chemist Dmitri Mendeleev (1834–1907) and the German chemist Julius Lothar Meyer (1830–1895), the **periodic law** (here given in its modern form) states that *when all the elements are arranged in order of increasing atomic numbers, elements with similar properties will occur at periodic (regularly recurring) intervals.*

periodic table

Printed on the inside front cover of this text is a *periodic table,* which is a graphical representation of the behavior described by the periodic law. A **periodic table** *is a graphical arrangement of the elements in order of increasing atomic number such that elements with similar properties are arranged in vertical columns.* In this periodic table, each element occupies a separate box. Within each box is given the symbol, atomic number, and atomic weight of the element, as shown in Figure 5-8.

groups

periods

Special chemical terminology exists for specifying the position (location) of an element within the periodic table. This terminology involves the use of the words *group* and *period.* **Groups** *in the periodic table are vertical columns of elements.* Notice, by referring to the periodic table inside the front cover, that groups are designated by Roman numerals and the letters A and B. (The group designation is given at the top of each group.) The elements with atomic numbers 8, 16, 34, 52, and 84 (O, S, Se, Te, and Po) constitute group VIA in the periodic table. **Periods** *in the periodic table are horizontal rows of elements.* For identification purposes, the periods are numbered sequentially starting at the top of the table. The elements H, Li, Na, K, Rb, Cs, and Fr (all members of group IA — the group on the extreme left of the periodic table) are, respectively, members of periods 1, 2, 3, 4, 5, 6, and 7. The location of any element in the periodic table is thus specified by giving its group number and period number. Thus, the element gold, with an atomic number of 79, belongs to group IB and period 6.

In later chapters we will find that a periodic table contains much more information

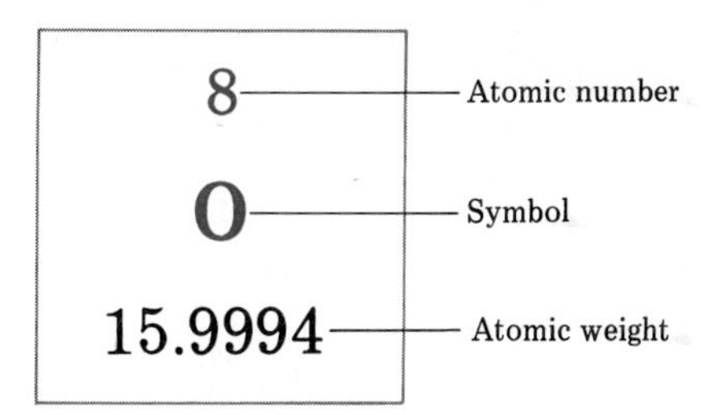

Figure 5-8. Arrangement of information within the periodic table.

about the elements than their symbols, atomic numbers, and atomic weights, which are explicitly shown. Information about the arrangement of electrons in an atom is "coded" into the table as is information concerning some physical and chemical property trends. Indeed, the periodic table is considered the single most useful study aid available for organizing information about the elements.

For many years after the formulation of the periodic law and the periodic table, both were considered to be empirical. The law worked and the table was very useful, but there was no explanation available for the law or for why the periodic table had the shape that it did. It is now known that the theoretical basis for both the periodic law and table lies in electronic theory. The properties of the elements repeat themselves in a periodic manner, because the arrangement of electrons about the nucleus of an atom follows a periodic "pattern." The theoretical foundations for the periodic table will be discussed further in Sections 9.9 and 9.10.

5.9 Ions: Charged Atoms or Groups of Atoms

Under certain circumstances a neutral atom may acquire either a positive or a negative charge. This charge results from the loss or gain of electrons by the atom. Neutrality of atoms is the result of the number of protons (positive charges) being equal to the number of electrons (negative charges). Loss or gain of electrons destroys this proton–electron balance and leaves a net charge on the atom.

ion

The charged particles formed from loss or gain of electrons are called ions. An **ion** *is an atom (or group of atoms) that is electrically charged as the result of an excess or deficiency of electrons.* If one or more electrons are gained by an atom, a negatively charged ion is produced; excess negative charge is present since electrons now outnumber protons. Loss of one or more electrons by an atom results in the formation of a positively charged ion; more protons are now present than electrons resulting in excess positive charge. Note that the excess positive charge associated with a positive ion is never caused by proton gain but rather always from electron loss. Having the number of protons remain constant and decreasing the number of electrons will result in net positive charge. The number of protons (which determines the identity of an element) never changes during ion formation.

The charge on an ion is directly correlated with the number of electrons lost or gained. Loss of one, two, or three electrons gives, respectively, ions with a +1, +2, and +3 charge. Similarly, a gain of one, two, or three electrons gives, respectively, ions with −1, −2, and −3 charges. (Ions that have lost or gained more than three electrons are very seldom encountered — see Sec. 10.4.)

The notation for denoting the charge on an ion is a superscript placed to the right of the elemental symbol. Some examples of ion symbols are

Positive ions: Na^+, Ba^{2+}, K^+, Al^{3+}, Ca^{2+}

Negative ions: Cl^-, O^{2-}, F^-, S^{2-}, N^{3-}

Note that a lone plus or minus sign is used to denote a charge of one, instead of using $^{1+}$ or $^{1-}$. Also note that in multiply charged ions the number precedes the charge sign, that is, the notation for a charge of plus two is $^{2+}$ rather than $^{+2}$.

EXAMPLE 5.7

Give the symbol for each of the following ions.

(a) The ion formed when a calcium atom loses two electrons.
(b) The ion formed when a nitrogen atom gains three electrons.

SOLUTION

(a) A neutral calcium atom contains twenty protons and twenty electrons. A calcium ion would still contain twenty protons, but would have only eighteen electrons since two electrons were lost.

$$\begin{aligned} 20 \text{ protons} &= 20 + \text{charges} \\ 18 \text{ electrons} &= \underline{18 - \text{charges}} \\ \text{Net charge} &= 2+ \end{aligned}$$

The symbol for the ion is Ca^{2+}.
(b) The atomic number of nitrogen is 7. Thus, seven protons and seven electrons are present in a neutral nitrogen atom. A gain of three electrons raises the electron count to ten.

$$\begin{aligned} 7 \text{ protons} &= 7 + \text{charges} \\ 10 \text{ electrons} &= \underline{10 - \text{charges}} \\ \text{Net charge} &= 3- \end{aligned}$$

The symbol for the ion is N^{3-}.

EXAMPLE 5.8

Calculate the number of protons, neutrons, and electrons in the following ions.

(a) $^{23}_{11}Na^{+}$ (b) $^{18}_{8}O^{2-}$ (c) $^{27}_{13}Al^{3+}$

SOLUTION

(a) The numbers of protons and neutrons are the same as in a neutral atom of sodium and are calculated as in Example 5.2.

$$\begin{aligned} \text{Number of protons} &= \text{atomic number} = 11 \\ \text{Number of neutrons} &= (\text{mass number}) - (\text{atomic number}) \\ &= 23 - 11 = 12 \end{aligned}$$

The number of electrons in a neutral Na atom is eleven. The charge on the ion of +1 indicates the loss of one electron.

$$\text{Number of electrons} = 11 - 1 = 10$$

(b) The atomic number is 8 and the mass number is 18.

$$\text{Number of protons} = \text{atomic number} = 8$$

$$\text{Number of neutrons} = (\text{mass number}) - (\text{atomic number})$$
$$= 18 - 8 = 10$$

A neutral oxygen atom would have eight electrons. The charge on the ion of −2 indicates a gain of two electrons.

$$\text{Number of electrons} = 8 + 2 = 10$$

Remember, a negative charge on an ion always indicates the gain of electrons (excess negative charge) and a positive charge on an ion always indicates the loss of electrons (deficiency of negative charge).

(c) The atomic number is 13 and the mass number is 27.

$$\text{Number of protons} = \text{atomic number} = 13$$
$$\text{Number of neutrons} = (\text{mass number}) - (\text{atomic number})$$
$$= 27 - 13 = 14$$
$$\text{Number of electrons} = 13 \text{ (the atomic number)} - 3 \text{ (electrons lost)} = 10$$

Information about why atoms gain or lose electrons and the factors that determine how many electrons are lost or gained will be presented in Chapter 10.

A final point about ions is that their chemical properties are very unlike those of the neutral atoms from which they are derived. For example, water solutions containing Na^+ ion are very stable even though the element sodium (neutral Na) reacts explosively with water. Reasons for this will also be presented in Chapter 10.

Learning Objectives

After completing this chapter you should be able to

- Explain current scientific thought concerning atoms as stated in the postulates of the atomic theory of matter (Sec. 5.1).
- Understand the difference between homoatomic and heteroatomic molecules and identify the type of substance in which each would be found (Sec. 5.2).
- Interpret a correctly written formula in terms of the number of elements and the number of atoms present (Sec. 5.3).
- Name the three major subatomic particles that make up an atom, tell where each is located within the atom, and indicate the electrical charge and relative mass associated with each particle (Sec. 5.4).
- Define atomic number and mass number and know how to determine the number of protons, neutrons, and electrons present in an atom given these two numbers (Sec. 5.5).
- Define what isotopes are and be able to write the symbol for an isotope (Sec. 5.6).
- Calculate the atomic weight of an element from the isotopic masses and percentage abundances of its isotopes (Sec. 5.7).
- Understand and state the periodic law (Sec. 5.8).
- Understand the rationale behind the organization of the periodic table. Relate the terms period and group to the periodic table. List the general information given on the periodic table (Sec. 5.8).
- Write a correct symbol for an ion (including charge) given the number of protons, neutrons, and electrons present (Sec. 5.9).

Terms and Concepts for Review

The new terms or concepts defined in this chapter are

atom (Sec. 5.1)
atomic theory of matter (Sec. 5.1)
molecule (Sec. 5.2)
homoatomic molecule (Sec. 5.2)

heteroatomic molecule (Sec. 5.2)
chemical formula (Sec. 5.3)
subatomic particle (Sec. 5.4)
proton (Sec. 5.4)
electron (Sec. 5.4)
neutron (Sec. 5.4)
nucleus (Sec. 5.4)
atomic number (Sec. 5.5)
mass number (Sec. 5.5)
isotope (Sec. 5.6)
atomic weight (Sec. 5.7)
periodic law (Sec. 5.8)
periodic table (Sec. 5.8)
group (Sec. 5.8)
period (Sec. 5.8)
ion (Sec. 5.9)

Questions and Problems

Atoms and Molecules

5-1 Indicate whether each of the following statements is *true* or *false*. If a statement is false, change it to make it true. (Such changes should involve more than merely making the statement the negative of itself.)

a. Molecules must contain three or more atoms.
b. The atom is the limit of chemical subdivision for an element.
c. All compounds have molecules as their basic structural unit.
d. A molecule of a compound must be heteroatomic.
e. A molecule of an element may be homoatomic or heteroatomic depending on which element is involved.
f. The limit of chemical subdivision for a molecular compound is a molecule.
g. There is only one kind of molecule for any given molecular substance.
h. Heteroatomic molecules do not maintain the properties of their constituent elements.
i. Only one kind of atom may be present in a homoatomic molecule.
j. The main difference between molecules of elements and molecules of compounds is the number of atoms they contain.

Chemical Formulas

5-2 On the basis of its formula classify each of the following substances as element or compound.

a. KNO_3 **b.** CO **c.** Co **d.** $COCl_2$
e. $CoCl_2$ **f.** S_8 **g.** Hf

5-3 From the information given about one molecule of each of the following molecular compounds write the chemical formula for the compound.

a. Cortisone molecule: contains 21 carbon atoms, 28 hydrogen atoms, and 5 oxygen atoms.
b. Caffeine molecule: contains 8 carbon atoms, 10 hydrogen atoms, 4 nitrogen atoms, and 2 oxygen atoms.
c. Ozone molecule: contains 3 oxygen atoms.

5-4 How many atoms of hydrogen are represented in each of the following formulas.

a. HNO_3 **b.** H_3PO_4 **c.** $NaHCO_3$
d. NH_3 **e.** NH_4CN **f.** $(NH_4)_3PO_4$
g. $Al(H_2PO_4)_3$ **h.** $C_{10}H_{22}$

5-5 How many atoms of each kind are represented by the following formulas.

a. SO_3 **b.** $COCl_2$ **c.** NH_4ClO_3
d. Fe_2O_3 **e.** $Ba(NO_3)_2$ **f.** $Al_2(SO_4)_3$
g. $Ca(C_2H_3O_2)_2$ **h.** $(NH_4)_3PO_4$

Protons, Neutrons, and Electrons

5-6 Indicate which subatomic particle (proton, neutron, or electron) correctly matches each of the following statements. More than one particle may be used as an answer.

a. Possesses a positive charge.
b. Has no charge.
c. Has a mass slightly greater than that of a proton.
d. Has a charge equal but opposite in sign to that of an electron.
e. Is located in the nucleus.
f. Has a negative charge.
g. Can be called a nucleon.
h. Has the lowest mass of the three particles.
i. Is located outside the nucleus.
j. Has a relative mass of 1839 if the mass of an electron is 1.

5-7 Explain why atoms are considered to be the fundamental building blocks of matter even though smaller particles (subatomic particles) exist.

Atomic Numbers, Mass Numbers, and Isotopes

5-8 Write complete symbols ($^{A}_{Z}$symbol), with the help of the periodic table, for atoms with the following characteristics.

a. Contains 51 protons and 70 neutrons.
b. Contains 19 protons and 21 neutrons.
c. Contains 30 electrons and 34 neutrons.
d. Contains 1 proton and 0 neutrons.

5-9 Write complete symbols ($^{A}_{Z}$symbol), using the periodic table when needed, for each of the following atoms.

a. The nitrogen isotope containing 8 neutrons.
b. The chromium isotope with a mass number of 50.
c. An isotope with 45 protons and 58 neutrons.
d. An isotope with 91 neutrons and a mass number of 155.

5-10 In what ways are isotopes of an element alike? In what ways are they different?

5-11 Write the symbol for an isotope of $^{77}_{34}Se$ that

a. Contains two more neutrons.
b. Contains two fewer neutrons.

5-12 The following symbols represent different atoms.

$^{50}_{25}X$, $^{53}_{30}X$, $^{48}_{26}X$, $^{48}_{24}X$, $^{47}_{26}X$, $^{56}_{26}X$, $^{49}_{25}X$

a. Which of the above are isotopes of $^{48}_{25}X$?
b. Which of the above have the same mass number as $^{48}_{25}X$?
c. Which of the above contain the same number of nucleons as $^{48}_{25}X$?
d. Which of the above contain the same number of neutrons as $^{48}_{25}X$?

5-13 What restrictions on the numbers of protons, neutrons, and electrons apply to each of the following?
a. Atoms of different isotopes of the same element.
b. Atoms of a particular isotope of an element.
c. Atoms of different elements with the same mass numbers.

5-14 The notations $^{15}_{7}N$ and ^{15}N may be used interchangeably. Why is this allowable? Could the notations $^{15}_{7}N$ and $_{7}N$ be used interchangeably without loss of meaning? Why?

Atomic Weights

5-15 A certain isotope of zinc ($^{91}_{40}Zr$) is 7.5754 times as heavy as $^{12}_{6}C$. What is the atomic mass of this isotope in amu units?

5-16 The masses of four objects are as follows.

Object A	3.00 g
Object B	5.00 g
Object C	8.00 g
Object D	9.00 g

Using these masses, construct relative mass scales with the following reference points.
a. Object B = 2.00 bobs
b. Object C = 4.00 bibs
c. Object D = 7.00 bebs

5-17 How many times heavier, on the average, is an atom of calcium than an atom of sodium?

5-18 Suppose it was decided to redefine the atomic weight scale by choosing as an arbitrary reference point an amu value of 20.0000 to represent the naturally occurring mixture of nickel isotopes. What would be the atomic weight of the following elements on the new atomic weight scale?
a. carbon **b.** silver **c.** gold

5-19 Using the data found in Table 5-2, calculate the atomic weights of the following elements.
a. chlorine **b.** copper **c.** carbon
d. titanium

5-20 Naturally occurring boron is a mixture of two isotopes: $^{10}_{5}B$ (10.013 amu) and $^{11}_{5}B$ (11.009 amu). Given that the atomic weight of boron is 10.811 amu, which one of the two isotopes is most abundant in nature?

5-21 Using the data found in Table 5-2, calculate how many $^{17}_{8}O$ and $^{18}_{8}O$ atoms you would find in a sample of oxygen containing 1.0×10^6 atoms.

Periodic Table

5-22 Identify, with the help of the periodic table on the inside front cover, each of the following elements.
a. located in period 4 and group IA
b. located in group VIA and period 2
c. located in period 3 and group IIIA
d. located in group VA and period 6

Ions

5-23 Why does an atom that loses electrons become positively charged?

5-24 What would be the charge, if any, on particles with the following subatomic makeups?
a. 18 protons, 20 neutrons, 18 electrons
b. 8 protons, 9 neutrons, 10 electrons
c. 11 protons, 12 neutrons, 10 electrons
d. 15 protons, 16 neutrons, 18 electrons

5-25 What is the difference in meaning associated with the notations Al and Al^{3+}?

5-26 Calculate the numbers of protons, neutrons, and electrons in each of the following ions.
a. $^{43}_{20}Ca^{2+}$ **b.** $^{39}_{19}K^{+}$ **c.** $^{15}_{7}N^{3-}$ **d.** $^{81}_{35}Br^{-}$

5-27 Write the complete symbol (includes *Z*, *A*, and charge) for ions containing the following numbers of subatomic particles.
a. 26 protons, 30 neutrons, 23 electrons
b. 17 protons, 18 neutrons, 18 electrons
c. 30 protons, 34 neutrons, 28 electrons
d. 7 protons, 7 neutrons, 10 electrons
e. 1 proton, 1 neutron

6 Compounds: Their Formulas and Names

6.1 The Law of Definite Proportions

In Chapter 4 we saw that compounds are pure substances with the following characteristics.

1. They are a chemical combination of two or more elements.
2. They can be broken down into constituent elements by chemical, but not physical, means.
3. They have a definite, constant elemental composition.

In this section we consider further the fact that compounds have a definite composition.

The composition of a compound can be determined by decomposing a weighed amount of the compound into its elements and then determining the masses of the individual elements. Alternatively, the mass of a compound formed by the combination of known masses of elements will also lead to its composition.

law of definite proportions

Studies of composition data for many compounds have led to the conclusion that the percentage of each element present in a given compound does not vary. This conclusion has been formalized into a statement known as the **law of definite proportions:** *In a pure compound, the elements are always present in the same definite proportion by mass.*

Let us consider how composition data obtained from decomposing a compound are used to illustrate the law of definite proportions. Samples of the compound of *known* mass are decomposed. The masses of the constituent elements present in each sample are then obtained. This experimental information (elemental masses and the original sample mass) is then used to calculate the percentage composition of the samples. Within the limits of experimental error, the calculated percentages for any given element present turn out to be the same, validating the law of definite proportions. Example 6.1 gives actual decomposition data for a compound and shows how it is treated to verify the law of definite proportions.

An alternate way of examining the constancy of composition found for compounds

EXAMPLE 6.1

Four samples of ammonia (NH_3) of differing mass and from different sources are individually decomposed to yield ammonia's constituent elements (nitrogen and hydrogen). The results of the decomposition experiments are as follows.

	Sample Mass Before Decomposition (g)	Mass of Nitrogen Produced (g)	Mass of Hydrogen Produced (g)
Sample 1	1.840	1.513	0.327
Sample 2	2.000	1.644	0.356
Sample 3	27.60	22.69	4.91
Sample 4	87.40	71.86	15.54

Show that these data are consistent with the law of definite proportions.

SOLUTION

Calculating the percent nitrogen in each sample will be sufficient to show whether the data are consistent with the law. All the nitrogen percentages should come out the same if the law is obeyed.

$$\text{Percent nitrogen} = \frac{\text{mass of nitrogen obtained}}{\text{total sample mass}} \times 100$$

Sample 1

$$\% \text{ N} = \frac{1.513 \not{g}}{1.840 \not{g}} \times 100 = 82.228261\% \quad \text{(calculator answer)}$$

$$= 82.23\% \quad \text{(correct answer, four significant figures)}$$

Sample 2

$$\% \text{ N} = \frac{1.644 \not{g}}{2.000 \not{g}} \times 100 = 82.2\% \quad \text{(calculator answer)}$$

$$= 82.20\% \quad \text{(correct answer)}$$

Sample 3

$$\% \text{ N} = \frac{22.69 \not{g}}{27.60 \not{g}} \times 100 = 82.210145\% \quad \text{(calculator answer)}$$

$$= 82.21\% \quad \text{(correct answer)}$$

Sample 4

$$\% \text{ N} = \frac{71.86 \not{g}}{87.40 \not{g}} \times 100 = 82.21968\% \quad \text{(calculator answer)}$$

$$= \mathbf{82.22\%} \quad \text{(correct answer)}$$

Note that the percentages are close to being equal but are not identical. This is due to measuring uncertainties and rounding in the original experimental data. All of the percentages are the same to three significant figures. The difference lies in the fourth significant digit. Recall from Section 2.2 that the last of the significant digits in a number (the fourth one here) has uncertainty in it. The four percentages above are considered to be the same within experimental error.

An alternate way of treating the given data to illustrate the law of definite proportions involves calculating the mass ratio between nitrogen and hydrogen. This ratio should be the same for each sample; otherwise all of the samples are not the same compound.

Sample 1

$$\frac{\text{mass N}}{\text{mass H}} = \frac{1.513 \not{g}}{0.327 \not{g}} = 4.6269113 \quad \text{(calculator answer)}$$

$$= \mathbf{4.63} \quad \text{(correct answer, three significant figures)}$$

Sample 2

$$\frac{\text{mass N}}{\text{mass H}} = \frac{1.644 \not{g}}{0.356 \not{g}} = 4.6179775 \quad \text{(calculator answer)}$$

$$= \mathbf{4.62} \quad \text{(correct answer)}$$

Sample 3

$$\frac{\text{mass N}}{\text{mass H}} = \frac{22.69 \not{g}}{4.91 \not{g}} = 4.6211813 \quad \text{(calculator answer)}$$

$$= \mathbf{4.62} \quad \text{(correct answer)}$$

Sample 4

$$\frac{\text{mass N}}{\text{mass H}} = \frac{71.86 \not{g}}{15.54 \not{g}} = 4.6241956 \quad \text{(calculator answer)}$$

$$= \mathbf{4.624} \quad \text{(correct answer, four significant figures)}$$

Again, slight differences in the ratios are caused by measuring errors and rounding in the original data.

involves looking at the mass ratios in which elements combine to form compounds. Let us consider the reaction between the elements calcium and sulfur to produce the compound calcium sulfide. Suppose an attempt is made to combine various masses of sulfur with a fixed mass of calcium. A set of possible experimental data for this attempt

Table 6-1. Data Illustrating the Law of Definite Proportions

Mass of Ca Used (g)	Mass of S Used (g)	Mass of CaS Formed (g)	Mass of Excess Unreacted Sulfur (g)	Ratio in Which Substances React
55.6	44.4	100.0	none	1.25
55.6	50.0	100.0	5.6	1.25
55.6	100.0	100.0	55.6	1.25
55.6	200.0	100.0	155.6	1.25
111.2	88.8	200.0	none	1.25

is given in the first four lines of Table 6-1. Notice that, regardless of the mass of S present, only a certain amount, 44.4 grams, reacts with the 55.6 grams of Ca. The excess S is left over in an unreacted form. The data therefore illustrate that Ca and S will react in only one fixed mass ratio (55.6/44.4 = 1.25) to form CaS. This fact is consistent with the law of definite proportions. Note also that if the amount of Ca used is doubled (line 5 of Table 6-1), the amount of S with which it reacts also doubles (compare lines 1 and 5 of the table). Nevertheless, the ratio in which the substances react (111.2/88.8) still remains 1.25.

6.2 "Natural" and "Synthetic" Compounds

Approximately five million chemical compounds are now known, with more being characterized daily. No end appears to be in sight as to the number of compounds that can and will be prepared in the future. At present, approximately 7000 new chemical substances are registered every week with Chemical Abstracts Service, a clearing house for new information concerning chemical substances.

Many compounds, perhaps the majority now known, are not naturally occurring substances. These synthetic (laboratory-produced) compounds are legitimate compounds and should not be considered "second class" or "unimportant" because they lack the distinction of being "natural." Many of the plastics, synthetic fibers, and prescription drugs now in common use are synthetic materials produced from controlled chemical change carried out on an industrial scale.

We have noted that chemists may produce compounds not found in nature. The reverse is also true. Nature is capable of making many compounds, especially those found in living systems, that chemists are not yet able to prepare in the laboratory.

There is a middle-ground area also. Many compounds that exist in nature may also be synthetically produced in the laboratory. A fallacy exists in the thinking of some people concerning these compounds that have "dual origins." A belief still persists that there is a difference between compounds prepared in the laboratory and samples of the same compounds found in nature. This is not true for pure samples of compound. The message of the law of definite proportions is that all pure samples of a compound, regardless of their origin, have the same composition. Since compositions are the same, properties will also be the same. There is no difference, for example, between a laboratory-prepared vitamin and a "natural" vitamin, if both are pure samples of the same vitamin, despite frequent claims to the contrary.

6.3 Classification Systems for Compounds

ionic compounds

molecular compounds

In Section 5.2 we learned that there are two distinct categories of compounds: ionic and molecular. **Ionic compounds** *have structures based upon infinite three-dimensional arrangements of positive and negative ions.* Ionic compounds do not have a simple fixed group that can be identified as a basic structural unit for the compound. **Molecular compounds** *contain separate fixed groupings of atoms called molecules.*

How does a student determine whether a given compound is composed of ions or made up of molecules. To make this decision a student needs to classify the elements present in the compound as metals or nonmetals. Once this is done the following guidelines apply.

1. Most *molecular* compounds contain only nonmetallic elements.
2. Most *ionic* compounds contain both metallic and nonmetallic elements.

Note that both of these guidelines begin with the word *most* rather than the word *all*. These two generalizations do have exceptions to them. They are, however, still widely used because of the great number of cases in which they do give the right classification. Specifics concerning the limitations of these guidelines will be given in Section 10.6.

Classification of a compound as molecular or ionic is dependent on the types of elements present — metals or nonmetals. What is a metal? What is a nonmetal? This is a topic we have not yet considered.

metals

nonmetals

The classification of an element as a metal or a nonmetal, one of the earliest classification schemes developed for elements, is based on selected general physical properties of the elements. **Metals** *are elements having the characteristic properties of luster, thermal and electric conductivity, and malleability.* Copper, silver, gold, and lead — all familiar metals — are readily recognized as having these properties. All metals, except for mercury and gallium, are solids at room temperature. The two exceptions are liquids. **Nonmetals** *are elements characterized by the absence of the properties of luster, thermal and electric conductivity, and malleability.* Many nonmetals (oxygen, hydrogen, nitrogen, etc.) are gases at room temperature; others are powdery solids (sulfur, phosphorus).

The majority of the elements are metals. Only 22 elements are nonmetals; the rest (84) are metals. It is not necessary to memorize which elements are nonmetals and which are metals as this information is obtainable from a periodic table. As can be seen in Figure 6-1, the location of an element in the periodic table correlates directly with its classification as a metal or nonmetal. Note the step-like dark line in the periodic table that separates the metals from the nonmetals. Note also that the location of hydrogen presents an exception; it is a nonmetal.

Some metals are more metallic than others; that is, they exhibit the properties associated with metals to a greater degree than do other elements. For example, the ability to conduct electricity varies considerably among the metals. In general, the more metallic elements are found on the left side of the periodic table and toward the bottom. The less metallic elements are toward the upper right of the table; that is, elements become less metallic as we move from left to right in the table and as we move from bottom to top. Using these trends we would predict (correctly) that sodium ($Z = 11$) is more metallic than lithium ($Z = 3$), which is located directly above sodium

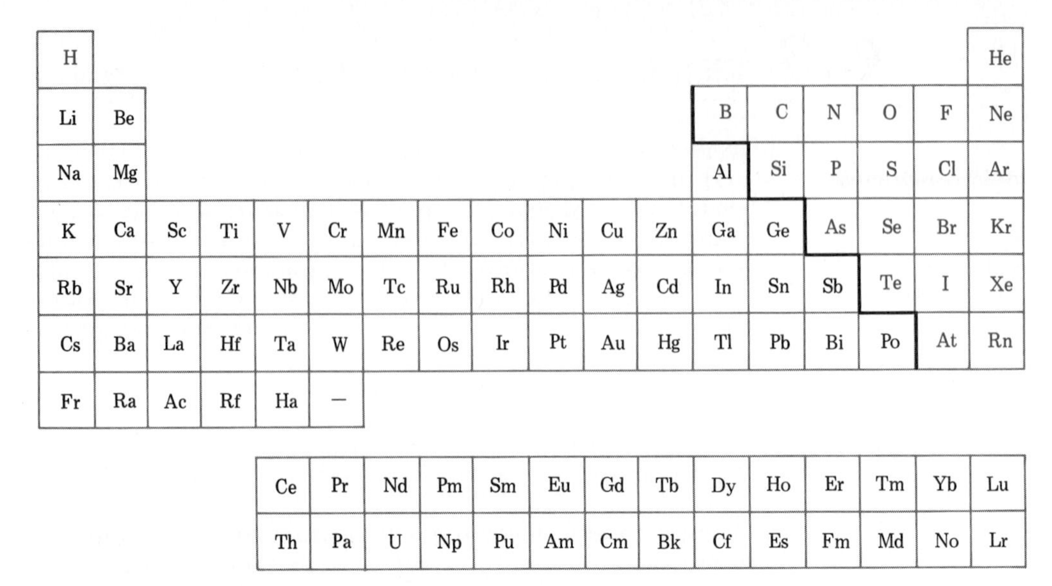

Figure 6-1. Periodic table showing the location of nonmetals (in color) and metals.

in the table, and that cesium ($Z = 55$) is much more metallic than osmium ($Z = 76$), which is located a number of positions to the right of cesium in the table.

Elements adjacent to the step-like dividing line between metals and nonmetals exhibit some of the properties of both metals and nonmetals. Sometimes these elements are considered a separate category of elements called *metalloids*.

EXAMPLE 6.2

Classify each of the following compounds as molecular or ionic.

(a) NaCl **(b)** H_2O **(c)** K_2O **(d)** CF_4

SOLUTION

We will use the periodic table (Fig. 6-1) to determine whether each of the elements present in a compound is a metal or nonmetal and the two guidelines given at the start of Section 6.3.

(a) Sodium (Na) is a metal. Chlorine (Cl) is a nonmetal. The compound is thus an ionic compound.

(b) Both hydrogen (H) and oxygen (O) are nonmetals. Thus, water is a molecular compound.

(c) A metal (potassium) and a nonmetal (oxygen) are present in this compound. The presence of both types of elements is a characteristic of ionic compounds.

(d) Carbon and fluorine are both in the nonmetal area of the table. Thus, CF_4 is a molecular compound.

binary compound

ternary compound

A second classification scheme for compounds, of use to us later in this chapter, is based on the number of elements present in a compound. A **binary compound** *is a compound containing just two different elements.* The compounds NH_3, H_2O, CO_2, and NaCl are all binary compounds. Any number of atoms of the two elements may be present in a molecule or formula unit of a binary compound, but only two elements may be present. A **ternary compound** *is a compound containing three different elements.* Such compounds, although not as common as binary compounds, are nevertheless frequently encountered. The compounds H_2SO_4, $NaNO_3$, and $POCl_3$ are all examples of ternary compounds. Compounds containing more than three kinds of elements are known. We will not, however, worry about specific classification terms for such compounds. We will, however, occasionally encounter such compounds.

6.4 Charges on Monoatomic Ions

In Section 5.9 ion formation was discussed. Atoms of some elements lose electrons to form positive ions, while atoms of other elements exhibit just the opposite behavior, gaining electrons to form negative ions.

Are there any guidelines concerning which elements tend to lose electrons and which atoms tend to gain electrons? The answer is yes. In general, *nonmetals tend to gain electrons; metals tend to lose electrons.*

In this section we are particularly concerned with *how many* electrons a metal loses or a nonmetal gains. This determines the magnitude of the charge on the ion. Knowing charge magnitude for ions is a prerequisite for writing formulas for compounds containing ions (Sec. 6.5) and for naming the compounds (Sec. 6.6).

To help systematize the process of learning ionic charges for monoatomic ions, they are classified into three categories.

1. Ions of nonmetallic elements.
2. Ions of metallic elements where the metal forms only one type of ion.
3. Ions of metallic elements where the metal forms more than one type of ion.

fixed-charge metals

variable-charge metals

Note that we have two categories of metal ions. In general, all metals lose electrons when forming ions. Some metals (those in category 2), called fixed-charge metals, always exhibit the same behavior in ion formation, that is, they always lose the same number of electrons. Thus, **fixed-charge metals** *form only one type of ion, which always has the same magnitude of charge.* Other metals (those in category 3), called variable-charge metals, do not always lose the same number of electrons upon ion formation. Thus, **variable-charge metals** *form more than one type of ion with the ion types differing in charge.* For example, the metal iron sometimes forms a Fe^{2+} ion and other times a Fe^{3+} ion.

Figure 6-2 shows the most common nonmetallic element ions (category 1) and their charges. The information about nonmetallic ion charge is presented in Figure 6-2 in the format of a periodic table because the charges on these ions can be related to position in the periodic table. Observe that the charge on the ions increases from -1 to -2 to -3 as we move to the left in the table (from group VIIA to VIA to VA). The numbers -1, -2, and -3 can be obtained by subtracting the number 8 from the group number. (The significance of the number 8 is discussed in Sec. 10.3.)

Fixed-charge metals also form ions whose charge can be related to position in the

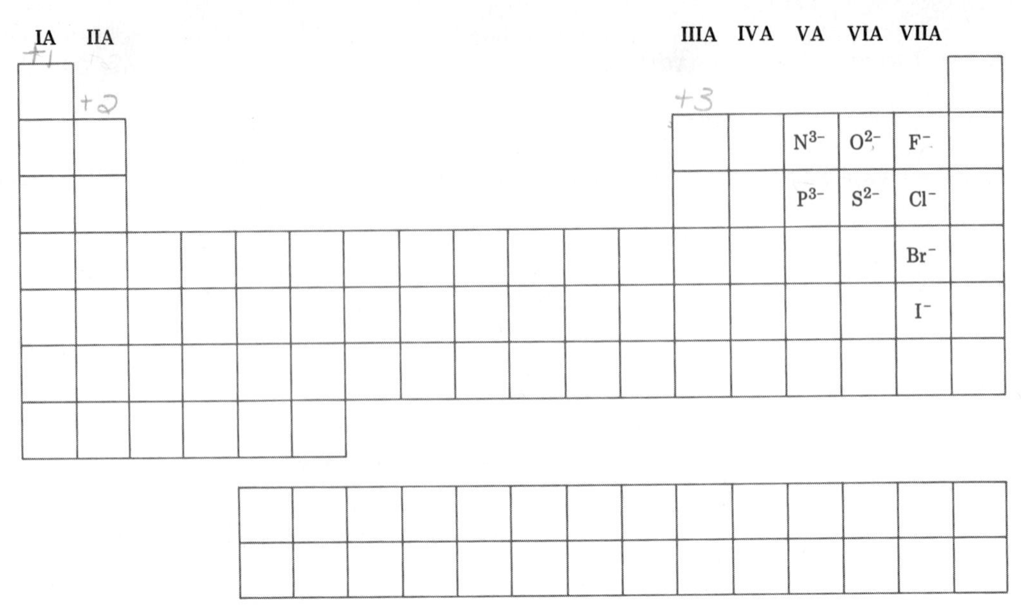

Figure 6-2. Periodic table showing common nonmetallic ions (in color).

periodic table. Looking at Figure 6-3, we see that all group IA elements form +1 ions. The charges for ions of elements in groups IIA and IIIA are +2 and +3, respectively. Group numbers and charge also directly correlate for Zn, Cd, and Ag, the other fixed-charge metallic ions.

Learning ionic charges for common variable-charge metals is more difficult than for the other types of ions. It is not possible to predict the magnitude of the charge from the metal's position in the periodic table. Table 6-2 gives the charges for the most commonly encountered variable-charge metal ions. From Table 6-2 note that the charges of +2 and +3 are a common combination. It is not, however, the only combination found. There are two examples of a +2 and +4 pairing (Pb and Sn), a +1 and +2 pairing (Cu), and a +1 and +3 pairing (Au).

Table 6-2. Common Variable-Charge Metallic Element Ions and Their Charges

Element	Ions Formed
Chromium	Cr^{2+}, Cr^{3+}
Cobalt	Co^{2+}, Co^{3+}
Copper	Cu^{+}, Cu^{2+}
Gold	Au^{+}, Au^{3+}
Iron	Fe^{2+}, Fe^{3+}
Lead	Pb^{2+}, Pb^{4+}
Manganese	Mn^{2+}, Mn^{3+}
Tin	Sn^{2+}, Sn^{4+}

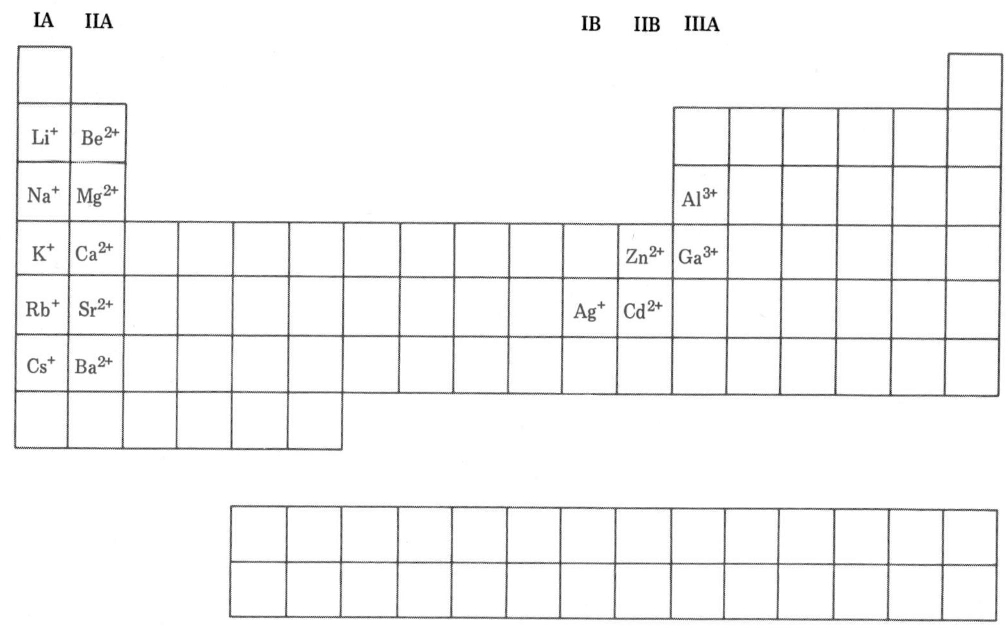

Figure 6-3. Periodic table showing common fixed-charge metallic ions and their charges (in color).

6.5 Formulas for Binary Ionic Compounds

Our discussions about ions to this point (Secs. 5.9, 6.3, and 6.4) have focused on the loss or gain of electrons by isolated individual atoms. Metal atoms lose electrons; nonmetal atoms gain electrons. In reality, loss and gain of electrons are "partner processes"; that is, one does not occur without the other also occurring. Ion formation occurs only when atoms of two elements are present—an element to donate electrons (electron loss) and an element to accept electrons (electron gain). The electrons lost by the one element are the same ones gained by the other element. Thus, positive and negative ion formations always occur together. The mutual attraction of the positive and negative ions that result from the *electron transfer* constitutes the force that holds the ions together as an ionic compound. Again, no atom will lose electrons unless other atoms are available to accept them.

Electron loss always equals electron gain in an electron-transfer process. Consequently, ionic compounds are always neutral; no net charge is present. The total positive charge present on the ions that have lost electrons always is exactly counterbalanced by the total negative charge on the ions that have gained electrons.

The procedures for writing formulas for ionic compounds are based on this required total charge balance between positive and negative ions. *The ratio in which positive and negative ions combine is that ratio which achieves charge neutrality for the resulting compound.* This combining ratio is always directly dependent on the charges on the combining ions.

The correct combining ratio when magnesium ions (Mg^{2+}) and chlorine ions (Cl^{-})

combine is 1 to 2. Two Cl^- ions (each of minus one charge) will be required to balance the charge on a single Mg^{2+} ion.

$$\begin{array}{rl} Mg^{2+}: & (\text{charge of } +2) \times (1 \text{ ion}) = +2 \\ 2(Cl^-): & (\text{charge of } -1) \times (2 \text{ ions}) = -2 \\ \hline & \text{Net charge} = \quad 0 \end{array}$$

From another viewpoint, magnesium atoms have all lost two electrons and chlorine atoms have all gained one electron. Two chlorine atoms, each accepting one electron, will be required to accept the two electrons from one magnesium atom. From either viewpoint, the correct combining ratio is one magnesium atom to two chlorine atoms. The formula of the resulting compound is written as $MgCl_2$. The subscript 2 indicates the presence of two chlorine atoms (ions).

In writing formulas for ionic compounds, such as $MgCl_2$, the positive ion is always written first followed by the negative ion. Subscripts indicate the number of ions of each kind, with the subscript 1 being understood. The charges on the ions are *not shown* in the formula. Knowledge of charges is necessary to determine the formula, but once it is determined the charges are not explicitly written.

Example 6-3 gives further illustrations of the procedures involved in determining correct combining ratios between ions and in writing correct ionic formulas from the combining ratios.

EXAMPLE 6.3

Determine the formula for the resulting ionic compound when each of the following pairs of elements (a metal and a nonmetal) interact.

(a) Li and Br **(b)** K and N **(c)** Al and O
(d) Fe and S (Fe is present as Fe^{2+})

SOLUTION

(a) The metal lithium always forms a Li^+ ion (see Fig. 6-3). The nonmetal bromine always forms a Br^- ion (see Fig. 6-2). The resulting ions, Li^+ and Br^-, will combine in a 1:1 ratio since that simple combination will cause the total charge to add up to zero.

$$\begin{array}{rl} Li^+: & (\text{charge of } +1) \times (1 \text{ ion}) = +1 \\ Br^-: & (\text{charge of } -1) \times (1 \text{ ion}) = -1 \\ \hline & \textbf{Net charge} = \quad 0 \end{array}$$

The formula of the compound is, thus, simply **LiBr.** Note again that the positive ion is always written first in the formula.

(b) The metal potassium always forms a +1 ion (see Fig. 6-3). (Any element found in group IA of the periodic table will always form a +1 ion.) Figure 6-2 shows that nitrogen forms ions that carry a charge of −3. Combining K^+ and N^{3-} ions in a 1:1 ratio will not work, since the charge on the resulting entity would be a −2, [(+1) + (−3) = −2], rather than the desired zero. The charge will add up to zero only when three K^+ ions are present for every N^{3-} ion.

$$\begin{array}{rl} 3(K^+): & (\text{charge of } +1) \times (3 \text{ ions}) = +3 \\ N^{3-}: & (\text{charge of } -3) \times (1 \text{ ion}) = -3 \\ \hline & \textbf{Net charge} = \quad 0 \end{array}$$

The formula of the compound is $\mathbf{K_3N}$. The subscript 3 indicates the presence of the 3 K^+ ions.

(c) The combining ratio for this combination of elements is slightly more difficult to visualize than in the previous two exercises. Aluminum forms a +3 ion (Fig. 6-3). Oxygen forms a −2 ion (Fig. 6-2). The simplest satisfactory combining ratio for these ions is two Al^{3+} ions and three O^{2-} ions, giving the formula $\mathbf{Al_2O_3}$. With this 2:3 combining ratio we have a total of six positive charges and six negative charges per formula unit of compound.

$$\begin{array}{rl} 2(Al^{3+}): & (\text{charge of } +3) \times (2 \text{ ions}) = +6 \\ 3(O^{2-}): & (\text{charge of } -2) \times (3 \text{ ions}) = -6 \\ \hline & \textbf{Net charge} = \ \ 0 \end{array}$$

Note that 6 (the total positive or negative charge) is the lowest common multiple of the numbers 2 and 3.

(d) Iron is a variable-charge metal, forming both Fe^{2+} and Fe^{3+} ions (see Table 6-2). Note that it was specified that we are dealing with Fe^{2+} in the problem statement. Without this information we would not have known whether to use Fe^{2+} or Fe^{3+} in working the problem. Sulfur, like oxygen, forms a −2 ion (Fig. 6-2). These two ions, Fe^{2+} and S^{2-}, combine in a 1:1 ratio. Iron gives up two electrons. Sulfur accepts two electrons. The formula is, thus, **FeS**. What about the formula Fe_2S_2? Would it not also be equally correct? The answer is no. The correct formula contains the *simplest* ratio of ions present. A 2:2 ratio (Fe_2S_2) can be converted into the smaller whole-number 1:1 ratio by dividing both subscripts by 2, giving the formula FeS.

6.6 Nomenclature of Binary Ionic Compounds

A study of chemical history indicates that over the years numerous methods for naming chemical compounds have been devised. In many of the early systems compounds were named very arbitrarily. Today procedures are much more systematic. Currently most chemists throughout the world use a set of systematic rules developed by the International Union of Pure and Applied Chemistry (IUPAC). We will use the IUPAC rules in this text. An IUPAC committee periodically meets to revise and update these nomenclature rules to accommodate any newly discovered situations.

For purposes of naming we will classify binary ionic compounds into two categories.

1. Binary ionic compounds containing a *fixed-charge* metal.
2. Binary ionic compounds containing a *variable-charge* metal.

The basic rule for naming both categories of compounds is the same. However, naming compounds in the second category requires the use of a subrule in addition to the basic rule.

Compounds Containing a Fixed-Charge Metal

Names for binary ionic compounds in which the metal is a fixed-charge metal are easily assigned. *The full name of the metallic element is given first, followed by a separate word containing the stem of the nonmetallic element name and the suffix -ide.*

Table 6-3. Names of Some Common Nonmetal Ions

Element	Stem	Name of Ion	Formula
Bromine	brom-	bromide	Br^-
Carbon	carb-	carbide	C^{4-}
Chlorine	chlor-	chloride	Cl^-
Fluorine	fluor-	fluoride	F^-
Hydrogen	hydr-	hydride	H^-
Iodine	iod-	iodide	I^-
Nitrogen	nitr-	nitride	N^{3-}
Oxygen	ox-	oxide	O^{2-}
Phosphorus	phosph-	phosphide	P^{3-}
Sulfur	sulf-	sulfide	S^{2-}

Thus, to name the compound NaF we start with the name of the metal (sodium), follow it with the stem of the name of the nonmetal (fluor-), and then add the suffix -ide. The name becomes *sodium fluoride.*

The stem of the name of the nonmetal is always the first few letters of the nonmetal's name, that is, the name of the nonmetal with its ending chopped off. Table 6-3 gives the stem part of the name for the most common nonmetallic elements. The name of the metal ion present is always exactly the same as the name of the metal itself. The metal's name is never shortened. Example 6-4 illustrates further the use of the rule for naming binary ionic compounds containing a fixed-charge metal.

EXAMPLE 6.4

Name the following binary ionic compounds, all of which contain a fixed-charge metal.

(a) BeO **(b)** AlF_3 **(c)** K_3N **(d)** Na_2S

SOLUTION

The general pattern for naming compounds of this type is

Name of metal + stem of name of nonmetal + ide

(a) The metal is beryllium and the nonmetal is oxygen. Thus, the compound name is beryllium <u>ox</u>ide (the stem of the nonmetal name is underlined).
(b) The metal is aluminum and the nonmetal is fluorine. Hence, the name is aluminum <u>fluor</u>ide. Note that no mention is made in the name of the subscript 3 found after the symbol for fluorine in the formula. *The name of an ionic compound never contains any reference to formula subscript numbers.* Since aluminum is a fixed-charge metal, there is only one ratio in which aluminum and fluorine may combine. Thus, just telling the names of the elements present in the compound is adequate nomenclature.
(c) Potassium (K) and nitrogen (N) are present in the compound. Its name is potassium <u>nitr</u>ide.
(d) This compound is named sodium <u>sulf</u>ide.

Compounds Containing a Variable-Charge Metal

In naming binary ionic compounds containing a variable-charge metal, the charge on the metal ion must be incorporated into the name of the compound. The magnitude of the charge on the metal ion is indicated by using a Roman numeral, inside parentheses, placed immediately after the name of the metal. This Roman numeral is considered to be part of the metal's name, as shown by the following examples.

Fe^{2+}: iron(II) ion

Fe^{3+}: iron(III) ion

Au^{+}: gold(I) ion

Note that the magnitude of the Roman numeral is always the same as the magnitude of the charge on the metal ion.

The chlorides of Fe^{2+} and Fe^{3+} ($FeCl_2$ and $FeCl_3$, respectively), are named iron(II) chloride and iron(III) chloride. Without the use of Roman numerals (or some equivalent system) these two compounds would have the same name, an unacceptable situation.

Sometimes, when given the formula of an ionic compound containing a variable-charge metal, a student is uncertain about the charge on the metal ion. When this is the case, the charge on the nonmetal ion is used to calculate the charge on the metal ion. This technique, as well as Roman numeral use in compound names, is illustrated in Example 6.5.

EXAMPLE 6.5

Name the following binary ionic compounds, all of which contain a variable-charge metal ion.

(a) FeO **(b)** Fe_2O_3 **(c)** PbO_2 **(d)** AuCl

SOLUTION

We will need to indicate the magnitude of the charge on the metal ion, using Roman numerals, in the names of each of these compounds.

(a) Let us assume you are uncertain as to the metal ion charge in this compound, a quantity you need to know to determine the Roman numeral to be used in the name. The metal ion charge is easily calculated using the following procedure.

$$(\text{Iron charge}) + (\text{oxygen charge}) = 0$$

The oxide ion has a -2 charge (Fig. 6-2). Therefore,

$$(\text{Iron charge}) + (-2) = 0$$

Solving, we get

$$\text{Iron charge} = +2$$

Therefore, the iron ion present is Fe^{2+} and the name of the compound is **iron(II) oxide.**

(b) For charge balance in this compound we have the equation

$$2(\text{Iron charge}) + 3(\text{oxygen charge}) = 0$$

Note that we have to take into account the number of each kind of ion present (2 and 3 in this case). Oxide ions carry a −2 charge (Fig. 6-2). Therefore,

$$2(\text{Iron charge}) + 3(-2) = 0$$

$$2(\text{Iron charge}) = +6$$

$$\text{Iron charge} = +3$$

Here we note that we are interested in the charge on a single iron ion (+3) and not in the total positive charge present (+6). Since Fe^{3+} ions are present, the compound is named **iron(III) oxide**. As is the case for all ionic compounds, the name does not contain any reference to the numerical subscripts in the compound's formula.

(c) The charge balance equation is

$$(\text{Lead charge}) + 2(\text{oxygen charge}) = 0$$

Substituting a charge of −2 into the equation for oxygen, we get

$$(\text{Lead charge}) + 2(-2) = 0$$

$$\text{Lead charge} = +4$$

The compound is thus named **lead(IV) oxide**.

(d) This compound is **gold(I) chloride**. Chloride ions carry a −1 charge (Table 6-2). Since the compound contains ions in a 1 to 1 ratio, the gold ions must have a +1 charge to counterbalance the −1 charge of the chloride ions. Hence, Au^{+} ions are present.

There is an older method for indicating the charge on metal ions that uses suffixes rather than Roman numerals. This system is more complicated and less precise than the Roman numeral system and fortunately is being abandoned. It is mentioned here because it is still occasionally encountered (especially on the labels of bottles of chemi-

Table 6-4. Comparison of Preferred and Old System Names for Selected Metal Ions

Element	Ions	Preferred Name	Old System
Copper	Cu^{+}	copper(I)	cuprous
	Cu^{2+}	copper(II)	cupric
Iron	Fe^{2+}	iron(II)	ferrous
	Fe^{3+}	iron(III)	ferric
Tin	Sn^{2+}	tin(II)	stannous
	Sn^{4+}	tin(IV)	stannic
Lead	Pb^{2+}	lead(II)	plumbous
	Pb^{4+}	lead(IV)	plumbic
Gold	Au^{+}	gold(I)	aurous
	Au^{3+}	gold(III)	auric

cals). In this system when a metal has two common ionic charges, the ending -ous is used for the ion of lower charge and the suffix -ic for the ion of higher charge. Table 6-4 compares the two systems for the metals where the old system is most often encountered.

6.7 Polyatomic Ions

monoatomic ions

polyatomic ion

To this point in the text all references to and comments about ions have involved monoatomic ions. **Monoatomic ions** *are single atoms that have lost or gained electrons.* Such ions are very common and very important. Another large and important category of ions, called polyatomic ions, exists. A **polyatomic ion** *is a group of atoms tightly bound together that have acquired a charge.* Numerous ionic compounds exist in which the positive or negative ion (sometimes both) is polyatomic. Polyatomic ions are very stable species, generally maintaining their identity during chemical reactions.

Polyatomic ions are not molecules. They never occur alone as do molecules. Instead, they are always found associated with other ions, ions of opposite charge. Polyatomic ions are *pieces* of compounds, not compounds. Compounds require the presence of both positive and negative ions and are neutral overall. Polyatomic ions are always charged species.

Table 6-5 gives the names and formulas of some of the more common polyatomic ions. The table is organized around key elements (other than oxygen) present in the polyatomic ions. Note, from the table, that almost all the polyatomic ions listed contain oxygen atoms. The names, but not necessarily the formulas, of some of these common polyatomic ions, should be familiar to you. Many of these ions are found in commercial products. Examples are fertilizers (phosphates, sulfates, nitrates), baking powder and soda (bicarbonates), and building materials (carbonates, sulfates).

There is no easy way of learning the common polyatomic ions. Memorization is required. The charges and formulas for the various polyatomic ions cannot be related easily to the periodic table as was the case for many of the monoatomic ions. In Table 6-5 the most frequently encountered polyatomic ions are in color. Their formulas should definitely be memorized. Your instructor may want you also to memorize others. The inability to recognize the presence of polyatomic ions (both by name and formula) in a compound is a major stumbling block for many chemistry students. Effort must be put forth to avoid this obstacle.

Note from Table 6-5 the following items concerning polyatomic ions.

1. Most of the ions have a negative charge, which can vary from −1 to −3. Only two positive ions are listed in the table: NH_4^+ (ammonium) and H_3O^+ (hydronium).
2. Two of the polyatomic ions, OH^- (hydroxide) and CN^- (cyanide), have names ending in -ide. These names represent exceptions to the rule that the suffix *-ide* be reserved for use in naming binary ionic compounds (Sec. 6.6).
3. A number of -ate, -ite pairs of ions exist, for example, SO_4^{2-} (sulfate) and SO_3^{2-} (sulfite). The ion in the pair with the higher number of oxygens is always the *-ate* ion. The *-ite* ion always contains one less oxygen than the *-ate* ion.
4. A number of pairs of ions exist where one member of the pair differs from the other by having a hydrogen atom present, for example, CO_3^{2-} (carbonate) and HCO_3^- (bicarbonate or hydrogen carbonate). In such pairs, the charge on the hydrogen-containing ion is always one less than that on the other ion.

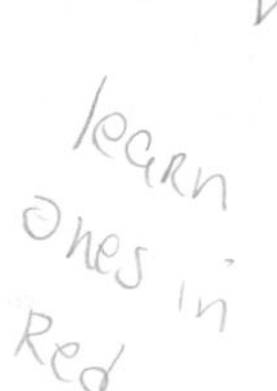

Table 6-5. Formulas and Names of Some Common Polyatomic Ions

Key Element Present	Formula	Name of Ion
Nitrogen	NO_3^-	nitrate
	NO_2^-	nitrite
	NH_4^+	ammonium
Sulfur	SO_4^{2-}	sulfate
	HSO_4^-	bisulfate or hydrogen sulfate
	SO_3^{2-}	sulfite
	HSO_3^-	bisulfite or hydrogen sulfite
Phosphorus	PO_4^{3-}	phosphate
	HPO_4^{2-}	hydrogen phosphate
	$H_2PO_4^-$	dihydrogen phosphate
	PO_3^{3-}	phosphite
Carbon	CO_3^{2-}	carbonate
	HCO_3^-	bicarbonate or hydrogen carbonate
	$C_2O_4^{2-}$	oxalate
	$C_2H_3O_2^-$	acetate
	CN^-	cyanide
Chlorine	ClO_4^-	perchlorate
	ClO_3^-	chlorate
	ClO_2^-	chlorite
	ClO^-	hypochlorite
Hydrogen	H_3O^+	hydronium
	OH^-	hydroxide
Metals	MnO_4^-	permanganate
	CrO_4^{2-}	chromate
	$Cr_2O_7^{2-}$	dichromate

6.8 Formulas for Ionic Compounds Containing Polyatomic Ions

Formulas for ionic compounds containing polyatomic ions are determined in the same way as those for ionic compounds containing monoatomic ions (Sec. 6.5). The basic rule is still the same: The total positive and negative charge present must add up to zero.

Two conventions not encountered previously in formula writing often arise when writing polyatomic-ion-containing formulas. They are

1. When more than one polyatomic ion of a given kind is required in a formula, the polyatomic ion is enclosed in parentheses and a subscript placed outside the parentheses is used to indicate the number of polyatomic ions needed.
2. To preserve the identity of polyatomic ions, the same elemental symbol may be used more than once in a formula.

Example 6.6 contains examples illustrating the use of both of these new conventions.

EXAMPLE 6.6

Determine the formulas for the ionic compounds containing the following pairs of ions.

(a) K^+ and OH^- **(b)** Na^+ and SO_4^{2-} **(c)** Ca^{2+} and NO_3^-
(d) NH_4^+ and CN^-

SOLUTION

(a) Combining these two ions in a 1:1 ratio will balance the charge. The formula of the compound is thus KOH.
(b) In order to equalize the total positive and negative charge, we need two sodium ions (+1 charge) for each sulfate ion (−2 charge). We indicate the presence of two Na^+ ions with the subscript 2 following the symbol of this ion. The formula of the compound is Na_2SO_4. The convention that the positive ion is always written first in the formula still holds when polyatomic ions are present.
(c) Two nitrate ions (−1 charge) are required to balance the charge on one calcium (+2 charge). Since more than one polyatomic ion is needed, the formula will contain parentheses, $Ca(NO_3)_2$. The subscript 2 outside the parentheses indicates two of what is inside the parentheses. If parentheses were not used, the formula would appear to be $CaNO_{32}$, which is not intended and which actually conveys false information. The formula $Ca(NO_3)_2$ indicates a formula unit containing one Ca atom, two N atoms, and six O atoms; the formula $CaNO_{32}$ indicates a formula unit containing one Ca atom, one N atom, and thirty-two O atoms. Verbally the correct formula, $Ca(NO_3)_2$, would be read as "C-A" (pause) "N-O-three-taken-twice."
(d) In this compound both ions are polyatomic, a perfectly legal situation. Since the ions have equal but opposite charges, they will combine in a 1:1 ratio. The formula is thus NH_4CN. No parentheses are needed, since we only need one polyatomic ion of each type in a formula unit. Parentheses are used only when two or more polyatomic ions of a given kind are present in a formula unit. What is different about this formula is the appearance of the symbol for the element nitrogen (N) at two locations in the formula. This could be prevented by combining the two nitrogens, giving the formula N_2H_4C. Combining is not, however, done in situations like this because the identity of any polyatomic ions present is lost in the resulting formula. The formula N_2H_4C does not convey the message that NH_4^+ and CN^- ions are present. The formula NH_4CN does convey this message. Thus, in writing formulas that contain polyatomic ions we always maintain the identities of these ions even if it means having the same elemental symbol at more than one location in the formula.

Often students can construct with ease formulas for ionic compounds given the ions present (Ex. 6.6) but have trouble with the reverse process — recognizing the ions present given the formula of an ionic compound. Students having problems with this reverse process usually are not familiar enough with the common polyatomic ions to recognize them by sight. This "sight recognition" is basic to working from a formula back to ions. Example 6.7 illustrates this "reverse-thinking" process.

EXAMPLE 6.7

Indicate the number of ions present of each type in each of the following ionic compounds.

(a) Na_3PO_4 **(b)** $KClO$ **(c)** $K_2Cr_2O_7$ **(d)** $(NH_4)_2CO_3$

SOLUTION

Binary ionic compounds do not contain enough different elements for polyatomic ions to be present. Conversely, when three or more elements are present in an ionic compound the presence of polyatomic ions should be expected.

(a) We recognize the grouping PO_4 as being characteristic of the phosphate ion. A total of four ions are present in one formula unit of this compound: three Na^+ ions and one PO_4^{3-} ion.

(b) The grouping ClO is found in the hypochlorite ion (ClO^-) (see Table 6-5). Only two ions are present in one formula unit of this compound, one K^+ ion and one ClO^- ion.

(c) The monoatomic K^+ ion is present. The nonpotassium part of the formula (Cr_2O_7) we suspect is a polyatomic ion. Consulting Table 6-5, we find it is the dichromate ion ($Cr_2O_7^{2-}$). Three ions are present in a formula unit: two K^+ ions and one $Cr_2O_7^{2-}$ ion.

(d) The parentheses present quickly indicate that two NH_4 units (the ammonium ion, NH_4^+) are present. The other part of the formula (CO_3) also represents a polyatomic ion, the carbonate ion (CO_3^{2-}). Thus, three ions, all polyatomic, are present in one formula unit of this ionic compound.

6.9 Nomenclature for Ionic Compounds Containing Polyatomic Ions

The names of ionic compounds containing polyatomic ions are derived in a manner similar to that for binary ionic compounds (Sec. 6.6). The rule for naming binary ionic compounds, it should be recalled, is: Give the name of the metallic element first (including, when needed, a Roman numeral indicating ion charge), and then as a separate word give the stem of the nonmetallic element name to which the suffix *-ide* is appended.

For our present situation, *if the polyatomic ion is positive, its name is substituted for that of the metal. If the polyatomic ion is negative, its name is substituted for the nonmetal stem plus -ide.* In the case where both positive and negative ions are polyatomic, dual substitution occurs and the resulting name includes just the names of the polyatomic ions. Example 6.8 illustrates the use of these rules.

EXAMPLE 6.8

Name the following compounds, which contain one or more polyatomic ions.

(a) K_2CO_3 **(b)** $Co(NO_3)_3$ **(c)** $Fe_2(SO_4)_3$ **(d)** $(NH_4)_3PO_4$

SOLUTION

(a) The positive ion present is the potassium ion (K^+). The negative ion is the polyatomic carbonate ion (CO_3^{2-}). The name of the compound is **potassium carbonate.** As in naming binary ionic compounds, subscripts in the formula are not incorporated into the name.

(b) The positive ion present is cobalt and the negative ion is the nitrate ion. Since cobalt is a variable-charge metal, a Roman numeral must be used to indicate ionic charge. In this case, the Roman numeral is (III). The fact that cobalt ions carry a +3 charge is deduced by noting that there are three nitrate ions present, each of which carries a −1 charge. The charge on the single cobalt ion present must be a +3 in order to counterbalance the total negative charge of −3. The name of the compound, thus, is **cobalt(III) nitrate.**

(c) The positive ion present is iron(III). The negative ion is the polyatomic sulfate ion (SO_4^{2-}). The name of the compound is **iron(III) sulfate.** The determination that iron is present as iron(III) involves the following calculation dealing with charge balance.

$$2(\text{Iron charge}) + 3(\text{sulfate charge}) = 0$$

The sulfate charge is −2. Therefore,

$$2(\text{Iron charge}) + 3(-2) = 0$$

$$2(\text{Iron charge}) = +6$$

$$\textbf{Iron charge} = \mathbf{+3}$$

(d) Both the positive and negative ions in this compound are polyatomic—the ammonium ion (NH_4^+) and the phosphate ion (PO_4^{3-}). The name of the compound is simply the combination of the names of the two polyatomic ions: **ammonium phosphate.**

6.10 Formulas for Molecular Compounds

Section 6.9 completes our discussion of formulas and names for ionic compounds. It is now time to turn our attention to the other major category of compounds: molecular compounds.

Writing formulas for molecular compounds, knowing what the combining elements are, is not as simple a task as it was for ionic compounds. Writing such formulas requires a knowledge of electron arrangements about the atoms—a topic not yet discussed. Consequently, we will delay consideration of molecular compound formula writing until Chapter 10.

Even though we cannot yet write formulas for molecular compounds from "scratch," this does not prevent us from learning how to name such compounds when the formula is known. The next two sections of this chapter deal with this topic.

6.11 Nomenclature for Binary Molecular Compounds

Table 6-6. Greek Numerical Prefixes

Greek Prefix	Number
Mono-	1
Di-	2
Tri-	3
Tetra-	4
Penta-	5
Hexa-	6
Hepta-	7
Octa-	8
Ennea-[a]	9
Deca-	10

[a]The prefix ennea- is preferred to the Latin nona- by the IUPAC, but nona- is still frequently used.

In this section we consider the rules for naming binary molecular compounds, the simplest type of molecular compound. Binary molecular compounds have two nonmetallic elements present (Sec. 6.3). The two nonmetals present are named in the order they appear in the formula. The name of the first nonmetal is used in full. The name of the second nonmetal is treated as was the nonmetal in binary ionic compounds, that is, the stem of the name is used and the suffix *-ide* is added.

Besides the names of the elements present, names for binary molecular compounds contain an additional feature. The number of atoms of each element present in a molecule of the compound is explicitly incorporated into the name of the compound through the use of Greek numerical prefixes. The prefixes precede the names of both nonmetals. The use of these prefixes is in direct contrast to the procedures used for naming ionic compounds. For ionic compounds, formula subscripts are not mentioned in the name.

Prefix use is needed in naming binary molecular compounds because numerous different compounds exist for many pairs of nonmetallic elements. For example, all of the following nitrogen–oxygen molecular compounds exist: NO, NO_2, N_2O, N_2O_3, N_2O_4, and N_2O_5. The prefixes used are the standard Greek numerical prefixes, which are given in Table 6-6 for the numbers 1 through 10. Example 6-9 shows how these prefixes are used in binary molecular compound nomenclature.

EXAMPLE 6.9

Name the following binary molecular compounds.

(a) N_2O **(b)** N_2O_3 **(c)** PCl_5 **(d)** P_4S_{10} **(e)** CCl_4

SOLUTION

The names of each of these compounds will consist of two words with the words having the following general formats.

First word: (prefix) + (full name of the first element)

Second word: (prefix) + (stem of name of second element) + (-ide)

(a) The elements present are nitrogen and oxygen. The two portions of the name, before adding Greek numerical prefixes, are *nitrogen* and *oxide.* Adding the prefixes gives *dinitrogen* (two nitrogen atoms are present) and *monoxide* (one oxygen atom is present). (When an element name begins with a vowel, the *a* or *o* at the end of the Greek prefix is dropped for ease of pronunciation — monoxide instead of monooxide.) The name of this compound is, thus, dinitrogen monoxide.

(b) The elements present are again nitrogen and oxygen. This time the two portions of the name are *dinitrogen* and *trioxide,* which are combined to give the name dinitrogen trioxide.
(c) When there is only one atom of the first element present, it is common to omit the prefix *mono-* for that element. Following this guideline, we have for the name of this compound phosphorus pentachloride.
(d) The prefix for four atoms is *tetra-* and for ten atoms *deca-*. This compound, thus, has the name tetraphosphorus decasulfide.
(e) Omitting the initial *mono-* (see part c), we name this compound carbon tetrachloride.

There is one standard exception to the use of Greek numerical prefixes in naming binary molecular compounds. Binary compounds with hydrogen listed as the first element in the formula are named without prefix use. Thus, the compounds H_2S and HCl are named hydrogen sulfide and hydrogen chloride, respectively.

A few binary molecular compounds have names completely unrelated to the IUPAC rules for naming such compounds. They have "common" names, which were coined prior to the development of the systematic rules. At one time, in the early history of chemistry, all compounds had common names. With the advent of systematic nomenclature, most common names were discontinued. A few, however, have persisted and are now officially accepted. The most "famous" example of this is the compound H_2O, which has the systematic name dihydrogen monoxide (or hydrogen oxide if prefixes are not used). Neither of these names is ever used; H_2O is known as *water,* a name that is not going to be changed. Another very common example of "common" nomenclature is the compound NH_3, which is *ammonia.* Table 6-7 gives additional examples of compounds for which common names are used in preference to systematic names. The "common" name exceptions are actually very few when compared to the total number of compounds named by systematic rules.

Writing formulas for binary molecular compounds given their names is a very easy task. The Greek prefixes in the names of such compounds tell you exactly how many atoms of each kind are present. For example, the compounds dinitrogen pentoxide

Table 6-7. Selected Binary Molecular Compounds That Have Common Names

Compound Formula	Accepted Common Name
H_2O	water
H_2O_2	hydrogen peroxide
NH_3	ammonia
N_2H_4	hydrazine
CH_4	methane
C_2H_6	ethane
PH_3	phosphine
AsH_3	arsine

and carbon dioxide have, respectively, the formulas N_2O_5 (di- and penta-) and CO_2 (mono- and di-). Remember, the prefix mono- is usually dropped at the start of a name, as in carbon dioxide.

6.12 Nomenclature for Acids

Many molecular hydrogen-containing compounds dissolve in water to give solutions with properties markedly different from those of the compounds that were dissolved. These solutions, which we will discuss in detail in Chapter 14, are called *acids*.

Not all hydrogen-containing molecular compounds are acids. Those that are can be recognized by their formulas, which have hydrogen written as the first element in the formula. The compounds HCl, H_2S, H_2SO_4, and HNO_3 are all acids. The compounds NH_3, CH_4, PH_3, and H_2O are not generally considered acids. Note that the element H is not written first in these formulas, except for H_2O, which is an exception to the rule.

Because of differences in the properties of acids and the anhydrous (without water) compounds from which they are produced, the acids and the anhydrous compounds are given different names. Acid nomenclature is derived from the names of the parent anhydrous compounds (which we have learned to name in this chapter). Because of this close relationship, we will also learn to name acids at this time, even though a detailed discussion of acids and their properties does not come until Chapter 14.

For nomenclatural purposes, acids will be classified into two categories.

1. *Nonoxyacids.* Water solutions of molecular compounds composed of hydrogen and one or more nonmetals other than oxygen.
2. *Oxyacids.* Water solutions of molecular compounds made up of hydrogen, some nonmetal, and oxygen.

Nonoxyacids, with one exception (HCN), are binary compounds. They are named by modifying the name of the parent anhydrous compounds as follows.

1. The word *hydrogen* is replaced with the prefix hydro-.
2. The suffix *-ide* on the stem of the name of the nonmetal is replaced with the suffix -ic.
3. The word **acid** is added to the end of the name (as a separate word).

By these rules, the compound hydrogen chloride (HCl) in water solution is called hydrochloric acid. A water solution of hydrogen selenide (H_2Se) is called hydroselenic acid, and HCN (hydrogen cyanide) dissolved in water becomes hydrocyanic acid. Binary molecular compounds containing hydrogen and one or more nonmetals other than oxygen may, thus, be named in two ways: as a pure compound and as a water solution (acid). Table 6-8 contrasts this dual naming system for selected other compounds.

Oxyacids are ternary molecular compounds, all with the two elements hydrogen and oxygen in common. Formulas for common oxyacids include H_2SO_4, HNO_3, H_3PO_4, and H_2CO_3.

Oxyacid molecules, when dissolved in water, break apart to give ions. The positive ions produced are H^+ ions, the species characteristic of all acids. The negative species produced are polyatomic ions, the same polyatomic ions we dealt with in Sections 6.7 through 6.9. Even though ions are produced in solution, oxyacids are not ionic compounds. They are molecular compounds that interact with water to produce ions.

Table 6-8. The Dual Naming System for Molecular Compounds Containing Hydrogen and One or More Nonmetals Other Than Oxygen

Formula	Name as a Pure Compound	Name of Water Solution
HF	hydrogen fluoride	hydrofluoric acid
HBr	hydrogen bromide	hydrobromic acid
HI	hydrogen iodide	hydroiodic acid
H_2S	hydrogen sulfide	hydrosulfuric acid[a]

[a] For acids involving sulfur, "ur" from sulfur is reinserted in the acid name for pronunciation reasons.

Names for oxyacids are derived from the names of the polyatomic ions produced when the acid molecules break into ions in solution. Polyatomic ion names are modified to give oxyacid names in the following manner.

1. When the polyatomic ion produced in solution has a name ending in *-ate*, the *-ate* ending is dropped, the suffix **-ic** is added in its place, and then the word **acid** is added.
2. When the polyatomic ion produced in solution has a name ending in *-ite*, the *-ite* ending is dropped, the suffix **-ous** is added in its place, and then the word **acid** is added.

Example 6.10 illustrates the use of these rules.

EXAMPLE 6.10

Name the following ternary molecular compounds as acids.

(**a**) HNO_3 (**b**) $HClO_3$ (**c**) HNO_2 (**d**) HClO

SOLUTION

(**a**) The polyatomic ion present in solutions of this acid is the nitrate ion (NO_3^-). Removing the *-ate* ending from the word nitrate, replacing it with the suffix *-ic*, and then adding the word *acid* gives the name nitric acid for the name of this oxyacid.
(**b**) The name base this time is the polyatomic ion chlorate (ClO_3^-). Removing the *-ate* suffix and adding an *-ic* suffix and the word *acid* produces the oxyacid name of chloric acid.
(**c**) The polyatomic ion produced when this compound dissolves in water is nitrite (NO_2^-). Note that two closely related (and often confused) polyatomic ions exist: nitr*ite* (NO_2^-) and nitr*ate* (NO_3^-). We are dealing here with nitr*ite*. For polyatomic ions whose names end in *-ite*, the *-ite* suffix is replaced by an *-ous* suffix and then the word *acid* is added. This compound is, thus, nitrous acid.
(**d**) The name for the polyatomic ion ClO^-, hypochlorite, is our starting point this time. The *-ite* ending becomes *-ous* and with the addition of the word acid we have the compound's name, hypochlorous acid. Note that if a polyatomic ion name contains the prefixes *per-* or *hypo-* they carry over to the acid name, as is the case here.

EXAMPLE 6.11

Give the formulas for the following oxyacids.

(a) sulfuric acid **(b)** perchloric acid **(c)** chlorous acid

SOLUTION

(a) The *-ic* ending of the acid name indicates that the formula will contain a polyatomic ion whose name ends in *-ate*. (Note that we are going through the reverse reasoning process to that used in Ex. 6.10.) The sulfur-containing polyatomic ion whose name ends in -ate is sulfate (SO_4^{2-}). The hydrogen present, for formula-writing purposes, is considered to be H^+ ions. Combining these ions in the right ratio to give a neutral compound (Sec. 6.8) gives the formula H_2SO_4 for sulfuric acid.
(b) The *-ic* ending again indicates the presence of an *-ate* polyatomic ion. This time it will be a *per- . . . -ate*, indeed, the perchlorate ion. This ion has the formula ClO_4^- (see Table 6-5). Combining H^+ and ClO_4^- ions in the correct ratio, a 1:1 ratio, gives the formula $HClO_4$ for perchloric acid.
(c) The *-ous* ending in the acid name indicates that an *-ite* polyatomic ion is present, the chlorite ion with the formula ClO_2^- (Table 6-5). Combining H^+ and ClO_2^- ions in a 1:1 ratio gives the formula $HClO_2$ for chlorous acid.

6.13 Nomenclature Rules: A Summary

Within this chapter various sets of nomenclature rules have been presented. Students sometimes have problems deciding which set of rules to use in a given situation. This dilemma can be avoided, if, when confronted with the request to name a compound, the student goes through the following reasoning pattern.

1. Decide, first, whether the compound is ionic (metal + nonmetal) or molecular (nonmetals only).
2. If the compound is ionic, then classify it as binary or ternary and use the rules appropriate for that classification. If the compound is ternary ionic, it will always contain polyatomic ions.
3. If the compound is molecular, classify it as an acid or nonacid. An acid must have the element hydrogen present (written first in the formula) and the compound must be in water solution. If the compound is a nonacid, name it according to the rules for binary molecular compounds.
4. If the compound is an acid, further classify it as a nonoxyacid or an oxyacid and then use the appropriate nomenclature rules.

Figure 6-4 summarizes in diagrammatic form the steps involved in deciding how to name a compound.

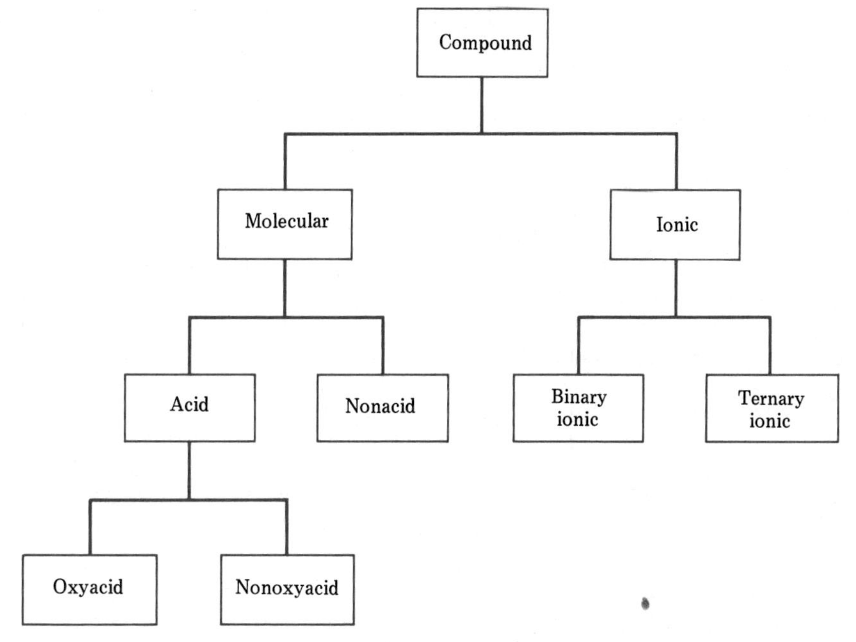

Figure 6-4. Classifying a compound for the purpose of naming it.

EXAMPLE 6.12

Name the following compounds.

(a) K_3P **(b)** N_2O **(c)** $FeSO_4$ **(d)** $HClO_4$ (in solution)
(e) H_2S (in solution)

SOLUTION

(a) This is a binary compound containing a metal (K) and a nonmetal (P). The presence of both a metal and nonmetal indicates that the compound is an ionic one. Its name is thus **potassium phosphide**. Remember that Greek numerical prefixes are never used in naming ionic compounds. Their use is limited to molecular compounds. The name tripotassium phosphide is incorrect. Note that no Roman numeral is needed for the K since it is a fixed-charge metal (Fig. 6-3).
(b) This is a binary compound containing two nonmetals. Therefore, it is a molecular compound. It cannot be an acid since it does not contain hydrogen. Using the rules for binary molecular compounds, we name the compound **dinitrogen monoxide**.
(c) Iron (Fe) is a metal. Sulfur (S) and oxygen (O) are nonmetals. This is a ternary ionic compound; it is ionic since both a metal and nonmetals are present. Since it is a ternary compound, polyatomic ions must be present. The polyatomic ion is the sulfate ion (SO_4^{2-}). The metal ion is iron(II). The +2 charge on the iron counterbalances the −2 charge on the sulfate ions. The compound name is **iron(II) sulfate**.
(d) The compound is an oxyacid. The polyatomic ion on which the name is based is ClO_4^- (perchlorate). Changing the *-ate* ending to *-ic* yields the name **perchloric acid**.

(e) This compound is a nonoxyacid. The pure compound would be named hydrogen sulfide. Using the prefix *hydro-* and the suffix *-ic* transforms this name into hydrosulfuric acid.

Learning Objectives

After completing this chapter you should be able to

- State the law of definite proportions, and show how composition data for compounds is treated to verify this law (Sec. 6.1).
- Distinguish between ionic and molecular compounds; between binary and ternary compounds (Sec. 6.3).
- Given the name or symbol, state the charge(s) on common monoatomic ions (Sec. 6.4).
- Write formulas for binary ionic compounds given the combining elements (Sec. 6.5).
- Given the formula of a binary ionic compound write its name and vice versa (Sec. 6.6).
- Recognize the names and symbols of common polyatomic ions (Sec. 6.7).
- Write formulas for ionic compounds containing polyatomic ions given the identity of the combining ions (Sec. 6.8).
- Given the formula of an ionic compound containing polyatomic ions, write its name and vice versa (Sec. 6.10).
- Recognize acids from other molecular compounds, distinguish between nonoxyacids and oxyacids, and name both types of acids given the formula or vice versa (Sec. 6.11).

Terms and Concepts for Review

The new terms or concepts defined in this chapter are

law of definite proportions (Sec. 6.1)
ionic compound (Sec. 6.3)
molecular compound (Sec. 6.3)
metal (Sec. 6.3)
nonmetal (Sec. 6.3)
binary compound (Sec. 6.3)
ternary compound (Sec. 6.3)
fixed-charge metal (Sec. 6.4)
variable-charge metal (Sec. 6.4)
monoatomic ion (Sec. 6.7)
polyatomic ion (Sec. 6.7)
nonoxyacid (Sec. 6.12)
oxyacid (Sec. 6.12)

Questions and Problems

Law of Definite Proportions

6-1 Two different samples of a pure compound (containing elements A and B) were analyzed with the following results:

Sample I: 17.35 g of compound yielded 10.03 g of A and 7.32 g of B
Sample II: 22.78 g of compound yielded 13.17 g of A and 9.61 g of B
Show that this data is consistent with the law of definite proportions.

6-2 It has been found experimentally that the elements X and Y react to produce two different compounds, depending upon conditions. Some sample experimental data are as follows.

Experiment	Grams of X	Grams of Y	Grams of Compound
1	3.37	8.90	12.27
2	0.561	1.711	2.272
3	26.9	71.0	97.9

Which two of the three experiments produced the same compound?

6-3 The elemental composition, by weight, of ammonia (NH_3) is 82.2% nitrogen (N) and 17.8% hydrogen (H). Suppose you mixed 82.2 g of N with 25.0 g of H. What is the maximum amount of NH_3 that can be formed from this mixture? Which element is in excess, and how much H will remain after completion of the chemical reaction?

6-4 Criticize, from a scientific viewpoint, the statement "The best pure sugar comes from Hawaii."

Classification Systems for Compounds

6-5 Classify the following elements as metals or nonmetals.

a. $_{15}P$ **b.** $_{4}Be$ **c.** $_{36}Kr$ **d.** $_{31}Ga$
e. $_{64}Gd$ **f.** $_{28}Ni$ **g.** $_{34}Se$ **h.** $_{78}Pt$

6-6 Identify the following compounds as ionic or molecular. Also state whether the smallest basic grouping present is a molecule or a formula unit.

a. NaI **b.** CO_2 **c.** Be_3N_2 **d.** $FeCl_3$
e. NH_3 **f.** AlF_3 **g.** P_4O_{10} **h.** Cu_2O

6-7 Classify each of the following compounds as binary or ternary.

a. N_2O_5 **b.** CO **c.** $(NH_4)_2S$
d. SiF_4 **e.** Na_2SO_4 **f.** SO_3
g. $Fe(CN)_2$ **h.** $AlPO_4$

Ionic Charge

6-8 Indicate the number of electrons lost or the number of electrons gained when each of the following atoms forms an ion.

a. $_{13}Al$ **b.** $_{7}N$ **c.** $_{9}F$ **d.** $_{47}Ag$
e. $_{11}Na$ **f.** $_{16}S$ **g.** $_{35}Br$ **h.** $_{26}Fe$

6-9 Classify each of the following metals as a fixed-charge metal or a variable-charge metal.

a. $_{50}Sn$ **b.** $_{3}Li$ **c.** $_{30}Zn$ **d.** $_{27}Co$
e. $_{29}Cu$ **f.** $_{20}Ca$ **g.** $_{12}Mg$ **h.** $_{82}Pb$

Formulas for Binary Ionic Compounds

6-10 Write formulas for the compounds formed from the following ions.

a. Na^+ and O^{2-} **b.** Ca^{2+} and Cl^-
c. Al^{3+} and I^- **d.** Zn^{2+} and N^{3-}
e. Ag^+ and Br^- **f.** Fe^{2+} and S^{2-}
g. Cu^+ and P^{3-} **h.** Ba^{2+} and F^-

6-11 Write a correct formula for the binary ionic compound formed by combining

a. lithium and bromine
b. beryllium and nitrogen
c. iron (the +3 ion) and oxygen
d. aluminum and phosphorus
e. gold (the +1 ion) and fluorine
f. magnesium and iodine
g. calcium and sulfur
h. silver and chlorine

6-12 Identify the ions present (by symbol and charge) in each of the following binary ionic compounds. In parts (e) through (g) X stands for the symbol of one element.

a. Na_2S **b.** Li_3N **c.** $AlCl_3$ **d.** Fe_3N_2
e. XCl_3 **f.** Na_2X **g.** Al_2X_3 **h.** XO_2

Nomenclature for Binary Ionic Compounds

6-13 When must a Roman numeral be used as part of an ionic compound's name?

6-14 Name the following binary ionic compounds.

a. $BeCl_2$ **b.** K_2S **c.** ZnO **d.** $NaCl$
e. Co_2O_3 **f.** Cu_2O **g.** $SnCl_4$ **h.** PbO

6-15 Write formulas for the following binary ionic compounds.

a. potassium iodide
b. silver oxide
c. calcium fluoride
d. barium phosphide
e. gallium nitride
f. iron(II) bromide
g. cadmium chloride
h. copper(II) sulfide

6-16 What is the charge on the metal ion in each of the following binary ionic compounds?

a. Au_2O **b.** CoS **c.** Ag_3P **d.** AlN
e. Cr_2O_3 **f.** $FeCl_3$ **g.** MnI_2 **h.** SnO_2

Polyatomic Ions

6-17 Give names for the following polyatomic ions.

a. NO_3^- **b.** SO_4^{2-} **c.** OH^-
d. CN^- **e.** CO_3^{2-} **f.** PO_4^{3-}
g. HCO_3^- **h.** MnO_4^-

6-18 Write formulas (including charge) for the following polyatomic ions.

a. chlorate **b.** hydronium **c.** nitrite
d. sulfite **e.** ammonium **f.** acetate
g. perchlorate **h.** chromate

6-19 Indicate which of the following compounds contain polyatomic ions and identify the polyatomic ion by name if present.

a. Al_2S_3 **b.** $Ca_3(PO_4)_2$ **c.** $ZnSO_4$
d. KNO_3 **e.** NH_4Cl **f.** KCN
g. $K_2Cr_2O_7$ **h.** $NaClO$

6-20 Each of the following formulas contains polyatomic ions within parentheses. Show your understanding of the use of parentheses by indicating how many atoms are present in a formula unit of each compound.

a. $(NH_4)_2S$ **b.** $Ba(NO_3)_2$ **c.** $Al_2(SO_4)_3$
d. $Ca(OH)_2$

Formulas for Ionic Compounds Containing Polyatomic Ions

6-21 Write formulas for the compounds formed between these positive and negative ions.

a. Ba^{2+} and NO_3^-
b. Fe^{3+} and OH^-
c. Al^{3+} and CO_3^{2-}
d. Na^+ and ClO_4^-
e. K^+ and PO_4^{3-}
f. NH_4^+ and SO_4^{2-}
g. Co^{2+} and $H_2PO_4^-$
h. NH_4^+ and $C_2H_3O_2^-$

6-22 Write formulas for the ionic compound formed by combining

a. potassium ion and perchlorate ion.
b. ammonium ion and chloride ion.
c. calcium ion and sulfate ion.
d. copper(II) ion and hydroxide ion.
e. lead(IV) ion and phosphate ion.
f. silver ion and dichromate ion.
g. gold(I) ion and acetate ion.
h. iron(III) ion and oxalate ion.

6-23 Each of the following compounds contains a polyatomic ion whose identity has been "masked" by use of the symbol X in place of a nonmetallic element symbol. What is the charge on the polyatomic ion in each case?

a. $AgXO_3$ **b.** Na_3XO_4 **c.** $Al(XO_2)_3$
d. $Ca(XO)_2$

Nomenclature for Ionic Compounds Containing Polyatomic Ions

6-24 Name the following compounds.
a. $CuSO_4$ **b.** NH_4NO_3 **c.** $BaCO_3$
d. $CrHPO_4$ **e.** $Zn(C_2H_3O_2)_2$ **f.** Na_3PO_4
g. Na_3PO_3 **h.** $KMnO_4$

6-25 Write formulas for the following compounds.
a. cobalt(III) sulfate
b. ammonium cyanide
c. potassium bicarbonate
d. sodium hydrogen carbonate
e. lead(II) hydroxide
f. silver carbonate
g. gold(III) nitrate
h. aluminum phosphate

Nomenclature for Binary Molecular Compounds

6-26 Name the following binary molecular compounds.
a. SF_4 **b.** Cl_2O **c.** P_4O_6
d. CBr_4 **e.** H_2S (as pure compound **f.** ClO_2
g. NH_3 **h.** H_2O

6-27 Write formulas for the following binary molecular compounds.
a. iodine monochloride
b. nitrogen trichloride
c. sulfur hexafluoride
d. oxygen difluoride
e. carbon dioxide
f. silicon tetrachloride
g. sulfur dioxide
h. hydrogen peroxide

Nomenclature for Acids

6-28 Classify the following as oxy- or nonoxyacids.
a. HCl **b.** HClO **c.** $HClO_2$ **d.** H_2SO_4
e. H_2S **f.** HNO_3 **g.** HI **h.** H_3PO_4

6-29 Name the following compounds as acids.
a. HBr **b.** $HClO_2$ **c.** H_2SO_4
d. H_2SO_3 **e.** H_3PO_4 **f.** HCN
g. HCl **h.** H_2Te

6-30 Write formulas for the following acids.
a. acetic acid
b. phosphorous acid
c. carbonic acid
d. perchloric acid
e. hydroiodic acid
f. hydrosulfuric acid
g. oxalic acid
h. permanganic acid

Nomenclature: General Problems

6-31 Classify the following compounds as (1) binary ionic, (2) ternary ionic, (3) binary molecular, (4) nonoxyacid, or (5) oxyacid.
a. Na_3P **b.** NF_3 **c.** K_3PO_4 **d.** HClO
e. LiOH **f.** H_2SO_4 **g.** NH_4Cl **h.** P_4O_{10}

6-32 Write formulas for the following compounds.
a. phosphorus pentafluoride
b. gold(III) bromide
c. sodium chromate
d. hydrofluoric acid
e. lithium phosphide
f. copper(I) sulfate
g. nitrous acid
h. ammonium sulfite

6-33 Name the following compounds.
a. HI (pure compound) **b.** $Fe(NO_3)_3$
c. Rb_3N **d.** $Pb(CO_3)_2$
e. N_2O **f.** Cu_2S
g. Al_2O_3 **h.** Mn_2O_3

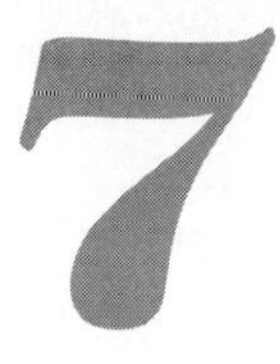

Chemical Calculations I: The Mole Concept and Chemical Formulas

7.1 Formula Weights

This chapter is the first of two chapters dealing with "chemical arithmetic," that is, with the quantitative relationships between elements and compounds. The emphasis in this chapter will be on quantitative relationships that involve chemical formulas. Emphasis in Chapter 8 will be on quantitative relationships that involve chemical equations.

Our entry point into the realm of "chemical arithmetic" will be a discussion of the quantity called *formula weight.* Formula weights play a role in almost all chemical calculations and will be used extensively in later sections of this chapter and in succeeding chapters.

formula weight

Once the formula of a substance has been established, it is a simple matter to calculate its formula weight. The **formula weight** *of a substance is the sum of the atomic weights of the atoms in its formula.* Formula weights, like the atomic weights from which they are calculated, are relative masses based on the $^{12}_{6}C$ relative mass scale (Sec. 5.7). They may be calculated for compounds (both molecular and ionic) and for elements that exist in molecular form.

The term molecular weight is used interchangeably with formula weight by many chemists when reference is made to substances that contain discrete molecules. It is incorrect, however, to use the term molecular weight when talking about ionic substances, since molecules are not their basic structural unit.

Example 7.1 illustrates how formula weights are calculated. You will note as you follow through this example that significant figures are back to "haunt" us again. Significant figures are a consideration in any calculation that involves experimental data or numbers obtained from such data. Atomic weights, the basis for formula-weight calculations, fall in this latter category.

EXAMPLE 7.1

Calculate formula weights for the following substances.

(a) CO_2 (b) $BeSO_4$ (c) $(NH_4)_3PO_4$ (d) P_4

SOLUTION

Formula weights are obtained simply by adding the atomic weights of the constituent elements, counting each atomic weight as many times as the symbol for the element occurs in the formula.

(a) A molecule of CO_2 (carbon dioxide) contains three atoms: one atom of C and two atoms of O. The formula weight, the collective mass of these three atoms, is calculated as follows.

$$1 \text{ atom C} \times \left(\frac{12.011 \text{ amu}}{1 \text{ atom C}}\right) = 12.011 \text{ amu}$$

$$2 \text{ atoms O} \times \left(\frac{15.9994 \text{ amu}}{1 \text{ atom O}}\right) = 31.9988 \text{ amu}$$

$$\text{Formula weight} = 44.0098 \text{ amu} \quad \text{(calculator answer)}$$

$$= 44.010 \text{ amu} \quad \text{(correct answer)}$$

The conversion factors in the calculation were derived from the atomic weights listed on the inside front or back cover of the text. The atomic weight, the mass of an average atom, is 12.011 amu for C and 15.9994 amu for O. The calculator answer, 44.0098 amu, had to be rounded down one significant figure, since the atomic weight of C is known only to the thousandths decimal place. (Recall the rules for significant figures in addition, Sec. 2.3.)

Often conversion factors are not explicitly shown in a formula-weight calculation as we did above; the calculation is simplified as follows.

$$\text{C:} \quad 1 \times 12.011 \text{ amu} = 12.011 \text{ amu}$$

$$\text{O:} \quad 2 \times 15.9994 \text{ amu} = 31.9988 \text{ amu}$$

$$\text{Formula weight} = 44.0098 \text{ amu} \quad \text{(calculator answer)}$$

$$= 44.010 \text{ amu} \quad \text{(correct answer)}$$

(b) Similarly, for $BeSO_4$ (beryllium sulfate) we calculate the formula weight as

$$\text{Be:} \quad 1 \times 9.01218 \text{ amu} = 9.01218 \text{ amu}$$

$$\text{S:} \quad 1 \times 32.06 \text{ amu} = 32.06 \text{ amu}$$

$$\text{O:} \quad 4 \times 15.9994 \text{ amu} = 63.9976 \text{ amu}$$

$$\text{Formula weight} = 105.06978 \text{ amu} \quad \text{(calculator answer)}$$

$$= 105.07 \text{ amu} \quad \text{(correct answer)}$$

Note how the atomic weight of S severely limits the precision of the formula weight. Why does the periodic table not give a more precise atomic weight for S, you may ask. A more fundamental question is "Why does the preciseness of atomic weights of the elements vary as much as it does?" Some elements have atomic weights known to 0.00001 amu; others are known to only 0.1 amu. A major factor in atomic weight preciseness is the constancy of the abundance percentages for the isotopes of an element. Abundance percentages fluctuate more for some elements than for others; the less the fluctuation, the greater the atomic weight precision. (The fluctuation is a relative matter; none of the fluctuations are very large on an absolute scale.) The atomic weights given in the periodic table are stated to current precision limits.

(c) The formula for this compound contains parentheses. Improper interpretation of parentheses (see Sec. 5.3) is a common error made by students doing formula-weight calculations. In this formula, $(NH_4)_3PO_4$, the subscript 3 outside the parentheses affects all symbols inside the parentheses. Thus, we have

$$\begin{aligned}
\text{N:}&\quad 3 \times 14.0067 \text{ amu} = 42.0201 \text{ amu}\\
\text{H:}&\quad 12 \times 1.0079 \text{ amu} = 12.0948 \text{ amu}\\
\text{P:}&\quad 1 \times 30.97376 \text{ amu} = 30.97376 \text{ amu}\\
\text{O:}&\quad 4 \times 15.9994 \text{ amu} = 63.9976 \text{ amu}\\
&\text{Formula weight} = 149.08626 \text{ amu} \quad \text{(calculator answer)}\\
&= 149.0863 \text{ amu} \quad \text{(correct answer)}
\end{aligned}$$

(d) This is a very simple problem. We multiply the atomic weight of P by 4 to get the formula weight.

$$\begin{aligned}
\text{P:}&\quad 4 \times 30.97376 \text{ amu} = 123.89504 \text{ amu} \quad \text{(calculator answer)}\\
&\text{Formula weight} = 123.8950 \text{ amu} \quad \text{(correct answer)}
\end{aligned}$$

The point of this example is that elements that exist in molecular form have both a formula weight and an atomic weight. Numerically, these two weights are not the same.

In Example 7.1, all the formula weights were calculated purposely to the maximum number of significant figures possible. This was done to emphasize the fact that the number of significant figures in "input" atomic weights determines the preciseness of calculated formula weights.

In most chemical calculations you will not need maximum formula-weight preciseness. Other numbers that enter into the calculation will usually restrict the answer to three or four significant figures. Thus, in most problems, formula weights rounded to three or four significant figures may be used. Occasionally, however, formula weights of maximum precision will be required.

Formula weights, relative masses on the ${}^{12}_{6}C$ relative mass scale, may be used for mass comparisons. For example, using the formula weights calculated in Example 7.1 we can make the following comparison statements.

1. One formula unit of $BeSO_4$ is 2.3874 times heavier than one molecule of CO_2 (105.07/44.010 = 2.3874).
2. One formula unit of $(NH_4)_3PO_4$ is 3.3876 times heavier than one molecule of CO_2 (149.0863/44.010 = 3.3876).
3. One formula unit of $(NH_4)_3PO_4$ is 1.4189 times heavier than one formula unit of $BeSO_4$ (149.0863/105.07 = 1.4189).

7.2 Percentage Composition

percentage composition

A useful piece of information about a compound is its percentage composition. **Percentage composition** *specifies the percentage by mass of each element present in a compound.* For instance, the percentage composition of water is 88.81% oxygen and 11.19% hydrogen.

Percentage compositions are frequently used to compare compound compositions. For example, the compounds gold(III) iodide (AuI_3), gold(III) nitrate [$Au(NO_3)_3$], and gold(I) cyanide (AuCN) contain, respectively, 34.10%, 51.43%, and 88.33% gold by mass. If you were given the choice of receiving a gift of one pound of one of these three gold compounds, which one would you choose?

We can calculate the percentage composition for a compound from its chemical formula. Example 7.2 illustrates how this is done.

EXAMPLE 7.2

Determine the percentage composition (to two decimal places) of sodium bicarbonate (baking soda) from its formula $NaHCO_3$.

SOLUTION

First, we calculate the formula weight for $NaHCO_3$, using atomic weights rounded to the hundredths decimal place.

$$\begin{array}{ll} \text{Na:} & 1 \times 22.99\text{ amu} = 22.99\text{ amu} \\ \text{H:} & 1 \times 1.01\text{ amu} = 1.01\text{ amu} \\ \text{C:} & 1 \times 12.01\text{ amu} = 12.01\text{ amu} \\ \text{O:} & 3 \times 16.00\text{ amu} = \underline{48.00\text{ amu}} \\ & \text{Formula weight} = 84.01\text{ amu} \end{array}$$

The mass percent of each element in the compound is found by dividing the mass contribution of each element, in amu, by the total mass (formula weight), in amu, and multiplying by 100.

$$\%\text{ element} = \frac{\text{mass of element in one formula unit}}{\text{formula weight}} \times 100$$

Finding the percentages, we have

$$\%\text{ Na:} \quad \frac{22.99\text{ amu}}{84.01\text{ amu}} \times 100 = 27.36579\% \quad \text{(calculator answer)}$$

$$= 27.37\% \quad \text{(correct answer)}$$

$$\%\text{ H:} \quad \frac{1.01\text{ amu}}{84.01\text{ amu}} \times 100 = 1.2022378\% \quad \text{(calculator answer)}$$

$$= 1.20\% \quad \text{(correct answer)}$$

$$\% \text{ C:} \quad \frac{12.01 \text{ amu}}{84.01 \text{ amu}} \times 100 = 14.295917\% \quad \text{(calculator answer)}$$

$$= 14.30\% \quad \text{(correct answer)}$$

$$\% \text{ O:} \quad \frac{48.00 \text{ amu}}{84.01 \text{ amu}} \times 100 = 57.136055\% \quad \text{(calculator answer)}$$

$$= 57.14\% \quad \text{(correct answer)}$$

To check our work we can add the percentages of all the parts. They, of course, have to total 100. (On occasion, round-off errors may not cancel and totals such as 99.99% or 100.01% may be obtained.)

$$27.37\% + 1.20\% + 14.30\% + 57.14\% = 100.01\%$$

Percentage compositions may also be calculated from mass data obtained from compound decomposition or compound synthesis experiments. Example 7.3 shows how decomposition data are treated to yield percentage composition.

EXAMPLE 7.3

Iron pills, for iron-deficiency anemia, usually contain iron in the form of iron(II) sulfate ($FeSO_4$). A 0.3473 g sample of iron(II) sulfate is decomposed into its elemental constituents to give 0.1277 g of iron, 0.07330 g of sulfur, and 0.1463 g of oxygen. What is the percentage composition of iron(II) sulfate?

SOLUTION

The percentage by mass of each element present is found by dividing its mass by the total sample mass and multiplying by 100.

$$\% \text{ element} = \frac{\text{mass of element}}{\text{total sample mass}} \times 100$$

Finding the percentages, we have

$$\% \text{ Fe} = \frac{0.1277 \text{ g}}{0.3473 \text{ g}} \times 100 = 36.769364\% \quad \text{(calculator answer)}$$

$$= 36.77\% \quad \text{(correct answer)}$$

$$\% \text{ S} = \frac{0.07330 \text{ g}}{0.3473 \text{ g}} \times 100 = 21.105672\% \quad \text{(calculator answer)}$$

$$= 21.11\% \quad \text{(correct answer)}$$

$$\% \text{ O} = \frac{0.1463 \text{ g}}{0.3473 \text{ g}} \times 100 = 42.124964\% \quad \text{(calculator answer)}$$

$$= 42.12\% \quad \text{(correct answer)}$$

Checking our work, we see that the percentages do add up correctly.

$$36.77\% + 21.11\% + 42.12\% = 100.00\%$$

7.3 The Mole: The Chemist's Counting Unit

Two common methods exist for specifying the quantity of material in a sample of a substance: (1) in terms of units of *mass* and (2) in terms of units of *amount*. We measure *mass* by using a balance (Sec. 3.2). Common mass units are gram, kilogram, and pound. For substances that consist of discrete units, we can specify the *amount* of substance present by indicating the number of units present — 12, 27, 113, and so on.

Both units of mass and units of amount are used on a daily basis by all of us. We work well with this dual system. Sometimes it doesn't matter which type of unit is used; other times one system is preferred over the other. When buying potatoes at the grocery store we can decide on quantity in either mass units (10 lb bag, 20 lb bag, etc.) or amount units (9 potatoes, 15 potatoes, etc.). When buying eggs, amount units are used almost exclusively — 12 eggs (1 dozen), 24 eggs (2 dozen), and so on. We do not ordinarily buy 2 pounds of eggs; eggs tend to break when weighed. On the other hand, peanuts and grapes are almost always purchased in weighed quantities. It is impractical to count the number of grapes in a bunch. Very few people go to the store with the idea that they will buy 117 grapes.

In chemistry, as in everyday life, both the mass and amount methods of specifying quantity find use. Again, the specific situation dictates the method used. In laboratory work, practicality dictates working with quantities of known mass (12.3 g, 0.1365 g, etc.). (Counting out a given number of atoms for a laboratory experiment is somewhat impractical, since we cannot see individual atoms.)

In performing chemical calculations, after the laboratory work has been done, it is often useful (even necessary) to think of quantities of substances present in terms of numbers of atoms, molecules, or ions present. A problem exists when this is done — very, very large numbers are always encountered. Any macroscopic-sized sample of a chemical substance contains many trillions of atoms, molecules, or ions.

In order to cope with this "large number problem" chemists have found it convenient to use a special counting unit when counting. Employment of such a unit should not surprise you as specialized counting units are used in many areas. The two most common counting units in use are *dozen* and *pair*. Other more specialized counting units exist. For example, at the stationery store paper is sold by the *ream* (500 sheets), pencils by the *gross* (144), and stencils by the *quire* (24).

mole

The chemist's counting unit is called a *mole*. What is unusual about the mole is its magnitude. A **mole** *is* 6.02×10^{23} *objects*. The extremely large size of the mole is necessitated by the extremely small size of atoms, molecules, and ions. Use of a traditional counting unit, such as a dozen, would be, at best, only a slight improvement over counting atoms singly.

$$6.02 \times 10^{23} \text{ atoms} = 5.02 \times 10^{22} \text{ dozen atoms} = 1 \text{ mole atoms}$$

Note how the use of the mole counting unit decreases very significantly the magnitude of numbers encountered. The number 1 represents 6.02×10^{23} objects, the number 2 double that number of objects. (Why the number 6.02×10^{23} was chosen as the counting unit rather than some other number will be discussed in Sec. 7.4. A more formal definition of the mole will also be presented in that section.)

Avogadro's number

The number 6.02×10^{23} also has a special name. **Avogadro's number** *is the name given to the numerical value* (6.02×10^{23}) *associated with a mole.* This designation honors Amedeo Avogadro (1776–1856), an Italian physicist, whose pioneering work on gases later proved to be valuable in determining the number of particles present in a given mass of a substance.

In solving mathematical problems dealing with the number of atoms, molecules, or ions present in a given amount of material, Avogadro's number becomes part of the conversion factor used to relate number of particles present to moles present.

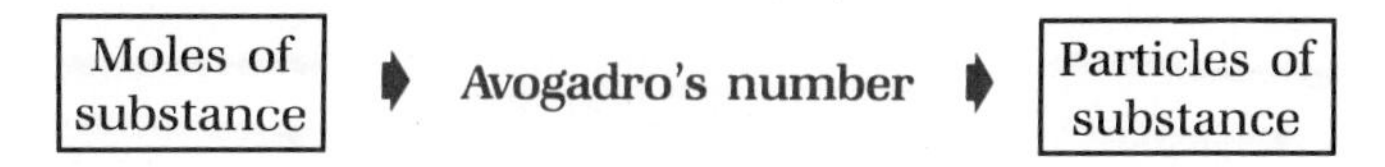

From the definition

$$1 \text{ mole} = 6.02 \times 10^{23} \text{ objects}$$

two conversion factors may be derived.

$$\frac{1 \text{ mole}}{6.02 \times 10^{23} \text{ objects}} \quad \text{and} \quad \frac{6.02 \times 10^{23} \text{ objects}}{1 \text{ mole}}$$

Example 7.4 illustrates the use of particle–mole conversion factors.

EXAMPLE 7.4

How many objects are there in each of the following quantities?

(a) 2.00 moles of hydrogen molecules (H_2)
(b) 1.78 moles of sulfur atoms
(c) 0.345 mole of sodium chloride (NaCl) formula units
(d) 3.00 moles of elephants

SOLUTION

We will use dimensional analysis (Sec. 3.2) in solving each of the parts of this problem. All of the parts are similar in that we are given a certain number of moles of substance and want to find the number of particles contained in the given number of moles. All parts may be classified as moles-to-particles problems and each solution will involve the use of Avogadro's number.

(a) The given quantity is 2.00 moles of H_2 molecules and the desired quantity is numbers of hydrogen molecules.

$$2.00 \text{ moles } H_2 = ? \; H_2 \text{ molecules}$$

The setup, by dimensional analysis, involves only one conversion factor.

$$2.00 \text{ moles } H_2 \times \left(\frac{6.02 \times 10^{23}\ H_2 \text{ molecules}}{1 \text{ mole } H_2}\right) = 1.204 \times 10^{24}\ H_2 \text{ molecules}$$
(calculator answer)

$$= \mathbf{1.20 \times 10^{24}\ H_2 \text{ molecules}}$$
(correct answer)

(b) The given quantity is 1.78 moles of sulfur atoms and the desired quantity is the actual number of sulfur atoms present.

1.78 moles S = ? S atoms

The setup, with the same conversion factor used in part (a), is

$$1.78 \text{ moles S} \times \left(\frac{6.02 \times 10^{23} \text{ S atoms}}{1 \text{ mole S}}\right) = 1.07156 \times 10^{24} \text{ S atoms}$$
(calculator answer)

$$= \mathbf{1.07 \times 10^{24} \text{ S atoms}}$$
(correct answer)

(c) Similarly, the setup this time is

$$0.345 \text{ mole NaCl} \times \left(\frac{6.02 \times 10^{23} \text{ NaCl formula units}}{1 \text{ mole NaCl}}\right)$$

$$= 2.0769 \times 10^{23} \text{ NaCl formula units}$$ (calculator answer)

$$= \mathbf{2.08 \times 10^{23} \text{ NaCl formula units}}$$ (correct answer)

(d) Use of the mole as a counting unit is usually found only in a chemical context. Technically, however, any type of object can be counted in units of moles. One mole denotes 6.02×10^{23} objects; it does not matter what the objects are — even elephants. Just as we can talk about 3 dozen elephants, we can talk about 3 moles of elephants, although the latter is an above average-sized herd of elephants.

$$3.00 \text{ moles elephants} \times \left(\frac{6.02 \times 10^{23} \text{ elephants}}{1 \text{ mole elephants}}\right) = 1.806 \times 10^{24} \text{ elephants}$$
(calculator answer)

$$= \mathbf{1.81 \times 10^{24} \text{ elephants}}$$
(correct answer)

It is somewhat unfortunate that the name mole was selected as the name for the chemist's counting unit because of its similarity to the word molecule. Students often think of the word mole as being an abbreviated form of the word molecule. That is not the case. The word mole comes from the Latin *moles*, which means "heap or pile." The word molecule is a diminutive form of moles meaning "the smallest piece" of that heap or pile. A mole is a macroscopic amount, a heap or pile of objects, that can easily be seen. A molecule is an invisible particle too small to be seen with the naked eye.

In Example 7.4 we calculated the number of objects present in molar-sized samples ranging from 0.345 mole to 3.00 moles. Our answers were numbers carrying the exponents 10^{23} or 10^{24}. Numbers with these exponents are inconceivably large. The magnitude of Avogadro's number itself is so large that it is almost incomprehensible. There is nothing in our experience to relate to it. (When chemists count, they really count in "big" jumps.) Many attempts have been made to create word pictures of the vast size of Avogadro's number. Such pictures, however, really only hint at its magnitude since other large numbers must be used in the word pictures. Three such pictures are as follows.

1. Suppose a fraternity decided they would throw a large "pizza party." So as not to run out of food 1 mole of pizza pies was ordered. Two thousand students showed up at the party. It would take these 2000 students, each "snarfing" one pizza every 3 minutes, 2×10^{15} years to finish the stack of pizzas. If everyone living on this planet attended the party (4 billion pizza eaters), it would still take 5×10^{5} years to eat the pizza, just 500,000 years of non-stop pizza eating. (You have to wonder who the caterer was who supplied all the pizza. How many years ahead of time did the caterer have to start baking pizza in order to supply the fraternity's order?)
2. If each one of the 4 billion people on earth were made a millionaire (receiving 1 million dollar bills), we would still need 150 million other worlds, each inhabited with the same number of millionaires, in order to have Avogadro's number of dollar bills in circulation. (Where would we put all of the dollar bills?) One mole of dollar bills is a lot of money, enough to pay all the expenses of the United States Government for the next billion years or so (in inflated dollars).
3. Avogadro's number (1 mole) of chemistry textbooks the approximate thickness of this one (3 cm) piled on top of each other would reach to the moon and back 1.3×10^{10} times, 13 billion times. (Outer space would no longer be empty.)

7.4 The Mass of a Mole

How much does a mole weigh; that is, what is its mass? Are you uncertain about the answer to that question? Let us, then, consider a similar (but more familiar) question first: "How much does a dozen weigh?" Your response is now immediate. "A dozen what?" you reply. The mass of a dozen identical objects obviously depends on the identity of the object. For example, the mass of a dozen elephants will be "somewhat greater" than the mass of a dozen marshmallows. The mole, like the dozen, is a counting unit. Similarly, the mass of a mole of objects will depend on the identity of the object. Thus, the mass of a mole, *molar mass,* is not one set number; it varies, being different for each different chemical substance. This is in direct contrast to the *molar number,* Avogadro's number, which is the same for all chemical substances.

molar mass of an element

The **molar mass of an element,** *when the element is in atomic form, is a mass in grams that is numerically equal to the atomic weight of the element.* Thus, if we know the atomic weight of an element, we also know the mass of one mole of atoms of the element. The two quantities are numerically the same, differing only in units. For the elements carbon, oxygen, and sodium we can write the following mass-number relationships.

Mass of 1 carbon atom = 12.011 amu (atomic weight)
Mass of 1 mole of carbon atoms = 12.011 g

Mass of 1 oxygen atom = 15.9994 amu (atomic weight)
Mass of 1 mole of oxygen atoms = 15.9994 g

Mass of 1 sodium atom = 22.98977 amu (atomic weight)
Mass of 1 mole of sodium atoms = 22.98977 g

It is not a coincidence that the mass in grams of a mole of atoms of an element and the element's atomic weight are numerically equal. Avogadro's number was given the value that it has in order to cause this relationship to exist. Experimentally it was determined that when 6.02×10^{23} atoms of an element are present, molar masses (in grams) and atomic weights (in amu's) are numerically equal. Again, only when the chemist's counting unit has the value 6.02×10^{23} does this relationship hold. Use of the mass unit grams in specifying the molar mass is also of significance. Use of other mass units would require a different counting unit than 6.02×10^{23} for a numerical match between atomic weight and molar mass.

The molecular form of an element will have a different molar mass than will its atomic form. Consider the element chlorine, which is found in nature in the form of diatomic molecules (Cl_2). The mass of one mole of chlorine atoms (Cl) is different from the mass of one mole of chlorine molecules (Cl_2). Since there are two atoms in each molecule of chlorine, the molar mass of molecular chlorine is twice the molar mass of atomic chlorine. The following relationships hold for chlorine, whose atomic weight is 35.453 amu.

$$6.02 \times 10^{23} \text{ Cl atoms} = 1 \text{ mole Cl atoms} = 35.453 \text{ g Cl}$$

$$6.02 \times 10^{23} \text{ Cl}_2 \text{ molecules} = 1 \text{ mole Cl}_2 \text{ molecules} = 70.906 \text{ g Cl}_2$$

Note that one mole of molecular chlorine contains twice as many atoms as one mole of atomic chlorine; however, the number of discrete particles present is the same (Avogadro's number) in both cases. For atomic chlorine, atoms are considered to be the object counted; for molecular chlorine, molecules are considered to be the discrete particle counted. There are the same number of atoms in the former case as there are molecules in the latter case. This atomic–molecular chlorine situation is analogous to the difference between a dozen shoes and a dozen pairs of shoes. Both the mass and the actual number of shoes for the dozen pairs of shoes is double that for the dozen shoes.

The existence of some elements in molecular form becomes a source of error in chemical calculations if care is not taken to distinguish properly between atomic and molecular forms of the element. The statement "one mole of chlorine" is an ambiguous term. Does it mean one mole of chlorine atoms (Cl) or does it mean one mole of chlorine molecules (Cl_2)?

molar mass of a compound

The **molar mass of a compound** *is a mass in grams that is numerically equal to the formula weight of the compound.* Thus, for compounds a numerical equivalence exists between molar mass and formula weight, if the molar mass is specified in grams. When we add atomic weights to get the formula weight (in amu's) of a compound, we are simultaneously finding the mass of one mole of compound (in grams). The molar mass–formula weight relationships for the compounds water (H_2O), ammonia (NH_3), and barium chloride ($BaCl_2$) are

Mass of 1 H_2O molecule = 18.0152 amu (formula weight)
Mass of 1 mole of H_2O molecules = 18.0152 g

Mass of 1 NH_3 molecule = 17.0304 amu (formula weight)
Mass of 1 mole of NH_3 molecules = 17.0304 g

Mass of 1 $BaCl_2$ formula unit = 208.25 amu (formula weight)
Mass of 1 mole of $BaCl_2$ formula units = 208.25 g

It should now be very evident to you why the chemist's counting unit, the mole, has the value it does — 6.02×10^{23}. *Avogadro's number represents the experimentally determined number of atoms, molecules, or formula units contained in a sample of a pure substance with a mass in grams numerically equal to the atomic weight or formula weight of the pure substance.*

The numerical match between molar mass and atomic or formula weights makes the calculation of the mass of any given number of moles of a substance a very simple procedure. In solving problems of this type, the numerical value of the molar mass becomes part of the conversion factor used to convert from moles to grams.

Moles of substance → molar mass → Grams of substance

For example, for the compound CO_2, which has a formula weight of 44.0 amu, we can write the equality

$$44.0 \text{ g } CO_2 = 1 \text{ mole } CO_2$$

From this statement two conversion factors may be written.

$$\frac{44.0 \text{ g } CO_2}{1 \text{ mole } CO_2} \quad \text{and} \quad \frac{1 \text{ mole } CO_2}{44.0 \text{ g } CO_2}$$

Example 7.5 illustrates the use of gram–mole conversion factors like these.

EXAMPLE 7.5

What is the mass in grams of each of the following molar quantities of chemical substances?

(a) 1.37 moles of SO_2
(b) 7.89 moles of NaCl
(c) 4.33 moles of O_2 molecules
(d) 4.33 moles of O atoms

SOLUTION

We will use dimensional analysis to solve each of these problems. The relationship between molar mass and atomic or formula weight will serve as a conversion factor in the setup of each problem.

(a) The given quantity is 1.37 moles of SO_2 and the desired quantity is grams of SO_2. Thus,

$$1.37 \text{ moles } SO_2 = ? \text{ g } SO_2$$

The molecular weight of SO_2 is calculated to be 64.1 amu.

$$\begin{aligned} \text{S:} \quad & 1 \times 32.1 \text{ amu} = 32.1 \text{ amu} \\ \text{O:} \quad & 2 \times 16.0 \text{ amu} = \underline{32.0 \text{ amu}} \\ & \qquad\qquad\qquad\quad 64.1 \text{ amu} \end{aligned}$$

(The molecular weight need contain only three significant figures since the given quantity, 1.37 moles, contains only that number of significant figures.) Using the molecular weight, we can write the equality

$$64.1 \text{ g } SO_2 = 1 \text{ mole } SO_2$$

The dimensional analysis setup for the problem, with the gram–mole equation as a conversion factor, is

$$1.37 \text{ moles } SO_2 \times \left(\frac{64.1 \text{ g } SO_2}{1 \text{ mole } SO_2}\right) = 87.817 \text{ g } SO_2 \quad \text{(calculator answer)}$$

$$= 87.8 \text{ g } SO_2 \quad \text{(correct answer)}$$

(b) The given quantity is 7.89 moles of NaCl and the desired quantity is grams of NaCl.

$$7.89 \text{ moles NaCl} = ? \text{ g NaCl}$$

The calculated formula weight of NaCl is 58.5 amu. Thus,

$$58.5 \text{ g NaCl} = 1 \text{ mole NaCl}$$

With this relationship as a conversion factor, the setup for the problem becomes

$$7.89 \text{ moles NaCl} \times \left(\frac{58.5 \text{ g NaCl}}{1 \text{ mole NaCl}}\right) = 461.565 \text{ g NaCl} \quad \text{(calculator answer)}$$

$$= 462 \text{ g NaCl} \quad \text{(correct answer)}$$

(c) The given quantity is 4.33 moles of O_2 molecules. The desired quantity is grams of O_2 molecules. Thus,

$$4.33 \text{ moles } O_2 = ? \text{ g } O_2$$

We are dealing here with diatomic oxygen molecules (O_2) and not oxygen atoms. Thus, 32.0 amu, twice the atomic weight of oxygen, is the formula weight used in the mole–gram equality statement.

$$32.0 \text{ g } O_2 = 1 \text{ mole } O_2$$

With this relationship as a conversion factor, the setup becomes

$$4.33 \text{ moles } O_2 \times \left(\frac{32.0 \text{ g } O_2}{1 \text{ mole } O_2}\right) = 138.56 \text{ g } O_2 \quad \text{(calculator answer)}$$

$$= 139 \text{ g } O_2 \quad \text{(correct answer)}$$

(d) The given quantity is 4.33 moles of O atoms and the desired quantity is grams of O atoms. Thus,

$$4.33 \text{ moles O} = ? \text{ g O}$$

This problem differs from the previous one in that atoms, rather than molecules, of oxygen are being counted. The atomic weight of oxygen is 16.0 amu, and the mole–gram equality statement is

$$16.0 \text{ g O} = 1 \text{ mole O}$$

With this relationship as a conversion factor, the setup becomes

$$4.33 \text{ moles O} \times \left(\frac{16.0 \text{ g O}}{1 \text{ mole O}}\right) = 69.28 \text{ g O} \quad \text{(calculator answer)}$$

$$= 69.3 \text{ g O} \quad \text{(correct answer)}$$

The answer to part (c) is double the answer here. That is the way it should be. An oxygen molecule (O_2) has a mass twice that of an oxygen atom (O).

In Section 5.3 we defined the mole simply as

$$1 \text{ mole} = 6.02 \times 10^{23} \text{ objects}$$

mole

Although this statement conveys correct information (the value of Avogadro's number to three significant figures is 6.02×10^{23}), it is not the officially accepted definition for Avogadro's number. The official definition, which is mass based, is: *The* **mole** *is the amount of substance in a system that contains as many elementary units (atoms, molecules, or formula units) as there are* $^{12}_{6}C$ *atoms in exactly 12.00000 grams of* $^{12}_{6}C$. The value of Avogadro's number is, thus, an experimentally determined quantity (the number of atoms in exactly 12.00000 grams of $^{12}_{6}C$ atoms) rather than an exactly defined quantity. Its value is not even mentioned in the definition. The most up to date experimental value for Avogadro's number is 6.022045×10^{23}, which is consistent with our previous definition. In calculations we will never need such a precise value as the experimentally determined one. Most often three significant figures will suffice; occasionally four significant figures (6.022×10^{23}) will be needed for a calculation. But, remember, more significant figures are available if the need ever arises for their use.

7.5 Counting Particles by Weighing

In a laboratory situation it is often necessary to work with equal numbers of atoms of two different substances or twice as many atoms of one type than another, and so forth. Atoms are so small that it is impossible to count them one by one. How does the chemist resolve this requirement of "equal or proportional numbers of atoms"? The problem is solved by means of the concept of "counting by weighing," a concept closely related to molar mass (Sec. 7.4).

To illustrate the "counting by weighing" concept, let us compare the masses of two kinds of atoms — oxygen and nitrogen — when a varying equal number of atoms is present. (The choice of elements in the comparison is arbitrary; any two could be used.) From a table of atomic weights we determine that the mass of a single oxygen atom (on the average) is 16.0 amu and that of a single nitrogen atom (on the average) is 14.0 amu. Using these single atom masses and varying the equal number of atoms present from 1 to 2 to 5 to 100 to 1 billion to Avogadro's number generates the mass comparisons given in Table 7-1. Note that the mass ratio is identical in every case; also, that the numerical value of this ratio is the same as the ratio of the atomic weights of oxygen and nitrogen. What Table 7-1 establishes is that *the mass ratio of equal numbers of atoms of two elements is the same as the ratio of the atomic weights of the two elements.* This statement is true regardless of the elements involved in the comparison. The ratio between two other elements will not be 1.143 (as was the case for O and N); each pair of elements will have its own unique mass ratio determined by the atomic weights of the elements involved.

Turning the conclusion statement derived from Table 7-1 around (reversing it) gives an even more useful generalization (from a laboratory viewpoint): *If samples of two or*

Table 7-1. Mass Ratios for Varying Equal Numbers of Oxygen and Nitrogen Atoms

Atom Ratio	Mass Ratio of Equal Numbers of Atoms
$\frac{1 \text{ atom O}}{1 \text{ atom N}}$	$\frac{1 \times 16.00 \text{ amu}}{1 \times 14.00 \text{ amu}} = \frac{16.00}{14.00} = 1.143$
$\frac{2 \text{ atoms O}}{2 \text{ atoms N}}$	$\frac{2 \times 16.00 \text{ amu}}{2 \times 14.00 \text{ amu}} = \frac{32.00}{28.00} = 1.143$
$\frac{5 \text{ atoms O}}{5 \text{ atoms N}}$	$\frac{5 \times 16.00 \text{ amu}}{5 \times 14.00 \text{ amu}} = \frac{80.00}{70.00} = 1.143$
$\frac{100 \text{ atoms O}}{100 \text{ atoms N}}$	$\frac{100 \times 16.00 \text{ amu}}{100 \times 14.00 \text{ amu}} = \frac{160.0}{140.0} = 1.143$
$\frac{10^9 \text{ atoms O}}{10^9 \text{ atoms N}}$	$\frac{10^9 \times 16.00 \text{ amu}}{10^9 \times 14.00 \text{ amu}} = \frac{1.600 \times 10^{10}}{1.400 \times 10^{10}} = 1.143$
$\frac{6.022 \times 10^{23} \text{ atoms O}}{6.022 \times 10^{23} \text{ atoms N}}$	$\frac{6.022 \times 10^{23} \times 16.00 \text{ amu}}{6.022 \times 10^{23} \times 14.00 \text{ amu}} = \frac{9.635 \times 10^{24}}{8.431 \times 10^{24}} = 1.143$

more elements have mass ratios equal to their atomic weight ratios they must contain equal numbers of atoms. Thus, for example, since the atomic weight ratio between Au and Cu is 3.100 (197.0/63.55), any time samples of these elements are weighed out in a 3.100 to 1.000 ratio the samples will contain equal numbers of atoms. A 3.100 gram sample of Au contains the same number of atoms as a 1.000 gram sample of Cu. Similarly, 3.100 pound and 3.100 ton samples of Au contain, respectively, the same number of atoms as 1.000 pound and 1.000 ton samples of Cu. Any time, regardless of units (g, lb, kg, ton, etc.), Au and Cu are present in a 3.100 to 1.000 mass ratio, equal numbers of atoms are present.

Our comparisons (and generalizations) so far have involved atoms and elements. Parallel comparisons (and generalizations) exist for molecules (or formula units) and compounds. The generalizations for compounds differ from those for elements only in that formula weight has replaced the phrase atomic weight and molecules (or formula weight) has replaced atoms. The substitution of formula weights for atomic weights is valid because both are based on the same $^{12}_{6}C$ scale (Sec. 7.1). Therefore, the following samples all contain the same number of molecules.

$$18.0 \text{ g } H_2O \quad (\text{formula weight of } H_2O = 18.0 \text{ amu})$$

$$44.0 \text{ g } CO_2 \quad (\text{formula weight of } CO_2 = 44.0 \text{ amu})$$

$$64.1 \text{ g } SO_2 \quad (\text{formula weight of } SO_2 = 64.1 \text{ amu})$$

counting particles by weighing

We can now make the following generalization, which is applicable to both elements and compounds. **Samples of two or more pure substances (element or compound) found to have mass ratios equal to the ratios of their atomic or formula weights must contain identical numbers of particles (atoms, molecules, or formula units).** The following samples, therefore, contain the same number of particles.

27.0 g Al (atomic weight = 27.0 amu)

18.0 g H_2O (molecular weight = 18.0 amu)

The particles for water are molecules and the particles for aluminum are atoms.

We see, therefore, that to obtain samples of elements or compounds containing equal numbers of particles, we merely weigh out quantities (in any units) whose mass ratio is numerically equal to the ratio of the substance's atomic or molecular weights.

EXAMPLE 7.6

Determine whether each of the following pairs of samples contains the same number of particles.

(a) 160.3 g of Ca and 107.9 g of Al
(b) 60.07 g of H_2O and 48.66 g of NH_3
(c) 52.81 g of CO_2 and 10.81 g of Be

SOLUTION

(a) From the periodic table, the ratio of the atomic weights of the two elements is

$$\frac{\text{atomic weight Ca}}{\text{atomic weight Al}} = \frac{40.08 \text{ amu}}{26.98 \text{ amu}} = 1.4855448 \quad \text{(calculator answer)}$$

$$= 1.486 \quad \text{(correct answer)}$$

The ratio of the masses of the two given samples is

$$\frac{\text{mass Ca}}{\text{mass Al}} = \frac{160.3 \text{ g}}{107.9 \text{ g}} = 1.4856348 \quad \text{(calculator answer)}$$

$$= 1.486 \quad \text{(correct answer)}$$

Since the atomic weight ratio and the mass ratio are the same (to four significant figures), equal numbers of atoms are present (to four significant figures).

(b) The formula weight of H_2O is 18.02 amu and that of NH_3 is 17.03 amu. The ratio of formula weights is

$$\frac{\text{formula weight } H_2O}{\text{formula weight } NH_3} = \frac{18.02 \text{ amu}}{17.03 \text{ amu}} = 1.0581327 \quad \text{(calculator answer)}$$

$$= 1.058 \quad \text{(correct answer)}$$

The ratio of the masses of the two given samples is

$$\frac{\text{mass } H_2O}{\text{mass } NH_3} = \frac{60.07 \text{ g}}{48.66 \text{ g}} = 1.2344842 \quad \text{(calculator answer)}$$

$$= 1.234 \quad \text{(correct answer)}$$

An equal number of molecules is not present, since the formula weight ratio and the mass ratio have different values.

(c) Here we are comparing amounts of an element and a compound. For the comparison we will use the atomic weight of the element and the formula weight of the compound.

$$\frac{\text{formula weight } CO_2}{\text{atomic weight Be}} = \frac{44.02 \text{ amu}}{9.012 \text{ amu}} = 4.8845983 \quad \text{(calculator answer)}$$

$$= 4.885 \quad \text{(correct answer)}$$

The ratio of the given masses is

$$\frac{\text{mass } CO_2}{\text{mass Be}} = \frac{52.81 \text{ g}}{10.81 \text{ g}} = 4.8852914 \quad \text{(calculator answer)}$$

$$= 4.885 \quad \text{(correct answer)}$$

An equal number (to four significant figures) of particles is present. That is, the number of CO_2 molecules is the same as the number of Be atoms.

7.6 The Mole and Chemical Formulas

A chemical formula has two meanings or interpretations: (1) a microscopic level interpretation and (2) a macroscopic level interpretation.

The first of these two interpretations has been discussed previously in Section 5.3. At a microscopic level a chemical formula indicates the number of atoms of each element present in one molecule or formula unit of a substance. **The numerical subscripts in the formula give directly the number of atoms of the various elements present in one formula unit of the substance.** The formula C_2H_6, interpreted at the microscopic level, conveys the information that two atoms of C and six atoms of H are present in one molecule of C_2H_6.

microscopic level interpretation of a chemical formula

Now that the mole concept has been introduced, a macroscopic interpretation of formulas is possible. At a macroscopic level a chemical formula indicates the number of moles of atoms of each element present in one mole of a substance. **The numerical subscripts in the formula give directly the number of moles of atoms of the various elements present in one mole of the substance.** The designation "macroscopic" is given to this molar interpretation, since moles are "laboratory-sized" quantities of atoms. The formula C_2H_6, interpreted at the macroscopic level, conveys the information that two moles of C atoms and six moles of H atoms are present in one mole of C_2H_6.

macroscopic level interpretation of a chemical formula

It is now evident, then, that the subscripts in a formula always carry a dual meaning: "atoms" at the microscopic level and "moles of atoms" at the macroscopic level.

The validity of the molar interpretation for subscripts in a formula derives from the following line of reasoning. In x molecules of C_2H_6, where x is any number, there are $2x$ atoms of C and $6x$ atoms of H. Regardless of the value of x, there must always be two times as many C atoms as molecules and six times as many H atoms as molecules; that is,

$$\text{Number of molecules} = x$$

$$\text{Number of C atoms} = 2x$$

$$\text{Number of H atoms} = 6x$$

Now let x equal 6.02×10^{23}, the value of Avogadro's number. With this x value, the following statements are true.

$$\text{Number of molecules} = 6.02 \times 10^{23}$$

$$\text{Number of C atoms} = 2 \times (6.02 \times 10^{23}) = 1.204 \times 10^{24} \quad \text{(calculator answer)}$$

$$= 1.20 \times 10^{24} \quad \text{(correct answer)}$$

$$\text{Number of H atoms} = 6 \times (6.02 \times 10^{23}) = 3.612 \times 10^{24} \quad \text{(calculator answer)}$$

$$= 3.61 \times 10^{24} \quad \text{(correct answer)}$$

Since 6.02×10^{23} is equal to 1 mole, 1.20×10^{24} to 2 moles, and 3.61×10^{24} to 6 moles, these statements may be changed to read

Number of molecules = 1 mole

Number of C atoms = 2 moles

Number of H atoms = 6 moles

Thus, the mole ratio is the same as the subscript ratio: 2 to 6.

In calculations where the moles of a particular element within a compound are asked for, the subscript of that particular element in the chemical formula of the compound becomes part of the conversion factor used to convert from moles of compound to moles of element within the compound.

For example, again using C_2H_6 as our chemical formula, we can write as conversion factors the following.

$$\text{For C:} \quad \frac{\text{2 moles C atoms}}{\text{1 mole } C_2H_6 \text{ molecules}} \quad \text{or} \quad \frac{\text{1 mole } C_2H_6 \text{ molecules}}{\text{2 moles C atoms}}$$

$$\text{For H:} \quad \frac{\text{6 moles H atoms}}{\text{1 mole } C_2H_6 \text{ molecules}} \quad \text{or} \quad \frac{\text{1 mole } C_2H_6 \text{ molecules}}{\text{6 moles H atoms}}$$

Example 7.7 illustrates the use of conversion factors of this type in a problem-solving context.

EXAMPLE 7.7

How many moles of each type of atom are present in each of the following molar quantities?

(a) 0.753 moles of CO_2 molecules **(b)** 1.31 moles of P_4O_{10} molecules

SOLUTION

(a) One mole of CO_2 will contain one mole of C atoms and two moles of O atoms. From this statement we obtain the following conversion factors.

$$\left(\frac{\text{1 mole C atoms}}{\text{1 mole } CO_2 \text{ molecules}}\right) \quad \text{and} \quad \left(\frac{\text{2 moles O atoms}}{\text{1 mole } CO_2 \text{ molecules}}\right)$$

From the first of these conversion factors, the moles of C atoms present are calculated as follows.

$$0.753 \text{ mole } CO_2 \text{ molecules} \times \left(\frac{1 \text{ mole C atoms}}{1 \text{ mole } CO_2 \text{ molecules}}\right) = 0.753 \text{ mole C atoms}$$
(calculator answer and correct answer)

Similarly, by the second conversion factor, the moles of O atoms present are as follows.

$$0.753 \text{ mole } CO_2 \text{ molecules} \times \left(\frac{2 \text{ moles O atoms}}{1 \text{ mole } CO_2 \text{ molecules}}\right) = 1.506 \text{ moles O atoms}$$
(calculator answer)

$$= 1.51 \text{ moles O atoms}$$
(correct answer)

(b) Interpreting the formula P_4O_{10} in terms of moles, we obtain the following conversion factors.

$$\left(\frac{4 \text{ moles P atoms}}{1 \text{ mole } P_4O_{10} \text{ molecules}}\right) \quad \text{and} \quad \left(\frac{10 \text{ moles O atoms}}{1 \text{ mole } P_4O_{10} \text{ molecules}}\right)$$

The setup for calculating the moles of P atoms present is

$$1.31 \text{ moles } P_4O_{10} \text{ molecules} \times \left(\frac{4 \text{ moles P atoms}}{1 \text{ mole } P_4O_{10} \text{ molecules}}\right) = 5.24 \text{ moles P atoms}$$
(calculator answer and correct answer)

The setup for calculating the moles of O atoms present is

$$1.31 \text{ moles } P_4O_{10} \text{ molecules} \times \left(\frac{10 \text{ moles O atoms}}{1 \text{ mole } P_4O_{10} \text{ molecules}}\right) = 13.1 \text{ moles O atoms}$$
(calculator answer and correct answer)

If the question were asked "How many total moles of atoms are present," we could obtain the answer by adding the moles of P atoms to the moles of O atoms.

$$(5.24 \text{ moles atoms}) + (13.1 \text{ moles atoms}) = 18.34 \text{ moles atoms}$$
(calculator answer)

$$= 18.3 \text{ moles atoms}$$
(correct answer)

Alternatively, by noting that there are a total of 14 moles of atoms (the subscripts 4 and 10 added together give 14) present in 1 mole of P_4O_{10}, we could calculate the total moles of atoms present using the following setup.

$$1.31 \text{ moles } P_4O_{10} \text{ molecules} \times \left(\frac{14 \text{ moles atoms}}{1 \text{ mole } P_4O_{10} \text{ molecules}}\right) = 18.34 \text{ moles atoms}$$
(calculator answer)

$$= 18.3 \text{ moles atoms}$$
(correct answer)

7.7 The Mole and Chemical Calculations

In this section we combine the major points we've learned about moles in previous sections to produce a general approach to problem solving, which is applicable to a variety of types of chemical calculations.

The three quantities most often calculated in chemical problems are

1. The number of *particles* of a substance, that is, the number of atoms, molecules, or formula units.
2. The number of *moles* of a substance.
3. The number of *grams* (mass) of a substance.

These quantities are interrelated. The conversion factors dealing with these relationships, as previously noted, involve the concepts of (1) Avogadro's number, (2) molar mass, and (3) molar interpretation of chemical formula subscripts.

1. Avogadro's number (Sec. 7.3) provides a relationship between the number of particles of a substance and the number of moles of the same substance.

2. Molar mass (Sec. 7.4) provides a relationship between the number of grams of a substance and the number of moles of the same substance.

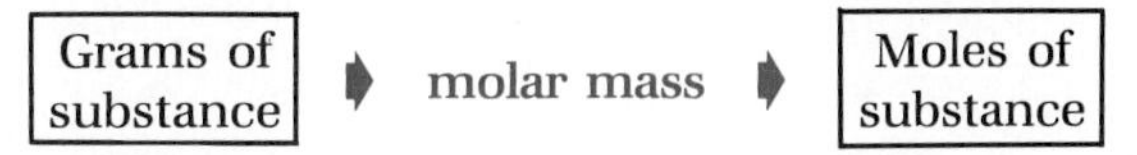

3. Molar interpretation of chemical formula subscripts (Sec. 7.6) provides a relationship between the number of moles of a substance and the number of moles of its component parts.

Chemical problems solved by the above three relationships fall into two general categories.

1. Calculations where information (moles, particles, or grams) is given about a particular substance and additional information is asked for concerning this *same substance*, and
2. Calculations where information (moles, particles, or grams) is given about a particular substance and information is asked for concerning a *related substance.*

Note the difference between the two problem types. In the first type of calculation the whole calculation deals with a single substance; two substances are involved in the second type of calculation.

Only two of the three relationships previously given are needed to solve problems of the first type. Figure 7-1 shows these relationships, arranged in a manner that will be very useful in determining the sequence of conversion factors needed to solve such problems. Figure 7-1 is just a combination of relationships we have used previously on an individual basis. Figure 7-2 shows the relationships needed to solve problems

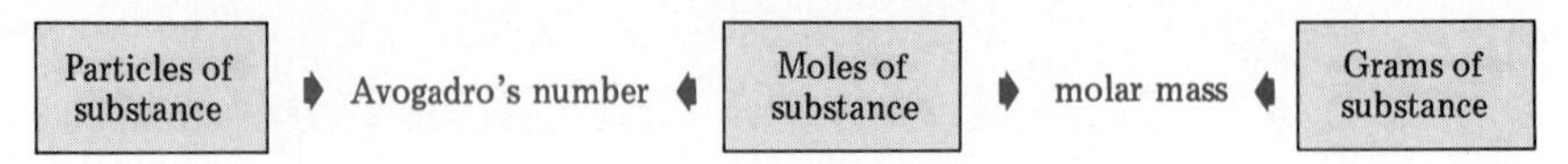

Figure 7-1. Quantitative relationships needed to solve problems involving a single substance.

where two different related substances are involved. Each of the boxes on the left of Figure 7-2 represents a quantity often given in a problem. This given quantity is the starting point for the calculation. The boxes on the right represent quantities of the related substance you may be asked to calculate in the problem. The arrows between boxes represent the concepts to be used in constructing conversion factors to effect the indicated changes. Note the central role that the mole plays in these relationships. Every single arrow touches at least one box labeled moles. The mole is the "heart" of chemical calculations.

Both Figures 7-1 and 7-2 are "road maps." We will see shortly that they are both outlines of the steps to be followed in setting up chemical problems using dimensional analysis. Examples 7.8 through 7.11 illustrate a few of the types of problems that can be solved with these two diagrams as "guides." Figure 7-1 will be useful in determining the problem "setups" in Examples 7.8 and 7.9, and Figure 7-2 will give us the sequence of conversion factors needed to set up the problems of Examples 7.10 and 7.11.

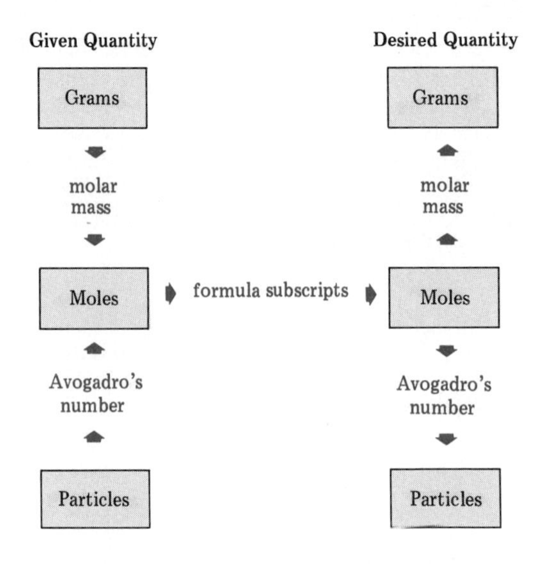

Figure 7-2. Quantitative relationship needed to solve problems involving two related substances.

EXAMPLE 7.8

Calculate the number of copper atoms present in a 37.83 g sample of copper.

SOLUTION

Step 1. The given quantity is 37.83 g of copper. The units of the desired quantity are atoms of copper.

$$37.83 \text{ g Cu} = ? \text{ atoms Cu}$$

Step 2. Since the given and desired quantities both involve the same substance, copper, we will use Figure 7-1 as a guide in setting up this problem.

Grams Cu ➧ molar mass ➧ Moles Cu ➧ Avogadro's number ➧ Particles Cu

This sequence of steps indicates that we should start with grams of Cu, change the grams of Cu to moles of Cu, and then change the moles of Cu to atoms (particles) of Cu.

Step 3. The setup, with dimensional analysis, for this sequence of conversions is

$$37.83\ \cancel{\text{g Cu}} \times \left(\frac{1\ \cancel{\text{mole Cu}}}{63.55\ \cancel{\text{g Cu}}}\right) \times \left(\frac{6.022 \times 10^{23}\ \text{atoms Cu}}{1\ \cancel{\text{mole Cu}}}\right)$$

g Cu ⟶ moles Cu ⟶ particles Cu

The number 63.55 used in the first conversion factor is the atomic weight of Cu. It was not given in the problem, but had to be obtained from a table of atomic weights. Avogadro's number, the basis for the second conversion factor, is quoted to four significant figures since the given data contain that number of significant figures.

Step 4. The solution, obtained by doing the arithmetic, is

$$\frac{37.83 \times 1 \times 6.022 \times 10^{23}}{63.55 \times 1}\ \text{atoms Cu} = 3.584772 \times 10^{23}\ \text{atoms Cu}$$

(calculator answer)

$$= 3.585 \times 10^{23}\ \text{atoms Cu}$$

(correct answer)

EXAMPLE 7.9

What is the mass, in grams, of a single oxygen atom?

SOLUTION

Step 1. The given quantity is 1 atom of O and the desired quantity is grams of O.

$$1\ \text{atom O} = ?\ \text{g O}$$

In the jargon of Figure 7-1 this is a particle-to-gram problem.

Step 2. The sequence of conversion factors for this problem is the exact reverse of that in the previous problem, where grams were given and particles were desired. The pathway is

Particles O ➧ Avogadro's number ➧ Moles O ➧ molar mass ➧ Grams O

Step 3. Translating this pathway into "conversion factor language" gives

$$1\ \cancel{\text{atom O}} \times \left(\frac{1\ \cancel{\text{mole O atoms}}}{6.02 \times 10^{23}\ \cancel{\text{O atoms}}}\right) \times \left(\frac{16.0\ \text{g O atoms}}{1\ \cancel{\text{mole O atoms}}}\right)$$

particles O ⟶ moles O ⟶ g O

Step 4. Collecting numerical terms, after cancellation of units, and doing the indicated arithmetic gives

$$\frac{1 \times 1 \times 16.0}{(6.02 \times 10^{23}) \times 1} \text{ g} = 2.6578073 \times 10^{-23} \text{ g O} \quad \text{(calculator answer)}$$

$$= \mathbf{2.66 \times 10^{-23} \text{ g O}} \quad \text{(correct answer)}$$

The number of significant figures in the answer is determined by the number of significant figures in Avogadro's number. The number of given atoms, one, is an exact (counted) number and therefore has unlimited significance (Sec. 2.2).

This calculation substantiates the generalization we have made a number of times in the text that atoms are very, very small. The mass of an oxygen atom, 2.66×10^{-23} g, written in decimal notation is

$$0.0000000000000000000000266 \text{ g}$$

EXAMPLE 7.10

Calculate the number of nitrogen atoms in 74.3 g of N_2O_5.

SOLUTION

Step 1. The given quantity is 74.3 g N_2O_5. The desired quantity is particles (atoms) of N.

$$\mathbf{74.3 \text{ g } N_2O_5 = ? \text{ atoms N}}$$

A very important difference exists between this problem and the previous two examples; here we are dealing with not one but two substances. We are given information about N_2O_5 and asked for information about a component of the N_2O_5, the nitrogen. Figure 7-2 will serve as a guide in setting up this problem.

Step 2. This is a gram-to-particle problem. The pathway appropriate for solving this problem is

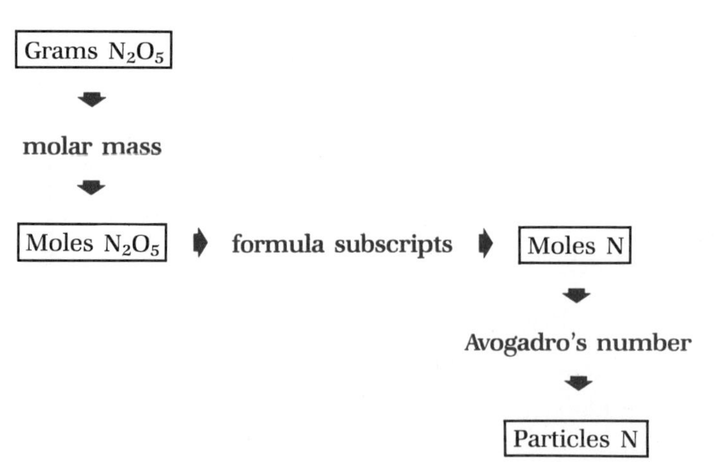

Step 3. The setup, with dimensional analysis, is

$$74.3\ \text{g}\ N_2O_5 \times \left(\frac{1\ \text{mole}\ N_2O_5}{108.0\ \text{g}\ N_2O_5}\right) \times \left(\frac{2\ \text{moles N atoms}}{1\ \text{mole}\ N_2O_5}\right) \times \left(\frac{6.02 \times 10^{23}\ \text{N atoms}}{1\ \text{mole N atoms}}\right)$$

$$\text{g}\ N_2O_5 \longrightarrow \text{moles}\ N_2O_5 \longrightarrow \text{moles N} \longrightarrow \text{atoms N}$$

The number 108.0 in the first conversion factor is the formula weight of N_2O_5. The number 2 in the second conversion factor comes from the chemical formula N_2O_5. Invoking the molar interpretation for the subscripts in a formula, we know that one mole of N_2O_5 contains two moles of N atoms and five moles of O atoms. (The information about O atoms is not needed in solving the problem.)

Step 4. Collecting numerical terms, after cancellation of units, gives

$$\frac{74.3 \times 1 \times 2 \times (6.02 \times 10^{23})}{108.0 \times 1 \times 1}\ \text{N atoms} = 8.2830741 \times 10^{23}\ \text{N atoms} \quad \text{(calculator answer)}$$

$$= \mathbf{8.28 \times 10^{23}\ N\ atoms} \quad \text{(correct answer)}$$

EXAMPLE 7.11

How many grams of gold contain the same number of atoms as 12.37 g of silver?

SOLUTION

This problem can be solved using two different dimensional-analysis approaches.

Method A:
We will work this problem in two parts. First, we will determine the number of silver atoms present in 12.37 g of Ag. Taking a like number of gold atoms, we will then determine their mass in grams.

Number of Ag Atoms
Step 1. This is a grams-to-particles problem involving only one substance.

$$\mathbf{12.37\ g\ Ag = ?\ atoms\ Ag}$$

Step 2. The pathway, from Figure 7-1, is

Grams Ag ➧ molar mass ➧ Moles Ag ➧ Avogadro's number ➧ Particles Ag

Step 3. By this pathway, the mathematical setup becomes

$$12.37\ \text{g Ag} \times \left(\frac{1\ \text{mole Ag}}{107.9\ \text{g Ag}}\right) \times \left(\frac{6.022 \times 10^{23}\ \text{atoms Ag}}{1\ \text{mole Ag}}\right)$$

$$\text{g Ag} \longrightarrow \text{moles Ag} \longrightarrow \text{particles Ag}$$

Step 4. Collecting numerical terms, after cancellation of units, gives

$$\frac{12.37 \times 1 \times (6.022 \times 10^{23})}{107.9 \times 1} = 6.9038128 \times 10^{22} \text{ atoms Ag} \quad \text{(calculator answer)}$$

$$= 6.904 \times 10^{22} \text{ atoms Ag} \quad \text{(correct answer)}$$

Grams of Gold

Step 1. This is a particle-to-gram calculation, just the opposite of the Ag calculation.

$$6.904 \times 10^{22} \text{ Au atoms} = ? \text{ g Au}$$

Step 2. The pathway, from Figure 7-1, is

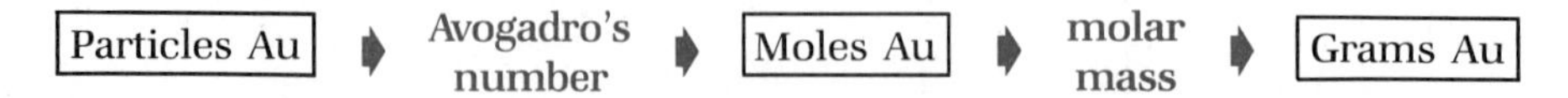

Step 3. By this pathway, the mathematical setup becomes

$$(6.904 \times 10^{22} \text{ atoms Au}) \times \left(\frac{1 \text{ mole Au}}{6.022 \times 10^{23} \text{ atoms Au}}\right) \times \left(\frac{197.0 \text{ g Au}}{1 \text{ mole Au}}\right)$$

particles Au ⟶ moles Au ⟶ g Au

Step 4. Doing the indicated arithmetic gives

$$\frac{(6.904 \times 10^{22}) \times 1 \times 197.0}{(6.022 \times 10^{23}) \times 1} \text{ g Au} = 22.58532 \text{ g Au} \quad \text{(calculator answer)}$$

$$= 22.59 \text{ g Au} \quad \text{(correct answer)}$$

Method B:

It is also possible to solve this problem by using a continuous "string" of conversion factors.

Step 1. We will consider this problem to be a grams-to-grams problem involving Ag and Au.

$$12.37 \text{ g Ag} = ? \text{ g Au}$$

Step 2. The pathway, from Figure 7-2, is

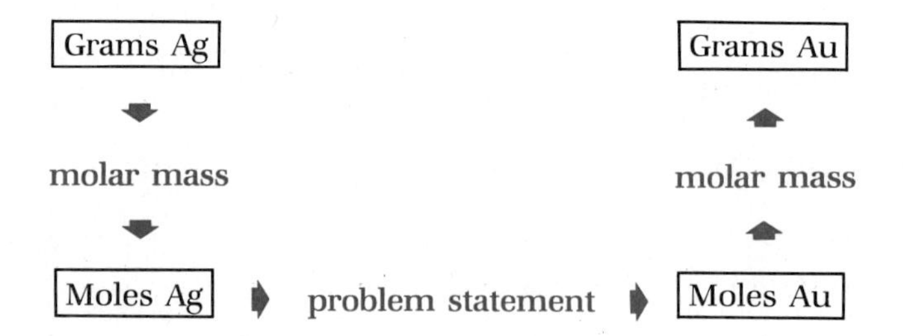

Step 3. The sequence of conversion factors obtained from this pathway is

$$12.37\ \cancel{\text{g Ag}} \times \left(\frac{1\ \cancel{\text{mole Ag}}}{107.9\ \cancel{\text{g Ag}}}\right) \times \left(\frac{1\ \cancel{\text{mole Au}}}{1\ \cancel{\text{mole Ag}}}\right) \times \left(\frac{197.0\text{ g Au}}{1\ \cancel{\text{mole Au}}}\right)$$

g Ag ⟶ moles Ag ⟶ moles Au ⟶ g Au

The middle conversion factor is derived from the statement of the problem itself, which indicates that equal numbers of gold and silver atoms are present. Equal numbers of atoms means a 1 to 1 ratio of moles.

Step 4. Doing the indicated arithmetic gives

$$\frac{12.37 \times 1 \times 1 \times 197.0}{107.9 \times 1 \times 1}\text{ g Au} = 22.584708\text{ g Au} \quad \text{(calculator answer)}$$

$$= 22.58\text{ g Au} \quad \text{(correct answer)}$$

Notice how much shorter this second method is compared to the first method.

7.8 The Determination of Empirical and Molecular Formulas

Chemical formulas provide a great deal of useful information about the substances they represent, and, as seen in previous sections of this chapter, are key entities in many types of calculations. How are formulas, themselves, determined? They are determined by calculation from experimentally obtained information.

Depending on the amount of experimental information available two types of chemical formulas may be obtained: an empirical formula or a molecular formula. The

empirical formula

empirical formula *(or simplest formula) gives the* SMALLEST *whole-number ratio of atoms present in a formula unit of a compound.* In empirical formulas the subscripts in the formula cannot be reduced to a simpler set of numbers by division with a small integer. The

molecular formula

molecular formula *(or true formula) gives the* ACTUAL *number of atoms present in a formula unit of a compound.*

For ionic compounds the empirical and molecular formulas are almost always the same, that is, the actual ratio of atoms present in a formula unit and the smallest ratio of atoms present are one and the same. For molecular compounds the two types of formulas may be the same, but frequently they are not the same. When they are not the same, the molecular formula is a multiple of the empirical formula.

$$\text{Molecular formula} = \begin{pmatrix}\text{whole-number}\\ \text{multiplier}\end{pmatrix} \times \begin{pmatrix}\text{empirical}\\ \text{formula}\end{pmatrix}$$

Table 7-2 contrasts the differences between the two types of formulas for selected compounds.

Percentage composition data (Sec. 7.2) may be used to determine empirical formulas. Examples 7.12 and 7.13 show the steps used in obtaining an empirical formula from such data.

Table 7-2. A Comparison of Empirical and Molecular Formulas for Selected Compounds

Compound	Empirical Formula	Molecular Formula	Whole-Number Multiplier
Dinitrogen tetrafluoride	NF_2	N_2F_4	2
Hydrogen peroxide	HO	H_2O_2	2
Sodium chloride	NaCl	NaCl	1
Benzene	CH	C_6H_6	6

EXAMPLE 7.12

Freon-12, a gas, was widely used as an aerosol can propellant until it was found that it may possibly cause damage to the ozone layer in the upper atmosphere. Its percentage composition is 9.90% carbon, 58.7% chlorine, and 31.4% fluorine. What is the empirical formula for the compound?

SOLUTION

The problem of calculating the empirical formula of a compound from percentage composition data may be broken down into three steps.

1. Express the composition of each element in the compound in grams.
2. Convert the grams of each element to moles of element.
3. Express the mole ratio between the elements in terms of *small whole numbers.*

Step 1. Percentage composition figures can easily be converted to masses (grams) by assuming that you have a sample size of 100 g. The mass of each element in this case becomes the same as the percentage figure.

Carbon: 9.90% of 100 g = 9.90 g

Chlorine: 58.7% of 100 g = 58.7 g

Fluorine: 31.4% of 100 g = 31.4 g

Step 2. The subscripts in a formula give the ratio of the number of moles of each element present in the compound (Sec. 7.6). We next convert the grams (from step 1) to moles.

$$9.90\ \text{g C} \times \left(\frac{1\ \text{mole C}}{12.0\ \text{g C}}\right) = 0.825\ \text{mole C} \quad \text{(calculator and correct answer)}$$

$$58.7\ \text{g Cl} \times \left(\frac{1\ \text{mole Cl}}{35.5\ \text{g Cl}}\right) = 1.6535211\ \text{moles Cl} \quad \text{(calculator answer)}$$

$$= 1.65\ \text{moles Cl} \quad \text{(correct answer)}$$

$$31.4\ \text{g F} \times \left(\frac{1\ \text{mole F}}{19.0\ \text{g F}}\right) = 1.6526316\ \text{moles F} \quad \text{(calculator answer)}$$

$$= 1.65\ \text{moles F} \quad \text{(correct answer)}$$

Thus, for every 0.825 moles of C there are 1.65 moles of both Cl and F, a 0.825 to 1.65 to 1.65 ratio.

Step 3. The subscripts in a formula are expressed as whole numbers, not as decimals. To obtain whole numbers from the decimals of step 2 we proceed as follows. Each of the numbers is divided by the smallest of the numbers.

$$\text{C:} \quad \frac{0.825}{0.825} = 1.00 \quad \text{(calculator and correct answer)}$$

$$\text{Cl:} \quad \frac{1.65}{0.825} = 2.00 \quad \text{(calculator and correct answer)}$$

$$\text{F:} \quad \frac{1.65}{0.825} = 2.00 \quad \text{(calculator and correct answer)}$$

The empirical formula thus becomes CCl_2F_2.

EXAMPLE 7.13

Aspirin contains 60.0% carbon, 4.48% hydrogen, and 35.5% oxygen. What is its empirical formula?

SOLUTION

We will proceed using the steps outlined in the previous example.

Step 1.

Carbon: 60.0% of 100 g = 60.0 g

Hydrogen: 4.48% of 100 g = 4.48 g

Oxygen: 35.5% of 100 g = 35.5 g

Step 2.

$$60.0\ \cancel{\text{g C}} \times \left(\frac{1\ \text{mole C}}{12.0\ \cancel{\text{g C}}}\right) = 5\ \text{moles C} \quad \text{(calculator answer)}$$

$$= 5.00\ \text{moles C} \quad \text{(correct answer)}$$

$$4.48\ \cancel{\text{g H}} \times \left(\frac{1\ \text{mole H}}{1.01\ \cancel{\text{g H}}}\right) = 4.4356436\ \text{moles H} \quad \text{(calculator answer)}$$

$$= 4.44\ \text{moles H} \quad \text{(correct answer)}$$

$$35.5\ \cancel{\text{g O}} \times \left(\frac{1\ \text{mole O}}{16.0\ \cancel{\text{g O}}}\right) = 2.21875\ \text{moles O} \quad \text{(calculator answer)}$$

$$= 2.22\ \text{moles O} \quad \text{(correct answer)}$$

Step 3. Dividing the three numbers from step 2 by the smallest of the three numbers gives

$$\text{C:} \quad \frac{5.00}{2.22} = 2.2522523 \quad \text{(calculator answer)} = 2.25 \quad \text{(correct answer)}$$

$$\text{H:} \quad \frac{4.44}{2.22} = 2 \quad \text{(calculator answer)} = 2.00 \quad \text{(correct answer)}$$

$$\text{O:} \quad \frac{2.22}{2.22} = 1 \quad \text{(calculator answer)} = 1.00 \quad \text{(correct answer)}$$

Frequently all whole numbers will result at this point (as was the case in Ex. 7.12) and the calculation is finished. Sometimes, however, as is the case here, we obtain small decimals, which may be recognized as simple fractions. Such decimals may be cleared by multiplying *all* numbers by a small whole number. Since 0.25, the decimal occurring in this problem, is one-fourth ($\frac{1}{4}$), it may be cleared by multiplying all numbers by 4.

$$\text{C:} \quad 2.25 \times 4 = 9.00$$

$$\text{H:} \quad 2.00 \times 4 = 8.00$$

$$\text{O:} \quad 1.00 \times 4 = 4.00$$

We now have a whole number ratio, and the desired empirical formula for aspirin is

$$C_9H_8O_4$$

The decimals that most commonly occur in empirical-formula calculations are fourths, 0.25 ($\frac{1}{4}$) and 0.75 ($\frac{3}{4}$), which are multiplied by 4 to clear; thirds, 0.33 ($\frac{1}{3}$) and 0.67 ($\frac{2}{3}$), which are multiplied by 3 to clear; and halves, 0.50 ($\frac{1}{2}$), which are multiplied by 2 to clear.

Experimental data about the composition of a compound need not always be in the form of percentage composition before an empirical formula can be calculated. For example, we might know the masses of the elements present in a known mass of compound (1.64 g C, 3.73 g O, etc.). Data of this type simplify an empirical-formula calculation. The grams are changed to moles, and we are ready to find the smallest mole ratio.

Calculation of the molecular formula of a compound requires *both* an elemental analysis (percentage composition or its equivalent) and information concerning the molecular weight of the compound. A number of experimental methods (not discussed in this text) exist for determining this latter quantity. The procedure followed for a molecular-formula calculation is identical to that for an empirical-formula calculation with an additional step required at the end. The percentage composition is used to calculate the empirical formula (as in Examples 7.12 and 7.13), and then the molecular weight and empirical formula are used to calculate the molecular formula, the new last step. Example 7.14 shows how this last step is accomplished.

EXAMPLE 7.14

A compound, known to have a molecular weight of 176.0 amu, is found to have an empirical formula of $C_3H_4O_3$. What is its molecular formula?

SOLUTION

The formula weight for the molecular formula is always a whole-number multiple of the formula weight for the empirical formula. This whole number relating the two formula weights is obtained by dividing the formula weight of the molecular formula by the formula weight of the empirical formula.

We know the formula weight of the molecular formula is 176.0 amu (it was given). We will need to calculate the formula weight of the empirical formula ($C_3H_4O_3$).

Formula weight for $C_3H_4O_3$:

$$\begin{array}{ll} \text{C:} & 3 \times 12.0 \text{ amu} = 36.0 \text{ amu} \\ \text{H:} & 4 \times 1.0 \text{ amu} = 4.0 \text{ amu} \\ \text{O:} & 3 \times 16.0 \text{ amu} = \underline{48.0 \text{ amu}} \\ & \qquad\qquad\qquad\quad 88.0 \text{ amu} \end{array}$$

Dividing the formula weight of the empirical formula into the formula weight for the molecular formula gives

$$\frac{176.0 \text{ amu}}{88.0 \text{ amu}} = 2.00 \quad \text{(whole-number multiplier)}$$

The molecular formula is determined by multiplying each of the subscripts in the empirical formula by 2.

$$(C_3H_4O_3)_2 = C_{3\times2}H_{4\times2}O_{3\times2} = C_6H_8O_6$$

Therefore, $C_6H_8O_6$ is the molecular formula.

Learning Objectives

After completing this chapter you should be able to

- Calculate the formula weight of a substance given its formula and a table of atomic weights (Sec. 7.1).
- Calculate the percentage composition of any component in a molecular or ionic compound given its formula and a table of atomic weights (Sec. 7.2).
- Interpret the mole as a counting unit and calculate the number of particles in a given number of moles of a substance (Sec. 7.3).
- Interpret the mole as a variable mass unit and calculate the mass of a given number of moles of a substance (Sec. 7.4).
- Recognize the mass ratio between substances of which equal numbers of particles are present (Sec. 7.5).
- Interpret the subscripts in a chemical formula in terms of the number of moles of the various elements present in one mole of substance (Sec. 7.6).
- Given the chemical formula and one of the following pieces of information about a pure substance, calculate any other item: (1) mass of the substance, (2) moles of the substance, (3) number of particles of the substance, (4) atoms of any element of the substance, (5) mass of any element in the substance, or (6) moles of any element in the substance (Sec. 7.7).
- Calculate an empirical formula for a substance from percentage composition information (Sec. 7.8).

- Calculate a molecular formula for a substance from its empirical formula and formula weight information (Sec. 7.8).

Terms and Concepts for Review

The new terms or concepts defined in this chapter are

formula weight (Sec. 7.1)
percentage composition (Sec. 7.2)
mole (Secs. 7.3 and 7.4)
Avogadro's number (Sec. 7.3)
molar mass of an element (Sec. 7.4)
molar mass of a compound (Sec. 7.4)
counting by weighing (Sec. 7.5)
microscopic level interpretation of a chemical formula (Sec. 7.6)
macroscopic level interpretation of a chemical formula (Sec. 7.6)
empirical formula (Sec. 7.8)
molecular formula (Sec. 7.8)

Questions and Problems

Formula Weights

7-1 Calculate the formula weight, to the closest 0.01 amu, of each of the following substances. Obtain the needed atomic weights from the inside front or back cover.
a. $NaHCO_3$ (baking soda)
b. $C_{12}H_{22}O_{11}$ (table sugar)
c. $(NH_4)_2SO_4$ (a fertilizer)
d. $C_6H_8O_6$ (vitamin C)
e. $C_3H_5N_3O_9$ (an explosive, nitroglycerin)
f. $Ca(C_6H_{12}NSO_3)_2$ (an artificial sweetening agent, calcium cyclamate)

7-2 Ethyl butanoate, a compound responsible in part for the characteristic odor and taste of pineapple, has a formula weight of 116.18 amu and the formula $C_6H_{12}O_y$. What number does y stand for in this formula?

Percentage Composition

7-3 Calculate the percentage composition for each of the following compounds. (Round off all atomic weights to 0.01 amu before using them.)
a. $Ca(NO_3)_2$ (used in fireworks)
b. $C_{20}H_{30}O$ (vitamin A)
c. NH_3 (ammonia)
d. $C_{14}H_9Cl_5$ (a banned insecticide, DDT)
e. $Pb(C_2H_5)_4$ (a gasoline additive)
f. $C_{10}H_{14}N_2$ (nicotine)

7-4 When 100.0 g of $C_6H_{16}N_2$, one of the raw materials for the production of nylon, is decomposed, 62.01 g of C, 13.87 g of H, and 24.11 g of N are produced. Using this information, calculate the percentage composition for $C_6H_{16}N_2$.

7-5 Decomposition of a sample of a compound yields 19.43 g of Na, 13.55 g of S, and 27.04 g of O. What is the percentage composition for this compound?

7-6 A compound that is 47.49% sulfur by weight contains two sulfur atoms per molecule. What is the formula weight of the compound?

7-7 Without doing a mathematical calculation, indicate which one of the following compounds contains the greatest percent of carbon by mass: CO, CO_2, Na_2CO_3, or CH_4. Explain your reasoning.

7-8 Which of the following compounds, each of which has value as a fertilizer, is potentially the richest source of nitrogen on a mass basis: $(NH_4)_2SO_4$ (ammonium sulfate), $(NH_4)_3PO_4$ (ammonium phosphate), $(NH_2)_2CO$ (urea), NH_4NO_3 (ammonium nitrate), or $(NH_2)_2CNH$ (guanidine).

The Mole as a Counting Unit

7-9 How many molecules or formula units are present in each of the following amounts of substance?
a. 1.00 mole of H_2O molecules
b. 2.00 moles of NaCl formula units
c. 2.67 moles of CO_2 molecules
d. Avogadro's number of O_2 molecules
e. Avogadro's number of CO_2 molecules
f. Avogadro's number of $Al(NO_3)_3$ formula units

7-10 What would be the value of Avogadro's number if it were defined as the number of particles in a sample of a pure substance with a mass in pounds (rather than grams) numerically equal to the atomic or molecular weight of the substance?

Molar Mass

7-11 How much does 1 mole of each of the following substances weigh in grams? (Round off atomic weights to the nearest 0.01 amu before using.)
a. CO_2 **b.** $BaCl_2$ **c.** PbI_2 **d.** $Al_2(SO_4)_3$
e. N_2O_5 **f.** O_2

7-12 What is the mass, in grams, of each of the following quantities of matter?
a. 2.00 moles O atoms
b. 2.00 moles O_2 molecules
c. 3.25 moles CO molecules
d. 4.63 moles NaCl formula units
e. 0.0300 mole H_2O molecules
f. 2.37 moles NH_4CN formula units

Counting by Weighing

7-13 In which of the following pairs of quantities of substances do the members of the pair contain equal numbers of molecules?
a. 18.02 g of H_2O, 17.03 g of NH_3
b. 66.02 g of CO_2, 27.03 g of H_2O

c. 88.04 g of CO_2, 35.76 g of NH_3
d. 29.44 g of CO_2, 30.78 g of NO_2

Molar Interpretation of Chemical Formulas

7-14 Construct conversion factors that relate
a. The numbers of atoms of hydrogen, phosphorus, and oxygen to one molecule of H_3PO_4 (microscopic level interpretation of a formula).
b. The numbers of moles of atoms of hydrogen, phosphorus, and oxygen to one mole of H_3PO_4 molecules (macroscopic level interpretation of a formula).

7-15 How many moles of each type of atom are present in each of the following molar quantities?
a. 2.00 moles CH_4 molecules
b. 3.00 moles $(NH_4)_2SO_4$ formula units
c. 0.00300 mole H_2O molecules
d. 1.50 moles $BaCl_2$ formula units
e. 3.25 moles O_2 molecules
f. 1.13 moles Al_2S_3 molecules

7-16 How many moles of atoms are present in the following amounts of substance?
a. 1 mole of P_4 molecules
b. 0.50 mole of $Ca(NO_3)_2$ formula units
c. 2.0 moles of P_4O_{10} molecules
d. 3.75 moles of K_2S formula units
e. Avogadro's number of CO_2 molecules
f. 6.02×10^{23} H_2O molecules

7-17 How many sulfur atoms are present in the following quantities of substance?
a. 0.50 mole of Na_2SO_4 formula units
b. 2.0 moles of S_8 molecules
c. 0.15 mole of H_2S molecules
d. 1.50 moles of H_2SO_4 molecules
e. Avogadro's number of Al_2S_3 formula units
f. Avogadro's number of K_2S formula units

7-18 Indicate how many moles of molecules of each of the following substances could be prepared from 4.00 moles of C and an unlimited amount of other elements.
a. CO b. CO_2 c. C_2H_4
d. C_3H_6 e. $C_6H_{12}O_6$ f. $C_{20}H_{30}O$

Moles-to-Grams-to-Particles Problems

7-19 Determine the number of atoms in each of the following.
a. 9.012 g Be b. 79.90 g Br c. 3.00 g O
d. 7.67 g Fe e. 10.0 g S f. 100.0 g Al

7-20 Determine the number of atoms of oxygen in each of the following.
a. 10.0 g CO b. 10.0 g CO_2
c. 10.0 g NH_4NO_3 d. 10.0 g P_4O_{10}
e. 10.0 g Na_2SO_4 f. 10.0 g OF_2

7-21 Determine the mass, in grams, of each of the following.
a. 6.02×10^{23} atoms Cu
b. 3.00×10^{23} atoms N
c. 1.00×10^{11} atoms Be
d. 113 molecules NH_3
e. 2.23×10^{30} molecules H_2O
f. 3.76×10^{20} molecules H_2SO_4

7-22 Determine the number of moles of substance present in each of the following quantities.
a. 1.00×10^{-6} g He
b. 2.00×10^{6} g He
c. 30.0 g He
d. 145 atoms He
e. 4.00×10^{10} atoms He
f. 4.00×10^{23} atoms He

7-23 Determine the number of grams of S present in each of the following quantities.
a. 10.0 g H_2SO_4
b. 100.0 g SO_2
c. 1000.0 g $Al_2(SO_4)_3$
d. 1 atom of S
e. 1.37 moles CS_2 molecules
f. 3.00×10^{20} molecules S_4N_4

7-24 Norepinephrine is a substance that plays a key role in the transport of messages along nerves in the human body. Its formula is $C_8H_{11}O_3N$. In a 24.00 g sample of this compound,
a. How many moles of compound are present?
b. How many molecules of compound are present?
c. How many moles of hydrogen atoms are present?
d. How many atoms of carbon are present?
e. How many grams of oxygen are present?
f. How many total atoms are present?

7-25 How many grams of Be would contain the same number of atoms as there are in 3.50 moles of Li?

7-26 How many grams of Al would contain the same number of atoms as there are in 10.0 g of Cu?

7-27 How much of each of the following quantities would you need to obtain 1 million (1.00×10^6) atoms of P?
a. moles PH_3 molecules
b. formula units of $Ca_3(PO_4)_2$
c. molecules of P_4
d. grams of P
e. moles of P_4O_{10} molecules
f. grams of $P_3N_3Cl_6$

7-28 At the grocery store you buy a 25 lb bag (25.00 lb) of table sugar ($C_{12}H_{22}O_{11}$) for $7.99.
a. How many moles of table sugar did you buy?
b. How many molecules of table sugar did you buy?
c. How much is the cost of the sugar on a per molecule basis, that is, what is the cost of a single molecule?

d. How many molecules of sugar could you buy for a penny?

7-29 If a molecule of X has a mass of 1.792×10^{-22} g, what is the formula weight of X (in amu)?

7-30 Calculate the atomic weight of element Y given that 2.33×10^{14} atoms of Y have a mass of 4.59×10^{16} amu. What is the identity of Y?

7-31 Consider 10.00 g samples of each of the following compounds: NH_3, $P_3N_3Cl_6$, HN_3, and $S_4N_4H_4$. Which sample contains

a. The greatest number of total atoms?
b. The least number of total atoms?
c. The greatest number of moles of molecules?
d. The greatest mass, in grams, of nitrogen?
e. The least mass, in grams, of nitrogen?

Empirical and Molecular Formulas

7-32 Each of the following is a correctly written molecular formula. In each case write the empirical formula for the substance.

a. $P_3N_3Cl_6$ **b.** P_4O_{10} **c.** H_2S
d. $(NH_4)_2SO_4$ **e.** $C_6H_{12}O_6$ **f.** Cl_2O_7

7-33 Determine the molecular formulas of compounds with the following empirical formulas and formula weights.

a. CH_2, 28 amu **b.** HO, 34 amu
c. CH_2, 56 amu **d.** NO_2, 92 amu
e. NF_3, 71 amu **f.** C_5H_7N, 162 amu

7-34 Explain why C_8H_{18} must be a molecular formula, whereas C_9H_{20} could be either a molecular formula or an empirical formula.

7-35 Convert each of the following molar ratios between elements in a compound to whole-number molar ratios.

a. 1.00 to 1.67 **b.** 1.25 to 1.75
c. 2.00 to 1.50 **d.** 1.00 to 1.25 to 2.50
e. 1.50 to 2.50 to 3.00 **f.** 2.67 to 1.25 to 1.33

7-36 Determine the empirical formulas of compounds with the following percentage composition.

a. 88.8% O, 11.2% H
b. 45.5% Ni, 54.5% Cl
c. 52.9% Pt, 16.8% P, 30.4% O
d. 54.88% Cr, 45.12% S
e. 43.6% P, 56.4% O
f. 19.8% C, 2.50% H, 66.1% O, 11.6% N

7-37 A gaseous compound of boron and hydrogen is found to have a percentage composition of 81.06% B and 18.94% H by mass. By an independent experiment the formula weight of the compound is determined to be 53.3 amu. What are the empirical and molecular formulas for the compound?

7-38 A 20.00 g sample of the food flavor enhancer, monosodium glutamate (MSG), when decomposed in the laboratory yielded 2.72 g Na, 7.10 g C, 0.960 g H, 1.66 g N, and 7.56 g O. What is the empirical formula of MSG?

7-39 Isopropyl alcohol (rubbing alcohol) is a compound that contains only carbon (60.0%), hydrogen (13.3%), and oxygen. Its molecular and empirical formulas are identical. What is the formula weight of isopropyl alcohol?

7-40 Determine the empirical formulas for compounds containing each of the following.

a. 3.0×10^{30} atoms Fe, 6.0×10^{30} atoms Cr, 1.2×10^{31} atoms O
b. 0.36 mole Ba, 0.36 mole C, 1.08 mole O
c. 1.66 g Cs, 0.65 g Cr, 0.70 g O
d. 3.2 g S, 1.20×10^{23} atoms O
e. 1.81×10^{23} atoms H, 0.30 mole O atoms, 10.65 g Cl
f. 5.266 g Ca, 8.430 g S, 6.304 g O

8 Chemical Calculations II: Calculations Involving Chemical Equations

8.1 The Law of Conservation of Mass

Chemical reactions always involve changing one or more substances into one or more new different substances (Sec. 4.4). Chemical change is a union, separation, or rearrangement of atoms to give new substances. Studies of countless chemical reactions, over a period of two hundred years, have shown that there is no detectable change in the quantity of matter during an ordinary chemical reaction. This generalization concerning chemical reactions has been formalized into a statement known as the **law of conservation of mass:** *Mass is neither created nor destroyed in any ordinary chemical reaction.* To demonstrate the validity of this law the masses of all reactants (substances that react together) and all products (substances formed) in a chemical reaction are carefully determined. It is found that the sum of the masses of the products is always the same as the sum of the masses of the reactants.

law of conservation of mass

Consider, as an illustrative example of this law, the reaction of known masses of the elements beryllium (Be) and oxygen (O) to form the compound beryllium oxide (BeO). Experimentally it is found that 36.03 grams of Be will react *exactly* with 63.97 grams of O. After the reaction no Be or O remains in elemental form; the only substance present is the product BeO, combined Be and O. When this product is weighed, its mass is found to be 100.00 grams, which is the sum of the masses of the reactants (36.03 g + 63.97 g = 100.00 g). It is also found that when the 100.00 grams of product BeO is heated to a high temperature in the absence of air, the BeO decomposes into Be and O, producing 36.03 grams of Be and 63.97 grams of O. Once again, no detectable mass change is observed; the mass of the reactants is equal to the mass of the products.

The law of conservation of mass is consistent with the postulates of atomic theory (Sec. 5.1). Since all reacting chemical substances are made up of atoms (postulate 1), each with its unique identity (postulate 2), and these atoms can neither be created nor destroyed in a chemical reaction but merely rearranged (postulate 4), it follows that the total mass after the reaction must equal the total mass before the reaction. We have the same number of atoms of each kind after the reaction as we started out with. An alternate way of stating the law of conservation of mass is "the total mass of reactants and total mass of products in a chemical reaction are always equal."

The law of conservation of mass applies to all ordinary chemical reactions, and there is no known case of a *measurable* change in total mass during an *ordinary* chemical reaction.* This law will be a "guiding principle" for the discussion later in this chapter of chemical equations and their use.

8.2 Writing Chemical Equations

chemical equation

A **chemical equation** *is a written statement that uses symbols and formulas instead of words to describe the changes that occur in a chemical reaction.* The following example shows the contrast between a word description of a chemical reaction and a chemical equation for the same reaction.

Word description: Magnesium oxide reacts with carbon to produce carbon monoxide and magnesium.

Chemical equation: $MgO + C \longrightarrow CO + Mg$

In the same light that chemical symbols are considered the *letters* of chemical language and formulas the *words* of the language, chemical equations may be considered the *sentences* of chemical language.

The conventions used in writing chemical equations are

1. The correct formulas of the *reactants* are always written on the *left* side of the equation.

$$MgO + C \longrightarrow CO + Mg$$

2. The correct formulas of the *products* are always written on the *right* side of the equation.

$$MgO + C \longrightarrow CO + Mg$$

3. The reactants and products are separated by an arrow pointing toward the products.

$$MgO + C \longrightarrow CO + Mg$$

4. Plus signs are used to separate different reactants or different products from each other.

$$MgO + C \longrightarrow CO + Mg$$

*In the last half century the law of conservation of mass has had to be qualified. Certain types of reactions that involve radioactive processes have been found to deviate from this law. In these processes, there is a conversion of a small amount of matter into energy rather than into another form of matter. A more general law incorporates this apparent discrepancy — the law of conservation of matter and energy. This law takes into account the fact that matter and energy are interconvertible. Note that the statement of the law of conservation of mass as given at the start of this discussion contains the phrase "ordinary chemical reaction." Radioactive processes are not considered to be ordinary chemical reactions.

In reading chemical equations, plus signs on the reactant side of the equation are taken to mean "reacts with," the arrow to mean "to produce," and plus signs on the product side to mean "and."

A *valid* chemical equation must satisfy two conditions.

1. *It must be consistent with experimental facts.* Only the reactants and products actually involved in a reaction are shown in an equation. For each of these substances an accurate formula must be used. For compounds, molecular rather than empirical formulas (Sec. 7.8) are always used. Elements in the solid and liquid states are represented in equations by the chemical symbol for the element. Elements that are gases at room temperatures are represented by the molecular form in which they actually occur in nature. Monoatomic, diatomic, and tetraatomic elemental gases are known.

Monoatomic:	He, Ne, Ar, Kr, Xe
Diatomic:	H_2, O_2, N_2, F_2, Cl_2, Br_2 (vapor), I_2 (vapor)
Tetraatomic:	P_4 (vapor), As_4 (vapor)*

2. *It must be consistent with the law of conservation of mass* (Sec. 8.1). There must be the same number of product atoms of each kind as there are reactant atoms of each kind since atoms are neither created nor destroyed in an ordinary chemical reaction. Equations that satisfy the conditions of this law are said to be *balanced.* Using the four conventions previously listed for writing equations does not guarantee a balanced equation. Section 8.3 considers the steps that must be taken to assure that an equation is balanced.

8.3 Balancing Chemical Equations

balanced chemical equation

A **balanced chemical equation** *has the same number of atoms of each element involved in the reaction on each side of the equation.* It is, thus, an equation consistent with the law of conservation of mass (Sec. 8.1).

coefficient

An unbalanced equation is brought into balance by addition of *coefficients* to the equation, with such coefficients adjusting the number of reactant and/or product molecules (or formula units) present. A **coefficient** *is a number placed to the left of the formula of a substance that changes the amount, but not the identity, of the substance.* In the notation 2 H_2O, the "2" on the left is a coefficient; 2 H_2O means two molecules of H_2O and 3 H_2O means three molecules of H_2O. Coefficients, thus, tell how many formula units of a given substance are present.

The following is a balanced chemical equation with the coefficients shown in **color.**

$$3Cu + 8HNO_3 \longrightarrow 3Cu(NO_3)_2 + 2NO + 4H_2O$$

The message of this balanced equation is: "Three Cu atoms react with eight HNO_3 molecules to produce three $Cu(NO_3)_2$ formula units and two NO molecules and four H_2O molecules." A coefficient of "1" in a balanced equation is not explicitly written; it

*The four elements listed as vapors are not gases at room temperature, but vaporize at slightly higher temperatures. The resultant vapors contain molecules with the formulas indicated. Even if these elements do not vaporize, they still may be represented with these formulas.

is considered to be understood. Both PCl_3 and H_3PO_3 have understood coefficients of 1 in the following balanced equation.

$$PCl_3 + 3\ H_2O \longrightarrow H_3PO_3 + 3\ HCl$$

A coefficient placed in front of a formula applies to the whole formula. In contrast, subscripts, also present in formulas, affect only parts of a formula.

coefficient (affects both the H and O)

$$2\ H_2O$$

subscript (affects only H)

The above notation denotes two molecules of H_2O; it also denotes a total of four H atoms and two O atoms.

We now proceed to the mechanics involved in determining the proper coefficients needed to balance a given equation. They will be introduced in the context of actually balancing two equations (Exs. 8.1 and 8.2). Both of these examples should be studied carefully, since detailed commentary is given in each concerning the "ins and outs" of balancing equations.

EXAMPLE 8.1

Balance the equation

$$KClO_3 \longrightarrow KCl + O_2$$

SOLUTION

Step 1. *Examine the equation and pick one element to balance first.* It is often convenient to start with the compound that contains the greatest number of atoms, whether a reactant or product, and "key in" on the element in that compound with the greatest number of atoms. Using this guideline, we select $KClO_3$ and the element oxygen within it.

We note that there are three oxygen atoms on the left side of the equation (in the $KClO_3$) and two oxygen atoms on the right (in the O_2). For the oxygen atoms to balance we will need six on each side; six is the lowest number that both two and three will divide into evenly. To obtain six atoms on each side of the equation we place the coefficient 2 in front of $KClO_3$ and the coefficient 3 in front of O_2.

$$2\ KClO_3 \longrightarrow KCl + 3\ O_2$$

We now have six oxygen atoms on each side of the equation.

$$2\ KClO_3\text{:}\quad 2 \times 3 = 6$$

$$3\ O_2\text{:}\quad 3 \times 2 = 6$$

Step 2. *Now pick a second element to balance.* We will balance the element K next. (In this particular equation it does not matter whether we balance K or Cl second.) The number of K atoms on the left side of the equation is two; the coefficient 2 in front of $KClO_3$ sets the K atom number at two. We will need two K atoms on the product side, instead of the one K atom now present. This is accomplished by placing the coefficient 2 in front of KCl.

$$2\ KClO_3 \longrightarrow 2\ KCl + 3\ O_2$$

Some students might try to balance the atoms of K at two by using the notation K_2Cl_2 instead of 2 KCl. This is incorrect. *Subscripts in a formula may never be altered during the balancing process.* The notation K_2Cl_2 denotes a formula unit containing four atoms, whereas the notation 2 KCl denotes two formula units, each containing two atoms. The experimental fact is that a formula unit of KCl contains two rather than four atoms. Again, no subscript in a formula may be changed during the process of balancing an equation.

The addition of the coefficient 2 in front of KCl completes the balancing process; all of the coefficients have been determined.

Step 3. *As a final check on the correctness of the balancing procedure, count atoms on each side of the equation.* The following table may be constructed from our balanced equation.

$$2\ KClO_3 \longrightarrow 2\ KCl + 3\ O_2$$

Atom	Left Side	Right Side
K	$2 \times 1 = 2$	$2 \times 1 = 2$
Cl	$2 \times 1 = 2$	$2 \times 1 = 2$
O	$2 \times 3 = 6$	$3 \times 2 = 6$

All elements are in balance: two K atoms on each side, two Cl atoms on each side, and six O atoms on each side.

Notice that the elemental oxygen present on the product side of the equation is written in the form of diatomic molecules (O_2). This is in accordance with the guideline given in Section 8.2 on the use of molecular formulas for elements that are gases at room temperature.

EXAMPLE 8.2

Balance the equation

$$C_2H_6 + O_2 \longrightarrow CO_2 + H_2O$$

SOLUTION

Step 1. *Examine the equation and pick one element to balance first.* The formula containing the most atoms is C_2H_6. We will balance the element H first. We have six H atoms on the left and two H atoms on the right. The two sides are brought into balance by placing the coefficient 3 in front of H_2O on the right side. We now have six H atoms on each side.

$$3\ H_2O{:}\quad 3 \times 2 = 6$$

Our equation now has the following appearance.

$$1\,C_2H_6 + O_2 \longrightarrow CO_2 + 3\,H_2O$$

In setting the H balance at six atoms we are setting the coefficient in front of C_2H_6 at 1. The 1 has been explicitly shown in the above equation to remind us that the C_2H_6 coefficient has been determined. (In the final balanced equation the 1 need not be shown.)

Step 2. *Now pick a second element to balance.* We will balance C next. It is always better to balance the elements that appear only in one reactant and one product before trying to balance any elements appearing in several formulas on one side of the equation. Oxygen, our other choice for an element to balance at this stage, appears in two places on the product side of the equation. The number of carbon atoms is already set at two on the left side of the equation.

$$1\,C_2H_6\text{:}\quad 1 \times 2 = 2$$

We obtain a balance of two carbon atoms on each side of the equation by placing the coefficient 2 in front of CO_2.

$$1\,C_2H_6 + O_2 \longrightarrow 2\,CO_2 + 3\,H_2O$$

Step 3. *Now pick a third element to balance.* Only one element is left to balance—oxygen. The number of oxygen atoms on the right side of the equation is already set at seven—four O atoms from the CO_2 and three O atoms from the H_2O.

$$2\,CO_2\text{:}\quad 2 \times 2 = 4$$

$$3\,H_2O\text{:}\quad 3 \times 1 = 3$$

To obtain seven O atoms on the left side of the equation we need a fractional coefficient, $3\frac{1}{2}$.

$$3\tfrac{1}{2}\,O_2\text{:}\quad 3\tfrac{1}{2} \times 2 = 7$$

The coefficient 3 gives six atoms, and the coefficient 4 gives eight atoms. The only way we can get seven atoms is by using $3\frac{1}{2}$.

All of the coefficients in the equation have now been determined.

$$1\,C_2H_6 + 3\tfrac{1}{2}O_2 \longrightarrow 2\,CO_2 + 3\,H_2O$$

Equations containing fractional coefficients are not considered to be written in their most conventional form. Although such equations are "mathematically" correct, they have some problems "chemically." The above equation indicates the need for $3\frac{1}{2}$ O_2 molecules among the reactants and a half an O_2 molecule does not exist as such. Step 4 shows how to take care of this "problem."

Step 4. *After all coefficients have been determined, clear any fractional coefficients that are present.* We can clear the fraction present in this equation, $3\frac{1}{2}$, by multiplying each of the coefficients in the equation by the factor 2.

$$2C_2H_6 + 7O_2 \longrightarrow 4CO_2 + 6H_2O$$

Now we have the equation in its conventional form. Note that *all of the coefficients* had to be multiplied by 2, not just the fractional one. It will always be the case that whatever is done to a fractional coefficient to make it a whole number must also be carried out on all of the other coefficients.

If a coefficient involving $\frac{1}{3}$ had been present in the equation, we would have multiplied by 3 instead of by 2.

Step 5. *As a final check on the correctness of the balancing procedure, count atoms on each side of the equation.* The following table may be constructed from our balanced equation.

$$2\,C_2H_6 + 7\,O_2 \longrightarrow 4\,CO_2 + 6\,H_2O$$

Atom	Left Side	Right Side
C	$2 \times 2 = 4$	$4 \times 1 = 4$
H	$2 \times 6 = 12$	$6 \times 2 = 12$
O	$7 \times 2 = 14$	$(4 \times 2) + (6 \times 1) = 14$

All atom counts balance. We have accomplished our task.

Some additional comments and guidelines concerning equations in general and the process of balancing in particular are

1. The coefficients in a balanced equation are always the *smallest set of whole numbers* that will balance the equation. We mention this because more than one set of coefficients will balance an equation. Consider the following three equations.

$$2\,H_2 + O_2 \longrightarrow 2\,H_2O$$

$$4\,H_2 + 2\,O_2 \longrightarrow 4\,H_2O$$

$$8\,H_2 + 4\,O_2 \longrightarrow 8\,H_2O$$

All three of these equations are mathematically correct; there are equal numbers of H and O atoms on each side of the equation. The first equation, however, is considered the conventional form because the coefficients used there are the smallest set of whole numbers that will balance the equation. The coefficients in the second equation are double those in the first, and the third equation has coefficients four times those of the first equation.

2. It is helpful to consider polyatomic ions as single entities in balancing an equation, provided they maintain their identity in the chemical reaction, that is, they appear on both sides of the equation in the same form. For example, in an equation where sulfate units are present (SO_4^{2-}), balance them as a unit rather than trying to balance S and O separately. The reasoning would be: "we have two sulfate on this side so we need two sulfate on that side, etc."

3. Always remember the difference between a coefficient in front of a formula and a subscript in a formula. Both are numbers, but they have very different meanings. The coefficient deals with the number of formula units of a substance and the subscript with the composition of the substance. *The subscripts in a formula are never changed* *in the process of balancing an equation.* Subscripts illustrate the law of definite proportions; coefficients relate to the law of conservation of mass.

4. You are not expected, at this point, to be able to write down the products for a chemical reaction given what the reactants are. After learning how to balance equations, students sometimes get the mistaken idea that they ought to be able to write down equations from "scratch." This is not so. You will need more "chemical knowledge" before attempting this task. At this stage, you should be able to balance simple equations given *all* of the reactants and products.

5. Some equations are much more difficult to balance than those you encounter in this chapter's examples and problem exercises. The procedures discussed here simply are not adequate for these more difficult equations. In Chapter 15 a more systematic method for balancing equations, specifically designed for these more difficult situations, will be presented.

6. It needs to be emphasized that the ultimate source of any chemical equation is experimental information. The products formed in a chemical reaction are facts learned by experiment. We cannot discover the products of a chemical reaction simply by writing an equation. The products are identified experimentally. Then they can be represented by an equation.

7. Finally, the only way to learn to balance equations is through practice. The problem set at the end of this chapter contains numerous equations for you to practice upon.

8.4 Special Symbols Used in Equations

In addition to the essential plus sign and arrow notation used in chemical equations, a number of optional symbols convey more information about a chemical reaction than just the chemical species involved. In particular, it is often useful to know the physical state of the substances involved in a chemical reaction. These optional symbols, which are given in Table 8-1, enable physical state to be specified.

Table 8-1. Symbols Used in Equations

Symbol	Meaning
Essential	
$\longrightarrow$	"to produce"
+	"reacts with" or "and"
Optional	
(s)	solid
(l)	liquid
(g)	gas
(aq)	aqueous solution (a substance dissolved in water)

The equations we balanced in Section 8.3 (Exs. 8.1 and 8.2) appear as follows when the optional symbols are included.

$$2\ KClO_3(s) \longrightarrow 2\ KCl(s) + 3\ O_2(g)$$

$$2\ C_2H_6(g) + 7\ O_2(g) \longrightarrow 4\ CO_2(g) + 6\ H_2O(g)$$

Additional examples of optional symbol use include the equations

$$NaCl(aq) + AgNO_3(aq) \longrightarrow AgCl(s) + NaNO_3(aq)$$

$$NaOH(aq) + HCl(aq) \longrightarrow NaCl(aq) + H_2O(l)$$

The optional symbols in these two equations indicate that both reactions take place in aqueous solution. In the first reaction one of the products (AgCl) is insoluble, being present in the solution as a solid. In the second reaction, the product (NaCl) is soluble and thus remains in solution.

8.5 Chemical Equations and the Mole Concept

The coefficients in a balanced chemical equation, like the subscripts in a chemical formula (Sec. 7.6), have two levels of interpretation — a microscopic level of meaning and a macroscopic level of meaning.

The first of these two interpretations, the microscopic level, has been used in the previous sections of this chapter. At this level, a balanced chemical equation gives the relative number of formula units of the various reactants and products involved in a chemical reaction. *The coefficients in the equation give directly the numerical relationships among formula units consumed* (*used up*) *and/or produced in the chemical reaction.* Interpreted at the microscopic level, the equation

$$4\ NH_3 + 5\ O_2 \longrightarrow 4\ NO + 6\ H_2O$$

conveys the information that four molecules of NH_3 react with five molecules of O_2 to produce four molecules of NO and six molecules of H_2O.

molar interpretation of coefficients

At the macroscopic level of interpretation, a level not used up to this time, chemical equations are used to relate mole-sized quantities of reactants and products to each other. At this level, *the coefficients in the equation give the fixed molar ratios between substances consumed and/or produced in the chemical reaction.* Interpreted at the macroscopic level, the equation

$$4\ NH_3 + 5\ O_2 \longrightarrow 4\ NO + 6\ H_2O$$

conveys the information that four moles of NH_3 react with five moles of O_2 to produce four moles of NO and six moles of H_2O.

The validity of the molar interpretation of coefficients in an equation can be derived very straightforwardly with the microscopic level of interpretation as a starting point.

A balanced chemical equation remains valid (mathematically correct) when each of its coefficients is multiplied by the same number. (If molecules react in a 3 to 1 ratio, they will also react in a 6 to 2 or 9 to 3 ratio.) Thus, using the previous equation and multiplying by *y*, where *y* is any number, we have

$$4y\ NH_3 + 5y\ O_2 \longrightarrow 4y\ NO + 6y\ H_2O$$

The situation where y is equal to 6.02×10^{23} is of particular interest because 6.02×10^{23} is equal to 1 mole. Using y is equal to 1 mole, we have by substitution

$$4 \text{ moles } NH_3 + 5 \text{ moles } O_2 \longrightarrow 4 \text{ moles } NO + 6 \text{ moles } H_2O$$

Thus, as with the subscripts in formulas, the coefficients in equations carry a dual meaning: "number of formula units" at the microscopic level and "moles of formula units" at the macroscopic level.

The coefficients in an equation may be used to generate conversion factors used in problem solving. Numerous conversion factors are obtainable from a single balanced equation. Consider the balanced equation

$$P_4O_{10} + 6\,H_2O \longrightarrow 4\,H_3PO_4$$

Three mole-to-mole relationships are obtainable from this equation.

1 mole of P_4O_{10} produces 4 moles of H_3PO_4

6 moles of H_2O produces 4 moles of H_3PO_4

1 mole of P_4O_{10} reacts with 6 moles of H_2O

From these three macroscopic level relationships six conversion factors can be written.

From the first relationship:

$$\left(\frac{1 \text{ mole } P_4O_{10}}{4 \text{ moles } H_3PO_4}\right) \quad \text{and} \quad \left(\frac{4 \text{ moles } H_3PO_4}{1 \text{ mole } P_4O_{10}}\right)$$

From the second relationship:

$$\left(\frac{6 \text{ moles } H_2O}{4 \text{ moles } H_3PO_4}\right) \quad \text{and} \quad \left(\frac{4 \text{ moles } H_3PO_4}{6 \text{ moles } H_2O}\right)$$

From the third relationship:

$$\left(\frac{1 \text{ mole } P_4O_{10}}{6 \text{ moles } H_2O}\right) \quad \text{and} \quad \left(\frac{6 \text{ moles } H_2O}{1 \text{ mole } P_4O_{10}}\right)$$

All chemical equations are the source of numerous conversion factors. The more reactants and products there are in the equation the greater the number of derivable conversion factors.

Conversion factors obtained from equations are used in many different types of calculations. Example 8.3 illustrates some very simple applications of their use. Section 8.6 then explores more complicated problem-solving situations in which they are used.

EXAMPLE 8.3

The rusting of iron may be represented by the equation

$$4\,Fe + 3\,O_2 \longrightarrow 2\,Fe_2O_3$$

(a) How many moles of Fe_2O_3 are produced when 2.68 moles of O_2 are consumed?
(b) How many moles of Fe are needed to produce 1.35 moles of Fe_2O_3?
(c) How many moles of Fe are needed to react with 7.21 moles of O_2?

SOLUTION

All three parts of this problem are one-step mole-to-mole calculations. In each case the needed conversion factor is derived from the coefficients of the chemical equation.

Moles of substance A	➧ equation coefficients ➧	Moles of substance B

(a) ***Step 1.*** The given quantity is 2.68 moles of O_2 and the desired quantity is moles of Fe_2O_3.

$$2.68 \text{ moles } O_2 = ? \text{ moles } Fe_2O_3$$

Step 2. The conversion factor needed to convert from moles O_2 to moles Fe_2O_3 is derived from the coefficients of O_2 and Fe_2O_3 in the balanced equation. The equation tells us that 3 moles of O_2 produces 2 moles of Fe_2O_3. From this relationship two conversion factors are obtainable.

$$\left(\frac{3 \text{ moles } O_2}{2 \text{ moles } Fe_2O_3}\right) \quad \text{and} \quad \left(\frac{2 \text{ moles } Fe_2O_3}{3 \text{ moles } O_2}\right)$$

We will use the second of these conversion factors in solving the problem. The setup is

$$2.68 \cancel{\text{moles } O_2} \times \left(\frac{2 \text{ moles } Fe_2O_3}{3 \cancel{\text{moles } O_2}}\right)$$

We used the second of the two conversion factors because it had moles of O_2 in the denominator, a requirement for the units moles of O_2 to cancel.

Step 3. Collecting numerical terms, after cancellation of units, gives

$$\left(\frac{2.68 \times 2}{3}\right) \text{ moles } Fe_2O_3 = 1.7866667 \text{ moles } Fe_2O_3 \quad \text{(calculator answer)}$$

$$= 1.79 \text{ moles } Fe_2O_3 \quad \text{(correct answer)}$$

Note that the coefficients in the equation enter directly into the numerical calculation. Having a correctly balanced equation, thus, is of vital importance. Using an unbalanced or misbalanced equation as a source of a conversion factor will lead to a wrong numerical answer.

(b) ***Step 1.*** The given quantity is 1.35 moles of Fe_2O_3 and the desired quantity is moles of Fe.

$$1.35 \text{ moles } Fe_2O_3 = ? \text{ moles Fe}$$

Here we are given information about a product and desire information about a reactant. This is just the opposite situation to that in part (a). The problem is solved, however, in an identical way to part (a).

Step 2. The equation indicates that the fixed mole ratio between Fe_2O_3 and Fe is 2 to 4. The two conversion factors derived from this 2 to 4 ratio are

$$\left(\frac{4 \text{ moles Fe}}{2 \text{ moles } Fe_2O_3}\right) \quad \text{and} \quad \left(\frac{2 \text{ moles } Fe_2O_3}{4 \text{ moles Fe}}\right)$$

We will use the first of these conversion factors in solving the problem. The setup is

$$1.35 \cancel{\text{moles } Fe_2O_3} \times \left(\frac{4 \text{ moles Fe}}{2 \cancel{\text{moles } Fe_2O_3}}\right)$$

Step 3. Collecting numerical terms, after cancellation of units, gives

$$\left(\frac{1.35 \times 4}{2}\right) \text{moles Fe} = 2.7 \text{ moles Fe} \quad \text{(calculator answer)}$$

$$= 2.70 \text{ moles Fe} \quad \text{(correct answer)}$$

(c) ***Step 1.*** Both the given quantity (O_2) and desired quantity (Fe) are reactants.

$$7.21 \text{ moles } O_2 = ? \text{ moles Fe}$$

Step 2. Molar relationships obtained from an equation are not required to always involve one reactant and one product, as was the case in parts (a) and (b). Molar relationships involving only reactants or only products are often needed and used. In this problem we will need the molar relationship between the two reactants, Fe and O_2, which is 4 to 3. With this ratio the conversion factor

$$\left(\frac{4 \text{ moles Fe}}{3 \text{ moles } O_2}\right)$$

can be constructed, which is used in the setup of the problem as follows.

$$7.21 \cancel{\text{moles } O_2} \times \left(\frac{4 \text{ moles Fe}}{3 \cancel{\text{moles } O_2}}\right)$$

Step 3. Collecting numerical terms, after cancellation of units, gives

$$\left(\frac{7.21 \times 4}{3}\right) \text{moles Fe} = 9.6133333 \text{ moles Fe} \quad \text{(calculator answer)}$$

$$= 9.61 \text{ moles Fe} \quad \text{(correct answer)}$$

In terms of significant figures, the numbers in conversion factors obtained from equation coefficients are considered exact numbers. Thus, since 7.21 contains three significant figures, the answer to this problem should also contain three significant figures.

8.6 Calculations Based on Equations

If the information contained in a chemical equation is combined with the concepts of molar mass (Sec. 7.4) and Avogadro's number (Sec. 7.3), many useful types of chemical calculations may be carried out.

In a typical chemical equation-based calculation, information is given about one

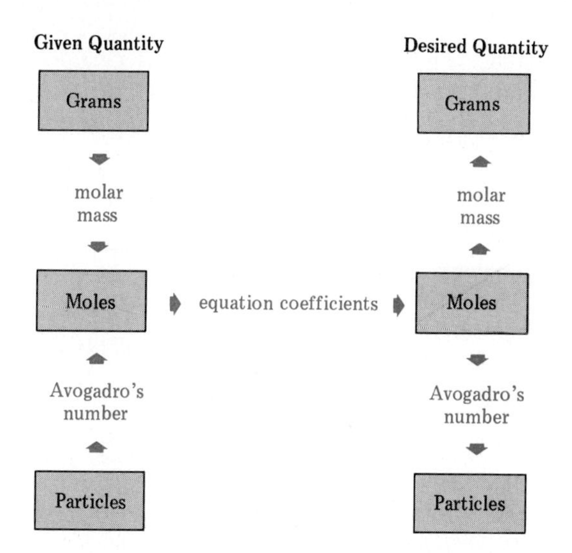

Figure 8-1. Conversion factor relationships needed for solving chemical equation-based problems.

reactant or product of a reaction (number of grams, moles, or particles) and information is requested concerning another reactant or product of the same reaction. The substances involved in such a calculation may both be reactants, may both be products, or may be one of each.

The conversion factor relationships needed to solve problems of the above general type are given in Figure 8-1. This diagram should seem very familiar to you, for it is almost identical to the figure in the last chapter (Fig. 7-2), which you have worked with repeatedly. Only one difference exists between the two diagrams. In the Chapter 7 diagram the subscripts in a chemical formula were listed as the basis for relating moles of given and desired substances to each other. In this new diagram, these same two quantities are related using the coefficients of a balanced chemical equation.

The most common type of chemical equation-based calculation is a mass–mass (gram-to-gram) type problem. In such problems the mass of one substance involved in a chemical reaction (either a reactant or product) is given and information is requested about the mass of another of the substances involved in the reaction (either a reactant or product). Situations requiring the solving of problems of this type are frequently encountered in laboratory settings. For example, a chemist has available so many grams of a certain chemical and wants to know how many grams of another substance can be produced from it, or how many grams of a third substance are needed to react with it, etc. Examples 8.4 and 8.5 are both problem-solving situations of the gram–gram type.

EXAMPLE 8.4

The natural gas burned to provide heat in many homes is predominantly methane, CH_4. Methane burns (reacts with O_2) as shown by the following equation.

$$CH_4 + 2\,O_2 \longrightarrow CO_2 + 2\,H_2O$$

How many grams of water (as vapor) will be produced when 48.3 g of CH_4 is burned?

SOLUTION

Step 1. The given quantity is 48.3 g of CH_4. The desired quantity is grams of H_2O.

$$48.3 \text{ g } CH_4 = ? \text{ g } H_2O$$

Step 2. This problem is of the gram-to-gram type. The pathway to be used in solving this type of problem, in terms of Figure 8-1, is

Step 3. The dimensional-analysis setup for this pathway is

$$48.3 \text{ g } CH_4 \times \left(\frac{1 \text{ mole } CH_4}{16.0 \text{ g } CH_4}\right) \times \left(\frac{2 \text{ moles } H_2O}{1 \text{ mole } CH_4}\right) \times \left(\frac{18.0 \text{ g } H_2O}{1 \text{ mole } H_2O}\right)$$

$$\text{g } CH_4 \longrightarrow \text{moles } CH_4 \longrightarrow \text{moles } H_2O \longrightarrow \text{g } H_2O$$

The 16.0 g in the first conversion factor is the molar mass of CH_4, the 2 and 1 in the second conversion factor are the coefficients, respectively, of H_2O and CH_4 in the balanced chemical equation, and the 18.0 g in the third conversion factor is the molar mass of H_2O.

Step 4. The solution, obtained from combining all of the numerical factors, is

$$\left(\frac{48.3 \times 1 \times 2 \times 18.0}{16.0 \times 1 \times 1}\right) \text{g } H_2O = 108.675 \text{ g } H_2O \quad \text{(calculator answer)}$$

$$= 109 \text{ g } H_2O \quad \text{(correct answer)}$$

EXAMPLE 8.5

Aluminum and sulfur react at high temperatures to form aluminum sulfide as shown by the following equation.

$$2\,Al + 3\,S \longrightarrow Al_2S_3$$

What mass, in grams, of Al is needed to completely react with 38.7 g of S?

SOLUTION

Step 1. Here we are given information about one reactant (38.7 g S) and asked to calculate information about the other reactant (? g Al).

$$38.7 \text{ g S} = ? \text{ g Al}$$

Step 2. This problem, like Example 8.4, is a gram-to-gram problem. The pathway used in solving it will be the same, which, in terms of Figure 8-1, is

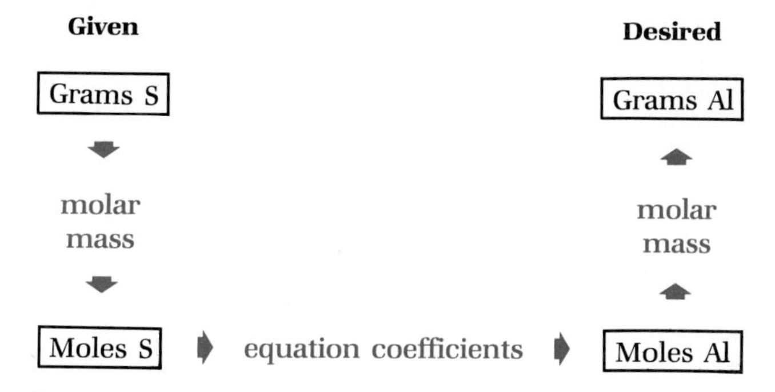

Step 3. The dimensional-analysis setup is

$$38.7 \cancel{\text{g S}} \times \left(\frac{1 \cancel{\text{mole S}}}{32.1 \cancel{\text{g S}}}\right) \times \left(\frac{2 \cancel{\text{moles Al}}}{3 \cancel{\text{moles S}}}\right) \times \left(\frac{27.0 \text{ g Al}}{1 \cancel{\text{mole Al}}}\right)$$

$$\text{g S} \longrightarrow \text{moles S} \longrightarrow \text{moles Al} \longrightarrow \text{g Al}$$

The chemical equation is the "bridge" that enables us to go from S to Al. The numbers in the second conversion factor are coefficients from this equation.

Step 4. The solution, obtained from combining all of the numerical factors in the setup, is

$$\left(\frac{38.7 \times 1 \times 2 \times 27.0}{32.1 \times 3 \times 1}\right) \text{g Al} = 21.700935 \text{ g Al} \quad \text{(calculator answer)}$$

$$= 21.7 \text{ g Al} \quad \text{(correct answer)}$$

Gram-to-gram problems (Exs. 8.4 and 8.5) are not the only type of problem for which the coefficients in a balanced equation can be used to relate quantities of two substances. As further examples of equation coefficient use in problem solving, consider Example 8.6 (a mole-to-gram problem) and Example 8.7 (a particle-to-gram problem).

EXAMPLE 8.6

Decomposition of $KClO_3$ serves as a convenient laboratory source of small amounts of oxygen gas. The reaction is

$$2\ KClO_3 \longrightarrow 2\ KCl + 3\ O_2$$

How many grams of $KClO_3$ must be decomposed in order to produce 1.73 moles of O_2?

SOLUTION

Step 1. The given quantity is 1.73 moles of O_2 and the desired quantity is grams of $KClO_3$.

$$1.73 \text{ moles } O_2 = ? \text{ grams of } KClO_3$$

Step 2. This is a mole-to-gram problem. The pathway used to solve such a problem is, according to Figure 8-1,

Given **Desired**

Grams $KClO_3$

↑ molar mass ↑

Moles O_2 → equation coefficients → Moles $KClO_3$

Step 3. The dimensional-analysis setup is

$$1.73 \text{ moles } O_2 \times \left(\frac{2 \text{ moles } KClO_3}{3 \text{ moles } O_2}\right) \times \left(\frac{122.6 \text{ g } KClO_3}{1 \text{ mole } KClO_3}\right)$$

$$\text{moles } O_2 \longrightarrow \text{moles } KClO_3 \longrightarrow \text{g } KClO_3$$

The number 122.6 in the second conversion factor is the formula weight of $KClO_3$.

Step 4. The solution, obtained from combining all of the numbers in the manner indicated in the setup, is

$$\left(\frac{1.73 \times 2 \times 122.6}{3 \times 1}\right) \text{g } KClO_3 = 141.39867 \text{ g } KClO_3 \quad \text{(calculator answer)}$$

$$= 141 \text{ g } KClO_3 \quad \text{(correct answer)}$$

EXAMPLE 8.7

Both water and sulfur dioxide are products from the reaction of sulfuric acid with copper as shown by the following equations.

$$2\, H_2SO_4 + Cu \longrightarrow SO_2 + 2\, H_2O + CuSO_4$$

How many grams of H_2O will be produced when 5 billion (5.00×10^9) SO_2 molecules are produced?

SOLUTION

Although a calculation of this type will not have a lot of practical significance, it will test your understanding of the problem-solving relationships under discussion in this section of the text.

Step 1. We are given a certain number of particles (molecules) and asked to find the number of grams of a related substance.

$$5.00 \times 10^9 \text{ molecules } SO_2 = ?\text{ g } H_2O$$

Step 2. This is a particle-to-gram problem. Even though SO_2 and H_2O are both products, we can still work this problem in a manner similar to previous problems. The coefficients in a balanced equation relate reactants to products, reactants to reactants and *products to products*. The pathway for this problem (see Fig. 8-1) is

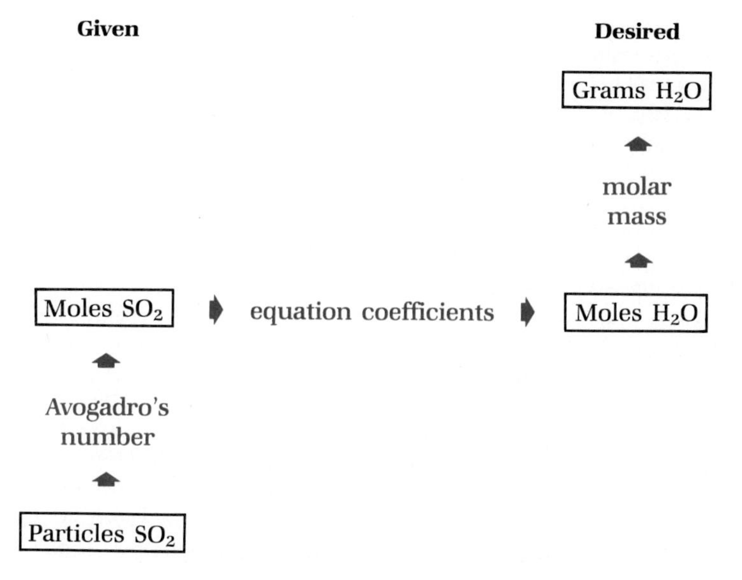

Step 3. The dimensional-analysis setup is

$$5.00 \times 10^9 \text{ molecules } SO_2 \times \left(\frac{1 \text{ mole } SO_2}{6.02 \times 10^{23} \text{ molecules } SO_2}\right) \times \left(\frac{2 \text{ moles } H_2O}{1 \text{ mole } SO_2}\right) \times \left(\frac{18.0 \text{ g } H_2O}{1 \text{ mole } H_2O}\right)$$

molecules SO_2 ⟶ moles SO_2 ⟶ moles H_2O ⟶ g H_2O

Step 4. The solution, obtained by combining all of the numerical factors in the setup, is

$$\left(\frac{(5.00 \times 10^9) \times 1 \times 2 \times 18.0}{(6.02 \times 10^{23}) \times 1 \times 1}\right) \text{ g } H_2O = 2.9900332 \times 10^{-13} \text{ g } H_2O \quad \text{(calculator answer)}$$

$$= 2.99 \times 10^{-13} \text{ g } H_2O \quad \text{(correct answer)}$$

8.7 The Limiting Reactant Concept

When a chemical reaction is carried out in a laboratory or industrial setting, the reactants are not usually present in the exact molar ratios specified in the balanced chemical equation for the reaction. Most often, on purpose, excess quantities of one or more of the reactants are present.

Numerous reasons exist for having some reactants present in excess. Sometimes such a procedure will cause a reaction to occur more rapidly. For example, large amounts of oxygen make combustible materials burn faster. Sometimes an excess of one reactant will insure that another reactant, perhaps a very expensive one, is completely consumed. (Reactions do not always go to completion in the way that theory predicts they should, that is, the reactants are not completely converted to products—for reasons to be discussed in Chapter 16.)

When one or more reactants are present in excess, the excess will not react because there is not enough of the other reactant to react with it. The reactant *not* in excess, thus, limits the amount of product(s) formed and is called the limiting reactant. The **limiting reactant** *is the reactant completely consumed by a reaction with an excess of other reactants.*

limiting reactant

The concept of a limiting reactant plays a major role in chemical calculations of certain types. It must be thoroughly understood. Let us consider some simple but analogous nonchemical examples of a "limiting reactant" before going to sample limiting reactant calculations.

Suppose we have a vending machine that contains forty 25¢ candy bars and that we have 27 quarters. In this case only 27 candy bars may be purchased. The quarters are the limiting reactant. The candy bars are present in excess. Suppose we have ten slices of cheese and eighteen slices of bread and we want to make as many cheese sandwiches as possible using one slice of cheese and two slices of bread per sandwich. The eighteen slices of bread limit us to nine sandwiches; one slice of cheese is left over. Bread is the limiting reactant in this case. A more complicated example of a limiting reactant involves the assembling of an eight-page "hand-out" laboratory experiment, as illustrated in Figure 8-2. It should be apparent to you, after study of this figure, that the graph paper limits the number of complete handouts to nineteen.

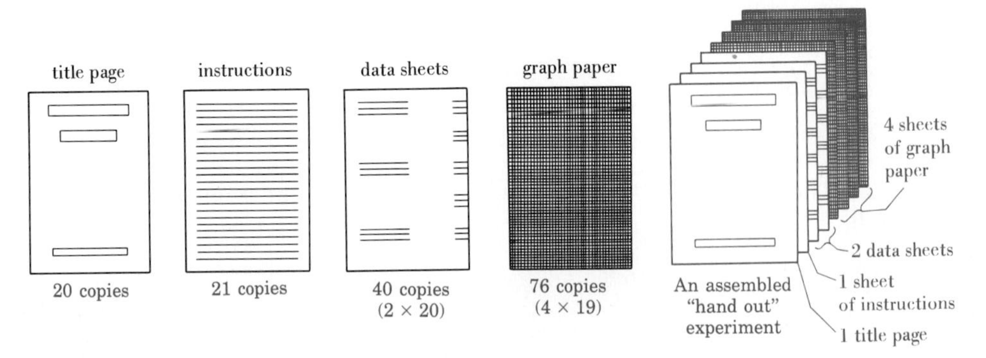

Figure 8-2. Assembling a "hand out" experiment—an analogy to the limiting reactant concept in a chemical reaction. [Adapted from *General Chemistry*, 3rd Edition, by Ralph H. Petrucci. New York: Macmillan Publishing Co., Inc., 1982.]

Now let us proceed to chemical calculations that involve a limiting reactant. *Whenever you are given the quantities of two or more reactants in a chemical reaction, it is necessary for you to determine which of the given quantities is the limiting reactant.* Determining the limiting reactant can be accomplished by the following procedure.

1. Determine the number of *moles* of each of the reactants present.
2. Calculate the number of moles of product *each* of the molar amounts of reactant would produce if it were the only reactant amount given. If more than one product is formed in the reaction, you need do this mole calculation for only one of the products.
3. The reactant that produces the *least number* of moles of product is the limiting reactant.

EXAMPLE 8.8

Silver tarnishes in the presence of hydrogen sulfide (H_2S), a gas that originates from the decay of food, because of the reaction

$$4\,Ag + 2\,H_2S + O_2 \longrightarrow 2\,Ag_2S + 2\,H_2O$$

The black product Ag_2S is the "tarnish." If 25.00 g of Ag, 5.00 g of H_2S, and 4.00 g of O_2 are present in a reaction mixture, which is the limiting reactant?

SOLUTION

Step 1. We first determine how many moles of each reactant are present. This will involve simple gram-to-mole calculations.

Grams reactant ➧ molar mass ➧ Moles reactant

For Ag:

$$25.00\,\cancel{g\,Ag} \times \left(\frac{1\text{ mole Ag}}{107.9\,\cancel{g\,Ag}}\right) = 0.23169601\text{ mole Ag} \quad \text{(calculator answer)}$$

$$= 0.2317\text{ mole Ag} \quad \text{(correct answer)}$$

For H_2S:

$$5.00\,\cancel{g\,H_2S} \times \left(\frac{1\text{ mole }H_2S}{34.1\,\cancel{g\,H_2S}}\right) = 0.14662757\text{ mole }H_2S \quad \text{(calculator answer)}$$

$$= 0.147\text{ mole }H_2S \quad \text{(correct answer)}$$

For O_2:

$$4.00\,\cancel{g\,O_2} \times \left(\frac{1\text{ mole }O_2}{32.0\,\cancel{g\,O_2}}\right) = 0.125\text{ mole }O_2$$

(calculator and correct answer)

Step 2. Next, we determine how many moles of product we can get from each of the previously determined number of moles of reactant. In this particular problem we have two products: Ag_2S and H_2O. It is sufficient to calculate the moles of either Ag_2S or H_2O. The decision as to which one is arbitrary; we will choose H_2O. The calculation is a mole-to-mole calculation that involves the use of the coefficients in the balanced equation.

Moles reactant → equation coefficients → Moles H_2O

For Ag: $0.2317 \text{ mole Ag} \times \left(\frac{2 \text{ moles } H_2O}{4 \text{ moles Ag}}\right) = 0.11585 \text{ mole } H_2O$ (calculator answer)

$= 0.1158 \text{ mole } H_2O$ (correct answer)

For H_2S: $0.147 \text{ mole } H_2S \times \left(\frac{2 \text{ moles } H_2O}{2 \text{ moles } H_2S}\right) = 0.147 \text{ mole } H_2O$ (calculator and correct answer)

For O_2: $0.125 \text{ mole } O_2 \times \left(\frac{2 \text{ moles } H_2O}{1 \text{ mole } O_2}\right) = 0.25 \text{ mole } H_2O$ (calculator answer)

$= 0.250 \text{ mole } H_2O$ (correct answer)

Step 3. The limiting reactant is the reactant that will produce the least number of moles of H_2O. Looking at the numbers just calculated we see that Ag will be the limiting reactant.

In determining the limiting reactant in this problem we broke the process into two calculational steps. We could also have worked the problem in a continuous sequence, taking the grams of each reactant and finding directly the moles of product H_2O.

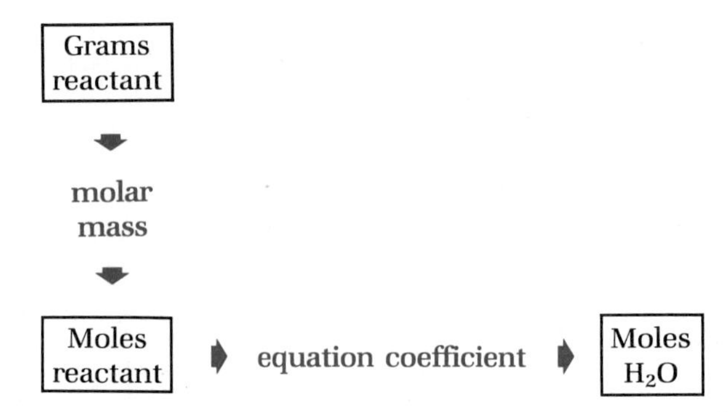

For Ag the continuous setup would be

$$25.00 \text{ g Ag} \times \left(\frac{1 \text{ mole Ag}}{107.9 \text{ g Ag}}\right) \times \left(\frac{2 \text{ moles } H_2O}{4 \text{ moles Ag}}\right) = 0.11584801 \text{ mole } H_2O$$

(calculator answer)

$$= 0.1158 \text{ mole } H_2O$$

(correct answer)

Once the limiting reactant has been determined, the amount of that reactant present becomes the starting point for any further calculations about the chemical reaction under consideration. Example 8.9 illustrates this point.

EXAMPLE 8.9

Ammonia (NH_3) reacts with oxygen as shown in the following equation.

$$4\,NH_3 + 3\,O_2 \longrightarrow 2\,N_2 + 6\,H_2O$$

How many grams of H_2O may be formed from a reaction mixture containing 35.0 g of NH_3 and 50.0 g of O_2?

SOLUTION

First, we must determine the limiting reactant since specific amounts of both reactants are given in the problem. We will use a continuous, rather than two-step, calculation to obtain the moles of product H_2O each of the reactants will form if it were the only reactant amount given.

Grams reactant → molar mass → Moles reactant → equation coefficients → Moles H_2O

For NH_3:

$$35.0\,\text{g}\,\cancel{NH_3} \times \left(\frac{1\,\cancel{\text{mole}\,NH_3}}{17.0\,\cancel{\text{g}\,NH_3}}\right) \times \left(\frac{6\,\text{moles}\,H_2O}{4\,\cancel{\text{moles}\,NH_3}}\right) = 3.0882353\,\text{moles}\,H_2O$$

(calculator answer)

$= 3.09$ moles H_2O (correct answer)

For O_2:

$$50.0\,\text{g}\,\cancel{O_2} \times \left(\frac{1\,\cancel{\text{mole}\,O_2}}{32.0\,\cancel{\text{g}\,O_2}}\right) \times \left(\frac{6\,\text{moles}\,H_2O}{3\,\cancel{\text{moles}\,O_2}}\right) = 3.125\,\text{moles}\,H_2O$$

(calculator answer)

$= 3.12$ moles H_2O (correct answer)

Thus, NH_3 is the limiting reactant since fewer moles of H_2O can be produced from it (3.09 moles) than from the O_2 (3.12 moles).

We can now calculate the grams of H_2O formed in the reaction using the 3.09 moles of H_2O (formed from our limiting reactant) as our starting factor. This calculation will be a simple one-step mole-to-gram calculation.

Moles H_2O → molar mass → Grams H_2O

$$3.09\,\cancel{\text{moles}\,H_2O} \times \left(\frac{18.0\,\text{g}\,H_2O}{1\,\cancel{\text{mole}\,H_2O}}\right) = 55.62\,\text{g}\,H_2O$$

(calculator answer)

$= 55.6$ g H_2O (correct answer)

8.8 Yields: Theoretical, Actual, and Percent

The amount of product isolated in a pure form from a chemical reaction is almost always less than the amount predicted using a limiting reactant calculation. A number of factors contribute to this situation. Some product is almost always lost in the process of its isolation and purification and in such mechanical operations as transferring materials from one container to another. Also, in many reactions *side reactions* occur that lead to the formation of small amounts of extraneous products. Reactants consumed in these side reactions obviously will not end up in the form of the desired product. The net effect of all of this is that the actual quantities of product isolated are less, sometimes far less, than the theoretically possible amount.

Product loss is specified in terms of *percent yield,* a measure of the ratio between the actual yield and the theoretical yield.

percent yield

$$\textbf{Percent yield} = \frac{\text{actual yield}}{\text{theoretical yield}} \times 100$$

actual yield

theoretical yield

The **actual yield** *is the amount of product actually obtained at the end of the experiment.* It is always an experimentally determined number; it cannot be calculated. The **theoretical yield** *is the maximum amount of product that can be produced from the starting amounts of reactants if no losses of any kind occurred.* It is always a calculated number obtained from a limiting reactant calculation.

EXAMPLE 8.10

A mixture of 80.0 g of Cr_2O_3 and 8.00 g of C is used to produce elemental Cr by the reaction

$$Cr_2O_3 + 3\,C \longrightarrow 2\,Cr + 3\,CO$$

(a) What is the theoretical yield of Cr that can be obtained from the reaction mixture?
(b) The actual yield is 21.7 g Cr. What is the percent yield for the reaction?

SOLUTION

(a) The limiting reactant must be determined before the theoretical yield can be calculated. Recalling the procedures of Examples 8.8 and 8.9 for determining the limiting reactant, we calculate the number of moles of Cr that can be produced from each individual reactant amount using gram-to-mole type calculations.

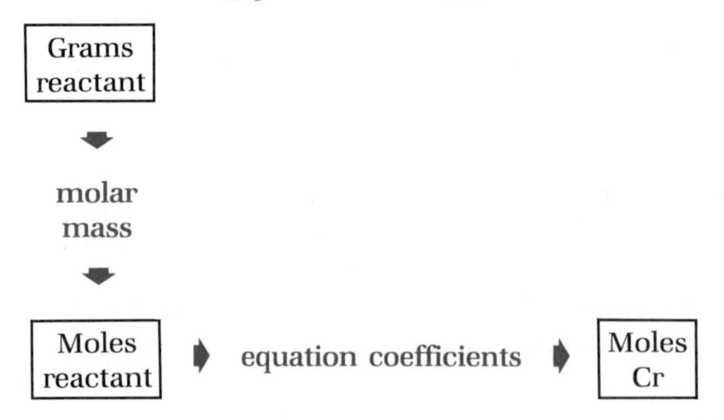

For Cr_2O_3:

$$80.0 \text{ g } \cancel{Cr_2O_3} \times \left(\frac{1 \text{ } \cancel{\text{mole } Cr_2O_3}}{152.0 \text{ g } \cancel{Cr_2O_3}}\right) \times \left(\frac{2 \text{ moles Cr}}{1 \text{ } \cancel{\text{mole } Cr_2O_3}}\right) = 1.0526316 \text{ moles Cr}$$

(calculator answer)

$= 1.05$ moles Cr

(correct answer)

For C:

$$8.00 \text{ g } \cancel{C} \times \left(\frac{1 \text{ } \cancel{\text{mole C}}}{12.0 \text{ g } \cancel{C}}\right) \times \left(\frac{2 \text{ moles Cr}}{3 \text{ } \cancel{\text{moles C}}}\right) = 0.44444444 \text{ mole Cr}$$

(calculator answer)

$= 0.444$ mole Cr

(correct answer)

Our calculations show that C is the limiting reactant.

The maximum number of grams of Cr obtainable from the limiting reactant, that is, the theoretical yield, can now be calculated. It is done using a one-step mole-to-gram setup.

Moles Cr → molar mass → Grams Cr

$$0.444 \text{ } \cancel{\text{mole Cr}} \times \left(\frac{52.0 \text{ g Cr}}{1 \text{ } \cancel{\text{mole Cr}}}\right) = 23.088 \text{ g Cr} \quad \text{(calculator answer)}$$

$= 23.1$ g Cr (correct answer)

(b) The percent yield is obtained by dividing the actual yield by the theoretical yield and multiplying by 100.

$$\text{Percent yield} = \frac{\text{actual yield}}{\text{theoretical yield}} \times 100 = \frac{21.7 \text{ } \cancel{\text{g Cr}}}{23.1 \text{ } \cancel{\text{g Cr}}} \times 100 = 93.939394\%$$

(calculator answer)

$= 93.9\%$

(correct answer)

8.9 Simultaneous and Consecutive Reactions (optional)

The concepts presented in previous sections of this chapter are easily adaptable to problem-solving situations that involve two or more chemical reactions. In some cases the two or more chemical reactions occur simultaneously (at the same time) and in other cases they occur consecutively (one right after the other). Example 8.11 deals with a pair of simultaneous reactions, and Example 8.12 deals with three consecutive reactions.

EXAMPLE 8.11

A mixture of gaseous fuel has the composition 83.1% methane (CH_4) and 16.9% ethane (C_2H_6). The combustion of this mixture produces CO_2 and H_2O as the only products. The combustion equations are

$$CH_4 + 2\,O_2 \longrightarrow CO_2 + 2\,H_2O$$

$$2\,C_2H_6 + 7\,O_2 \longrightarrow 4\,CO_2 + 6\,H_2O$$

How many moles of CO_2 would be produced from the combustion of 72.0 g of this gaseous fuel mixture?

SOLUTION

In solving this problem we will need to carry out two parallel calculations. In the one calculation we will determine the moles of CO_2 produced from the CH_4 component of the mixture and in the other the moles of CO_2 produced from the C_2H_6 component. Then we will add together the answers from the two parallel calculations to get our final answer, the total moles of CO_2 produced. The sequence of conversion factors for each setup is derived from the following pathway.

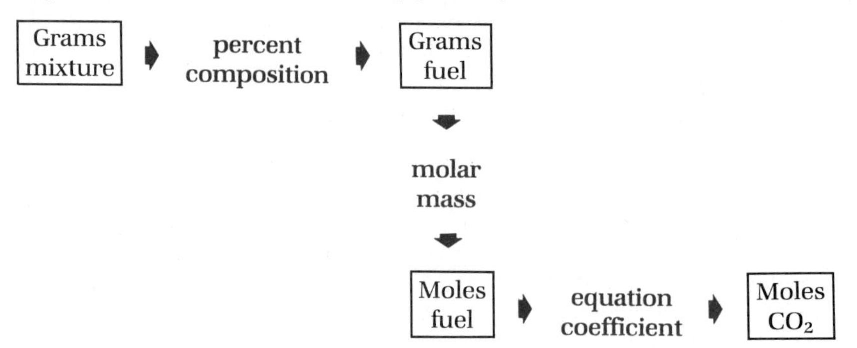

The moles of CO_2 produced from the CH_4 is given by the following setup.

$$72.0\ \cancel{\text{g mixture}} \times \left(\frac{83.1\ \cancel{\text{g CH}_4}}{100\ \cancel{\text{g mixture}}}\right) \times \left(\frac{1\ \cancel{\text{mole CH}_4}}{16.0\ \cancel{\text{g CH}_4}}\right) \times \left(\frac{1\ \text{mole CO}_2}{1\ \cancel{\text{mole CH}_4}}\right)$$

$$= 3.7395 \text{ moles } CO_2 \quad \text{(calculator answer)}$$

$$= 3.74 \text{ moles } CO_2 \quad \text{(correct answer)}$$

The moles of CO_2 produced from the C_2H_6 is given by the following parallel setup.

$$72.0\ \cancel{\text{g mixture}} \times \left(\frac{16.9\ \cancel{\text{g C}_2\text{H}_6}}{100\ \cancel{\text{g mixture}}}\right) \times \left(\frac{1\ \cancel{\text{mole C}_2\text{H}_6}}{30.0\ \cancel{\text{g C}_2\text{H}_6}}\right) \times \left(\frac{4\ \text{moles CO}_2}{2\ \cancel{\text{moles C}_2\text{H}_6}}\right)$$

$$= 0.8112 \text{ mole } CO_2 \quad \text{(calculator answer)}$$

$$= 0.811 \text{ mole } CO_2 \quad \text{(correct answer)}$$

The first conversion factor in each setup is derived from the given percentage of that compound in the fuel mixture. Use of percentages as conversion factors was covered in Section 3.4.

The total moles of CO_2 produced is the sum of the moles of CO_2 in the individual reactions.

$$3.74 \text{ moles } CO_2 + 0.811 \text{ mole } CO_2 = 4.551 \text{ mole } CO_2 \quad \text{(calculator answer)}$$

$$= 4.55 \text{ moles } CO_2 \quad \text{(correct answer)}$$

EXAMPLE 8.12

An older method for preparation of nitric acid (HNO_3) involves the following sequence of three reactions.

$$\text{Reaction (1):} \quad 4\,NH_3 + 5\,O_2 \longrightarrow 4\,NO + 6\,H_2O$$

$$\text{Reaction (2):} \quad 2\,NO + O_2 \longrightarrow 2\,NO_2$$

$$\text{Reaction (3):} \quad 3\,NO_2 + H_2O \longrightarrow 2\,HNO_3 + NO$$

Assuming an excess of O_2 and H_2O as reactants, how many grams of HNO_3 can be produced from 244 g NH_3?

SOLUTION

The key substances in this set of reactions, from a calculational point of view, are the nitrogen-containing species, NH_3, NO, NO_2, and HNO_3. Note that the nitrogen-containing species produced in the first and second reactions (NO and NO_2, respectively) are the reactants for the second and third reactions, respectively.

$$NH_3 \xrightarrow[\text{reaction (1)}]{} NO \xrightarrow[\text{reaction (2)}]{} NO_2 \xrightarrow[\text{reaction (3)}]{} HNO_2$$

We can solve this problem using a single multiple-step setup. The sequence of conversion factors needed is that for a gram-to-gram problem with two additional intermediate mole-to-mole steps added.

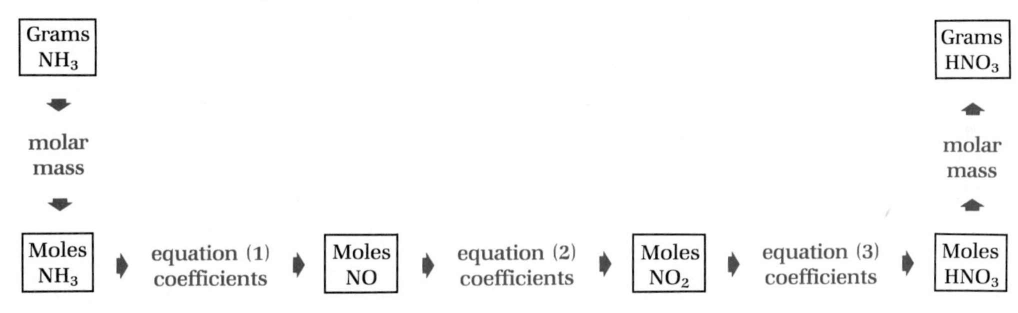

The dimensional-analysis setup is

$$244 \text{ g } NH_3 \times \left(\frac{1 \text{ mole } NH_3}{17.0 \text{ g } NH_3}\right) \times \left(\frac{4 \text{ moles } NO}{4 \text{ moles } NH_3}\right) \times \left(\frac{2 \text{ moles } NO_2}{2 \text{ moles } NO}\right)$$

grams NH_3 ⟶ moles NH_3 ⟶ moles NO ⟶ moles NO_2

$$\times \left(\frac{2 \text{ moles } HNO_3}{3 \text{ moles } NO_2}\right) \times \left(\frac{63.0 \text{ g } HNO_3}{1 \text{ mole } HNO_3}\right)$$

⟶ moles HNO_3 ⟶ g HNO_3

$= 602.82353 \text{ g } HNO_3$ (calculator answer)

$= 603 \text{ g } HNO_3$ (correct answer)

An alternate approach to solving this problem would involve setting up a separate calculation for each of the equations. As a first step, the number of moles of NO produced in the first reaction would be calculated. In the second step, one would determine the moles of NO_2 obtained if all of the NO produced in the first reaction entered into the second reaction. In the final step, one would determine the grams of HNO_3 derivable from the NO_2 produced in the second reaction. The answer obtained from this three-step method is the same as that obtained from the multiple-step method.

Learning Objectives

After completing this chapter you should be able to

- State the law of conservation of mass and describe how atomic theory accounts for this law (Sec. 8.1).
- Understand the conventions used in writing chemical equations (Secs. 8.2 and 8.4).
- Recognize whether an equation is balanced or not, and balance a chemical equation given the formulas of all the reactants and products (Sec. 8.3).
- Interpret coefficients in balanced chemical equations in terms of moles (Sec. 8.5).
- Use a balanced chemical equation and other appropriate information to calculate the quantities of reactants consumed and/or products produced in a chemical reaction (Sec. 8.6).
- Determine the limiting reactant in a chemical reaction given the mass of two or more reactants and the balanced chemical equation for the reaction (Sec. 8.7).
- Calculate the percent yield in a chemical reaction from appropriate data (Sec. 8.8).
- Calculate from appropriate data quantities of reactants consumed or products produced in chemical systems involving two or more simultaneous reactions or a series of consecutive reactions (Sec. 8.9) [optional].

Terms and Concepts for Review

The new terms or concepts defined in this chapter are

law of conservation of mass (Sec. 8.1)
chemical equation (Sec. 8.2)
balanced chemical equation (Sec. 8.3)
coefficient (Sec. 8.3)
molar interpretation of coefficients (Sec. 8.5)
limiting reactant (Sec. 8.7)
actual yield (Sec. 8.8)
theoretical yield (Sec. 8.8)
percent yield (Sec. 8.8)

Questions and Problems

The Law of Conservation of Mass

8-1 Consider a hypothetical reaction in which A and B are reactants and C and D are products. If 10 g of A completely reacts with 7 g of B to produce 14 g of C, how many g of D will be produced?

Chemical Equations

8-2 The formulas used in equations for gaseous elements should reflect their molecular makeup. Which gases (or vapors) should be written as
- **a.** monoatomic molecules
- **b.** diatomic molecules
- **c.** tetraatomic molecules

8-3 What do the symbols in parentheses stand for in the following balanced equations?
- **a.** $Mg(s) + 2\,HCl(aq) \longrightarrow MgCl_2(aq) + H_2(g)$
- **b.** $SO_2(g) + H_2O(l) \longrightarrow H_2SO_3(aq)$
- **c.** $PbO_2(s) + 2\,H_2(g) \longrightarrow Pb(s) + 2\,H_2O(l)$

8-4 What symbol is used in a chemical equation to represent the phrase "to produce"?

8-5 What is meant by a *balanced* chemical equation?

8-6 Identify each of the following equations as *balanced* or *unbalanced.*
- **a.** $CaCO_3 \longrightarrow CaO + CO_2$
- **b.** $2\,KClO_3 \longrightarrow 2\,KCl + 3\,O_2$
- **c.** $2\,Na + 2\,H_2O \longrightarrow 2\,NaOH + H_2$
- **d.** $2\,C_3H_8O + 8\,O_2 \longrightarrow 6\,CO_2 + 8\,H_2O$
- **e.** $2\,(NH_4)_2S + Pb(NO_3)_2 \longrightarrow PbS + 2\,NH_4NO_3$
- **f.** $2\,HNO_3 + Ca(OH)_2 \longrightarrow Ca(NO_3)_2 + 2\,H_2O$

8-7 Balance the following equations.
- **a.** $N_2 + O_2 \longrightarrow NO$
- **b.** $SO_2 + O_2 \longrightarrow SO_3$
- **c.** $H_2O \longrightarrow H_2 + O_2$
- **d.** $TiCl_4 + H_2O \longrightarrow TiO_2 + HCl$
- **e.** $Al + O_2 \longrightarrow Al_2O_3$
- **f.** $HNO_2 \longrightarrow N_2O_3 + H_2O$

8-8 Balance the following equations.
- **a.** $CH_4 + O_2 \longrightarrow CO_2 + H_2O$
- **b.** $C_2H_6 + O_2 \longrightarrow CO_2 + H_2O$
- **c.** $C_3H_8 + O_2 \longrightarrow CO_2 + H_2O$
- **d.** $C_4H_{10} + O_2 \longrightarrow CO_2 + H_2O$
- **e.** $C_5H_{12} + O_2 \longrightarrow CO_2 + H_2O$
- **f.** $C_6H_{14} + O_2 \longrightarrow CO_2 + H_2O$

8-9 Balance the following equations.
- **a.** $P_4O_{10} + H_2O \longrightarrow H_3PO_4$
- **b.** $Fe_2O_3 + C \longrightarrow Fe + CO_2$
- **c.** $Na_2NH + H_2O \longrightarrow NH_3 + NaOH$
- **d.** $NH_3 + O_2 \longrightarrow NO + H_2O$
- **e.** $Cl_2 + H_2O \longrightarrow HCl + HClO$
- **f.** $Na_2O_2 + H_2O \longrightarrow O_2 + NaOH$

8-10 Balance the following equations.
- **a.** $NaHCO_3 + H_2SO_4 \longrightarrow Na_2SO_4 + H_2O + CO_2$
- **b.** $PbO + NH_3 \longrightarrow Pb + N_2 + H_2O$
- **c.** $KClO_3 + HCl \longrightarrow KCl + ClO_2 + Cl_2 + H_2O$
- **d.** $CaC_2 + H_2O \longrightarrow C_2H_2 + Ca(OH)_2$
- **e.** $C_8H_{18}O_4 + O_2 \longrightarrow CO_2 + H_2O$
- **f.** $SO_2Cl_2 + HI \longrightarrow H_2S + H_2O + HCl + I_2$

8-11 Balance the following equations.
- **a.** $Ca(OH)_2 + HNO_3 \longrightarrow Ca(NO_3)_2 + H_2O$
- **b.** $BaCl_2 + (NH_4)_2SO_4 \longrightarrow BaSO_4 + NH_4Cl$
- **c.** $Fe(OH)_3 + H_2SO_4 \longrightarrow Fe_2(SO_4)_3 + H_2O$
- **d.** $Al + Sn(NO_3)_2 \longrightarrow Al(NO_3)_3 + Sn$
- **e.** $Na_2CO_3 + Mg(NO_3)_2 \longrightarrow MgCO_3 + NaNO_3$
- **f.** $Al(NO_3)_3 + H_2SO_4 \longrightarrow Al_2(SO_4)_3 + HNO_3$

Chemical Equations and the Mole Concept

8-12 Interpret the following balanced chemical equations in terms of the numbers of moles of reactants and products.
- **a.** $H_2O_2 + H_2S \longrightarrow 2\,H_2O + S$
- **b.** $4\,NH_3 + 3\,O_2 \longrightarrow 2\,N_2 + 6\,H_2O$

8-13 Write the twelve mole-to-mole conversion factors that can be derived from each of the following balanced equations.
- **a.** $3\,HNO_2 \longrightarrow 2\,NO + HNO_3 + H_2O$
- **b.** $N_2H_4 + 2\,H_2O_2 \longrightarrow N_2 + 4\,H_2O$

8-14 How many moles of the first listed product in each of the following equations could be obtained by reacting 0.750 mole of the first listed reactant with an excess of the other reactant?
- **a.** $H_2O_2 + H_2S \longrightarrow 2\,H_2O + S$
- **b.** $4\,NH_3 + 3\,O_2 \longrightarrow 2\,N_2 + 6\,H_2O$
- **c.** $Mg + 2\,HCl \longrightarrow MgCl_2 + H_2$
- **d.** $6\,HCl + 2\,Al \longrightarrow 3\,H_2 + 2\,AlCl_3$

8-15 Given the equation

$$4\,NH_3(g) + 5\,O_2(g) \longrightarrow 4\,NO(g) + 6\,H_2O(g)$$

answer the following.
- **a.** How many moles of O_2 are needed to produce 1.34 moles of NO?
- **b.** How many moles of H_2O will be produced from 0.789 mole of NH_3?
- **c.** How many moles of NH_3 are needed to react with 3.22 moles of O_2?
- **d.** How many moles of NO are produced when 0.763 mole of H_2O is produced?

Quantities of Reactants and Products

8-16 How many grams of the first reactant in each of the following equations would be needed to produce 3.00 moles of N_2?
- **a.** $4\,NH_3 + 3\,O_2 \longrightarrow 2\,N_2 + 6\,H_2O$
- **b.** $(NH_4)_2Cr_2O_7 \longrightarrow Cr_2O_3 + N_2 + 4\,H_2O$
- **c.** $N_2H_4 + 2\,H_2O_2 \longrightarrow N_2 + 4\,H_2O$
- **d.** $4\,C_3H_5O_9N_3 \longrightarrow 12\,CO_2 + 6\,N_2 + O_2 + 10\,H_2O$

8-17 How many grams of the first reactant in each of the following equations would be needed to produce 3.73 g of CO_2?
- **a.** $CS_2 + 3\,O_2 \longrightarrow CO_2 + 2\,SO_2$
- **b.** $CaCO_3 \longrightarrow CaO + CO_2$
- **c.** $Fe_3O_4 + CO \longrightarrow 3\,FeO + CO_2$
- **d.** $2\,C_8H_{18} + 25\,O_2 \longrightarrow 16\,CO_2 + 18\,H_2O$

8-18 A common method for producing oxygen gas (O_2) in the laboratory involves decomposing potassium chlorate ($KClO_3$) as shown by the following equation.

$$2\,KClO_3(s) \longrightarrow 2\,KCl(s) + 3\,O_2(g)$$

Based on this equation, how many grams of $KClO_3$ must be decomposed to produce
- **a.** 3.00 moles O_2
- **b.** 3.00 g O_2
- **c.** 2937 molecules O_2
- **d.** 7.27 moles KCl
- **e.** 0.627 g KCl
- **f.** 1.23×10^{20} formula units KCl

8-19 A mixture of hydrazine (N_2H_4) and hydrogen peroxide (H_2O_2) is used as a fuel for rocket engines. These two substances react with each other as shown by the equation

$$N_2H_4(l) + 2\,H_2O_2(l) \longrightarrow N_2(g) + 4\,H_2O(g)$$

- **a.** How many grams of N_2H_4 are needed to react with 0.453 mole of H_2O_2?
- **b.** How many grams of N_2 are obtained when 1.37 moles of N_2H_4 react?
- **c.** How many grams of H_2O are produced when 0.317 mole of N_2 is produced?
- **d.** How many grams of H_2O_2 must react in order to produce 19.6 moles of H_2O?

8-20 One way to remove gaseous carbon dioxide (CO_2) from the air in a spacecraft is to let cannisters of solid lithium hydroxide (LiOH) absorb it according to the following reaction.

$$CO_2(g) + 2\,LiOH(s) \longrightarrow Li_2CO_3(s) + H_2O(l)$$

a. How many grams of LiOH are required to absorb an astronaut's daily output of 925 g of CO_2 produced from breathing?
b. How many grams of by-product water are produced per day as a result of the reaction of the CO_2 from the astronaut's breathing with LiOH?

8-21 Pure silver metal results when silver carbonate is decomposed by heating as shown by the following equation.

$$2\,Ag_2CO_3(s) \longrightarrow 4\,Ag(s) + 2\,CO_2(g) + O_2(g)$$

a. How many moles of Ag_2CO_3 must be decomposed to produce 100.0 g of Ag?
b. How many grams of Ag_2CO_3 must be decomposed to produce 100.0 g of CO_2?
c. How many formula units of Ag_2CO_3 must be decomposed to produce 100.0 g of O_2?
d. How many grams of Ag_2CO_3 must be decomposed to produce 1 million (1.00×10^6) atoms of Ag?

Limiting Reactant

8-22 Magnesium nitride may be prepared by the direct reaction of the elements as shown by the equation

$$3\,Mg(s) + N_2(g) \longrightarrow Mg_3N_2(s)$$

For each of the following combinations of reactants, decide which is the limiting reactant.

a. 2.00 moles Mg, 0.500 mole N_2
b. 4.00 moles Mg, 4.00 moles N_2
c. 3.00 moles Mg, 1.25 moles N_2
d. 3.00 g Mg, 0.100 mole N_2
e. 3.00 g Mg, 3.00 g N_2
f. 20.00 g Mg, 7.00 g N_2

8-23 Under appropriate conditions nitrogen will react with hydrogen to produce ammonia as shown by the following equation.

$$N_2(g) + 3\,H_2(g) \longrightarrow 2\,NH_3(g)$$

How many grams of ammonia can be produced from the following amounts of reactants?

a. 3.0 g N_2, 5.0 g H_2 b. 30.0 g N_2, 10.0 g H_2
c. 50.0 g N_2, 8.00 g H_2 d. 56 g N_2, 12 g H_2

8-24 Sulfur dioxide (SO_2) can be produced from the reaction of hydrogen sulfide (H_2S) and oxygen as shown by the following equation.

$$2\,H_2S + 3\,O_2 \longrightarrow 2\,SO_2 + 2\,H_2O$$

a. How many grams of SO_2 can be produced from 70.0 g of H_2S and 125 g of O_2?
b. How many grams of excess reactant is left over after the reaction is complete?

8-25 The compound sodium cyanide, NaCN, can be prepared by the reaction

$$Na_2CO_3 + 4\,C + N_2 \longrightarrow 2\,NaCN + 3\,CO$$

If a mixture of 2.00 moles each of Na_2CO_3, C, and N_2 is allowed to react, how many grams of each reactant and product are present at the completion of the reaction?

Theoretical and Percent Yield

8-26 Aluminum and sulfur react at elevated temperatures to form aluminum sulfide as shown by the following equation.

$$2\,Al(s) + 3\,S(s) \longrightarrow Al_2S_3(s)$$

In a certain experiment, 125 g of Al_2S_3 was produced from 75.0 g of Al and an excess of S.

a. What is the theoretical yield of Al_2S_3?
b. What is the percent yield of Al_2S_3?

8-27 Chlorine gas is produced commercially by electrolyzing salt water solutions as shown by the following equation.

$$2\,NaCl(aq) \longrightarrow 2\,Na(s) + Cl_2(g)$$

If electrolysis of 2537 g of NaCl produces 1237 g of Cl_2, what is the percent yield?

8-28 Under appropriate conditions Zn and S react to produce ZnS as shown by the following equation.

$$Zn + S \longrightarrow ZnS$$

In a certain experiment using 30.7 g of Zn and an excess of S, a percent yield of 93.7% ZnS is obtained. What is the actual yield of ZnS in grams?

Simultaneous Reactions (optional)

8-29 A mixture of composition 47.3% magnesium carbonate ($MgCO_3$) and 52.7% calcium carbonate ($CaCO_3$) is heated until the carbonates are decomposed to oxides as shown by the following equations.

$$MgCO_3 \longrightarrow MgO + CO_2$$

$$CaCO_3 \longrightarrow CaO + CO_2$$

How many grams of CO_2 are produced from decomposition of 78.3 g of mixture?

8-30 A mixture of magnesium carbonate, $MgCO_3$, and magnesium hydroxide, $Mg(OH)_2$, has the composition 10.0% $MgCO_3$ and 90.0% $Mg(OH)_2$. Both components of the mixture dissolve in hydrochloric acid (HCl), with the dissolving reactions being

$$MgCO_3 + 2\,HCl \longrightarrow MgCl_2 + H_2O + CO_2$$

$$Mg(OH)_2 + 2\,HCl \longrightarrow MgCl_2 + 2\,H_2O$$

a. How many grams of HCl are required to dissolve a 20.0 g sample of the mixture?
b. How many grams of $MgCl_2$ are produced from dissolving a 50.0 g sample of the mixture?

Consecutive Reactions (optional)

8-31 How many grams of SO_2 can be obtained from 50.0 g of $KClO_3$ by the following two-step process?

$$2\ KClO_3(s) \longrightarrow 2\ KCl(s) + 3\ O_2(g)$$

$$S_8(s) + 8\ O_2(g) \longrightarrow 8\ SO_2(g)$$

8-32 NO_2 is a reddish brown gas that is a component of "smog." It is formed in a two-step process.

$$N_2(g) + O_2(g) \longrightarrow 2\ NO(g)$$

$$2\ NO(g) + O_2(g) \longrightarrow 2\ NO_2(g)$$

a. How many grams of N_2 are required to produce 135 g of NO_2?

b. How many grams of NO_2 are produced from 10.0 g of N_2 undergoing reaction?

8-33 In steelmaking a three-step process leads to the conversion of Fe_2O_3 (the iron-containing component of iron ore) to liquid iron.

$$3\ Fe_2O_3(s) + CO(g) \longrightarrow 2\ Fe_3O_4(s) + CO_2(g)$$

$$Fe_3O_4(s) + CO(g) \longrightarrow 3\ FeO(s) + CO_2(g)$$

$$FeO(s) + CO(g) \longrightarrow Fe(l) + CO_2(g)$$

How many grams of Fe can be produced from 500.0 g of Fe_2O_3?

9 The Electronic Structure of Atoms

9.1 The Energy of an Electron

In this chapter we consider how an atom's electrons are arranged about its nucleus. This is a most important subject since it is the arrangement of electrons about the nucleus that determines an element's chemical properties, that is, its behavior toward other substances.

Information already known, from Section 5.4, about the location of electrons within atoms includes

1. There are two parts (regions) to an atom: a nuclear region at the center of the atom containing all the neutrons and protons and an extranuclear region about that center containing all the electrons.
2. The extranuclear region of an atom is an extremely large region (compared to the nuclear region), mostly empty space, in which the electrons move rapidly about the nucleus.
3. The motion of the electrons in the extranuclear region determines the volume (size) of the atom.

More specific information concerning the behavior and arrangement of electrons within the extranuclear region of an atom is derived from a complex mathematical model for electron behavior called *quantum mechanics.* A formal discussion of quantum mechanics is beyond the scope of this course (and many other chemistry courses also) because of the rigorous mathematics involved. The answers obtained from quantum mechanics are, however, simple enough to be understood to a surprisingly large degree at the level of an introductory chemistry course. A consideration of these quantum-mechanical answers will enable us to develop a system for specifying electron arrangements around a nucleus and further to develop basic rules governing compound formation. This latter topic will occupy our attention in Chapter 10.

Present day quantum-mechanical theory describes the arrangement of an atom's electrons in terms of their energies. Indeed, the energy of an electron is its most important property when considering its behavior about the nucleus.

The energy of an electron is manifested primarily in its velocity. The higher its energy, the higher its average velocity. The faster an electron travels, the farther it tends to move from the nucleus with which it is associated.

quantized property

A most significant characteristic of an electron's energy is that it is a *quantized property*. A **quantized property** *is a property that can have only certain values, that is, not all values are allowed.* Since an electron's energy is quantized, an electron may have only certain specific energies.

Quantization is a phenomenon not commonly encountered in the macroscopic world. A condition somewhat analogous to quantization is encountered in the process of a person climbing a flight of stairs. In Figure 9-1a you see six steps between ground level and the level of the entrance door. As a person climbs these stairs there are only six permament positions he or she may occupy (with both feet together). Thus the person's position (height above ground level) is quantized; only certain heights are allowed. The opposite of quantization is continuous. A person climbing a ramp up to the entrance (Fig. 9-1b) would be able to assume a continuous set of heights above ground level; all values are allowed.

The energy of an electron determines its behavior about the nucleus. Since electron energies are quantized, only certain behavior patterns are allowed. Descriptions of these behavior patterns may be given in both semiquantitative and quantitative terms. We will consider both. Semiquantitative descriptions of electron arrangements involve the use of the terms shell, subshell, and orbital. Sections 9.2–9.4 consider the meaning of these terms and the mathematical interrelationships between them. Quantitative descriptions of electron behavior involve *quantum numbers*. Section 9.6 discusses such numbers.

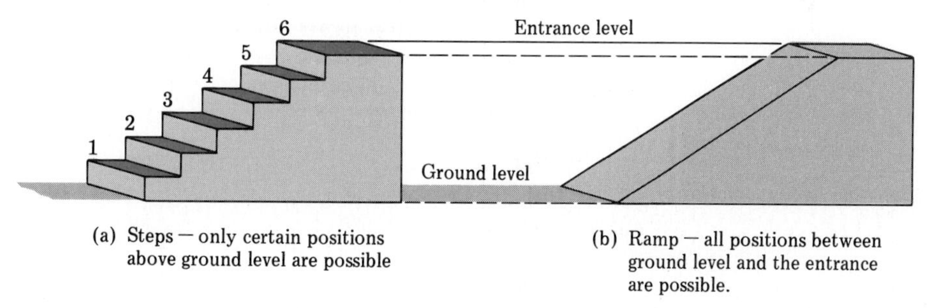

Figure 9-1. A stairway with quantized position levels versus a ramp with continuous position levels.

9.2 Electron Shells

It was mentioned in Section 9.1 that higher energy electrons will be found farther from the nucleus than lower energy electrons. Based on (energy)–(distance-from-the-nucleus) considerations, electrons may be grouped into shells or main energy levels. *A* ***shell*** *contains electrons that have approximately the same energy and that spend most of their time approximately the same distance from the nucleus.*

shell

Two different methods are used for identifying electron shells. An older method uses letters of the alphabet beginning with the letter *K* and then continuing sequentially. Shell *K* is the shell of electrons closest to the nucleus (the lowest in energy). The next closest shell is designated *L*, then come *M*, *N*, and so on. A more modern method designates shells by a number *n*, which may have the values 1, 2, 3, ... The lowest energy shell is assigned an *n* value of 1, the next higher 2, then 3, and so on. Comparing the two systems, we see that the $n = 1$ shell and the *K* shell are one and the same shell. Similarly, $n = 2$ and *L* are designations for the same shell. Only values of $n = 1$–7 (or letters *K*–*Q*) are used at present in designating electron shells. No known atom has electrons farther from the nucleus than the seventh main energy level ($n = 7$ or letter = *Q*).

The maximum number of electrons found in an electron shell varies; the higher the shell energy, the more electrons the shell may accommodate. The farther electrons are from the nucleus (a high energy shell), the greater the volume of space available for electrons; hence more electrons in the shell. (Conceptually, electron shells may be considered to be nested one inside another, somewhat analogous to the layers of flavors inside a jawbreaker or similar type of candy.)

The lowest energy shell ($n = 1$) accommodates a maximum of two electrons. In the second, third, and fourth shells, 8, 18, and 32 electrons are allowed, respectively. A very simple mathematical equation can be used to calculate the maximum number of electrons allowed in any given shell.

$$\text{Shell electron capacity} = 2n^2 \qquad (\text{where } n = \text{shell number})$$

For example, when $n = 4$ the value $2n^2 = 2(4)^2 = 32$, which is the number previously given for the number of electrons allowed in the fourth shell. Although a maximum

Table 9-1. *Important Characteristics of Electron Shells*

Shell	Number Designation (*n*)	Letter Designation	Electron Capacity ($2n^2$)
1st	1	*K*	$2 \times 1^2 = 2$
2nd	2	*L*	$2 \times 2^2 = 8$
3rd	3	*M*	$2 \times 3^2 = 18$
4th	4	*N*	$2 \times 4^2 = 32$
5th	5	*O*	$2 \times 5^2 = 50$[a]
6th	6	*P*	$2 \times 6^2 = 72$[a]
7th	7	*Q*	$2 \times 7^2 = 98$[a]

[a]The maximum number of electrons in this shell has never been attained in any element now known.

electron-occupancy level exists for each shell (group of electrons), these main energy levels may hold less than the allowable number of electrons in a given situation.

Table 9-1 summarizes concepts presented in this section concerning electron shells or electron main energy levels.

9.3 Electron Subshells

subshell

All electrons in a shell do not have the same energy. Energies are all close to each other in magnitude but not identical. The range of energies for electrons in a shell is due to the existence of electron *subshells* or electron *sub-energy levels.* A **subshell** *contains electrons that all have the same energy.*

The number of subshells within a shell varies. A shell contains the same number of subshells as its own shell number; that is,

$$\text{Number of subshells in a shell} = n \qquad (\text{where } n = \text{shell number})$$

Thus, each successive shell has one more subshell than the previous one. Shell 3 contains three subshells, shell 4 contains four subshells, and shell 5 contains five subshells.

Subshells are identified by a number and a letter. The number indicates the shell to which the subshell belongs. The letters are *s*, *p*, *d*, or *f* (all lowercase letters), which, in that order, denote subshells of increasing energy within a shell. The lowest energy subshell within a shell is always the *s* subshell, the next higher the *p* subshell, then the *d*, and finally the *f*. Shell 1 has only one subshell, the 1*s*. Shell 2 has two subshells, the 2*s* and 2*p*. The 3*s*, 3*p*, and 3*d* subshells are found in shell 3. Table 9-2 gives information concerning subshells for the first seven shells. Note that number–letter designations are not given for all the subshells in shells 5, 6, and 7, for these subshells are not needed to describe the electron arrangement for the 106 known elements. That they are not needed will become evident in Section 9.6.

The maximum number of electrons that a subshell may hold varies from two to fourteen depending on the type of subshell—*s*, *p*, *d*, or *f*. An *s* subshell can accommodate only two electrons. What shell the *s* subshell is located in does not affect the maximum electron-occupancy figure, that is, the 1*s*, 2*s*, 3*s*, 4*s*, 5*s*, 6*s*, and 7*s* subshells all have a maximum electron-occupancy figure of 2. Subshells of the *p*, *d*, and *f* types can accommodate a maximum of 6, 10, and 14 electrons, respectively. Again the maxi-

Table 9-2. Subshell Arrangements Within Shells

Shell number (*n*)	Subshells					
1	1*s*					
2	2*s*	2*p*				
3	3*s*	3*p*	3*d*			
4	4*s*	4*p*	4*d*	4*f*		
5	5*s*	5*p*	5*d*	5*f*	—	
6	6*s*	6*p*	6*d*	—	—	—
7	7*s*	—	—	—	—	—

Table 9-3. ***Distribution of Electrons Within Subshells***

Shell	Number of Subshells Within Shell	Maximum Number of Electrons Within Each Subshell				Maximum Number of Electrons Within Shell ($2n^2$)
		s	*p*	*d*	*f*	
$n = 1$ (*K*)	1	2				2
$n = 2$ (*L*)	2	2	6			8
$n = 3$ (*M*)	3	2	6	10		18
$n = 4$ (*N*)	4	2	6	10	14	32

mum number of electrons in these subshell types depends only on the subshell types and is independent of shell number. Table 9-3 summarizes the information just presented for shells 1 through 4. Notice the consistency between the numbers in columns 3 and 4 in Table 9-3. Within a shell, the sum of the subshell electron occupancies is the same as the shell electron occupancy ($2n^2$). For example, in shell 4, an *s* subshell containing 2 electrons, a *p* subshell containing 6 electrons, a *d* subshell containing 10 electrons, and an *f* subshell containing 14 electrons add up to a total of 32 electrons, which is the maximum occupancy of shell 4 as calculated by the $2n^2$ formula.

9.4 Electron Orbitals

orbital

The last and most basic of the three terms used in describing electron arrangements about nuclei is *orbital*. An **orbital** *is a region of space around a nucleus where an electron with a specific energy is most likely to be found.*

An analogy for the relationship between shells, subshells, and orbitals can be found in the physical layout of a high-rise condominium complex. A shell is the counterpart of a floor of the condominium in our analogy. Just as each floor will contain apartments (of varying size), a shell contains subshells (of varying size). Further, just as apartments contain rooms, subshells contain orbitals. An apartment is a collection of rooms; a subshell is a collection of orbitals. A floor of a condominium building is a collection of apartments; a shell is a collection of subshells.

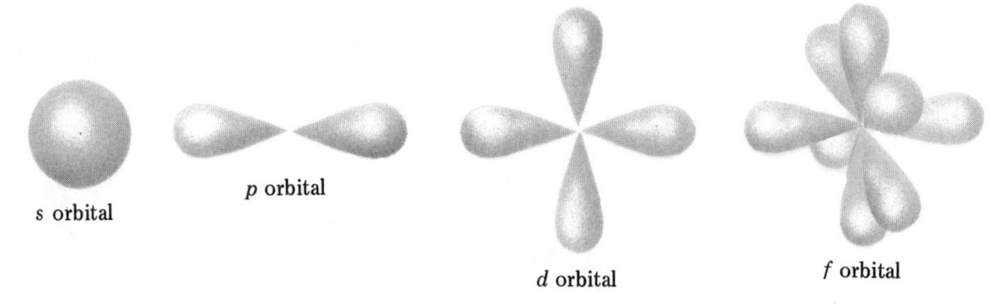

Figure 9-2. Shapes of atomic orbitals.

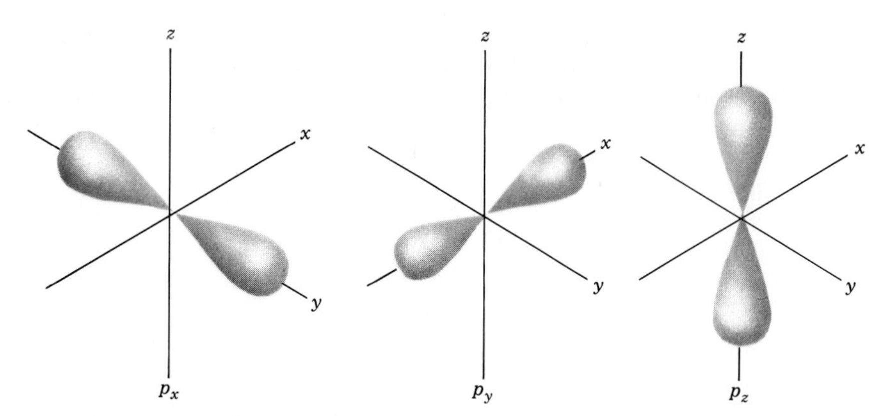

Figure 9-3. The orientation of the three orbitals in a *p* subshell.

Characteristics of orbitals, the rooms in our "electron apartment house," include

1. The number of orbitals in a subshell varies, being one for an *s* subshell, three for a *p* subshell, five for a *d* subshell, and seven for an *f* subshell.
2. The maximum number of electrons found in an orbital does not vary. It is always 2.
3. The notation used to designate orbitals is the same as that used for subshells. Thus, orbitals in the 4*f* subshell (there are seven of them) are called 4*f* orbitals.

We have already noted (Sec. 9.3) that all electrons in a subshell have the same energy. Thus all electrons in orbitals of the same subshell will have the same energy. This means *shell and subshell locations are sufficient to specify the energy of an electron.* This statement will be of great importance in the discussions of Section 9.5.

Orbitals have a definite size and shape related to the type of subshell in which they are found. (Remember, an orbital is a region of space. We are not talking about the size and shape of an electron, but rather the size and shape of a region of space where an electron is found.) At any given time an electron can be at only one point in an orbital, but because of its rapid movement throughout the orbital it "occupies" the entire orbital. An analogy would be the "definite" volume occupied by a rotating fan blade. Typical *s*, *p*, *d*, and *f* orbital shapes are given in Figure 9-2. Notice that the shapes increase in complexity in the order *s*, *p*, *d*, and *f*. Some of the more complex *d* and *f* orbitals have shapes related, but not identical, to those shown in Figure 9-2.

Orbitals within the same subshell differ mainly in orientation. For example, the three 2*p* orbitals look the same but are aligned in different directions — along the *x*, *y*, and *z* axes in a Cartesian coordinate system (see Fig. 9-3).

Orbitals of the same type found in different subshells (for example, 1*s*, 2*s*, and 3*s*) have the same general shape but differ in size (volume) (see Fig. 9-4).

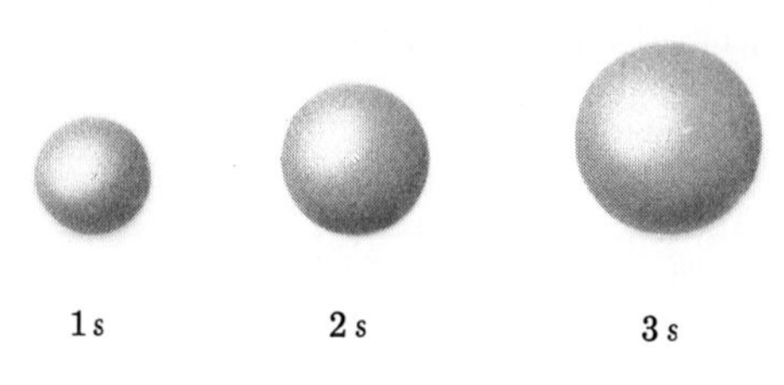

Figure 9-4. Size relationships among *s* orbitals in different subshells.

EXAMPLE 9.1

Determine the following for the fourth electron shell (fourth main electron energy level) of an atom.

(a) the number of subshells it contains
(b) the designation used to describe each subshell
(c) the number of orbitals in each subshell
(d) the maximum number of electrons that could be contained in each subshell
(e) the maximum number of electrons that could be contained in the shell

SOLUTION

(a) This whole problem deals with shell number 4. The number of subshells in a shell is the same as the shell number. Thus, shell 4 ($n = 4$) will contain four subshells.
(b) The lowest of the four subshells in terms of energy is designated 4*s*, the next higher 4*p*, the next 4*d*, and the final subshell (highest energy) is 4*f*.
(c) The number of orbitals in a given type (*s*, *p*, *d*, or *f*) of subshell is independent of the shell number. All ***s* subshells** (1*s*, 2*s*, 3*s*, etc.) contain one orbital, all *p* subshells three orbitals, all *d* subshells five orbitals, and all *f* subshells seven orbitals.
(d) The maximum number of electrons found in an orbital is always 2. Therefore the 4*s* subshell (one orbital) contains two electrons, the 4*p* subshell (three orbitals) six electrons, the 4*d* subshell (five orbitals) ten electrons, and the 4*f* subshell (seven orbitals) fourteen electrons.
(e) The maximum number of electrons in a shell is given by the formula $2n^2$, where n is the shell number. Since $n = 4$ in this problem, $2n^2 = 2 \times 4^2 = 32$. Alternatively, from part (*c*), we note that shell 4 contains 16 orbitals ($1 + 3 + 5 + 7$). Since each orbital can hold two electrons, the maximum number of electrons will be 32.

EXAMPLE 9.2

Characterize the similarities and differences between a 3*s* and a 3*d* orbital.

SOLUTION

Similarities. Both orbitals are located in shell 3 and therefore extend approximately the same distance from the nucleus; each orbital may accommodate a maximum of two electrons.
Differences. The orbitals belong to different subshells (*s* and *d*); they have different shapes (see Fig. 9-2); electrons contained within each orbital would possess different energies.

EXAMPLE 9.3

Characterize the similarities and differences between a 2*s* and a 5*s* orbital.

SOLUTION

Similarities. Both orbitals are in *s* subshells and therefore have the same general shape; each orbital may accommodate a maximum of two electrons.
Differences. They are in different shells (n = 2 and 5) and therefore have different sizes (5*s* is larger); an electron in the 5*s* orbital is of higher energy than one in the 2*s* orbital.

9.5 Writing Electron Configurations

electron configuration

An **electron configuration** *is a statement of how many electrons an atom has in each of its subshells.* Since subshells group electrons according to energy (Sec. 9.4), electron configurations indicate how many electrons an atom has of various energies.

Electron configurations are not written out in words; a shorthand system with symbols is used. Subshells containing electrons, listed in order of increasing energy, are designated using number–letter combinations (1*s*, 2*s*, 2*p*, etc.). A superscript following each subshell designation indicates the number of electrons in that subshell. The electron configuration for oxygen using this shorthand notation is

$1s^22s^22p^4$ (read "one-*s*-two, two-*s*-two, two-*p*-four")

An oxygen atom, thus, has an electron arrangement of two electrons in the 1*s* subshell, two electrons in the 2*s* subshell, and four electrons in the 2*p* subshell.

Aufbau principle

To find the electron configuration for an atom, a procedure called the *Aufbau principle* (German *aufbauen*, to build) is used. The **Aufbau principle** *states that electrons normally occupy the lowest energy subshell available.* This guideline brings order to what could be a very disorganized situation. Many orbitals exist about the nucleus of any given atom. Electrons do not occupy these orbitals in a random, haphazard fashion; a very predictable pattern, governed by the Aufbau principle, exists for electron orbital occupancy. Orbitals are filled in order of increasing energy.

Use of the Aufbau principle requires knowledge concerning the electron capacities of orbitals and subshells (which we already have — Sec. 9.4) and knowledge concerning the relative energies of subshells (which we now consider).

Figure 9-5 gives the order in which electron subshells about a nucleus acquire electrons. Note from studying this figure that the sequence of subshell filling is not as simple a pattern as you probably would have predicted. All subshells within a given shell do not necessarily have lower energies than all subshells of higher numbered shells. Because of energy overlaps, beginning with shell 4, one or more lower energy subshells of a specific shell have energies lower than the upper subshells of a preceding shell, and thus acquire electrons first. For example, the 4*s* subshell acquires electrons before the 3*d* subshell does (see Fig. 9-5). As another example, the *s* subshell of the sixth energy level fills before the *d* subshell of the fifth energy level or the *f* subshell of the fourth energy level (again refer to Fig. 9-5).

Aufbau diagram

The sequence in which subshells acquire electrons must be learned before electron configurations can be written. A useful mnemonic (memory) device, called an *Aufbau diagram*, helps considerably with this learning process. As can be seen from Figure 9-6, an **Aufbau diagram** *is simply a listing of subshells arranged in a specific manner.* All *s* subshells are located in column 1, all *p* subshells in column 2, and so on. Subshells belonging to the same shell are found in the same row. The order of subshell filling is

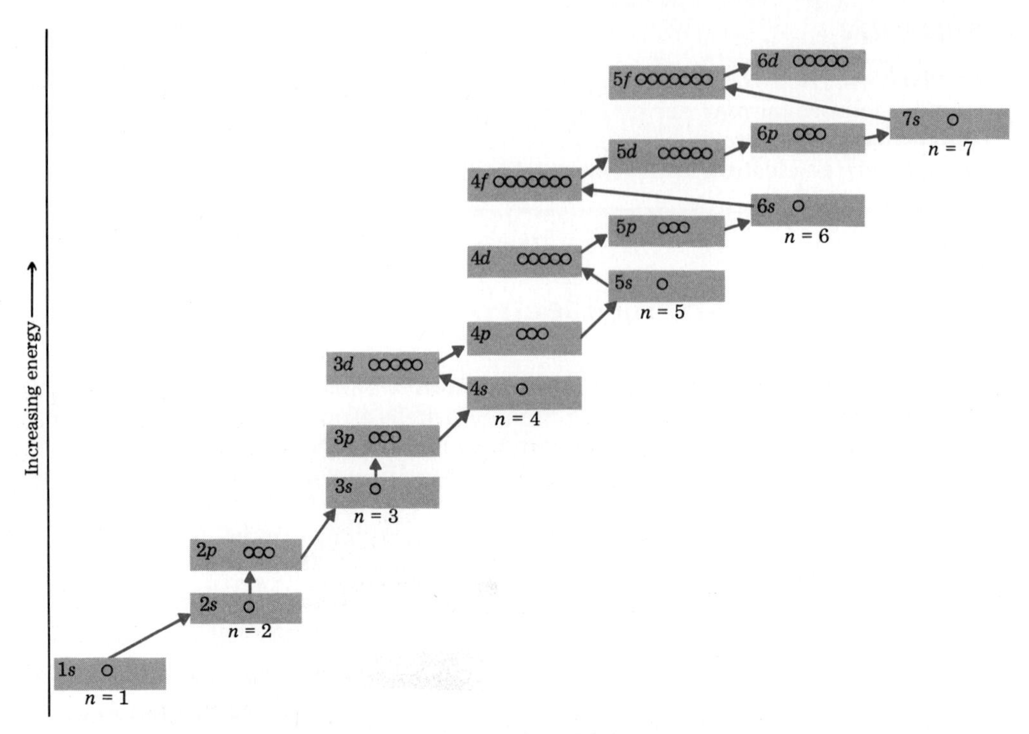

Figure 9-5. Relative energies and filling order for the electron subshells.

given by following the diagonal arrows, starting with the top one. The 1*s* subshell fills first. The second arrow points to (goes through) the 2*s* subshell. It fills next. The third arrow points to both the 2*p* and 3*s* subshells. The 2*p* fills first followed by the 3*s*. Any time a single arrow points to more than one subshell, start at the tail of the arrow and work to its head to determine proper filling sequence. The 3*p* subshell fills next, and so on. An Aufbau diagram is an easy way to catalog the information given in Figure 9-5.

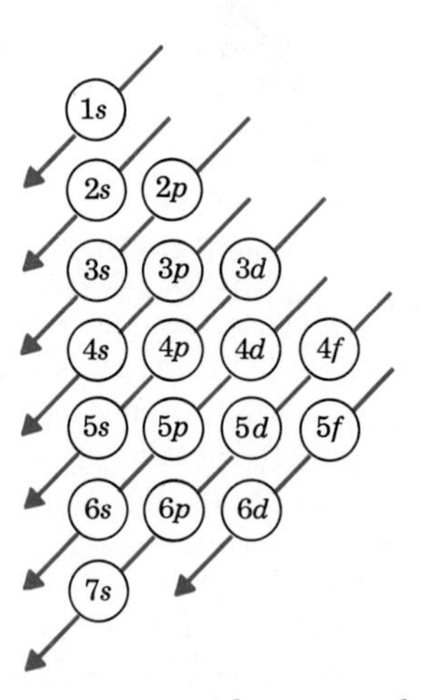

Figure 9-6. The Aufbau diagram — an aid to remembering subshell filling order.

We are now ready to write electron configurations. Let us systematically consider electron configurations for the first few elements in the periodic table.

Hydrogen ($Z = 1$) has only one electron, which goes into the 1s subshell, which has the lowest energy of all subshells. Hydrogen's electron configuration is written as

$$1s^1$$

Helium ($Z = 2$) has two electrons, both of which occupy the 1s subshell. (Remember, an s subshell contains one orbital and an orbital can accommodate two electrons.) Helium's electron configuration is

$$1s^2$$

For lithium ($Z = 3$), with three electrons, the third electron cannot enter the 1s subshell since its maximum capacity is two electrons. (All s subshells are completely filled with two electrons—Sec. 9.4.) The third electron is placed in the next highest energy subshell, the 2s. The electron configuration for lithium is, thus,

$$1s^22s^1$$

With beryllium ($Z = 4$) the additional electron is placed in the 2s subshell, which is now completely filled, giving beryllium the electron configuration

$$1s^22s^2$$

In boron ($Z = 5$) the 2p subshell, the subshell of next highest energy (check Fig. 9-5 or 9-6 to be sure) becomes occupied for the first time. Boron's electron configuration is

$$1s^22s^22p^1$$

A p subshell can accommodate six electrons since there are three orbitals within it (Sec. 9.4). The 2p subshell can thus accommodate the additional electrons found in C, N, O, F, and Ne. The electron configurations of these elements are

$$\text{C } (Z = 6): \quad 1s^22s^22p^2$$

$$\text{N } (Z = 7): \quad 1s^22s^22p^3$$

$$\text{O } (Z = 8): \quad 1s^22s^22p^4$$

$$\text{F } (Z = 9): \quad 1s^22s^22p^5$$

$$\text{Ne } (Z = 10): \quad 1s^22s^22p^6$$

With sodium ($Z = 11$) the 3s subshell acquires an electron for the first time.

$$\text{Na } (Z = 11): \quad 1s^22s^22p^63s^1$$

Note the pattern that is developing in the electron configurations we have written so far. Each element has an electron configuration the same as the one just before it with the addition of one electron.

Electron configurations for other elements are obtained by simply extending the principles we have just illustrated. Electrons are added to subshells, always filling one of lower energy before adding electrons to the next highest subshell, until the correct number of electrons have been accommodated.

EXAMPLE 9.4

Write the electronic configuration for

(a) arsenic ($Z = 33$) **(b)** gold ($Z = 79$)

SOLUTION

(a) The number of electrons in an arsenic atom is 33. Remember, the atomic number (Z) gives the number of electrons (Sec. 5.5). We will need to fill subshells, in order of increasing energy, until 33 electrons have been accommodated.

The $1s$, $2s$, and $2p$ subshells fill first, accommodating a total of ten electrons among them.

$$1s^2 2s^2 2p^6 \ldots$$

Next, according to Figure 9-5 or 9-6, the $3s$ fills and then the $3p$ subshell.

$$1s^2 2s^2 2p^6 3s^2 3p^6 \ldots$$

We have accommodated 18 electrons at this point. We still need to add 15 more electrons to get our desired number of 33.

The $4s$ subshell fills next, followed by the $3d$ subshell, giving us 30 electrons at this point.

$$1s^2 2s^2 2p^6 3s^2 3p^6 4s^2 3d^{10} \ldots$$

Note that the maximum electron population for d subshells is ten electrons.

Three more electrons are needed, which are added to the next higher subshell in energy, the $4p$.

$$1s^2 2s^2 2p^6 3s^2 3p^6 4s^2 3d^{10} 4p^3$$

The $4p$ subshell can accommodate six electrons, but we do not want it filled to capacity as this would give us too many electrons.

To double check that we have the correct number of electrons, 33, we add the superscripts in our final electron configuration:

$$2 + 2 + 6 + 2 + 6 + 2 + 10 + 3 = 33$$

The sum of the superscripts in any electron configuration should add up to the atomic number if the configuration is for a neutral atom.

(b) To write this configuration, we continue along the same lines as in part (a), remembering that the maximum electron subshell populations are $s = 2$, $p = 6$, $d = 10$, and $f = 14$.

To accommodate 79 electrons we will need to use all subshells through the $5d$ (as tabulated on the next page). The electron configuration for gold is, thus,

$$1s^2 2s^2 2p^6 3s^2 3p^6 4s^2 3d^{10} 4p^6 5s^2 4d^{10} 5p^6 6s^2 4f^{14} 5d^9$$

Subshell	Electrons in Subshell	Cumulative Electron Total
1*s*	2	2
2*s*	2	4
2*p*	6	10
3*s*	2	12
3*p*	6	18
4*s*	2	20
3*d*	10	30
4*p*	6	36
5*s*	2	38
4*d*	10	48
5*p*	6	54
6*s*	2	56
4*f*	14	70
5*d*	9	79

It should be noted that for a few elements in the middle of the periodic table the actual distribution of electrons within subshells differs slightly from that obtained using the Aufbau principle and Aufbau diagram. These exceptions are caused by very small energy differences between some subshells and are not important in the uses we shall make of electronic configurations.

Abbreviated electron configurations that give shell electron occupancy rather than subshell electron occupancy are sometimes written. Such configurations are simply a listing of numbers. Example 9.5 gives a comparison between a regular electron configuration (subshell occupancy) and an abbreviated electron configuration (shell occupancy).

EXAMPLE 9.5

The electron configuration for Cl is

$$1s^2 2s^2 2p^6 3s^2 3p^5$$

Write this configuration in an abbreviated form that gives only shell-occupancy information.

SOLUTION

Chlorine atoms have three electron shells that contain electrons.

Shell 1 contains two electrons: $1s^2$ 2

Shell 2 contains eight electrons: $2s^2 2p^6$ 8

Shell 3 contains seven electrons: $3s^2 3p^5$ 7

The abbreviated electron configuration for Cl is

$$2, 8, 7$$

In such notations the identity of each shell is not explicitly shown. It is assumed that you know that the listed numbers are ordered in terms of increasing shell number ($n = 1, 2, 3, \ldots$).

Chlorine has two electrons in the first shell, eight in the second shell, and seven in the third shell.

9.6 Quantum Numbers (optional)

We consider in this section a more rigorous approach to describing electron arrangements about atoms that compliments nicely the semiquantitative approach to electron arrangements just discussed (Secs. 9.1–9.5). We will not have to unlearn or cast aside any of the concepts about shells, subshells, and orbitals that we have already developed. We will simply "add to" these concepts.

quantum numbers

Quantum numbers are the foundation for more quantitative approaches to electron configurations. **Quantum numbers** *are numbers that describe the allowed energy states of electrons in orbitals.* A set of *four* quantum numbers is used in specifying the state of any given electron. Such a set of numbers may be considered a "numerical code" giving information about an electron and the orbital in which it is found. Identifying electron orbitals by sets of numbers is somewhat analogous to the United States government's system of identifying people by social security numbers. However, much, much more information is obtainable from a set of quantum numbers than from a social security number. Social security numbers for people are somewhat arbitrarily assigned. A set of quantum numbers for an electron is derived using very specific rules.

Principal Quantum Number

The first quantum number, n, is called the *principal quantum number.* Values of this quantum number must be positive and integral.

$$n = 1, 2, 3, 4, 5, 6, 7, \ldots$$

Values of n greater than 7 are not encountered in describing electrons in currently known atoms.

The function of the principal quantum number is to indicate the shell in which an electron is found. If $n = 3$ for an electron, it is found in an orbital located in shell 3. All orbitals and electrons within a given shell must have the same principal quantum number. By now you may have made the connection that a principal quantum number and the shell number discussed in Section 9.2 are one and the same.

principal quantum number

The **principal quantum number** *determines the size of the orbital in which an electron is located.* Orbital size and shell number are related (Sec. 9.2). As the value of n increases the size of the orbital increases.

Table 9-4. ***Relationship Between Principal and Secondary Quantum Number Values***

Value of n	Values of l	Total Number of l Values
1	0	1
2	0, 1	2
3	0, 1, 2	3
4	0, 1, 2, 3	4

Secondary Quantum Number

The second quantum number, l, is called the *secondary quantum number.* The value of n determines the allowable values of l. The value of l can be zero or any whole number up to and including $(n - 1)$; that is, the maximum value of l is one less than the value of n.

$$l = 0, 1, 2, \ldots, (n - 1)$$

Table 9-4 summarizes the relationship between principal and secondary quantum number values. Note that the number of l values for a given n is the same as the value of n. For example, when $n = 4$ there are four possible values for l.

The function of the secondary quantum number is to divide shells into subshells. The number of possible l values for a given n determines the number of subshells in a shell. Shell number 4 ($n = 4$) contains four subshells — the 4s ($l = 0$) subshell, the 4p ($l = 1$) subshell, the 4d ($l = 2$) subshell, and the 4f ($l = 3$) subshell. The matchup between subshell letter designations and secondary quantum number values given in the previous sentence holds in all situations and is formalized in Table 9-5.

All electrons found in a specific subshell, for example, the 3d subshell, must have the same n and l values.

secondary quantum number

The **secondary quantum number** *specifies the shape of the orbital in which an electron is found.* Subshell type is determined by the type of orbitals found within it. An s subshell contains s orbitals, a p subshell contains p orbitals, and so on. Shape is a differentiating factor between types of orbitals; s, p, d, and f orbitals have different shapes (Sec. 9.4). Hence, the statement that the secondary quantum number determines orbital shape.

Table 9-5. ***Relationship Between Secondary Quantum Number Values and Letter Designations for Subshells***

Value of l	Letter Designation for Subshell
0	s
1	p
2	d
3	f

Table 9-6. ***Relationship Between Secondary and Magnetic Quantum Number Values***

Value of l	Possible Values of m	Total Number of m Values
0 (s subshell)	0	1
1 (p subshell)	−1, 0, +1	3
2 (d subshell)	−2, −1, 0, +1, +2	5
3 (f subshell)	−3, −2, −1, 0, +1, +2, +3	7

Magnetic Quantum Number

The third quantum number, m, is called the *magnetic quantum number*. As with l, there are restrictions on the possible values of m. The values of m are integers that can range from the negative value of l through zero up to the positive value of l.

$$m = 0, \pm 1, \pm 2, \ldots, \pm l$$

For example, when $l = 2$ (a d subshell), m can have five possible values: −2, −1, 0, +1, +2. Table 9-6 shows the relationship between l and m values.

The function of the magnetic quantum number is to split subshells into individual orbitals. Note the values 1, 3, 5, and 7 in the last column of Table 9-6. These are numbers you have used before — the number of orbitals in subshells of various types (Sec. 9.4). The number of m values that are possible determines the number of orbitals in a subshell.

All electrons found in a specific orbital, a maximum of two, must have the same n, l, and m values.

magnetic quantum number

The **magnetic quantum number** *specifies the orientation in space of an orbital relative to other orbitals within the same subshell.* For example, the three p orbitals found in the 2p subshell, which differ from each other only in space orientation (see Fig. 9-3), all have the same n value ($n = 2$) and the same l value ($l = 1$) but differ in m values. One 2p orbital has an m value of −1, another a value of 0, and the third an m value of −1.

Spin Quantum Number

The fourth quantum number, s, is called the *spin quantum number*. The spin quantum number can have only one of two values, $-\frac{1}{2}$ or $+\frac{1}{2}$. Unlike the other three quantum numbers, which have interrelated values, the value of s is independent of other quantum number values.

The function of the spin quantum number is to differentiate between two electrons in the same orbital. If two electrons are in the same orbital they must have different spin quantum numbers.

spin quantum number

The **spin quantum number** *deals with a property of an electron itself, its spin, rather than with a characteristic of the orbital containing the electron.* Current quantum-mechanical theory indicates that an electron spins on its axis as it moves about the nucleus, much as the earth spins on its axis as it moves about the sun. Only two values for the spin quantum number results from there being only two directions for an electron to spin: clockwise or counterclockwise.

Table 9-7. General Expressions for the Allowed Values for Quantum Numbers

$n = 1, 2, 3, \ldots$
$l = 0, 1, 2, \ldots, (n - 1)$
$m = 0, \pm 1, \pm 2, \ldots, \pm l$
$s = +\frac{1}{2}, -\frac{1}{2}$

Quantum Numbers — A Summary

The mathematical restrictions that exist on the values quantum numbers may assume arise naturally from the solving of the equations of quantum theory. Table 9-7 summarizes, with general equations, the allowed values of each of the quantum numbers. Table 9-8 gives the actual allowed values for each of the four quantum numbers for electrons found in shells 1–4 of an atom.

Each of the quantum numbers conveys specific information about an electron's orbital or about the electron itself. Table 9-9 summarizes the informational aspect of quantum numbers.

9.7 Orbital Diagrams (optional)

The arrangement of electrons about a nucleus may be specified in terms of shell occupancy, subshell occupancy, or orbital occupancy. The notation for specifying electron arrangements in terms of shell occupancy (abbreviated electron configurations) and subshell occupancy (regular electron configurations) has been previously considered (Sec. 9.5). In this section we consider electron arrangements at the orbital level.

Three principles are used in specifying orbital occupancy for electrons.

1. The Aufbau principle.
2. The Pauli exclusion principle.
3. The principle of maximum multiplicity (Hund's rule).

Only one of these principles, the Aufbau principle, was needed to write electron arrangements in terms of shell or subshell occupancy. The increased number of parame-

Table 9-8. Specific Quantum Number Values for Electrons Found in Shells 1–4

n	l	Orbital Designation	m	Number of Orbitals	s
1	0	1s	0	1	$+\frac{1}{2}, -\frac{1}{2}$
2	0	2s	0	1	$+\frac{1}{2}, -\frac{1}{2}$
	1	2p	−1, 0, +1	3	$+\frac{1}{2}, -\frac{1}{2}$
3	0	3s	0	1	$+\frac{1}{2}, -\frac{1}{2}$
	1	3p	−1, 0, +1	3	$+\frac{1}{2}, -\frac{1}{2}$
	2	3d	−2, −1, 0, +1, +2	5	$+\frac{1}{2}, -\frac{1}{2}$
4	0	4s	0	1	$+\frac{1}{2}, -\frac{1}{2}$
	1	4p	−1, 0, +1	3	$+\frac{1}{2}, -\frac{1}{2}$
	2	4d	−2, −1, 0, +1, +2	5	$+\frac{1}{2}, -\frac{1}{2}$
	3	4f	−3, −2, −1, 0, +1, +2, +3	7	$+\frac{1}{2}, -\frac{1}{2}$

Table 9-9. Informational Aspects of Quantum Numbers

Name and Symbol of Quantum Number	Function of Quantum Number	Orbital Information Conveyed by Quantum Number
Principal (*n*)	locates the shell in which an electron is found	determines the size of the orbital in which the electron is found
Secondary (*l*)	divides shells into subshells	specifies the shape of the orbital in which the electron is found
Magnetic (*m*)	splits subshells into orbitals	specifies the orientation in space of an orbital relative to other orbitals in the same subshell
Spin (*s*)	differentiates between two electrons in the same orbital	gives no information about an orbital; specifies the relative spin of the electron itself

ters that must be dealt with at the orbital-occupancy level dictates the need for additional guidelines (principles). Electron arrangements written in terms of shell occupancy deal with only one quantum number, the principal quantum number. Specifying subshell occupancy for electrons requires looking at two quantum numbers — the principal and secondary quantum numbers. All four quantum numbers must be considered when electron arrangements are stated in terms of orbital occupancy.

We are familiar with only one of the three principles needed to write electron arrangements in terms of orbital occupancies. The Aufbau principle was used extensively in Section 9.5 in writing electron configurations. The other two principles are entirely new to us.

Pauli exclusion principle

The **Pauli exclusion principle** states: "*No two electrons in one atom may have the same set of four quantum numbers.*" To be in the same orbital, two electrons must have identical values for the first three quantum numbers, *n*, *l*, and *m*. The Pauli exclusion principle, thus, requires that two such electrons have different spin quantum numbers, that is, that the electrons possess opposite spins; otherwise all four quantum numbers would be the same. Electrons in the same orbital having opposite spins are said to have spins that are "paired." If there is only one electron in an orbital, its spin is said to be "unpaired."

principle of maximum multiplicity (Hund's rule)

The **principle of maximum multiplicity (Hund's rule)** states: "*When electrons are placed in a set of orbitals of equal energy* (*the orbitals of a subshell*), *the order of filling of the orbitals is such as to give the maximum number of unpaired electrons.*" The consequence of Hund's rule is that electrons will go into unoccupied orbitals of a subshell whenever possible, rather than pairing with electrons in partially occupied orbitals. Numerous applications of Hund's rule will be required in our discussion of orbital diagrams which follows.

orbital diagrams

Orbital diagrams *are diagrams that show the electron occupancy of each orbital about a nucleus.* In writing orbital diagrams, circles are used to represent orbitals and arrows with a single barb to denote electrons. The possible spin quantum numbers of an electron are represented by pointing a single-barbed arrow either up or down.

The orbital diagram for hydrogen, with its one electron, is

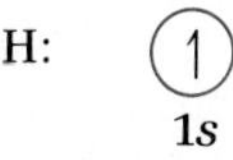

A helium atom contains two electrons, both of which occupy the 1*s* orbital. The Pauli exclusion principle requires that they have opposite spins, that is, that the spins are paired. The orbital diagram for helium is

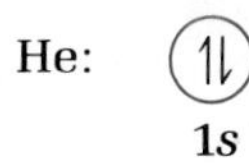

Orbital diagrams for the next two elements, lithium and beryllium, are written by reasoning similar to that for H and He. Again, the Pauli exclusion principle requires that the two electrons in the 2*s* orbital of Be be spin paired.

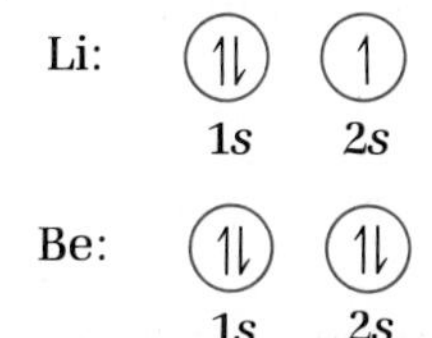

Boron has the electron configuration $1s^2 2s^2 2p^1$. The fifth electron in boron must enter a 2*p* orbital, since both the 1*s* and 2*s* orbitals are full. Boron's orbital diagram is

B: (↿⇂) (↿⇂) (↿)( )( )
1*s* 2*s* 2*p*

(Note, as with boron, that it is standard convention to show all orbitals of a given subshell even when some of them are empty.)

With carbon, element 6, we encounter the use of Hund's rule for the first time. Carbon has two electrons in the 2*p* subshell ($1s^2 2s^2 2p^2$). Do the two electrons go into the same orbital (paired spins) or do they go into separate equivalent orbitals (unpaired spins)? Hund's rule indicates that the latter is the case. The orbital diagram for carbon is

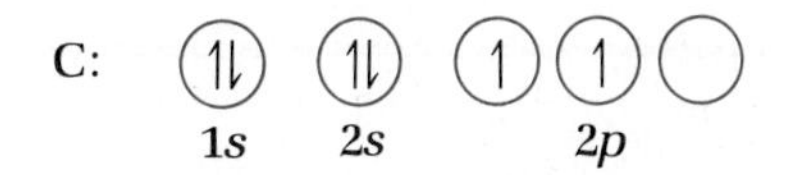

It does not matter which two of the three 2*p* orbitals of carbon are shown as containing electrons. All three orbitals have the same energy. Each of the following notations is equally correct.

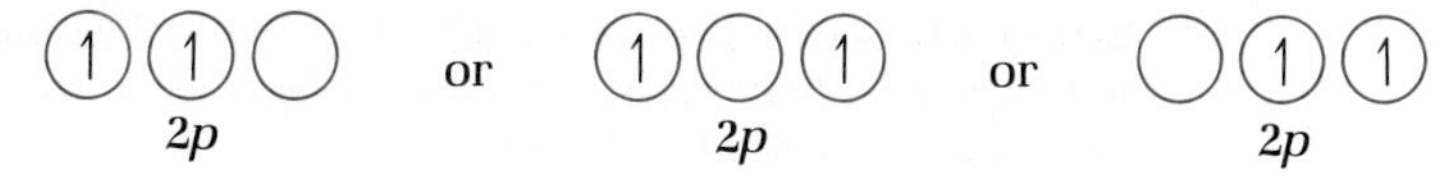

Nitrogen, with the electron configuration $1s^22s^22p^3$, contains three unpaired electrons.

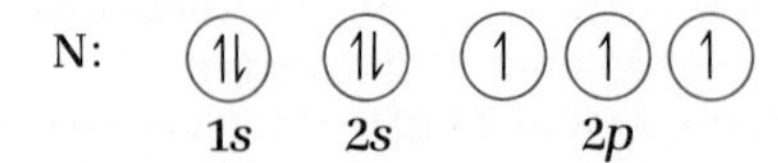

With oxygen ($1s^22s^22p^4$) two of the four $2p$ electrons must pair up, leaving two unpaired electrons. Fluorine, the next element in the periodic table, has only one unpaired electron. Finally, with neon, all of the electrons are paired up.

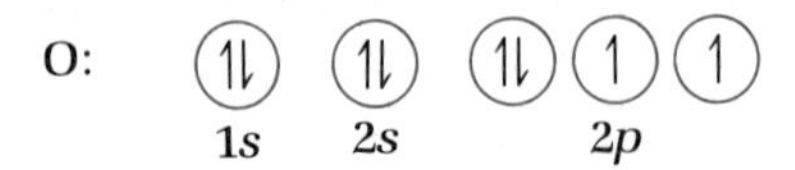

F: (↑↓) (↑↓) (↑↓)(↑↓)(↑)
  1s 2s 2p

Ne: (↑↓) (↑↓) (↑↓)(↑↓)(↑↓)
  1s 2s 2p

EXAMPLE 9.6

Write an orbital diagram for the electrons in the element manganese. The electron configuration for manganese is

$$1s^22s^22p^63s^23p^64s^23d^5$$

SOLUTION

All of the subshells of manganese are completely filled except for the $3d$. There are five orbitals in the $3d$ subshell and there are five electrons present. Following Hund's rule, we will put one electron in each of the five $3d$ orbitals, giving us five unpaired electrons.

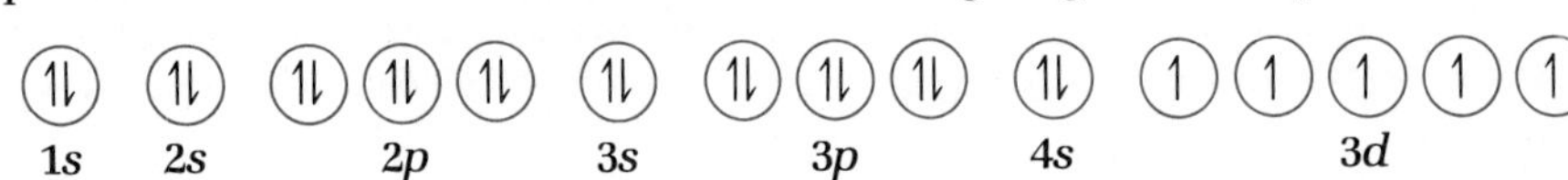

paramagnetic

diamagnetic

Atoms may be classified as *paramagnetic* or *diamagnetic* on the basis of unpaired electrons. A **paramagnetic** *atom has an electron arrangement containing one or more unpaired electrons.* The presence of unpaired electrons causes paramagnetic materials to be slightly attracted to a magnet. Measurement of paramagnetism provides experimental verification of the presence of unpaired electrons. A **diamagnetic** *atom has an electron arrangement in which all electrons are spin paired.*

9.8 The $(n + l)$ Rule (optional)

The relative energies of orbitals and subshells must be known before electron configurations or orbital diagrams can be determined. You are already aware that an Aufbau diagram (Fig. 9-6) is a very convenient mnemonic device for remembering the order in which orbitals and subshells acquire electrons.

$(n + l)$ rule

A second method, the $(n + l)$ rule is also available for determining the relative energies of orbitals and subshells. The **$(n + l)$ rule** *uses the sum of the values of the principal (n) and secondary (l) quantum numbers to predict relative energies.* There are two parts to the $(n + l)$ rule.

1. When comparing orbital energies, the orbital with the lowest sum for the principle quantum number, n, and the secondary quantum number, l, will be occupied first by electrons.
2. If two orbitals have the same $(n + 1)$ value, the one with the lower n value will be occupied first by electrons.

Let us use this rule to predict the relative energies of the $3d$ and $4s$ orbitals. (From the Aufbau diagram we already know that the $4s$ orbital receives electrons first.) A $3d$ orbital has $n = 3$ and $l = 2$ and thus an $(n + 1)$ sum of 5. The $4s$ orbital has $n = 4$ and $l = 0$ and an $(n + l)$ sum of 4. The $4s$ orbital has a lower $(n + l)$ sum and will thus fill first. Table 9-10 gives the $(n + l)$ values for various orbitals (subshells). Check for yourself that the $(n + 1)$ rule gives the same order of orbital filling as does the Aufbau diagram.

Table 9-10. Ordering of Orbital and Subshell Energies by the $(n + l)$ Rule

	n	l	$n + l$
$6d$	6	2	8
$5f$	5	3	8
$7s$	7	0	7
$6p$	6	1	7
$5d$	5	2	7
$4f$	4	3	7
$6s$	6	0	6
$5p$	5	1	6
$4d$	4	2	6
$5s$	5	0	5
$4p$	4	1	5
$3d$	3	2	5
$4s$	4	0	4
$3p$	3	1	4
$3s$	3	0	3
$2p$	2	1	3
$2s$	2	0	2
$1s$	1	0	1

EXAMPLE 9.7

Using the $(n + l)$ rule, show that
(a) The 3*s* orbital is lower in energy than the 3*p* orbital.
(b) The 4*f* orbital receives electrons before the 7*s* orbital does.

SOLUTION

(a) We examine the values of the *n* and *l* quantum numbers associated with each orbital.

Orbital	*n* Value	*l* Value	$(n + l)$ Value
3*s*	3	0	3
3*p*	3	1	4

Because $(n + l)$ is less for the 3*s*, it is of lower energy.
(b) Again, examining the values of the quantum numbers *n* and *l*, we find

Orbital	*n* Value	*l* Value	$(n + l)$ Value
4*f*	4	3	7
7*s*	7	0	7

The 4*f* and 7*s* orbitals both give an $(n + l)$ sum of 7. When two orbitals have the same $(n + l)$ sum, the *n* value determines the one of lower energy; the lower the *n* value the lower the energy. Hence, the 4*f* orbital receives electrons first.

9.9 Electron Configurations and the Periodic Law

A knowledge of electron configurations for the elements provides an explanation for the periodic law. Recall, from Section 5.8, that the periodic law points out that the properties of the elements repeat themselves in a regular manner when the elements are ordered in sequence of increasing atomic number. Those elements with similar chemical properties are found placed one under another in vertical columns (groups) in a periodic table.

Groups of elements have similar chemical properties because of similarities that exist in the electron configurations of the elements of the group. *Chemical properties repeat themselves in a regular manner among the elements because electron configurations repeat themselves in a regular manner among the elements.*

To illustrate this (similar-chemical-property)–(similar-electron-configuration) correlation, let us look at the electron configurations of two groups of elements known to have similar chemical properties.

We begin with the elements lithium, sodium, potassium, and rubidium — all members of group IA of the periodic table. The electron configurations for these similar-propertied elements are

$_{3}Li$: $1s^22s^1$

$_{11}Na$: $1s^22s^22p^63s^1$

$_{19}K$: $1s^22s^22p^63s^23p^64s^1$

$_{37}Rb$: $1s^22s^22p^63s^23p^64s^23d^{10}4p^65s^1$

We see that each of these elements has one outer electron (shown in color), the last one added by the Aufbau principle, in an *s* subshell. It is this similarity in outer shell electron arrangements that causes these elements to have similar chemical properties. It is found in general that elements with similar outer shell electron configurations have similar chemical properties.

Let us consider another group of elements known to have similar chemical properties: the elements fluorine, chlorine, bromine, and iodine of group VIIA of the periodic table. The electron configurations for these four elements are

$_{9}F$: $1s^22s^22p^5$

$_{17}Cl$: $1s^22s^22p^63s^23p^5$

$_{35}Br$: $1s^22s^22p^63s^23p^64s^23d^{10}4p^5$

$_{53}I$: $1s^22s^22p^63s^23p^64s^23d^{10}4p^65s^24d^{10}5p^5$

Once again similarities in electron configurations are readily apparent. This time the repeating pattern involves an outermost *s* and *p* subshell containing seven electrons (in color).

Section 10.2 will consider in depth the fact that the electrons most important in controlling chemical properties are those found in the outermost shell of an atom.

9.10 Electron Configurations and the Periodic Table

One of the strongest pieces of supporting evidence for the assignment of electrons to shells, subshells, and orbitals is the periodic table itself. The basic shape and structure of this table, which was determined many years before electrons were even discovered, is consistent with and can be explained by electron configurations. Indeed, the specific location of an element in the periodic table can be used to obtain information about its electron configuration.

distinguishing electron

The concept of *distinguishing electrons* is the key to obtaining "electron-configuration information" from the periodic table. The **distinguishing electron** *for an element is the last electron added to its electron configuration when the configuration is written by the Aufbau principle.* This last electron added is the one that causes an element's electron configuration to differ from that of the element immediately preceding it in the periodic table; hence, the term distinguishing electron.

As the first step in linking electron configurations to the periodic table let us analyze the general shape of the periodic table in terms of columns of elements. As shown in Figure 9-7, we have on the extreme left of the table two columns of elements, in the center an area containing ten columns of elements, to the right a block of six columns of elements, and at the bottom of the table, in two rows, fourteen columns of elements. These numbers of columns of elements in the various regions of the periodic table — 2, 6, 10, and 14 — are the same as the maximum numbers of electrons that the various

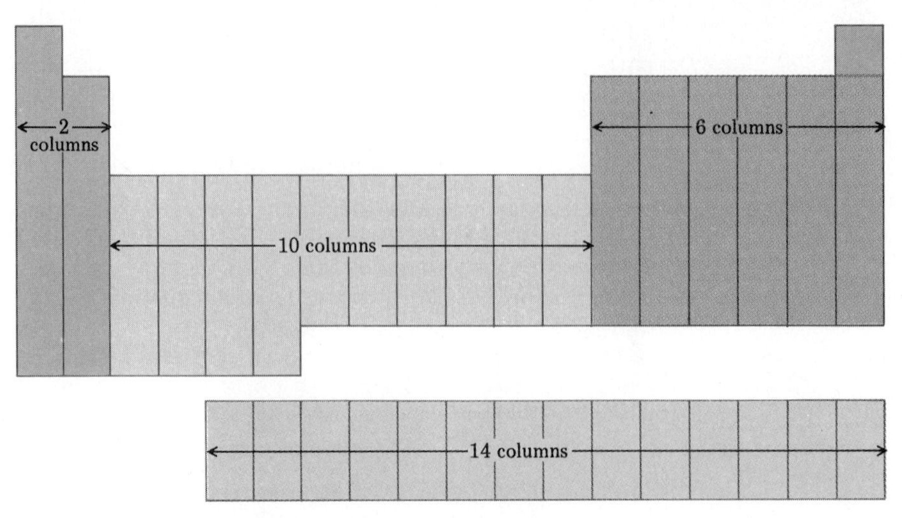

Figure 9-7. Structure of the periodic table in terms of columns.

types of subshells may accommodate. We will see shortly that this is a very significant observation; the number matchup is no coincidence. The various columnar regions of the periodic table are called the *s* area (two columns), the *p* area (six columns), the *d* area (ten columns), and the *f* area (fourteen columns) — (see Fig. 9-8).

For all elements located in the *s* area of the periodic table the distinguishing electron is always found in an *s* subshell. All *p*-area elements have distinguishing electrons in *p* subshells. Similarly, elements in the *d* and *f* areas of the periodic table have,

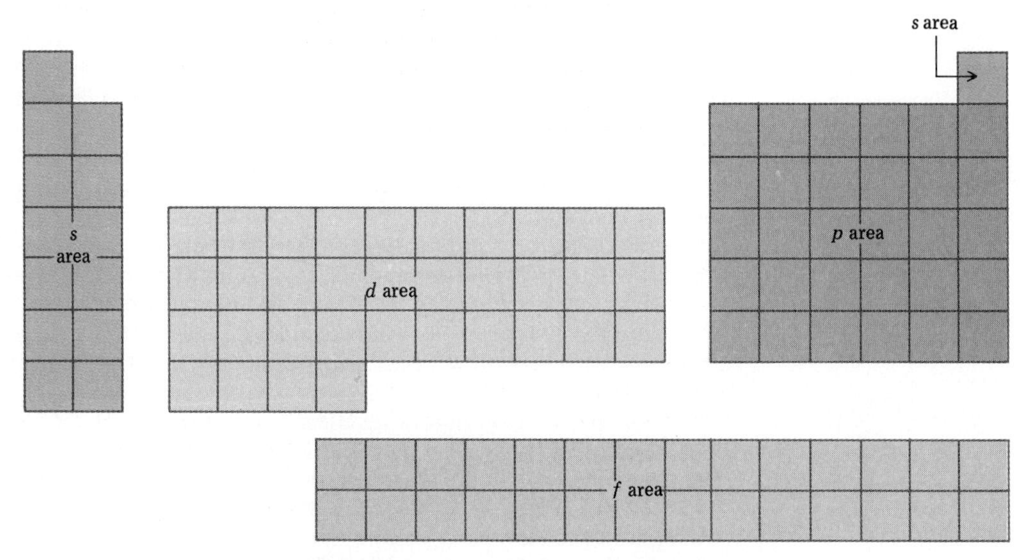

Figure 9-8. Areas of the periodic table.

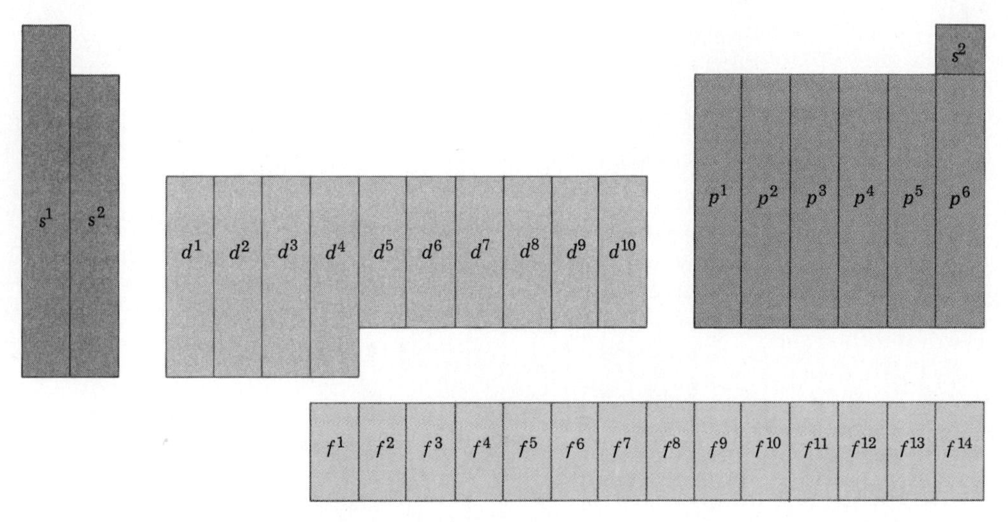

Figure 9-9. Extent of subshell filling as a function of periodic table location.

respectively, distinguishing electrons located in *d* and *f* subshells. Thus, the area location of an element in the periodic table may be used to determine the type of subshell that contains the distinguishing electron. Note that the element helium is considered to belong to the *s* rather than the *p* area of the periodic table even though its table position is on the right side. (The reason for this placement of helium will be explained in Sec. 10.3.)

The extent of filling of the subshell containing an element's distinguishing electron can also be determined from the element's position in the periodic table. All elements in the first column of a specific area contain only one electron in the subshell, all elements in the second column contain two electrons in the subshell, and so on. Thus, all elements in the first column of the *p* area (group IIIA) have an electron configuration ending in p^1. Elements in the second column of the *p* area (group IVA) have electron configurations ending in p^2 and so forth. Similar relationships hold in other areas of the table, as shown in Figure 9-9. A few exceptions to the above generalizations do exist in the *d* and *f* areas (because of irregular electron configurations — Sec. 9.5) but we will not be concerned with them in this text.

We can also use the periodic table to determine the shell in which the distinguishing electron is located. The relationship used involves the number of the period in which the element is found. In the *s* and *p* areas, the period number gives directly the shell number. In the *d* area, the period number minus one is equal to the shell number. (Remember that the 3*d* subshell is filled during the fourth period.) For similar reasons, in the *f* area the period number minus two equals the shell number. Thus, the subshell that contains the distinguishing electron for elements of period 6 may be the 6*s*, 6*p*, 5*d* (period number minus one), or 4*f* (period number minus two), depending on the location of the element in period 6. It must be remembered that even though the *f* area is located at the bottom of the table, it correctly belongs in periods 6 and 7. The complete matchup between period number and shell number for the distinguishing electron is given in the periodic table of Figure 9-10.

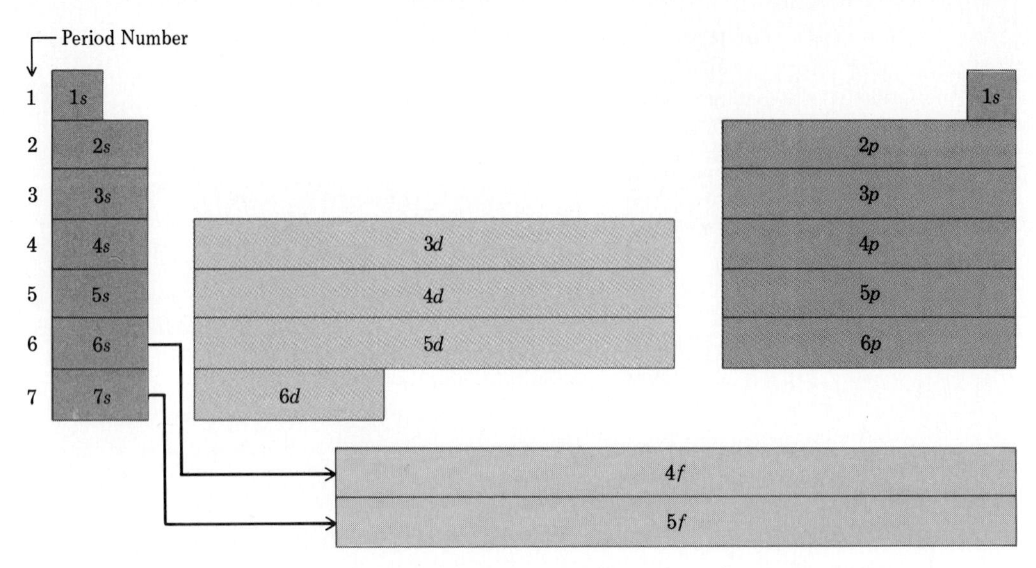

Figure 9-10. Relationships of period numbers to shell numbers for distinguishing electrons.

EXAMPLE 9.8

Using the periodic table and Figures 9-8, 9-9, and 9-10, determine the following for the elements magnesium ($Z = 12$), vanadium ($Z = 23$), and iodine ($Z = 53$).

(a) the type of subshell in which the distinguishing electron is found
(b) the extent of filling of the subshell containing the distinguishing electron
(c) the shell in which the subshell containing the distinguishing electron is found

SOLUTION

(a) Knowing the area of the periodic table in which an element is found is sufficient to determine the type of subshell in which the distinguishing electron is found.
Mg: Since this element is found in the *s* area of the periodic table, the distinguishing electron will be in an ***s* subshell.**
V: Since this element is found in the *d* area of the periodic table, the distinguishing electron will be in a ***d* subshell.**
I: Since this element is found in the *p* area of the periodic table, the distinguishing electron will be in a ***p* subshell.**
(b) The extent of filling of the subshell containing the distinguishing electron is determined by noting the column in the area that the element occupies.
Mg: Since this element is in the second column of the *s* area, the *s* subshell involved contains **two electrons** (s^2).
V: Since this element is in the third column of the *d* area, the *d* subshell involved contains **three electrons** (d^3).
I: Since this element is in the fifth column of the *p* area, the *p* subshell involved contains **five electrons** (p^5).
(c) The shell number of the subshell containing the distinguishing electron is obtained from the periodic number, sometimes directly and sometimes with modifications.

Mg: Since this element is in period 3, the *s* subshell involved is the 3*s*; therefore, the electron configuration for Mg ends in $3s^2$.
V: Since this element is in period 4, the *d* subshell involved is the 3*d* (period number minus one); therefore, the electron configuration for V ends in $3d^3$.
I: Since this element is in period 5, the *p* subshell involved is the 5*p*; therefore, the electron configuration for I ends in $5p^5$.

To write complete electron configurations the order of filling of the various electron subshells must be known. We have already considered two ways of obtaining this filling order — use of an Aufbau diagram (Sec. 9.5) and use of the $(n + l)$ rule (Sec. 9.8). This information may also be obtained directly from a periodic table. To obtain it, we merely follow a path of increasing atomic number through the table noting the various subshells as they are encountered. The results of this operation are summarized in Table 9-11.

Table 9-11. Subshell Filling Order from the Periodic Table

Atomic Number Range	Subshell Involved
1–2	1*s*
3–4	2*s*
5–10	2*p*
11–12	3*s*
13–18	3*p*
19–20	4*s*
21–30	3*d*
31–36	4*p*
37–38	5*s*
39–48	4*d*
49–54	5*p*
55–56	6*s*
57	5*d*
58–71	4*f*
72–80	5*d*
81–86	6*p*
87–88	7*s*
89	6*d*
90–103	5*f*
104–106	6*d*

EXAMPLE 9.9

Write the complete electron configuration of $_{80}$Hg using the periodic table as your sole guide.

SOLUTION

The needed information will be obtained by "working our way" through the periodic table. We will start at hydrogen and go from box to box, in order of increasing atomic number, until we arrive at position 80, mercury. Every time we traverse the *s* area we will fill an *s* subshell, the *p* area a *p* subshell, and so on. We will remember that in filling *s* and *p* subshells the shell number is given by the period number, that one must be subtracted from the period number for *d* subshells, and that a subtraction of two is required for *f* subshells.

Let us begin our journey through the periodic table. As we cross period 1 we encounter H and He, both 1*s* elements. We, thus, add the 1*s* electrons.

$$\text{Hg:}\quad \mathbf{1s^2} \ldots$$

In traversing period 2 we pass through the *s* area (elements 3 and 4) and the *p* area (elements 5–10) in that order. We add 2*s* and 2*p* electrons:

$$\text{Hg:}\quad 1s^2\mathbf{2s^2 2p^6} \ldots$$

Our trip through period 3 is very similar to that of period 2. Only *s*-area elements (11 and 12) and *p*-area elements (13–18) are encountered. The message is to add *s* and *p* electrons, 3*s* and 3*p* because we are in period 3.

$$\text{Hg:}\quad 1s^2 2s^2 2p^6 \mathbf{3s^2 3p^6} \ldots$$

In passing through period 4 we go through the *s* area (elements 19 and 20), the *d* area (elements 21–30), and the *p* area (elements 31–36). Electrons to be added are those in the 4*s*, 3*d* (period number minus one), and 4*p* subshells in that order.

$$\text{Hg:}\quad 1s^2 2s^2 2p^6 3s^2 3p^6 \mathbf{4s^2 3d^{10} 4p^6} \ldots$$

In period 5 we encounter scenery similar to that in period 4 — the *s* area (elements 37 and 38), the *d* area (elements 39–48), and the *p* area (elements 49–54). Hence, the 5*s*, 4*d* (period number minus one), and 5*p* subshells are filled in order.

$$\text{Hg:}\quad 1s^2 2s^2 2p^6 3s^2 3p^6 4s^2 3d^{10} 4p^6 \mathbf{5s^2 4d^{10} 5p^6} \ldots$$

Our journey ends in period 6. We go through the 6*s* area (elements 55 and 56), the 4*f* area (elements 58–71), and the 5*d* area (elements 57 and 72–80). We will completely fill the 6*s*, 4*f* (period number minus two), and 5*d* (period number minus one) subshells.

$$\text{Hg:}\quad \mathbf{1s^2 2s^2 2p^6 3s^2 3p^6 4s^2 3d^{10} 4p^6 5s^2 4d^{10} 5p^6 6s^2 4f^{14} 5d^{10}}$$

Again, let us mention that a few slightly irregular electronic configurations are encountered in the *d* and *f* areas of the periodic table. The generalizations in this section do not address this problem. We will be working mostly with *s* and *p* area elements in future chapters. There are no irregularities in electron configurations for these elements.

9.11 Classification Systems for the Elements

The elements can be classified in several ways. The two most common classification systems are

1. A system based on selected physical properties of the elements, in which the elements are described as *metals* or *nonmetals*. (This classification system has already been discussed in Sec. 6.3.)
2. A system based on the electron configurations of the elements, in which elements are described as *rare-gas, representative, transition,* or *rare-earth* elements.

The classification scheme based on electron configurations of the elements is depicted in Figure 9-11. The groupings of elements resulting from this classification scheme will be used in many discussions in subsequent chapters.

rare-gas elements

The **rare-gas elements** *are found in the far right column of the periodic table.* They are all gases at room temperature, and they have little tendency to form chemical compounds. With one exception, the distinguishing electron for a rare gas completes the *p* subshell. Therefore, they have electron configurations ending in p^6. The exception is helium, in which the distinguishing electron completes the first shell — a shell that has only two electrons. Helium's electron configuration is $1s^2$.

representative elements

The **representative elements** *are all of the elements of the s and p areas of the periodic table with the exception of the rare gases.* The distinguishing electron in these elements partially or completely fills an *s* subshell or partially fills a *p* subshell. The representative elements include most of the more common elements.

transition elements

The **transition elements** *are all of the elements of the d area of the periodic table.* The common feature in the electronic configurations of the transition elements is the presence of the distinguishing electron in a *d* subshell.

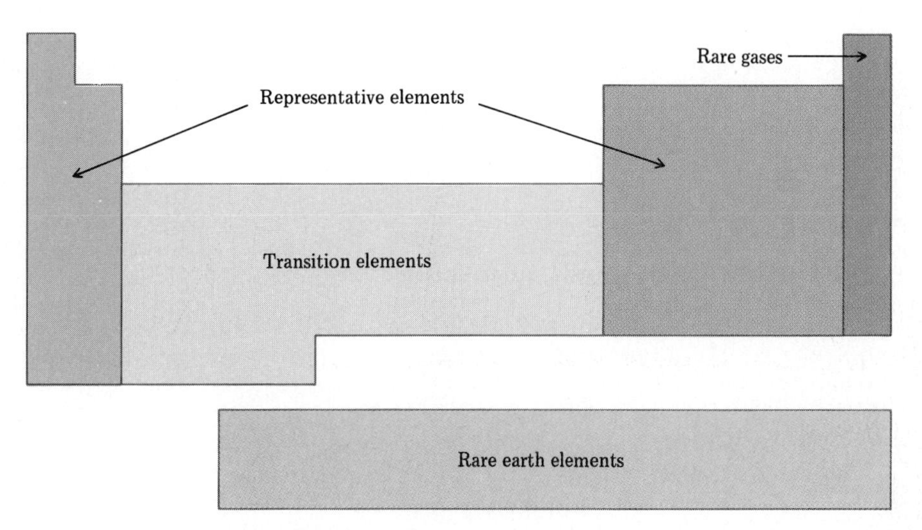

Figure 9-11. Elemental classification scheme based on the electron configurations of the elements.

rare-earth elements The **rare-earth elements** *are all of the elements of the f area of the periodic table.* The characteristic feature of their electronic configurations is the presence of the distinguishing electron in an *f* subshell. There is very little variance in the properties of these elements.

Learning Objectives

After completing this chapter you should be able to

- Explain what is meant by the phrase *quantized energy* (Sec. 9.1).
- Understand how the terms *shell, subshell,* and *orbital* are used in describing electron arrangements about nuclei and know the mathematical interrelationships between them (Secs. 9.2–9.4).
- Write the electron configuration for any element using the Aufbau principle (Sec. 9.5).
- List the names, symbols, and functions of the four quantum numbers and assign a physical meaning to each (Sec. 9.6) [optional].
- Draw orbital diagrams consistent with the Pauli exclusion principle and Hund's rule (Sec. 9.7) [optional].
- Use the $(n + l)$ rule to predict the order of filling of electron subshells (Sec. 9.8) [optional].
- Understand the relationship between the periodic law and electron configurations (Sec. 9.9).
- Use the periodic table to determine the extent of filling of the subshell containing the distinguishing electron and the shell in which this subshell is located for any given atom (Sec. 9.10).
- Use the periodic table as a guide in writing electron configurations (Sec. 9.10).
- Classify a given element as a rare gas, representative element, transition element, or rare-earth element (Sec. 9.11).

Terms and Concepts for Review

The new terms or concepts defined in this chapter are

quantized property (Sec. 9.1)
electron shell (Sec. 9.2)
electron subshell (Sec. 9.3)
electron orbital (Sec. 9.4)
electron configuration (Sec. 9.5)
Aufbau principle (Sec. 9.5)
Aufbau diagram (Sec. 9.5)
quantum number (Sec. 9.6)
principal quantum number (Sec. 9.6)
secondary quantum number (Sec. 9.6)
magnetic quantum number (Sec. 9.6)
spin quantum number (Sec. 9.6)
Pauli exclusion principle (Sec. 9.7)
principle of maximum multiplicity (Hund's rule) (Sec. 9.7)
orbital diagram (Sec. 9.7)
paramagnetic (Sec. 9.7)
diamagnetic (Sec. 9.7)
$(n + l)$ rule (Sec. 9.8)
distinguishing electron (Sec. 9.10)
rare gas (Sec. 9.11)
representative element (Sec. 9.11)
transition element (Sec. 9.11)
rare-earth element (Sec. 9.11)

Questions and Problems

Terminology Associated with Electron Arrangements

9-1 What does the term *quantization* mean and what property of an electron is quantized?

9-2 The following statements define or are closely related to the terms *shell, subshell,* and *orbital.* Match the terms to the appropriate statements.

a. In terms of electron capacity, this unit is the smallest of the three.
b. This unit may contain a maximum of two electrons.
c. This unit may contain as many or more electrons as/than either of the other two.
d. The term *subenergy level* is closely associated with this unit.
e. This unit may be designated by either a number or capital letter.
f. The formula $2n^2$ gives the maximum number of electrons that may occupy this unit.
g. This unit is designated in the same way as the orbitals that comprise it.

9-3 In each of the following sentences decide whether the phrase "dependent on" or "independent of" when placed in the blank makes the statement a correct statement.

a. The energy of a subshell is ______ the shell in which it is located.
b. The maximum number of electrons a subshell can accommodate is ______ the shell in which it is found.
c. The shape of an orbital is ______ the subshell in which it is found.
d. The size (volume) of a given type of orbital (*s, p,* etc.) is ______ the shell in which it is found.

e. The maximum number of electrons an orbital can accommodate is ______ the subshell in which it is found.

f. The maximum number of electrons an orbital can accommodate is ______ the shell in which it is found.

9-4 Indicate whether each of the following statements is *true* or *false*.

a. An orbital has a definite size and shape, which are related to the energy of the electrons it could contain.

b. The *M* and $n = 3$ shells are one and the same.

c. All of the orbitals in a subshell have the same energy.

d. All of the subshells in a shell have the same energy.

e. A 2*p* and 3*p* subshell would contain the same number of orbitals.

f. A *d* subshell always contains five orbitals.

g. An *s* orbital is shaped something like a four-leaf clover.

h. The $n = 3$ shell can accommodate a maximum of 18 electrons.

i. The fourth shell is made up of six subshells.

j. All subshells accommodate the same number of electrons.

9-5 Characterize the similarities and differences between

a. a 2*s* orbital and a 2*p* orbital

b. a 4*f* orbital and a 5*f* orbital

c. the third shell and the fourth shell

9-6 Give the maximum number of electrons that may occupy each of the following units.

a. 2*p* subshell **b.** 2*p* orbital
c. third shell **d.** *L* shell
e. 4*f* subshell **f.** 4*f* orbital

Writing Electron Configurations

9-7 For each of the following sets of subshells determine the one of lowest energy and the one of highest energy.

a. 1*s*, 2*s*, 3*s* **b.** 3*s*, 3*p*, 3*d*
c. 4*d*, 4*f*, 5*p* **d.** 3*d*, 5*p*, 7*s*

9-8 Write the complete electron configuration for each of the following atoms.

a. ${}_{8}O$ **b.** ${}_{35}Br$ **c.** ${}_{20}Ca$ **d.** ${}_{79}Au$
e. ${}_{50}Sn$ **f.** ${}_{87}Fr$ **g.** ${}_{15}P$ **h.** ${}_{30}Zn$

9-9 Identify the element that has each of the following electron configurations.

a. $1s^22s^22p^3$

b. $1s^22s^22p^63s^23p^64s^1$

c. $1s^22s^22p^63s^23p^64s^23d^{10}4p^4$

d. $1s^22s^22p^63s^23p^64s^23d^2$

9-10 What is wrong with each of the following attempts to write an electron configuration?

a. $1s^22s^3$ **b.** $1s^21p^62s^22p^6$
c. $1s^22s^22p^83s^23p^8$ **d.** $1s^22s^22p^63s^23d^{10}$

9-11 Write abbreviated electron configurations that give only shell occupancy rather than subshell occupancy in the following situations.

a. the element with the electron configuration $1s^22s^22p^63s^23p^64s^1$

b. the element with the electron configuration $1s^22s^22p^63s^23p^1$

c. the element ${}_{12}Mg$

d. the element ${}_{7}N$

Quantum Numbers (optional)

9-12 For each of the four quantum numbers give the following data.

a. name **b.** symbol **c.** function
d. information it conveys about an orbital

9-13 What are the allowed quantum number values in each of the following situations?

a. values of l when $n = 4$

b. values of m when $l = 3$

c. values of m when $n = 2$

d. values of s when $l = 2$

9-14 Give the values of n and l for the following subshells.

a. 2*s* **b.** 2*p* **c.** 4*f* **d.** 3*d* **e.** 6*p*
f. 7*s* **g.** 4*d* **h.** 5*f*

9-15 The following sets of four quantum numbers (n, l, m, s) each describe an electron in an atom. What are the shell and subshell designations for each set?

a. $(2, 0, 0, -\frac{1}{2})$ **b.** $(3, 1, -1, \frac{1}{2})$
c. $(3, 1, 1, -\frac{1}{2})$ **d.** $(5, 3, 2, -\frac{1}{2})$

9-16 There are six electrons in the 3*p* subshell. Below are the quantum numbers for three of the six electrons. Write the quantum numbers for the other three electrons.

$(3, 1, 0, -\frac{1}{2})$, $(3, 1, 0, +\frac{1}{2})$, $(3, 1, 1, -\frac{1}{2})$

9-17 Write the complete set of quantum numbers for all electrons that could populate the 3*d* subshell of an atom. (*Hint:* You will need ten sets of quantum numbers.)

9-18 Give the reason why each of the following sets of quantum numbers is an impossibility.

a. $(3, 3, 1, -\frac{1}{2})$ **b.** $(3, 0, 1, +\frac{1}{2})$
c. $(3, 0, 0, 1)$ **d.** $(3, -1, -1, -\frac{1}{2})$

9-19 List the values of all four quantum numbers for *each* electron in atoms of

a. ${}_{4}Be$ **b.** ${}_{10}Ne$

9-20 What is the maximum number of electrons that an atom can have with each of the following quantum number specifications?

a. $n = 4, \quad l = 0$

b. $n = 4, \quad l = 2$

c. $n = 3, \quad l = 2, \quad m = 2$

d. $n = 3,\ s = -\frac{1}{2}$
e. $n = 3,\ m = 1$
f. $n = 2,\ l = 0,\ m = 0,\ s = -\frac{1}{2}$

9-21 Using quantum numbers, explain why there is only one *s*-type orbital but three *p*-type orbitals in any shell (except the first shell).

Orbital Diagrams (optional)

9-22 Draw electron orbital diagrams for the following elements.
a. $_3$Li **b.** $_{22}$Ti **c.** $_{12}$Mg **d.** $_{16}$S
e. $_{19}$K **f.** $_9$F **g.** $_{33}$As

9-23 How many unpaired electrons are found in each of the following elements?
a. $_{20}$Ca **b.** $_{15}$P **c.** $_8$O **d.** $_{23}$V
e. $_{18}$Ar **f.** $_{55}$Cs

9-24 Predict whether each of the following atoms is paramagnetic or diamagnetic.
a. $_{12}$Mg **b.** $_6$C **c.** $_{30}$Zn **d.** $_{13}$Al
e. $_{10}$Ne **f.** $_{19}$K

The $(n + l)$ Rule (optional)

9-25 Using the $(n + l)$ rule, predict which subshell in each of the following groups is lower in energy.
a. 1*s*, 2*s*, 3*s* **b.** 3*s*, 3*p*, 4*p*
c. 6*s*, 7*p*, 4*d* **d.** 7*s*, 6*d*, 5*f*

The Periodic Table and Electron Configurations

9-26 Sketch an outline of the periodic table and indicate the following.
a. the *s* area **b.** the *p* area
c. the *d* area **d.** the *f* area

9-27 Indicate the position in the periodic table (by giving the symbol of the element) where each of the following occurs.
a. the 3*s* subshell begins filling
b. the 4*d* subshell begins filling
c. the 4*p* subshell becomes half-filled
d. the 5*p* subshell becomes completely filled
e. the fifth shell begins filling
f. the third shell becomes completely filled

9-28 Using only the periodic table, determine the electron configurations of the following elements.
a. $_{20}$Ca **b.** $_{33}$As **c.** $_{14}$Si **d.** $_{41}$Nb
e. $_{57}$La **f.** $_{86}$Rn

9-29 Using the periodic table as a guide indicate the number of
a. 3*p* electrons in a $_{16}$S atom
b. 3*d* electrons in a $_{29}$Cu atom
c. 4*s* electrons in a $_{37}$Rb atom
d. 4*d* electrons in a $_{78}$Pt atom
e. 4*p* electrons in a $_{15}$P atom
f. 3*d* electrons in a $_{30}$Zn atom

9-30 What is the identity of the subshell (2*s*, 3*p*, 4*f*, etc.) containing the distinguishing electron in each of the following atoms?
a. $_4$Be **b.** $_{34}$Se **c.** $_{44}$Ru **d.** $_{51}$Sb
e. $_{73}$Ta **f.** $_{86}$Rn

9-31 Each of the following is the outer subshell electron configuration after the distinguishing electron has been added. Using the periodic table, identify the elements with these configurations.
a. $6s^2$ **b.** $4d^4$ **c.** $3p^6$ **d.** $5f^{14}$
e. $6p^1$ **f.** $2p^6$

9-32 Based on the relationship between electron configurations and the periodic table, indicate which elements contain the electron characteristics below. In those cases where a series of elements have the indicated characteristics you need not write all the symbols but rather may write the atomic numbers of the first and last elements in the series, i.e., elements 70–83.
a. a total of 82 electrons
b. only three 5*p* electrons
c. two 7*s* electrons (note that more than one element qualifies)
d. a total of fifteen "*p*" electrons (they are not all in the same subshell)
e. only one electron in shell 7
f. a total of six "*s*" electrons (note that more than one element qualifies)

Classification Systems for the Elements

9-33 Formulate definitions for elements classified as rare gases, representative elements, transition elements, and rare-earth elements in terms of *s*, *p*, *d*, and *f* electrons.

9-34 Identify each of the following as a rare gas, representative element, transition element, or rare-earth element.
a. $_{54}$Xe **b.** $_{45}$Rh **c.** $_{37}$Rb **d.** $_{73}$Ta
e. $_{64}$Gd **f.** $_{15}$P **g.** $_{81}$Tl **h.** $_{100}$Fm

10 Chemical Bonding

10.1 Chemical Bonds

In both Chapters 5 (Sec. 5.2) and 6 (Sec. 6.3) we dealt with the fact that two types of compounds exist — ionic compounds and molecular compounds. Ionic compounds, we recall, have structures based on infinite three-dimensional arrangements of positive and negative ions and molecular compounds contain separate fixed groupings of atoms called molecules. Most ionic compounds contain both metallic and nonmetallic elements. Most molecular compounds contain only nonmetallic elements.

Why is it that some compounds contain ions whereas others have molecular building blocks? What is it that determines whether the interaction of two elements produces ions or molecules? In this chapter we consider answers to these questions. The answers are found in a consideration of the subject of *chemical bonding.*

chemical bonds

Chemical bonds *are the attractive forces that hold atoms or ions together in more complex aggregates.* The way in which atoms interact with each other to form chemical bonds is dictated by electron configuration. Thus, many of the concepts of Chapter 9 will be used in treating the subject of chemical bonding.

ionic bond

covalent bond

It is useful to classify chemical attractive forces (chemical bonds) into two categories — ionic bonds and covalent bonds. An **ionic bond** *is formed when one or more electrons are* TRANSFERRED *from one atom or group of atoms to another.* As suggested by its name, the ionic bond is particularly useful in describing the attractive forces in ionic compounds. A **covalent bond** *is formed when two atoms* SHARE *one or more electron pairs between them.* The covalent bond model is particularly useful in describing attractions between atoms in molecular compounds.

Prior to considering the details of these two bond models it is important to emphasize that the notions of ionic and covalent bonds are merely convenient concepts. Most bonds are not 100% ionic or 100% covalent. Instead, most bonds have at least some degree of both ionic and covalent character, that is, some degree of both the transfer and the sharing of electrons. But it is easiest to understand these intermediate bonds (the real bonds) by relating them to the pure or ideal bond types called ionic and covalent.

10.2 Valence Electrons and Electron-Dot Structures

Chemical bonds form as the result of the interaction (transfer or sharing) of electrons found in the combining atoms. Certain electrons, called valence electrons, are particularly important in determining the bonding characteristics of a given atom. For representative elements (Sec. 9.11) **valence electrons** *are those electrons in the outermost electron shell, that is, in the shell with the highest shell number* (n). These electrons will always be found in either s or p subshells. Note the restriction on the use of this definition; it applies only for representative elements. Most of the common elements are representative elements; hence the definition still finds much use. (We will not consider in this text the more complicated valence electron definitions for transition or rare-earth elements (Sec. 9.11); the presence of incompletely filled *inner d* or *f* subshells is the complicating factor in definitions for these elements.)

valence electrons

EXAMPLE 10.1

Determine the number of valence electrons present in atoms of each of the following elements.

(a) $_{20}Ca$ (b) $_{16}S$ (c) $_{35}Br$

SOLUTION

(a) The element calcium has two valence electrons, as can be seen by examining its electron configuration.

Number of valence electrons

$$1s^22s^22p^63s^23p^64s^2$$

Highest value of the electron shell number

The highest value of the electron shell number is $n = 4$. Only two electrons are found in shell 4, two electrons in the 4s subshell.

(b) The element sulfur has six valence electrons.

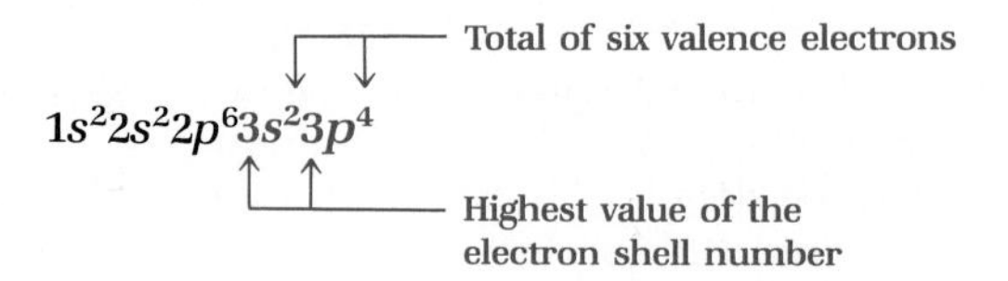

Electrons in two different subshells can simultaneously be valence electrons. The highest shell number is 3 and both the 3*s* and 3*p* subshells belong to shell number 3. Hence, all electrons in both subshells are valence electrons.

(**c**) The element bromine has seven valence electrons.

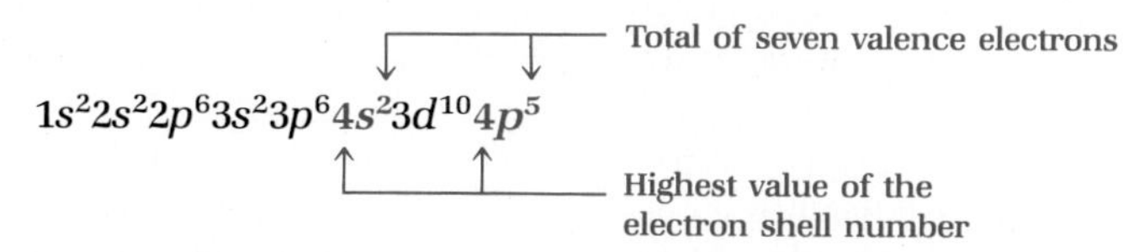

The 3*d* electrons are not counted as valence electrons because the 3*d* subshell is in shell 3 and shell 3 is not the shell with maximum *n* value. Shell 4 is the outermost shell, the shell with maximum *n* value.

The fact that the outermost electrons of atoms are those involved in bonding seems reasonable when it is remembered that the outermost electrons will be the first to come into close proximity when atoms collide — an event necessary before atoms can combine. Also, since these electrons are located the farthest from the nucleus, they are therefore the least tightly bound (attraction to the nucleus decreases with distance) and thus the most susceptible to change (transfer or sharing).

A shorthand system for designating numbers of valence electrons, which uses electron-dot structures, has been developed. Use of this system will make it easier, later in this chapter, to picture the role that valence electrons play in chemical bonding. An **electron-dot structure** *consists of an element's symbol with one dot for each valence electron placed about the elemental symbol.* Electron-dot structures for the first 20 elements, arranged as in the periodic table, are given in Figure 10-1. Note that the location of the dots is not critical. The following all have the same meaning.

electron-dot structure

Mg· Mg· ·Mg ·Mg Mg ·Mg·

Three important generalizations about valence electrons can be drawn from a study of the structures in Figure 10-1.

1. *Representative elements in the same group of the periodic table have the same number of valence electrons.* This should not be surprising to you. Elements in the same group in the periodic table have similar chemical properties as a result of having similar outershell electron configurations (Sec. 9.9). The electrons in the outermost shell are the valence electrons.
2. *The number of valence electrons for representative elements in a group is the same as the periodic table group number.* For example, the electron-dot structures for O

IA	IIA	IIA	IVA	VA	VIA	VIIA	Rare Gases
H·							·He·
Li·	·Be·	·Ḃ·	·Ċ·	·Ṅ:	:Ȯ:	:F:	:Ne:
Na·	·Mg·	·Ȧl·	·Si·	·Ṗ:	:S:	:Cl:	:Ar:
K·	·Ca·						

Figure 10-1. Electron-dot structures of selected elements.

and S, both members of group VIA, show six dots. Similarly, the electron-dot structures of H, Li, Na, and K, all members of group IA, show one dot.

3. *The maximum number of valence electrons for any representative element is eight.* All of the rare gases (Sec. 9.11), beginning with Ne, have the maximum number of eight electrons. They are the only elements with eight valence electrons. Helium, with only two valence electrons, is the exception in the rare gas family (obviously, an element with a grand total of two electrons cannot have eight valence electrons). Although shells with n greater than 2 are capable of holding more than eight electrons, they do so only when they are no longer the outermost shell and thus not the valence shell. For example, bromine (Ex. 10-1) has 18 electrons in its third shell; however, shell 4 is the valence shell in bromine.

10.3 The Octet Rule

A key concept in modern elementary bonding theory is that certain arrangements of valence electrons are more stable than others. The term "stable" as used here refers to the idea that a system (in this case an arrangement of electrons) does not easily undergo spontaneous change.

The valence-electron configurations possessed by the rare gases (He, Ne, Ar, Kr, Xe, and Rn — Sec. 9.11) are considered to be the *most stable of all valence-electron configurations.* All of the rare gases, except He, possess eight valence electrons, the maximum number possible. Helium's valence-electron configuration is $1s^2$. All of the other rare gases possess ns^2np^6 valence-electron configurations, where n has the maximum value found in the atom.

He: $1s^2$

Ne: $1s^22s^22p^6$

Ar: $1s^22s^22p^63s^23p^6$

Kr: $1s^22s^22p^63s^23p^64s^23d^{10}4p^6$

Xe: $1s^22s^22p^63s^23p^64s^23d^{10}4p^65s^24d^{10}5p^6$

Rn: $1s^22s^22p^63s^23p^64s^23d^{10}4p^65s^24d^{10}5p^66s^24f^{14}5d^{10}6p^6$

Each of the rare-gas valence-electron configurations, except for that of He, has the common characteristic of having the outermost s and p subshells *completely filled.*

The conclusion that an ns^2np^6 ($1s^2$ for He) configuration is the most stable of all valence-electron configurations is based on the chemical properties of the rare gases. The rare gases are the *most unreactive* of all the elements. They are the only elemental gases found in nature in the form of individual uncombined atoms. There are no known compounds of He, Ne, and Ar and only a very few compounds of Kr, Xe, and Rn. The rare gases appear to be "happy" the way they are. They have little or no desire to form bonds to other atoms.

Atoms of many elements that lack this very stable rare-gas valence-electron configuration tend to attain it in chemical reactions that result in compound formation. This observation has become known as the *octet rule* because of the eight valence electrons possessed by all of the rare gases except He. A formal statement of the **octet rule** is: *In*

octet rule

compound formation atoms of elements lose, gain, or share electrons in such a way as to produce a rare-gas electron configuration for each of the atoms involved.

Application of the octet rule to many different systems has shown that it has value in predicting correctly the observed combining ratios of atoms. For example, it explains why two hydrogen atoms rather than some other number are bonded to one oxygen atom in the molecular compound water. It explains why the formula of the ionic compound sodium chloride is NaCl rather than $NaCl_2$, $NaCl_3$, or Na_2Cl.

There are exceptions to the octet rule, but it is still used because of the large amount of information that it is able to correlate. It is particularly effective in explaining compound formation involving only representative elements. Often complications arise with transition and rare-earth elements because of the involvement of *d* and *f* electrons in the bonding.

10.4 Ionic Bonds

We have already encountered the topics of ions and ionic compounds (Secs. 5.9 and 6.3–6.9). Let us quickly recall some key pieces of information from these sections and then build on that foundation a discussion of ionic bonds.

1. Ionic compounds usually contain both metallic and nonmetallic elements (Sec. 6.3).
2. In binary ionic compounds the metallic element is always present as positive ions and the nonmetallic element as negative ions. Positive ions are formed as the result of the loss of electrons by metal atoms, and negative ions arise from the gain of electrons by nonmetal atoms (Sec. 6.4).
3. Electron loss and electron gain are not isolated processes: if one occurs so must the other. The electrons lost by metal atoms are the same ones gained by the nonmetal atoms. Electron loss is always equal to electron gain in any electron transfer process (Sec. 6.5).
4. In formula writing, the ratio in which positive and negative ions are combined is that ratio which achieves charge neutrality for the resulting compound (Sec. 6.5).

The formation of an ionic bond always involves the process of electron transfer. The attractions between the positive and negative ions formed as a result of the electron transfer process constitute ionic bonds.

A most simple example of ionic bonding is that which occurs between the elements sodium and fluorine in the compound NaF. Sodium atoms lose (transfer) one electron to fluorine atoms producing Na^+ and F^- ions. The ions combine in a 1 to 1 ratio to form the compound NaF.

Why do sodium atoms form Na^+ ions and not Na^{2+} or Na^- ions? Why do fluorine atoms form F^- ions rather than F^{2-} or F^+ ions? In general, why do metals always lose electrons to form positive ions and nonmetals always gain electrons to produce negative ions? What determines the specific number of electrons lost or gained in electron transfer processes?

Very simple and straightforward answers to these questions come from the octet rule (Sec. 10.3). *Atoms tend to gain or lose electrons until they have obtained an electron configuration that is the same as that of a rare gas.*

Consider the element sodium with the electron configuration

$$1s^22s^22p^63s^1$$

It can attain a rare-gas configuration by losing one electron (to give it the electron configuration of Ne) or by gaining seven electrons (to give it the electron configuration of Ar).

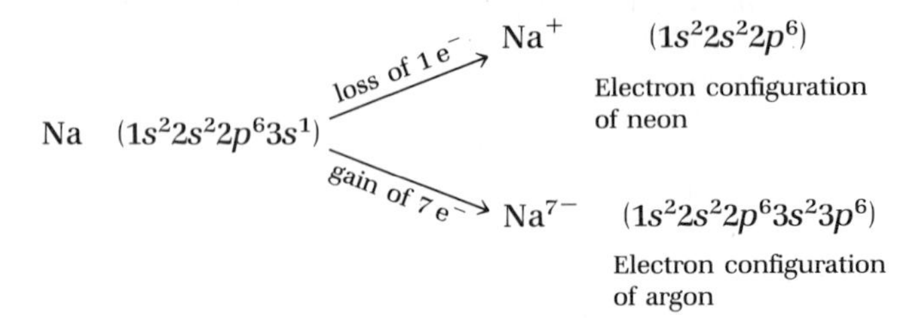

The first process, the loss of one electron, being more energetically favorable than the gain of seven electrons, is the process that occurs. The highly charged ions (such as Na^{7-}) that would result from the transfer of many electrons are not stable. Rarely are more than three electrons lost or gained by an atom during an electron transfer process.

Consider the element fluorine with the electron configuration

$$1s^22s^22p^5$$

It can attain a rare-gas configuration by losing seven electrons to give it a helium electron configuration ($1s^2$) or by gaining one electron to give it a neon electron configuration ($1s^22s^22p^6$). The latter occurs for the reason cited previously.

Considerations of the type we have just used for the elements sodium and fluorine lead to the following generalizations.

1. Metal atoms containing one, two, or three valence electrons (the metals in groups IA, IIA, and IIIA of the periodic table) tend to lose electrons to acquire a rare-gas configuration. The rare gas involved is the one preceding the metal in the periodic table.
2. Nonmetal atoms containing five, six, or seven valence electrons (the nonmetals in groups VA, VIA, and VIIA of the periodic table) tend to gain electrons to acquire a rare-gas configuration. The rare gas involved is the one following the nonmetal in the periodic table.

isoelectronic

An ion formed in the above manner with an electronic configuration the same as that of a rare gas is said to be *isoelectronic* with the rare gas. **Isoelectronic** *species contain the same number of electrons.* An atom and an ion or two ions may be isoelectronic. Numerous ions that are isoelectronic with a given rare gas exist as can be seen from the entries in Table 10-1.

It should be emphasized that an ion that is isoelectronic with a rare gas does not have the properties of the rare gas. It has not been converted into the rare gas. The number of protons in the nucleus of the isoelectronic ion is different from that in the rare gas. These points are emphasized by the comparison in Table 10-2 between Mg^{2+} and Ne, the rare gas with which Mg^{2+} is isoelectronic.

When learning how to name binary ionic compounds (Sec. 6.6) you were asked to memorize the charges associated with specific monoatomic metal and nonmetal ions

Table 10-1. ***Ions Isoelectronic with Selected Rare Gases***

Helium Structure		Neon Structure		Argon Structure	
H^-	$1s^2$	N^{3-}	$1s^2 2s^2 2p^6$	P^{3-}	$1s^2 2s^2 2p^6 3s^2 3p^6$
He		O^{2-}		S^{2-}	
Li^+		F^-		Cl^-	
Be^{2+}		Ne		Ar	
		Na^+		K^+	
		Mg^{2+}		Ca^{2+}	
		Al^{3+}			

(Figs. 6-2 and 6-3). To help you with this task, ionic charges were related to periodic table positions of the elements. The relationships given were

Group IA metals form +1 ions	Group VIIA nonmetals form −1 ions
Group IIA metals form +2 ions	Group VIA nonmetals form −2 ions
Group IIIA metals form +3 ions	Group VA nonmetals form −3 ions

We can now understand the basis for these relationships. Group IA metals are all located one periodic table position past a rare gas. Thus, they will each have one more electron than the preceding rare gas. This electron must be lost if a rare-gas configuration is to be obtained. Groups IIA and IIIA metals are two and three periodic table positions, respectively, beyond a rare gas. Consequently, two and three electrons, respectively, must be lost for these metals to become isoelectronic with a rare gas. The nonmetal ionic charge guidelines can be explained by similar reasoning. Only this time the periodic table positions are those immediately preceding the rare gases. Consequently electrons must be gained to attain rare-gas configurations. Elements in group IVA occupy unique positions relative to the rare gases. They are located equidistant between two rare gases. For example, the element carbon is four positions beyond He and four positions before Ne. Theoretically, ions with charges of plus or minus four could be formed by elements in this group, but in most cases the bonding that results is more adequately described by the covalent bond model to be discussed in Section 10.5.

The use of electron-dot structures helps in visualizing the formation of ionic compounds through electron-transfer processes. Let us consider the reaction between Na

Table 10-2. ***A Comparison of the Structure of a Mg^{2+} Ion and a Ne Atom, the Rare-Gas Atom Isoelectronic with the Ion***

	Ne Atom	Mg^{2+} Ion
Protons (in the nucleus)	10	12
Electrons (around the nucleus)	10	10
Atomic number	10	12
Charge	0	2+

(one valence electron) and F (seven valence electrons) to give NaF. This reaction can be represented as follows with electron-dot structures.

$$\text{Na}\cdot + \cdot\ddot{\underset{..}{\text{F}}}\!: \longrightarrow \left[\text{Na}^+ + :\ddot{\underset{..}{\text{F}}}\!:^-\right] \longrightarrow \text{NaF}$$

Loss of an electron empties the valence shell of Na. The next inner shell, which contains eight electrons, then becomes the valence shell.

$$\text{Na:} \quad 1s^22s^22p^63s^1 \longleftarrow \text{Original valence shell}$$
$$\text{Na}^+\text{:} \quad 1s^22s^22p^6 \longleftarrow \text{New valence shell}$$

The outer shell of F, after the gain of one electron, contains the desired eight valence electrons.

$$\text{F:} \quad 1s^22s^22p^5 \longleftarrow \text{Seven valence electrons}$$
$$\text{F}^-\text{:} \quad 1s^22s^22p^6 \longleftarrow \text{Eight valence electrons}$$

When Na (one valence electron) combines with O (six valence electrons), two Na atoms are needed to meet the need of two additional electrons for an oxygen atom.

$$2\,\text{Na}\cdot + :\dot{\underset{..}{\text{O}}}\!: \longrightarrow \left[2(\text{Na}^+) + :\ddot{\underset{..}{\text{O}}}\!:^{2-}\right] \longrightarrow \text{Na}_2\text{O}$$

A situation just opposite to that found in the formation of Na_2O occurs in the reaction between Mg (two valence electrons) and Cl (seven valence electrons). Here two Cl atoms are required to accommodate electrons transferred from one Mg atom.

$$\text{Mg}\!:\; + 2\,\cdot\ddot{\underset{..}{\text{Cl}}}\!: \longrightarrow \left[\text{Mg}^{2+} + 2(:\ddot{\underset{..}{\text{Cl}}}\!:^-)\right] \longrightarrow \text{MgCl}_2$$

EXAMPLE 10.2

Show the formation of the following ionic compounds using electron-dot structures.

(a) K_3N (b) BaO (c) Al_2O_3

SOLUTION

(a) K has one valence electron, which it would like to lose. N has five valence electrons and would, thus, like to acquire three more. Three K atoms will be required to supply enough electrons for one N atom.

$$3\,\text{K}\cdot + \cdot\dot{\underset{\cdot}{\text{N}}}\!: \longrightarrow \left[3(\text{K}^+) + :\ddot{\underset{..}{\text{N}}}\!:^{3-}\right] \longrightarrow \text{K}_3\text{N}$$

(b) Ba has two valence electrons and O has six valence electrons. Transfer of the two Ba valence electrons to an O atom will result in each atom having a rare gas configuration. Thus, these two elements combine in a 1 to 1 ratio.

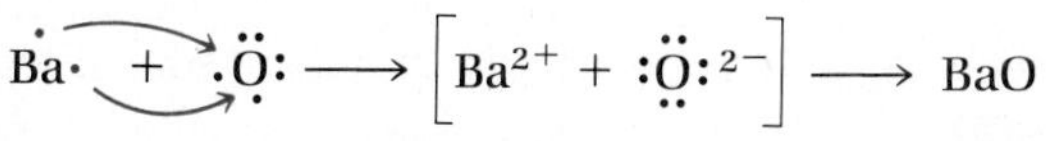

$$\text{Ba}\cdot\cdot + \cdot\ddot{\text{O}}: \longrightarrow [\text{Ba}^{2+} + :\ddot{\text{O}}:^{2-}] \longrightarrow \text{BaO}$$

(c) Al has three valence electrons, all of which need to be lost through electron transfer. O has six valence electrons and thus needs to acquire two more. Three O atoms are needed to accommodate the electrons given up by two Al atoms.

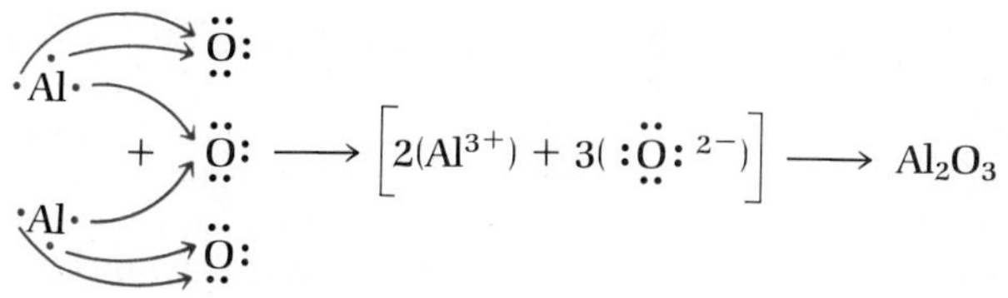

$$2\,\cdot\dot{\text{Al}}\cdot + 3\,\ddot{\text{O}}: \longrightarrow [2(\text{Al}^{3+}) + 3(:\ddot{\text{O}}:^{2-})] \longrightarrow \text{Al}_2\text{O}_3$$

10.5 Covalent Bonds

In binary ionic compounds, the two atoms involved in a given ionic bond are a metal and a nonmetal — atoms that are quite *dissimilar*. These dissimilar atoms are complimentary to each other; one atom (the metal) likes to lose electrons and the other atom (the nonmetal) likes to gain electrons. The net result is the transfer of one or more electrons (Sec. 10.4).

Covalent bonds are formed between *similar* or even *identical* atoms. Most often the atoms involved are two nonmetal atoms. It is not reasonable to suppose that one atom would give up electrons to another atom when the atoms are identical or very similar. The concept of *electron sharing* rather than electron transfer explains bonding between similar or identical atoms. In electron sharing two nuclei attract the same electrons with the resulting attractive forces holding the two nuclei together. The formation of a covalent bond always involves the process of electron sharing.

The hydrogen molecule (H_2) is the simplest covalent bonding situation that exists. Hydrogen, with just one 1*s* electron, needs one more electron to obtain a rare-gas configuration, that of helium ($1s^2$). Hydrogen atoms accomplish this by sharing their lone electron with another hydrogen atom, which in turn reciprocates, sharing its electron with the first hydrogen atom. The net result is the formation of an H_2 molecule. The two shared electrons in an H_2 molecule do "double duty," helping each of the H atoms achieve a rare-gas configuration.

Covalent bonds are represented using electron-dot structures in much the same way as we used them for ionic bonds. A pair of dots placed between the symbols of the bonded atoms indicates the shared pair of electrons. The electron-dot notation for H_2 is

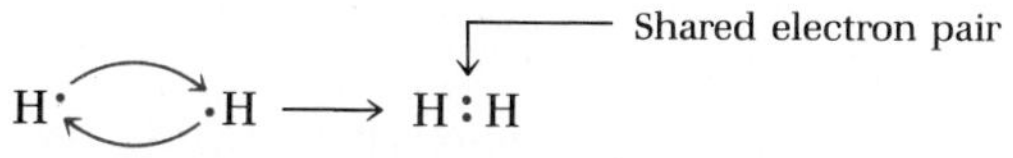

$$\text{H}\cdot + \cdot\text{H} \longrightarrow \text{H}:\text{H}$$

Shared electron pair

An alternate way of representing the shared electron pair in a covalent bond is to draw a dash between the symbols of the bonded atoms.

$$\text{H—H}$$

Both of the atoms in H_2 have access to the two electrons of the shared electron pair. The concept of overlap of orbitals helps visualize this. As shown in Figure 10-2a, suppose two H atoms are moving toward each other to form H_2 eventually. As long as the atoms are well separated, the 1*s* electrons on the two atoms are independent of each other. As the atoms get closer and closer together, the orbitals containing the electrons eventually overlap and create an orbital common to both atoms (Fig. 10-2b). When this happens, the two electrons move throughout the overlap region between the nuclei and are *shared* by both nuclei.

Two hydrogen atoms in an H_2 molecule sharing two electrons between them is a more stable situation than two separate H atoms, each with one electron. Thus, H atoms are always found in pairs (as H_2 molecules) in samples of elemental hydrogen.

Using the octet rule and electron-dot structures, let us consider some other simple molecular compounds where covalent bonding is present.

The element chlorine, located in group VIIA of the periodic table, has seven valence electrons. Its electron-dot structure is

$$\cdot\ddot{\underset{\cdot\cdot}{\text{Cl}}}:$$

Chlorine needs only one electron to achieve the octet of electrons that makes it isoelectronic with the rare gas argon. In ionic compounds, where it bonds to metals, the Cl receives the needed electron via electron transfer. When Cl combines with another nonmetal, a common situation, the octet of electrons is completed via electron sharing. Representative of the situation where chlorine obtains its eighth valence electron through an electron-sharing process are the molecules HCl, Cl_2, and BrCl, whose electron-dot structures are as follows.

$$\text{H}\cdot \;\; \cdot\ddot{\underset{\cdot\cdot}{\text{Cl}}}: \longrightarrow \text{H}:\ddot{\underset{\cdot\cdot}{\text{Cl}}}: \quad \text{or} \quad \text{H—}\ddot{\underset{\cdot\cdot}{\text{Cl}}}:$$

$$:\ddot{\underset{\cdot\cdot}{\text{Cl}}}\cdot \;\; \cdot\ddot{\underset{\cdot\cdot}{\text{Cl}}}: \longrightarrow :\ddot{\underset{\cdot\cdot}{\text{Cl}}}:\ddot{\underset{\cdot\cdot}{\text{Cl}}}: \quad \text{or} \quad :\ddot{\underset{\cdot\cdot}{\text{Cl}}}\text{—}\ddot{\underset{\cdot\cdot}{\text{Cl}}}:$$

$$:\ddot{\underset{\cdot\cdot}{\text{Br}}}\cdot \;\; \cdot\ddot{\underset{\cdot\cdot}{\text{Cl}}}: \longrightarrow :\ddot{\underset{\cdot\cdot}{\text{Br}}}:\ddot{\underset{\cdot\cdot}{\text{Cl}}}: \quad \text{or} \quad :\ddot{\underset{\cdot\cdot}{\text{Br}}}\text{—}\ddot{\underset{\cdot\cdot}{\text{Cl}}}:$$

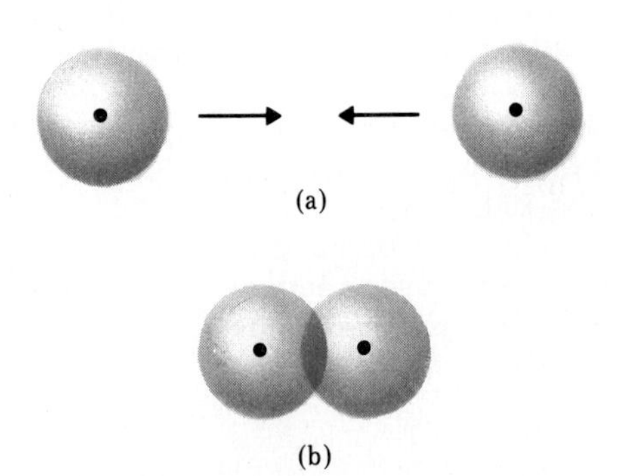

Figure 10-2. Overlap of orbitals during electron sharing.

Note that in each of these molecules Cl atoms have four pairs of electrons about them (an octet) and that only one of the four pairs is involved in electron sharing. The three pairs of electrons on each Cl atom not taking part in the bonding are called nonbonding electron pairs or unshared electron pairs.

The number of covalent bonds that an atom forms is equal to the number of electrons it needs to achieve a rare-gas configuration. Note that the chlorine atoms in HCl, Cl_2, and BrCl all formed one covalent bond. For chlorine, seven valence electrons plus one electron acquired by electron sharing (one bond) gives the eight valence electrons needed for a rare-gas electronic configuration. The elements oxygen, nitrogen, and carbon have, respectively, six, five, and four valence electrons. Therefore, these elements form, respectively, two, three, and four covalent bonds. The number of covalent bonds these three elements form is reflected in the formulas of their simplest hydrogen compounds — H_2O, NH_3, and CH_4. Electron-dot diagrams for these three molecules are as follows.

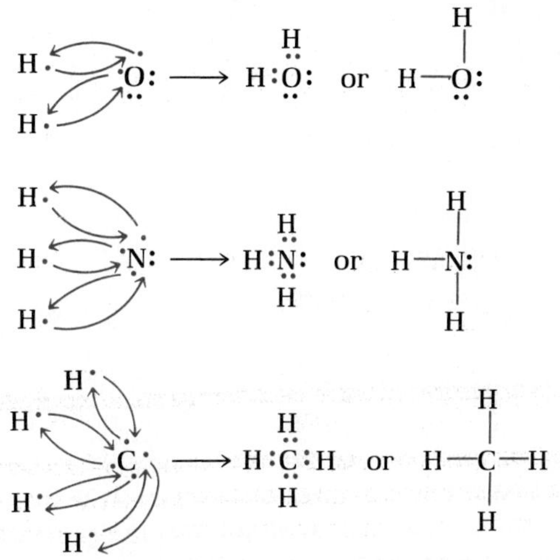

$$H\cdot \;\; H\cdot \;\; \cdot\underset{\cdot\cdot}{\dot{O}}: \longrightarrow H:\underset{\cdot\cdot}{\overset{H}{\ddot{O}}}: \quad \text{or} \quad H-\underset{\cdot\cdot}{\overset{H}{\overset{|}{O}}}:$$

$$H\cdot \;\; H\cdot \;\; H\cdot \;\; \cdot\underset{\cdot}{\dot{N}}: \longrightarrow H:\underset{H}{\underset{\cdot\cdot}{\overset{H}{\ddot{N}}}}: \quad \text{or} \quad H-\underset{H}{\underset{|}{\overset{H}{\overset{|}{N}}}}:$$

$$H\cdot \;\; H\cdot \;\; H\cdot \;\; H\cdot \;\; \cdot\underset{\cdot}{\dot{C}}\cdot \longrightarrow H:\underset{H}{\underset{\cdot\cdot}{\overset{H}{\ddot{C}}}}:H \quad \text{or} \quad H-\underset{H}{\underset{|}{\overset{H}{\overset{|}{C}}}}-H$$

Thus, we see that just as the octet rule was useful in determining the ratio of ions in ionic compounds, we can use it to predict formulas in covalent compounds. Example 10.3 gives additional illustrations of the use of the octet rule to determine formulas for molecular compounds.

EXAMPLE 10.3

Write electron-dot diagrams for the simplest binary compound formed from the following pairs of elements.

(a) phosphorus and bromine **(b)** hydrogen and sulfur
(c) oxygen and fluorine

SOLUTION

(a) Phosphorus is in group VA of the periodic table and thus has five valence electrons. It will, thus, want to form three covalent bonds. Bromine, in group VIIA of the periodic table, has seven valence electrons and will want to form only one covalent bond. Therefore, we have

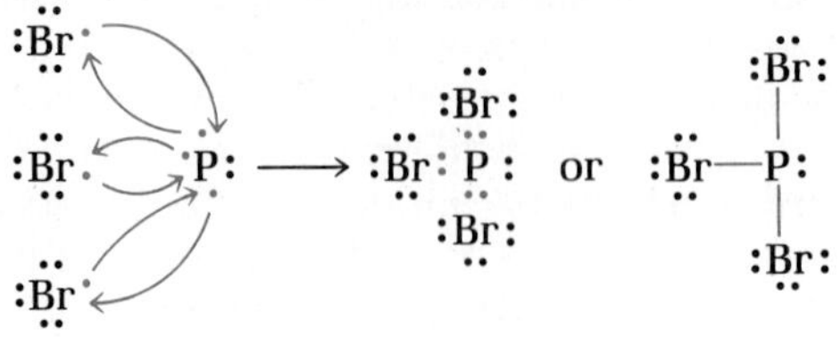

:Br· :Br· :Br· + ·P: ⟶ :Br:P:(Br)(Br) or :Br—P:(—Br)(—Br)

Each atom in PBr_3 has an octet of electrons, which is circled in color in the following diagram.

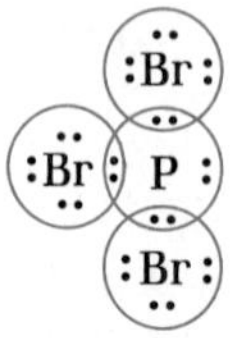

:Br:P:(Br)(Br)

(b) Sulfur has six valence electrons and hydrogen has one valence electron. Thus sulfur will form two covalent bonds $(6 + 2 = 8)$ and hydrogen will form one covalent bond $(1 + 1 = 2)$. Remember, that for H an "octet" is two electrons; the rare gas that hydrogen "mimics" is helium, which has only two valence electrons.

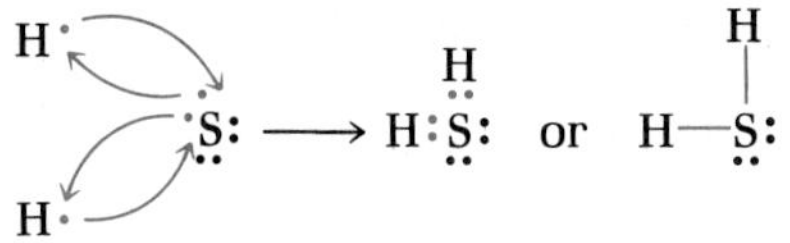

H· H· + ·S: ⟶ H:S:(H) or H—S:(—H)

(c) Oxygen, with six valence electrons, will form two covalent bonds, and fluorine, with seven valence electrons, will form only one covalent bond. The formula of the compound is thus OF_2, which has the following electron-dot structure.

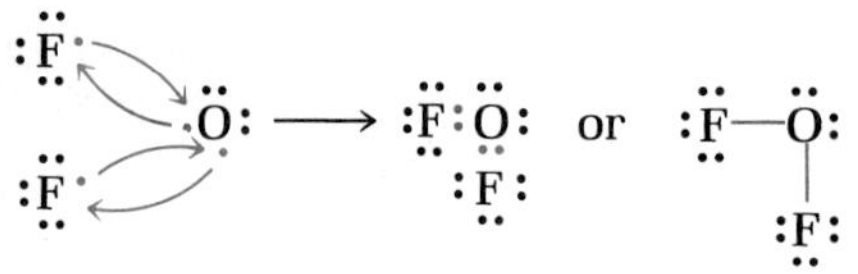

:F· :F· + ·O: ⟶ :F:O:(F) or :F—O:(—F)

10.6 Electronegativities and Bond Polarities

Most chemical bonds are not 100% ionic nor 100% covalent. Instead, most bonds have some characteristics of both types of bonding. We mentioned this fact in Section 10.1, as we started our discussion of chemical bonding.

Ionic and covalent bonding represent the two extremes of a broad continuum of bonding types. Between the extremes of complete transfer of electrons (ionic bonding) and equal sharing of electrons (covalent bonding) we find many intermediate cases where electrons are shared, but shared *unequally*.

Unequal sharing of electrons results from some atoms possessing a greater ability to attract bonding electrons to themselves than others. The greater the difference in electron-attracting ability between atoms involved in a bond the more unequal the sharing of bonding electrons becomes. The ultimate in unequal sharing is reached when one atom assumes complete control of the bonding electrons; electron transfer has occurred.

A pictorial representation of the continuum of bonding types possible is obtained by considering the distribution of bonding electrons about the nuclei in a diatomic molecule as a function of an increasing difference in the electron-attracting ability of the two nuclei. Such a representation is shown in Figure 10-3. Figure 10-3a shows the distribution of bonding electron density expected when the two atoms of the diatomic molecule are identical. (Note that an "electron cloud" is used to depict the bonding electrons, as in Sec. 5.4). The sharing of electrons must be equal since identical nuclei must affect the bonding electrons in the same way; hence the symmetrical electron density distribution. The situation depicted in Figure 10-3a corresponds to a 100% covalent bond. Note that this diagram is very similar to Figure 10-2b, which describes electron sharing in H_2, a situation where both nuclei are the same.

Any time two nuclei differ in their ability to attract a pair of bonding electrons unequal sharing results. Figure 10-3b shows the situation where such a difference is very small. Electron sharing will be "close" to being equal, but not exactly equal. Note that the electron distribution is no longer symmetrical. This means that the bonding electrons spend more time associated with the nucleus that has the greater electron-attracting ability. This will be the nucleus on the right side in Figure 10-3b, the side where the electron density is the largest.

Figure 10-3c depicts electron density distribution when a relatively large difference in electron-attracting ability exists between nuclei. The sharing of electrons here can be described as being "very unequal."

Finally, when the electron-attracting ability difference becomes very large, one

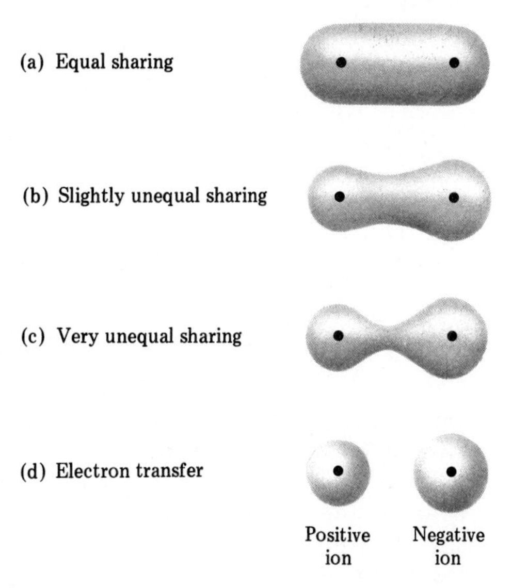

Figure 10-3. The continuum of bonding types.

atom "wins the battle." Electron transfer occurs, and the situation depicted in Figure 10-3d results. This situation corresponds to a 100% ionic bond.

electronegativity

An estimate of the "degree of unequalness" of electron sharing in a chemical bond can be obtained from the *electronegativity values* of the bonded atoms. What are electronegativity values? What is electronegativity? **Electronegativity** *is a measure of the relative attraction that an atom has for the shared electrons in a bond.* Electronegativity values are unitless numbers on a relative scale that are obtained from bond energies and other related experimental data. Electronegativities cannot be directly measured in the laboratory. The most widely accepted scale of electronegativity values is given in Figure 10-4. On this scale, fluorine, the most electronegative of all elements, has arbitrarily been assigned a value of 4.0 (the maximum on the scale) and serves as the reference element. *The higher the electronegativity of an element, the greater the electron-attracting ability of atoms of that element.*

For the representative elements, electronegativity increases from left to right across a period of the periodic table and decreases from top to bottom in a group of the periodic table. These two trends result in nonmetals generally having higher electronegativities than metals. This fact is consistent with our previous generalization (Sec. 6.4) that metals tend to lose electrons and nonmetals tend to gain electrons when an ionic bond is formed. Metals (low electronegativities, poor electron attractors) will give up electrons to nonmetals (high electronegativities, good electron attractors).

polarity

nonpolar covalent bond

polar covalent bond

The difference in the electronegativity values of the atoms in a bond is the key to predicting the *polarity* of that bond. **Polarity** *is a measure of inequality in the sharing of the bonding electrons.* The terms nonpolar and polar are used in describing bonds. A **nonpolar covalent bond** *is one where there is equal sharing of bonding electrons.* A **polar covalent bond** *is one in which sharing of bonding electrons is unequal.* The larger the difference in electronegativity between the atoms involved in a bond the more polar the bond. The "ultimate" in bond polarity is the ionic bond.

There are physical consequences to a bond being polar. In a polar covalent bond, the more electronegative atom, because of its greater electron-attracting ability, pulls the bonding electrons closer to it. The net effect of this is an unbalanced (nonsymmetrical) distribution of electron density in the bond (Fig. 10-3b and c). This results in a slight buildup of negative charge about the more electronegative atom. This additional electron density about the more electronegative atom comes from the other

H 2.1																	He —
Li 1.0	Be 1.5											B 2.0	C 2.5	N 3.0	O 3.5	F 4.0	Ne —
Na 0.9	Mg 1.2											Al 1.5	Si 1.8	P 2.1	S 2.5	Cl 3.0	Ar —
K 0.8	Ca 1.0	Sc 1.3	Ti 1.5	V 1.6	Cr 1.6	Mn 1.5	Fe 1.8	Co 1.8	Ni 1.8	Cu 1.8	Zn 1.6	Ga 1.6	Ge 1.8	As 2.0	Se 2.4	Br 2.8	Kr —
Rb 0.8	Sr 1.0	Y 1.2	Zr 1.4	Nb 1.6	Mo 1.8	Tc 1.9	Ru 2.2	Rh 2.2	Pd 2.2	Ag 1.9	Cd 1.7	In 1.7	Sn 1.8	Sb 1.9	Te 2.1	I 2.5	Xe —
Cs 0.7	Ba 0.9	57-71 1.1-1.2	Hf 1.3	Ta 1.5	W 1.7	Re 1.9	Os 2.2	Ir 2.2	Pt 2.2	Au 2.4	Hg 1.9	Tl 1.8	Pb 1.8	Bi 1.9	Po 2.0	At 2.2	Rn —
Fr 0.7	Ra 0.9																

Figure 10-4. Electronegativities of the elements.

element in the bond (the less electronegative element), which has acquired a slightly positive charge. This means that one end of the bond is negative *with respect* to the other end.

The unequal sharing of electrons in a covalent bond is often indicated by a notation that uses the lowercase Greek letter δ (delta). A δ− symbol, meaning a "partial negative charge," is placed above the relatively negative atom of the bond and a δ+ symbol, meaning a "partial positive charge," is placed above the relatively positive atom.

By "delta notation," the bond in hydrogen fluoride (HF) would be depicted as

$$\overset{\delta+}{\mathrm{H}}—\overset{\delta-}{\mathrm{F}}$$

Fluorine has an electronegativity of 4.0 and hydrogen's electronegativity is 2.1 (see Fig. 10-4). Since F is the more electronegative of the two elements, it dominates the electron-sharing process and draws the electrons closer to it. Hence, the F end of the bond has the δ− designation. Again, the δ− over the F atom indicates a "partial negative charge," that is, that the fluorine end of the molecule is negative with respect to the hydrogen end. The meaning of the δ+ over the H is that the H end of the molecule is positive with respect to the F end. Partial charges are always charges less than 1+ or 1−. Charges of 1+ and 1−, full charges, would result when an electron is transferred from one atom to another. With partial charges we are talking about an intermediate charge state between 0 and 1.

The message of this section is that there is no natural boundary between ionic and covalent bonding. Most bonds are a mixture of the pure ionic and pure covalent bonds, that is, unequal sharing of electrons occurs. Most bonds have both ionic and covalent characters. Nevertheless, it is still convenient to use the terms ionic and covalent in describing chemical bonds, based on the following guidelines, which relate to electronegativity differences.

1. Bonds between identical atoms (or nonidentical atoms of equal electronegativity), where there is zero difference in electronegativity between atoms, are called *nonpolar covalent bonds.*
2. Bonds where the electronegativity difference between atoms is greater than zero but less than 1.7 are called *polar covalent bonds.*
3. Bonds where the difference in electronegativity between atoms is 1.7 or greater are called *ionic bonds.*

We see, then, that the terms *ionic* and *covalent,* when applied to a particular bond,

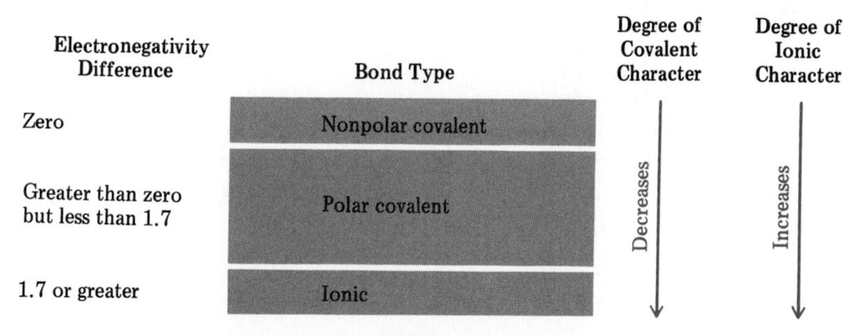

Figure 10-5. The relationships between bonding type, electronegativity difference, degree of covalent character, and degree of ionic character.

describe the predominant character of the bond. Almost all bonds have both ionic and covalent characters; we describe the bond in terms of the dominant type.

The above guidelines are useful in deciding whether to formulate the bonding description for a compound in terms of an ionic electron-dot structure (Sec. 10.4) or a covalent electron-dot structure (Sec. 10.5).

Figure 10-5 summarizes the relationships developed in this section among electronegativity differences, bond types, and bond terminology.

EXAMPLE 10.4

Indicate whether each of the following bonds is nonpolar covalent, polar covalent, or ionic. Also indicate the direction of polarity, if any, for bonds that involve a covalent designation, using delta notation ($\delta+$ and $\delta-$).

(a) ClF **(b)** MgO **(c)** HI

SOLUTION

(a) The electronegativities of Cl and F are 3.0 and 4.0, respectively. The electronegativity difference, obtained by subtracting the smaller electronegativity value from the larger, is

$$\text{Electronegativity difference} = 4.0 - 3.0 = 1.0$$

The bond is thus **polar covalent**. Bonds where the electronegativity difference is greater than zero but less than 1.7 are classified as polar covalent.

The fluorine end of the bond will be negative relative to the chlorine end, since F is the more electronegative of the two elements. Thus, the polarity direction is

$$\overset{\delta+}{\text{Cl}}—\overset{\delta-}{\text{F}}$$

(b) The electronegativity of oxygen is 3.5 and that of magnesium is 1.2. The electronegativity difference is

$$\text{Electronegativity difference} = 3.5 - 1.2 = 2.3$$

The term **ionic** is used to describe bonds when the electronegativity difference is 1.7 or greater. The polarity of the Mg—O bond can best be described in terms of ions—complete transfer of electrons.

$$[Mg^{2+}][O^{2-}]$$

(c) The electronegativity of iodine is 2.5 and that of hydrogen is 2.1. The electronegativity difference is 0.4. Since the electronegativity difference is greater than zero but less than 1.7 the bond is called a **polar covalent** bond.

In the polar covalent bond of concern here the I atom will have acquired a partial negative charge relative to the H atom since the I is the more electronegative of the two atoms. Thus, according to delta notation, the polarity is designated as

$$\overset{\delta+}{\text{H}}—\overset{\delta-}{\text{I}}$$

10.7 Multiple Covalent Bonds

single covalent bond

In our discussion of covalent bonding to this point (Secs. 10.5 and 10.6) all of the molecules chosen as examples to illustrate various aspects of such bonding have contained only *single covalent bonds.* A **single covalent bond** *is a bond where a single pair of electrons is shared (equally or unequally) between two atoms.*

multiple covalent bond

double covalent bond

triple covalent bond

Many molecules exist where two or three pairs of electrons must be shared between the same two atoms in order to provide a complete octet of electrons for each atom involved in the bonding. Such bonds are called multiple covalent bonds. **Multiple covalent bond** *is a collective term used to designate covalent bonds where two or three pairs of electrons are shared (equally or unequally) between the same two atoms.* Multiple covalent bonds may be further classified as double covalent bonds or triple covalent bonds. A **double covalent bond** *is a bond where two pairs of electrons are shared (equally or unequally) between the same two atoms.* A double covalent bond is stronger than a single covalent bond, but not twice as strong, because two electron pairs repel each other and cannot become fully concentrated between the two atoms. (Bond strength, the energy it takes to break a bond, can be measured experimentally.) A **triple covalent bond** *is a bond where three pairs of electrons are shared (equally or unequally) between the same two atoms.* A triple covalent bond is stronger than a single covalent bond or a double covalent bond, but not three times as strong as a single bond for the reason previously mentioned. Now let us consider some molecules where multiple bonding is present.

A diatomic N_2 molecule, the form in which nitrogen occurs in the atmosphere, contains a triple covalent bond. It is the simplest known triple covalent bond. A nitrogen atom has five valence electrons and thus needs three additional electrons to complete its octet.

$$\cdot\ddot{N}\cdot$$

In an N_2 molecule the only sharing that can take place is between the two nitrogen atoms. They are the only atoms present. Thus, to acquire a rare-gas electron configuration each nitrogen atom must share three of its electrons with the other nitrogen atom.

$$:\dot{N}\cdot \; \cdot\dot{N}: \longrightarrow :N:::N: \quad \text{or} \quad :N{\equiv}N:$$

Notice how all three shared electron pairs are placed in the area between the two nitrogen atoms in the above bonding diagrams. Note also that three lines are used to denote a triple covalent bond, paralleling the use of one line to denote a single covalent bond.

In "bookkeeping" electrons in an electron-dot structure to make sure that all atoms in the molecule have achieved their octet of electrons, *all* electrons in a multiple covalent bond are considered "to belong" to *both* of the atoms involved in the said bond. The "bookkeeping" for the N_2 molecule would be

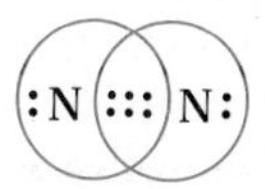

Each of the circles about an N atom contains eight valence electrons. Again, all of the electrons in a multiple covalent bond are considered to belong to each of the atoms in the bond. Circles are never drawn to include just some of the electrons in a multiple covalent bond.

A slightly more complicated molecule containing a triple covalent bond is the molecule C_2H_2 (acetylene). A carbon-carbon triple covalent bond is present as well as two carbon-hydrogen single covalent bonds. The arrangement of valence electrons in C_2H_2 is as follows.

H· ·C· → ←·C· ·H ⟶ H:C:::C:H or H—C≡C—H

The two atoms in a triple covalent bond are commonly the same element. They do not, however, have to be the same element. The molecule HCN contains a heteroatomic triple covalent bond.

H:C:::N: or H—C≡N:

As might be surmised from our discussion so far, C and N are the two representative elements most frequently encountered in triple covalent bonds.

Double covalent bonds are found in numerous molecules. A most common molecule that contains bonding of this type is carbon dioxide (CO_2). In fact, there are two carbon-oxygen double covalent bonds present in CO_2.

:Ö· ·C· ·Ö: ⟶ :Ö::C::Ö: or :Ö=C=Ö:

Note in the following diagram how the circles are drawn for the octet of electrons about each of the atoms in CO_2.

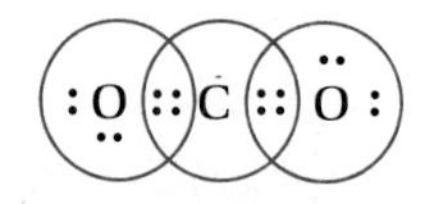

Not all elements can form multiple covalent bonds. There must be at least two vacancies in an atom's valence electron shell prior to bond formation if it is to participate in a multiple covalent bond. This requirement eliminates group VIIA elements (F, Cl, Br, I) and hydrogen from participating in such bonds. The group VIIA elements have seven valence electrons and one vacancy, and hydrogen has one valence electron and one vacancy. All bonds formed by these elements are single covalent bonds.

Double bonding becomes possible for elements needing two electrons to complete their octet, and triple bonding becomes possible when three or more electrons are needed to complete an octet. Note that the word possible was used twice in the previous sentence. Multiple bonding does not have to occur when an element has two or three or four vacancies in its octet; single covalent bonds can be formed instead. The "bonding behavior" of an element, when more than one behavior is possible, is determined by what other element or elements it is bonded to.

Let us consider the possible "bonding behaviors" for O (six valence electrons; two octet vacancies), N (five valence electrons; three octet vacancies), and C (four valence electrons; four octet vacancies).

To complete its octet by electron sharing an oxygen atom can form either one double bond or two single bonds.

$$:\ddot{O}- \text{ (with one vertical bond)} \qquad :\ddot{O}=$$

Two single bonds   One double bond

Nitrogen is a very versatile element relative to bonding. It can form single, double, or triple covalent bonds as dictated by the other atoms present in a molecule.

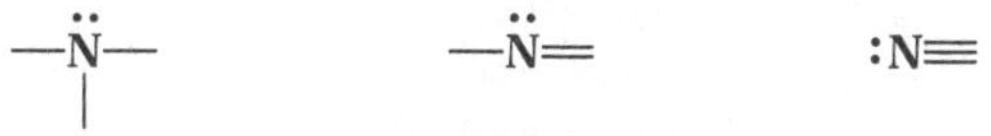

$$-\ddot{N}- \text{ (with one vertical bond)} \qquad -\ddot{N}= \qquad :N\equiv$$

Three single bonds   One single and one double bond   One triple bond

Note that in each of these bonding situations a nitrogen atom forms three bonds. A double bond counts as two bonds and a triple bond as three bonds. Since nitrogen has only five valence electrons, it must form three covalent bonds to complete its octet.

Carbon is an even more versatile element than nitrogen with respect to variety of types of bonding as illustrated by the following possibilities for bonding.

$$-C- \text{ (with two vertical bonds)} \qquad -C= \text{ (with one vertical bond)} \qquad =C= \qquad -C\equiv$$

Four single bonds   Two single bonds and one double bond   Two double bonds   One triple bond and one single bond

10.8 Coordinate Covalent Bonds

In the examples of covalent bonding encountered so far, whether the bonds be single, double, or triple, each of the bonded atoms has contributed an equal number of electrons to the bond — one each for a single covalent bond, two each for a double covalent bond, and three each for a triple covalent bond. A few molecules exist for which all of the covalent bonding within the molecule cannot be explained in this manner; instead the concept of coordinate covalency must be invoked.

coordinate covalent bond

A **coordinate covalent bond** *is a bond in which both electrons of a shared pair come from one of the two atoms.* Coordinate covalent bonding allows an atom that has two (or more) vacancies in its valence shell to share a pair of nonbonding electrons located on another atom.

The ammonium ion, NH_4^+, is an example of a species containing a coordinate covalent bond. The formation of a NH_4^+ ion can be viewed as resulting from the reaction of a hydrogen ion, H^+, with an ammonia molecule, NH_3. "Bookkeeping" on all of the valence electrons involved in this reaction, using × for nitrogen electrons and dots for hydrogen electrons, we get

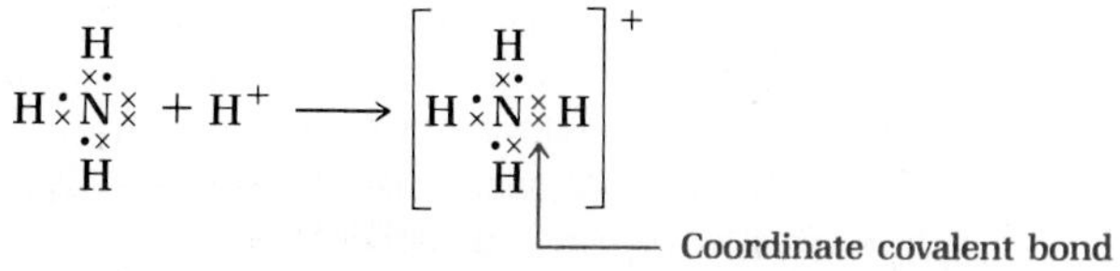

$$H{:}\overset{H}{\underset{H}{N}}{\times\atop\times} + H^+ \longrightarrow \left[H{:}\overset{H}{\underset{H}{N}}{:}H \right]^+$$

Coordinate covalent bond

A H^+ ion has no electrons, hydrogen having lost its only electron when it became an ion. The H^+ ion has two vacancies in its valence shell; that is, it needs two electrons to become isoelectronic with the rare gas helium. The nitrogen in NH_3 possesses a pair of nonbonding electrons. These electrons are used in forming the new nitrogen-hydrogen bond. The new species formed, the NH_4^+ ion, is charged since the H^+ was charged. The 1+ charge on the NH_4^+ ion is dispersed over the *entire* molecule; it is not localized on the "new" hydrogen atom.

Once a coordinate covalent bond is formed there is no way of distinguishing it from any of the other covalent bonds in a molecule; all electrons are identical regardless of their source. The main use of the concept of coordinate covalency is in helping to rationalize the existence of certain molecules and ions whose bonding-electron arrangement would otherwise present problems.

Single coordinate covalent bonds are most frequently encountered when "bookkeeping on electrons" in polyatomic ion and oxyacid structures. Another simple polyatomic ion containing a single coordinate covalent bond is the hydronium ion (H_3O^+). The formation of this ion can be visualized as resulting from the reaction of a water molecule with a H^+ ion.

$$\mathrm{H\overset{\times}{\cdot}\ddot{O}\!: \;(\text{with H below})} + H^+ \longrightarrow \left[\mathrm{H\overset{\times}{\cdot}\ddot{O}\!:\!H}\;(\text{with H below})\right]^+$$

Coordinate covalent bond

The molecule N_2O contains a single coordinate covalent bond.

$$\overset{\times}{_\times}\mathrm{N} \;\vdots\vdots\; \mathrm{N} : \mathrm{O}\overset{\times}{_\times}$$

The nitrogen-nitrogen triple bond in N_2O is a normal covalent bond; the nitrogen-oxygen bond is a coordinate covalent bond where both electrons are supplied by the nitrogen atom.

The concept of coordinate covalency must also be invoked in explaining certain multiple covalent-bonding situations. The triple covalent bond joining C and O in carbon monoxide (CO) is of the coordinate covalent type.

$$\overset{\times}{_\times}\mathrm{C}\overset{\times}{_\times}\!::\!\mathrm{O}\!:$$

Four of the six electrons in the triple bond may be considered to have come from the oxygen atom. Since C has only four valence electrons before bonding, it must share four electrons (from O) in order to achieve an octet of electrons. On the other hand, O has six valence electrons before bonding and thus needs to share only two electrons (from C).

Again, once a coordinate covalent bond is formed, it is no different from any other covalent bond. Electrons are electrons; they are all identical.

10.9 Resonance Structures

In Section 10.7 it was noted that, in general, triple covalent bonds are stronger than double covalent bonds, which in turn are stronger than single covalent bonds. **Bond strength** *is measured by the energy it takes to break a bond, that is, to separate bonded*

bond strength

bond length

atoms to give neutral particles. It may be determined experimentally. Another experimentally determinable parameter of bonds is bond length. **Bond length** *is the distance between the nuclei of bonded atoms.* A direct relationship exists between bond strength and bond length. It is found that as bond strength increases bond length decreases, that is, the stronger the bond the shorter the distance between the nuclei of the atoms of the bond. Thus, in general, triple covalent bonds are shorter than double covalent bonds, which are shorter than single covalent bonds.

Most electron-dot structures for molecules give bonding pictures consistent with available experimental bond-strength and bond-length information. However, there are some molecules for which a single electron-dot structure that is consistent with bond-strength and bond-length information cannot be written.

The molecule SO_2 is an example of a situation in which a single electron-dot structure does not adequately describe bonding. A plausible electron-dot structure for SO_2, in which the octet rule is satisfied for all three atoms, is

S O O: or S O O:

However, this structure suggests that one sulfur–oxygen bond, the double bond, should be stronger and shorter than the other sulfur–oxygen bond, the single bond. Experiment shows that this is not the case; both sulfur–oxygen bonds are equivalent with both bond-length and bond-strength characteristics intermediate between those for known sulfur–oxygen single and double bonds. An electron-dot diagram indicating this intermediate situation cannot be written.

resonance structures

The solution to the phenomenon in which no single electron-dot structure adequately describes bonding involves the use of two or more electron-dot structures, known as resonance structures, to represent the bonding in the molecule. **Resonance structures** *are two or more electron-dot structures for a molecule or ion that have the same arrangement of atoms, contain the same number of electrons, and differ only in the location of the electrons.*

Two resonance structures exist for an SO_2 molecule.

S O O: ⟷ :O S O or S O O: ⟷ O S O

A double-headed arrow is used to connect resonance structures. The only difference between the two SO_2 resonance structures is in the location of one pair of electrons. The positioning of this pair of electrons determines whether the oxygen atom on the right or the left is the oxygen atom involved in the double bond.

The actual bonding in a SO_2 molecule is said to be a *resonance hybrid* of the two contributing resonance structures. Beginning chemistry students frequently misinterpret the concept of a resonance hybrid. They incorrectly envision that a molecule, SO_2 in this case, is constantly changing (resonating) between various resonance structure forms. This is not the case. For example, SO_2 is not a mixture of two kinds of molecules, nor does a single type of molecule flip-flop back and forth between the two resonance forms. There is only one kind of SO_2 molecule in which the bonding is an average of that depicted by the resonance structures. SO_2 molecules exist "full time" in this average state. A mule, the offspring of a donkey and a horse, can be considered a hybrid of a donkey and a horse. However, it is not a horse at one instant and a donkey at another;

it is always a mule. Likewise, a molecule has only one real structure, which is different from any of the resonance structures; it has characteristics of each one of them but does not match any one of them exactly.

Sometimes three, four, or even more resonance structures can be drawn for a molecule. Again, such resonance structures must all contain the same number of electrons and have the same arrangement of atoms; the structures may differ only in the location of the electrons about the atoms.

10.10 Complex Electron-Dot Structures

The task of constructing electron-dot structures for molecules and polyatomic ions containing many electrons or for which several resonance hybrids must be used to describe the bonding can become quite frustrating if it is approached in a nonsystematic "trial-and-error" manner. Use of a systematic approach for electron-dot structure writing will enable a student to avoid most of this frustration. The following guidelines make the drawing of an electron-dot structure for any molecule or polyatomic ion that obeys the octet rule, even very complicated ones, a straightforward procedure.

Step 1. *Determine the arrangement of the atoms in the molecule or polyatomic ion.*

Determining which atom is the *central atom,* that is, which atom has the most other atoms bonded to it, is the key to determining the arrangement of atoms in a molecule or ion. Most other atoms present will be bonded to the central atom. For most binary molecular compounds the molecular formula is of help in deciding the identity of the central atom. The central atom is the atom that appears only once in the formula; for example, S is the central atom in SO_3, O is the central atom in H_2O, and P is the central atom in PF_3. For oxyacids, the central atom is the atom other than hydrogen or oxygen; for example, N is the central atom in HNO_3, and S is the central atom in H_2SO_4. In most oxyacids the oxygen atoms are bonded to the central atom and the hydrogen atoms are bonded to the oxygens. Carbon is the central atom in almost all ternary carbon-containing compounds, that is, HCN, $COCl_2$, etc. Hydrogen and fluorine are never the central atom.

Step 2. *Determine the total number of valence electrons present, that is, the total number of dots that must appear in the electron-dot structure.*

The total number of valence electrons is found by adding up the number of valence electrons each atom in the molecule or ion possesses. If the species is a polyatomic ion, add one electron for each unit of negative charge present or subtract one electron for each positive charge.

Step 3. *Determine the total number of valence electrons the molecule or ion would have to have in order for each atom to possess a rare-gas configuration.*

Each atom in the molecule or ion, except for hydrogen atoms, will need eight electrons. Hydrogen atoms need only two electrons to possess a rare-gas configuration.

Step 4. *Determine the number of bonding electron pairs needed, that is, the number of shared electron pairs.*

The number of bonding electron pairs needed is obtained by taking the total number of electrons needed for each atom to have a rare-gas configuration (the number from step 3), subtracting from it the number of valence electrons actually present in the molecule or ion (the number from step 2), and then dividing by two (to give electron pairs).

Step 5. *Write into the skeletal structure (the arrangement of atoms, step 1) the bonding electron pairs and then add nonbonding electron pairs as needed to satisfy the octet rule.*

Step 6. *Check the electron-dot structure to see that the total number of electrons (dots) present is the same as that calculated in step 2 and also check that each atom satisfies the octet rule.*

The three examples that follow illustrate the use of the above procedure. Although some steps in these examples might be done "by inspection" we will follow the outlined procedure step by step in order that you may become familiar with it.

EXAMPLE 10.5

Write the electron-dot structure for the oxyacid $HClO_3$.

SOLUTION

Step 1. Since this is an oxyacid the central atom will be Cl. Hydrogen or oxygen are never the central atom in an oxyacid. In this case the oxygens are attached to the Cl and the H atom to an oxygen atom. The arrangement of atoms is thus

```
O  Cl  O  H

    O
```

Step 2. Chlorine atoms have seven valence electrons, oxygen has six valence electrons, and hydrogen has one valence electron. The total number of valence electrons present in this molecule is 26.

$$\begin{aligned} 1\ \text{Cl:}\quad 1 \times 7 &= 7 \text{ valence electrons} \\ 3\ \text{O:}\quad 3 \times 6 &= 18 \text{ valence electrons} \\ 1\ \text{H:}\quad 1 \times 1 &= 1 \text{ valence electron} \\ \hline &\ 26 \text{ valence electrons} \end{aligned}$$

Step 3. The number of electrons needed for each atom in the molecule to have a rare-gas configuration is 34.

$$\begin{aligned} 1\ \text{Cl:}\quad 1 \times 8 &= 8 \text{ electrons} \\ 3\ \text{O:}\quad 3 \times 8 &= 24 \text{ electrons} \\ 1\ \text{H:}\quad 1 \times 2 &= 2 \text{ electrons} \\ \hline &\ 34 \text{ electrons} \end{aligned}$$

Remember, hydrogen needs only two electrons to achieve a rare-gas configuration.

Step 4. The number of electrons that will be involved in electron sharing is obtained by subtracting the number of valence electrons (step 2) from the number of electrons required to give each atom a rare-gas configuration (step 3).

$$34 \text{ electrons} - 26 \text{ electrons} = 8 \text{ bonding electrons}$$

The number of shared electron pairs is obtained by dividing the number of shared electrons by 2.

$$\frac{8 \text{ bonding electrons}}{2} = 4 \text{ shared electron pairs}$$

Step 5. We have four bonding locations in the molecule and four bonding pairs. Therefore, each bond is a single covalent bond.

```
O:Cl:O:H
  ..
  O
```

Now, adding enough nonbonding electron pairs to complete each atom's octet of electrons gives the structure

```
 ..  ..  ..
:O:Cl:O:H
 ..  ..  ..
   :O:
    ..
```

Step 6. Each of the atoms has a rare-gas configuration, as shown by the circles in color.

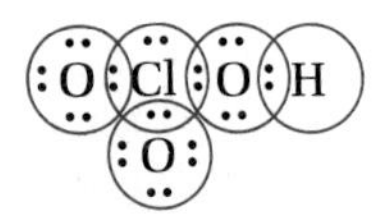

The total number of dots in this structure is 26, which matches the number calculated in step 2.

EXAMPLE 10.6

Write the electron-dot structure for the molecule SO_3.

SOLUTION

Step 1. The central atom is S; all of the O atoms will be attached to it. The atom arrangement is

```
O  S  O
   O
```

Step 2. Both S and O atoms have six valence electrons. The total number of valence electrons present is, thus, 24.

1 S:	1 × 6 =	6 valence electrons
3 O:	3 × 6 =	18 valence electrons
		24 valence electrons

Step 3. The number of electrons needed for each atom to have a rare-gas configuration is 32.

$$\begin{array}{lll} 1\,S: & 1 \times 8 = & 8 \text{ electrons} \\ 3\,O: & 3 \times 8 = & \underline{24 \text{ electrons}} \\ & & 32 \text{ electrons} \end{array}$$

Step 4. The number of electrons involved in bonding, that is, in electron sharing, is eight. Thus, four shared electron pairs are needed.

$$32 \text{ electrons} - 24 \text{ electrons} = 8 \text{ bonding electrons}$$
$$= 4 \text{ electron pairs}$$

Step 5. We have four bonding pairs and only three bonding "areas" in the molecule. Thus, one bond must be a double bond.

O:S::O
  ··
  O

There are three position choices, all equivalent, for location of the double bond. This is an indication that resonance structures exist for this molecule.

O:S::O ⟷ O:S:O ⟷ O::S:O
  ··   ::   ··
  O   O   O

We now add enough nonbonding electron pairs to complete the octet of electrons about each atom.

:Ö:S::Ö ⟷ :Ö:S:Ö: ⟷ Ö::S:Ö:
  :Ö: :Ö.   :Ö:

Step 6. Each of the atoms has a rare-gas configuration as shown by the circles in color in the following structures.

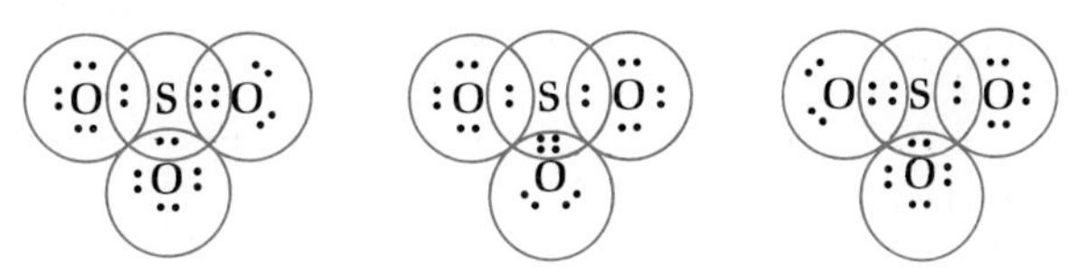

The total number of dots in each of the resonance structures is 24, which matches the number calculated in step 2.

EXAMPLE 10.7

Write the electron-dot structure for the polyatomic ion SO_4^{2-}, the sulfate ion.

SOLUTION

Step 1. Sulfur is the central atom according to the guidelines for determining central atoms.

```
    O
 O  S  O
    O
```

Step 2. Sulfur has six valence electrons as does each oxygen. There are two additional valence electrons present, since we are dealing with an ion of −2 charge.

$$\begin{aligned} 1\,S:&\quad 1 \times 6 = 6 \text{ valence electrons} \\ 4\,O:&\quad 4 \times 6 = 24 \text{ valence electrons} \\ \text{charge of } -2 =&\quad \underline{2 \text{ valence electrons}} \\ &\quad 32 \text{ valence electrons} \end{aligned}$$

If the polyatomic ion had had a positive charge instead of a negative one, we would have had to subtract valence electrons from the total instead of adding. A positive charge denotes loss of electrons and the electrons lost are valence electrons.

Step 3. The number of electrons needed for each atom to achieve a rare-gas configuration is

$$\begin{aligned} 1\,S:&\quad 1 \times 8 = 8 \text{ electrons} \\ 4\,O:&\quad 4 \times 8 = \underline{32 \text{ electrons}} \\ &\quad 40 \text{ electrons} \end{aligned}$$

Step 4. The number of electrons involved in bonding (electron sharing) is four electron pairs.

$$\begin{aligned} 40 \text{ electrons} - 32 \text{ electrons} &= 8 \text{ bonding electrons} \\ &= 4 \text{ electron pairs} \end{aligned}$$

Step 5. We have four bonding pairs and four bonding "areas" in which to put them. Thus single covalent bonding is sufficient to explain the bonding within this ion — one electron pair is assigned to each bonding area.

```
[     O     ]2−
[     ..    ]
[  O : S : O]
[     ..    ]
[     O     ]
```

We now add the nonbonding pairs.

```
[      :O:      ]2−
[  :O : S : O:  ]
[      :O:      ]
```

Step 6. Each of the atoms has a rare-gas configuration.

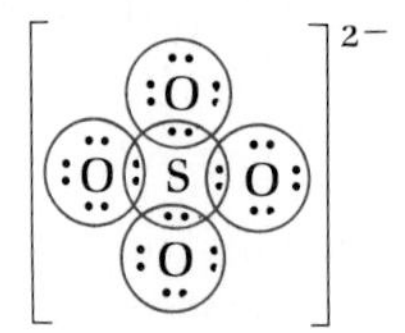

The total number of dots in each of the resonance structures is 32, which checks with the number of valence electrons calculated in step 2.

In this example we have shown the bonding within a polyatomic ion. This polyatomic ion, or any other one for that matter, is not a stable entity that exists alone. Polyatomic ions are parts of ionic compounds. The ion SO_4^{2-} would be found in ionic compounds such as Na_2SO_4, $CaSO_4$, $(NH_4)_2SO_4$, and $Pb(SO_4)_2$. Ionic compounds containing polyatomic ions thus offer an interesting combination of both ionic and covalent bonds: covalent bonding *within* the polyatomic ion and ionic bonding *between* it and ions of opposite charge.

The systematic approach to drawing electron-dot structures for molecules and polyatomic ions just illustrated does not take into account the origin of the electrons in a chemical bond, that is, which electron are contributed to the bond by which atoms. The system deals only with the total number of valence electrons. Thus, no distinction between normal covalent bonds and coordinate covalent bonds is made by the procedures of this system. This is acceptable. Each electron in a bond belongs to the bond as a whole. Electrons do not have labels of genealogy.

10.11 Molecular Polarity

Molecules as well as bonds can have polarity. As we shall see shortly, just because a molecule contains polar bonds does not mean that the molecule as a whole is polar. Molecular polarity depends on (1) the polarity of the bonds within a molecule and (2) the geometry of the molecule (when three or more atoms are present).

molecular geometry

Molecular geometry *gives the way in which atoms in a molecule are arranged in space relative to each other*. All molecules containing three or more atoms have characteristic three-dimensional shapes. For example, the triatomic CO_2 molecule is linear; that is, its three atoms lie in a straight line (Fig. 10-6a). On the other hand, a H_2O molecule, also a triatomic molecule, has a nonlinear or bent geometry (Fig. 10-6b). An NH_3 molecule has a trigonal pyramidal geometry, with the nitrogen atom at the apex and the hydrogen atoms at the base of the pyramid (Fig. 10-6c).

Determining the molecular polarity of a *diatomic* molecule is simple because only one bond is present. If that bond is nonpolar, the molecule is nonpolar; if the bond is polar, the molecule is polar.

With molecules containing more than one bond (triatomic molecules, tetraatomic molecules, and so on) the collective effect of individual bond polarities must be consid-

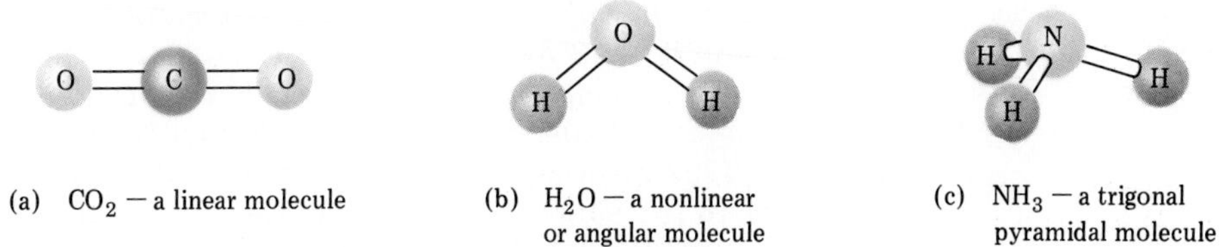

(a) CO_2 — a linear molecule (b) H_2O — a nonlinear or angular molecule (c) NH_3 — a trigonal pyramidal molecule

Figure 10-6. Molecular geometries of selected molecules.

ered. Molecular geometry plays an important role in determining this collective effect. In some instances, due to the symmetrical nature of the geometry of the molecule, the effects of the polar bonds are cancelled and a nonpolar molecule results.

Let us consider the polarities of three specific triatomic molecules — CO_2, H_2O, and HCN.

In the linear CO_2 molecule (Fig. 10-6a) both bonds are polar (oxygen is more electronegative than carbon). Despite the presence of these polar bonds, CO_2 molecules are *nonpolar.* The effects of the two polar bonds are cancelled out as a result of the O atoms being arranged symmetrically about the C atom. The shift of electronic charge toward one oxygen atom is exactly compensated by the shift of electronic charge toward the other. Thus, one end of the molecule is not negatively charged relative to the other end (a requirement for polarity); hence, the molecule is nonpolar. This cancellation of individual bond polarities is diagrammed as follows.

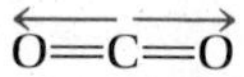

$$\overleftarrow{O{=}C}\overrightarrow{{=}O}$$

The two individual bond polarities (denoted by arrows), being of equal magnitude (each oxygen affects the C atom in the same way) but opposite in direction, cancel.

The nonlinear (angular) triatomic H_2O molecule (Fig. 10-6b) is polar. The bond polarities associated with the two hydrogen-oxygen bonds do not cancel each other because of the nonlinearity of the molecule.

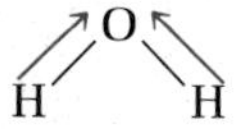

$$\begin{array}{ccc} & \nearrow O \nwarrow & \\ H \diagup & & \diagdown H \end{array}$$

As a result of their orientation, both bonds contribute to an accumulation of negative charge on the oxygen atom. The two bond polarities are equal in magnitude but not opposite in their direction.

The generalization that linear triatomic molecules are nonpolar and nonlinear triatomic molecules are polar, a generalization you might be tempted to make on the basis of our discussion of CO_2 and H_2O molecular polarities, is not valid. The linear molecule HCN, which is polar, invalidates this statement. Both bond polarities contribute to N acquiring a partial negative charge relative to H in HCN.

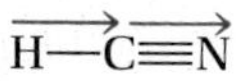

$$\overrightarrow{H{-}C}\overrightarrow{{\equiv}N}$$

(Note that the polarity arrows were drawn in the direction they were because N is more electronegative than C and C is more electronegative than H.)

As an analogy to the cancellation or noncancellation of bond polarities in a triatomic molecule, consider the effect that two forces operating on an object will have on its position. In Figure 10-7a the object does not move because both forces are of equal magnitude and exactly opposite in direction. In Figure 10-7b the block does move since the two forces, although equal in magnitude, are not opposite in direction. In Figure 10-7c, obviously, the block moves since both forces are in the same direction. The first situation (a) is somewhat analogous to the polarity situation in CO_2; the second (b), to the situation in H_2O; the third (c), to the situation in HCN. Only when there is "no movement" does nonpolarity result.

Molecules that contain four and five atoms commonly have, respectively, trigonal

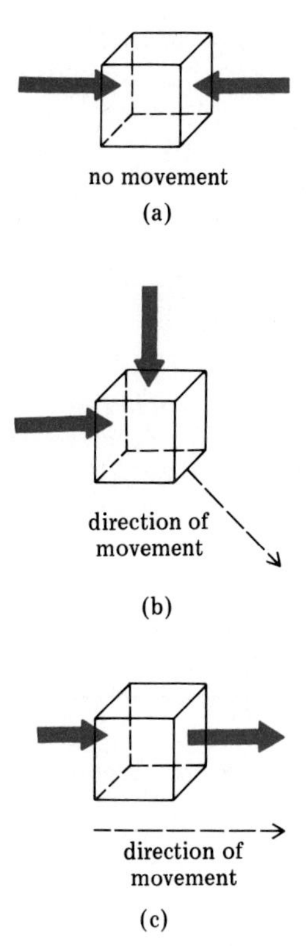

Figure 10-7. The effects of forces of various magnitudes and directions on the position of an object.

planar and tetrahedral geometries. The arrangement of atoms associated with these two geometries is

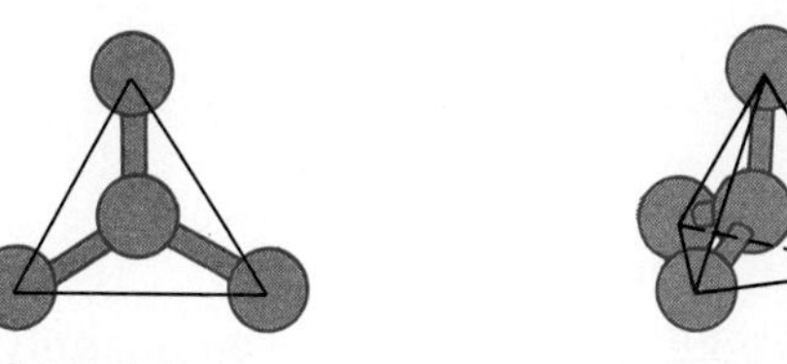

Trigonal planar and tetrahedral molecules in which all of the atoms attached to the central atom are identical, that is, molecules such as BF_3 (trigonal planar) and CH_4 (tetrahedral), are *nonpolar*. The individual bond polarities cancel as the result of the highly symmetrical arrangement of atoms about the central atom. (Proof that cancellation does occur involves some trigonometric considerations; it is not an obvious situation.)

If two or more kinds of atoms are attached to the central atom in a trigonal planar or tetrahedral molecule, the molecule will be polar. The high symmetry required for cancellation of the individual bond polarities is no longer present. For example, if one of the Cl atoms in CCl_4 (a nonpolar molecule) is replaced by a F atom a polar molecule

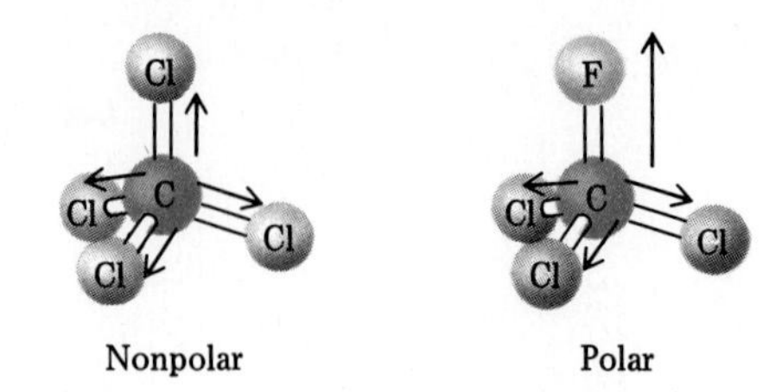

Figure 10-8. CCl_4 and CCl_3F, nonpolar and polar tetrahedral molecules.

results, even though the resulting CCl_3F is still a tetrahedral molecule. A C—F bond has a greater polarity than a C—Cl bond; F has an electronegativity of 4.0 and Cl has an electronegativity of only 3.0. Figure 10-8 contrasts the polar CCl_3F and nonpolar CCl_4 molecules, with a longer arrow being used to denote the increased polarity of the C—F bond.

EXAMPLE 10.8

Predict the polarity of each of the following molecules.

(a) NH_3 — trigonal pyramidal

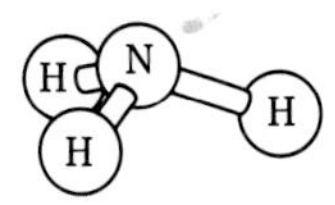

(b) H_2S — angular

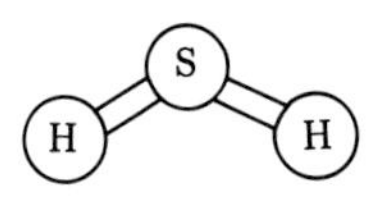

(c) N_2O — linear

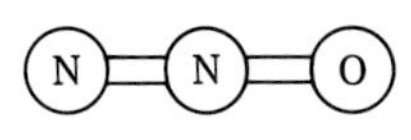

(d) SO_3 — trigonal planar

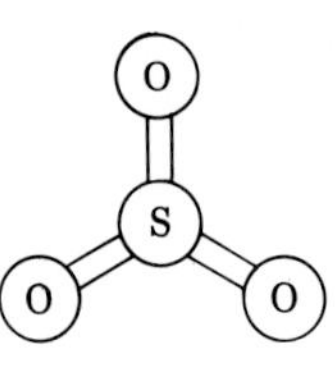

SOLUTION

Knowledge of molecular geometry, which is a given quantity for each molecule in this example, is a prerequisite for predicting molecular polarity.

(a) Noncancellation of the individual bond polarities in the trigonal pyramidal NH_3 molecule results in it being a polar molecule.

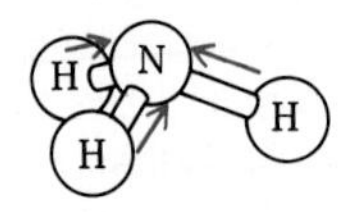

The bond polarity arrows all point toward the N atom because N is more electronegative than H.

(b) For the angular H_2S molecule, the shift in electron density in the polar sulfur-hydrogen bonds will be toward the S atom because S is more electronegative than H.

The H_2S molecule as a whole is polar due to the noncancellation of the individual S—H bond polarities.

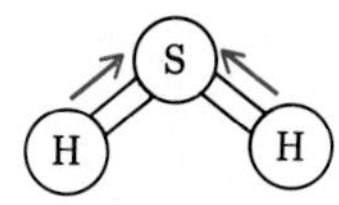

(c) The structure of the linear N_2O molecule is unsymmetrical; a nitrogen atom, rather than the O atom, is the central atom.

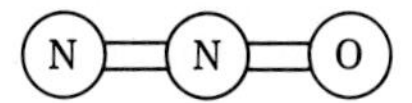

The N—N bond is nonpolar; the N—O bond is polar. The molecule as a whole is polar due to the polarity of the N—O bond.

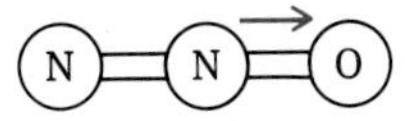

(d) SO_3 is a trigonal planar molecule. Three resonance structures (previously given in Ex. 10.6) are required to depict the bonding in this molecule. All of the S—O bonds, which are equivalent, are polar since S and O differ in electronegativity. However, the molecule as a whole is nonpolar because of the symmetrical arrangement of like O atoms about the central S atom which causes the individual bond polarities to cancel.

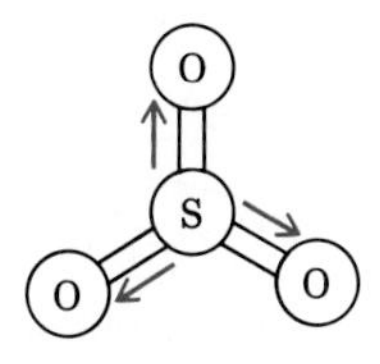

In later chapters we will see that molecular polarity is an important factor in determining the magnitude of some physical properties, for example, boiling point, as well as some chemical properties for molecular substances.

stop here
of final

10.12 Predicting Molecular Geometries (optional)

In this section we consider a theory, called valence-shell electron-pair repulsion theory (VSEPR theory for short), which will enable us to predict molecular geometries. It will also give us some insights about why molecules adopt the shapes that they do.

VSEPR theory

The basic principle behind **VSEPR theory** (pronounced "vesper") is very simple: *The* SETS *of valence shell electron pairs about an atom (whether bonding or nonbonding) tend to orient themselves so as to minimize the repulsions between them.* (Remember that all electrons carry a negative charge and like charges repel each other.) Let us apply this concept to various numbers of *sets* of electron pairs about an atom.

Two *sets* of electron pairs, to be as far apart as possible from each other, will be found on opposite sides of a nucleus, that is, at 180° angles to each other. Such an arrangement is called a *linear* arrangement of electron sets.

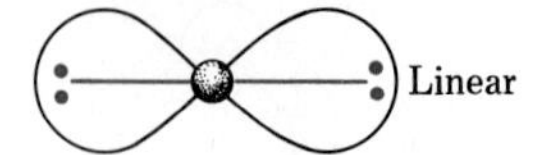

Three *sets* of electron pairs are as far apart as possible when they are found at the corners of an equilateral triangle. In such an arrangement, they are separated by angles of 120°, giving a *trigonal planar* arrangement of electron sets.

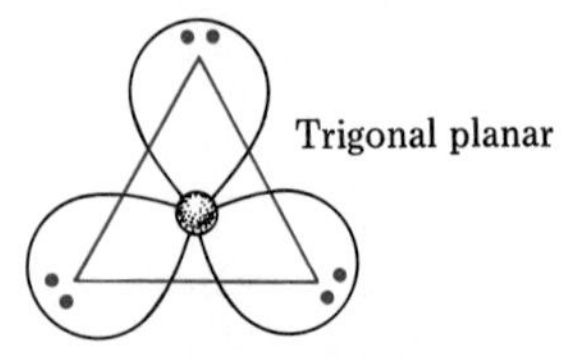

The arrangement that minimizes repulsions between four *sets* of electron pairs is *not* a square planar arrangement as you might first predict. Instead, a *tetrahedral* arrangement of electron pairs occurs. A tetrahedron is a four-sided solid, all four sides being identical equivalent triangles. It has the form of a pyramid with a triangular base.

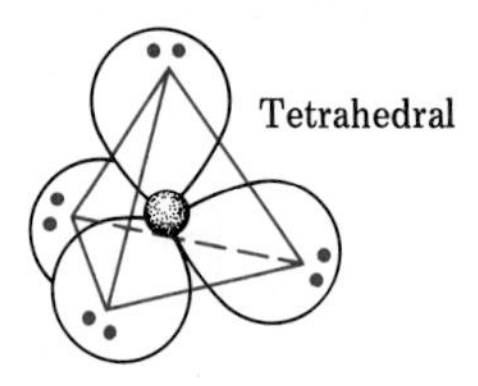

In the preceding paragraphs we have talked about the geometrical arrangements that would minimize the repulsions between *sets* of electron pairs with the word *set* always being italicized whenever it was mentioned. What is a *set* of electron pairs? A set of electron pairs may be any one of four things.

1. Three bonding pairs grouped together — a triple covalent bond.
2. Two bonding pairs grouped together — a double covalent bond.
3. One bonding pair — a single covalent bond.
4. One nonbonding pair.

The following guidelines make the determination of molecular geometry by VSEPR theory a straightforward procedure.

Step 1. *Draw the electron-dot structure for the molecule or polyatomic ion.*

The electron-dot structure does not directly give us information about geometry. It is always two dimensional; geometries involve three-dimensional considerations. The electron-dot structure serves as a source of information about numbers of valence electrons.

Step 2. *From the electron-dot structure determine the number of* SETS *of electron pairs that are present about the* CENTRAL ATOM *in the molecule or polyatomic ion.*

For molecules where the bonding is consistent with the octet rule (four electron pairs) a limited number of ways exist for grouping the electron pairs into sets. Table

Table 10-3. Various Electron Pair Sets That Can Exist About an Atom Having Eight Valence Electrons[a]

Groupings that give four sets of electron pairs

1. Four single bonds
2. Three single bonds and one nonbonding pair
3. Two single bonds and two nonbonding pairs

Groupings that give three sets of electron pairs

1. One double bond and two single bonds
2. One double bond, one single bond, and one nonbonding pair

Groupings that give two sets of electron pairs

1. Two double bonds
2. One triple bond and one single bond

[a] In listing the possibilities for grouping electron pairs into sets we did not consider cases where only one bond is present; for example, one single bond and three nonbonding pairs. If only one bond is present we have a diatomic molecule. All diatomic molecules have the same geometry; the two atoms lie along a straight line. We do not need a theory to predict that; common sense gives us the correct answer.

10-3 gives all the possible ways in which electron pairs may be grouped into sets. Note from the information in Table 10-3 that all of the electron pairs in a bond, whether it be a single, double, or triple bond, are counted as one *set* of electron pairs.

Step 3. *Determine the arrangement of the electron sets about the* CENTRAL ATOM *that minimizes electron repulsion.*

This is an easy step. The arrangement of electron sets is determined solely by the number of sets present. The arrangements, as previously discussed, are tetrahedral for four electron sets, trigonal planar for three electron sets, and linear for two electron sets.

Step 4. *Describe the shape of the molecule in terms of the positions of the atoms (the positions of the bonding electron sets).*

The molecular shape may or may not be the same as the electron-set geometry. If no nonbonding electron pairs are present the two are the same. If nonbonding pairs are present, molecular geometry differs from electron-set geometry. Molecular shape describes the arrangement of atoms, not the arrangement of electron pairs.

Let us illustrate the distinction between molecular geometry and electron-set geometry with an example. In a H_2O molecule there are four sets of electron pairs about the central O atom — two sets involved in single bonds and two sets present as nonbonding electron pairs. The four sets of electron pairs are arranged tetrahedrally.

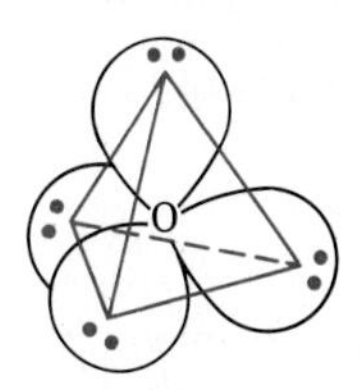

The molecular geometry is said to be nonlinear or angular.

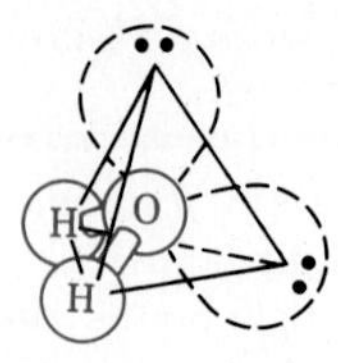

Note how only atoms are considered in coining a word to describe the molecular geometry.

Table 10-4 summarizes the relationships among number of sets of electron pairs, number of nonbonding electron pairs, electron-set geometry, and molecular geometry. In this figure note particularly the three cases where electron-set geometry and molecular geometry are different, and the terminology used to describe molecular geometry in these cases.

The two examples that follow illustrate the use of the above procedure for applying VSEPR theory to determine molecular geometry. Example 10.9 deals with molecules whose central atoms possess no nonbonding valence shell electron pairs. Molecular geometry and electron-set geometry are the same in such a situation. Example 10.10 deals with molecules whose central atoms do possess nonbonding electron pairs. For these molecules molecular geometry and electron-set geometry will always be different.

EXAMPLE 10.9

What is the molecular shape of each of the following molecules or polyatomic ions?

(a) CCl_4 **(b)** CO_2 **(c)** SO_4^{2-}

SOLUTION

(a) ***Step 1.*** The electron-dot structure for CCl_4 is

```
      ..
     :Cl:
  ..  ..  ..
 :Cl: C :Cl:
  ..  ..  ..
     :Cl:
      ..
```

Step 2. The central atom, C, has four sets of electron pairs; each set is an electron pair involved in a single covalent bond. No nonbonding valence shell electron pairs are present on the carbon atom.

Step 3. The arrangement of four sets of electron pairs about a central atom is always tetrahedral.

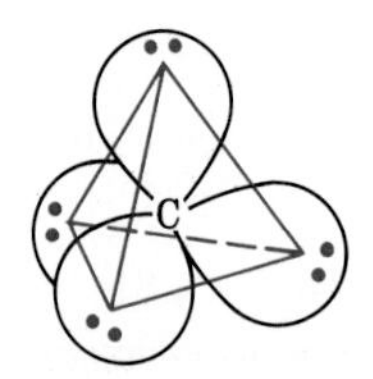

*Table 10-4. **The Relations Between Electron-Set Geometries About a Central Atom and Molecular Geometries***

Number of Sets of Electron Pairs	Number of Nonbonding Electron Pairs	Number of Bonds	Electron-Set Geometry	Molecular Geometry
4	0	4	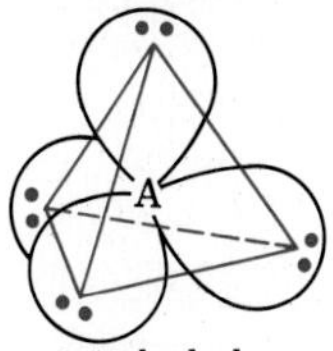 tetrahedral	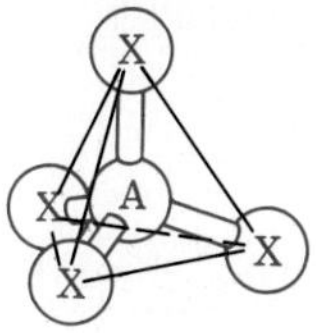 tetrahedral
4	1	3	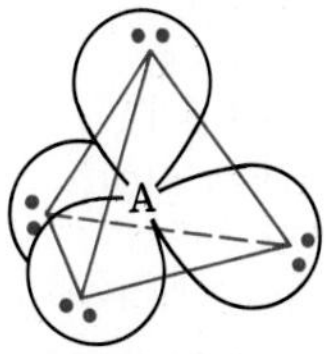 tetrahedral	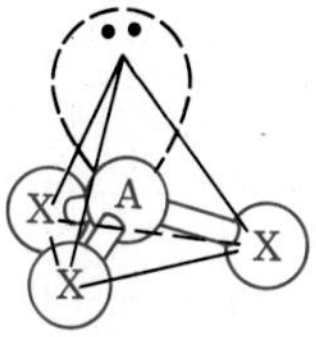 trigonal pyramidal
4	2	2	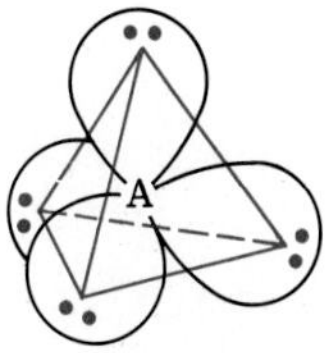 tetrahedral	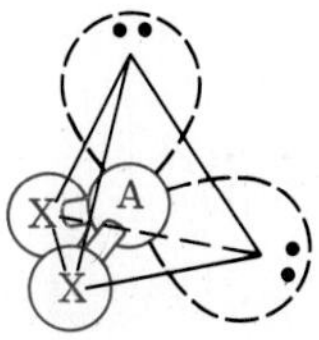 nonlinear or angular
3	0	3	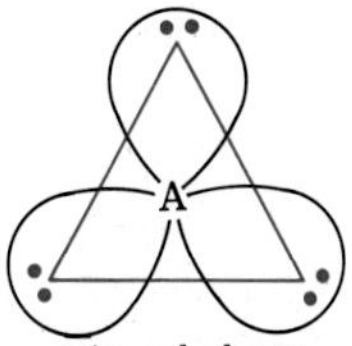 trigonal planar	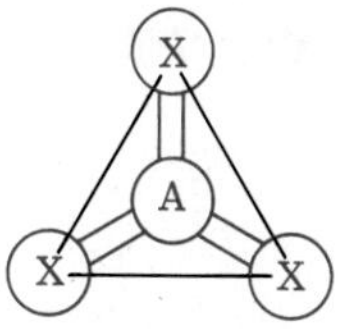 trigonal planar
3	1	2	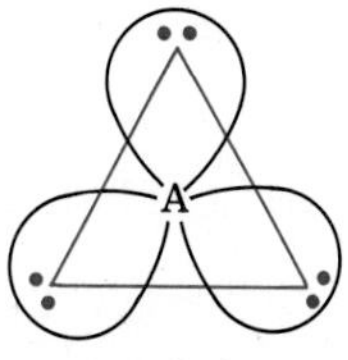 trigonal planar	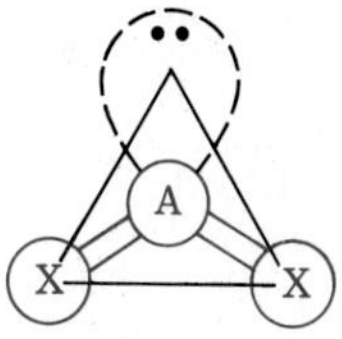 nonlinear or angular
2	0	2	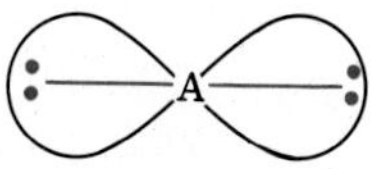 linear	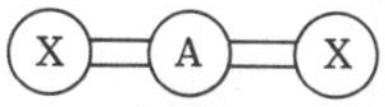 linear

Step 4. Since each electron set is bonding the molecular shape will also be tetrahedral.

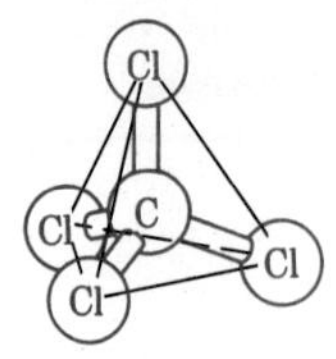

(b) ***Step 1.*** The electron-dot structure for CO_2 is

$$:\ddot{O}::C::\ddot{O}:$$

Step 2. The central atom, C, has two sets of electron pairs. Each set contains four electrons, that is, each set is involved in a double bond. No nonbonding electron pairs are present on the carbon atom.

Step 3. The arrangement of two electron sets about a central atom is always linear.

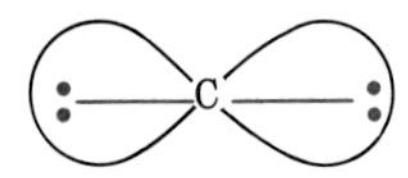

Step 4. Since each electron set is a bonding set, the molecular shape will also be linear.

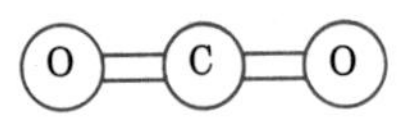

(c) ***Step 1.*** The electron-dot structure for SO_4^{2-} is

$$\left[\begin{array}{c} :\ddot{O}: \\ :\ddot{O}:\ddot{S}:\ddot{O}: \\ :\ddot{O}: \end{array}\right]^{2-}$$

Step 2. The central atom, S, has four sets of electron pairs, all of which are involved in bonding.

Step 3. The arrangements of four electron pairs about a central atom is always tetrahedral.

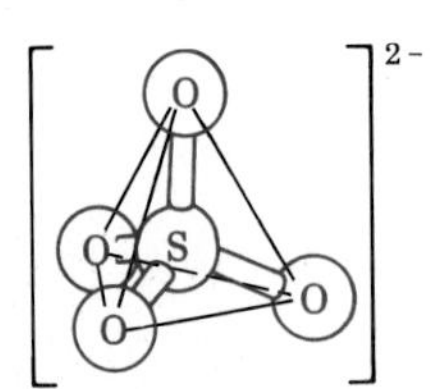

Step 4. Since each electron set is a bonding set the molecular geometry will be the same as the electron-set geometry: tetrahedral.

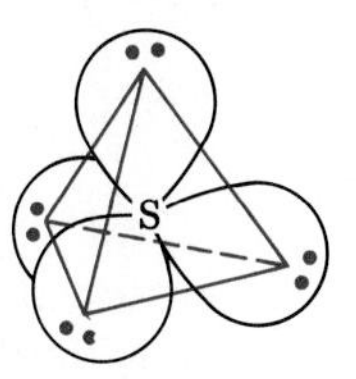

EXAMPLE 10.10

What is the molecular geometry of each of the following molecules?

(a) PF_3 **(b)** SO_2

SOLUTION

(a) ***Step 1.*** The electron-dot structure for PF_3 is

```
 ..  ..  ..
:F : P : F:
 ..  ..  ..
    :F:
     ..
```

Step 2. Four electron sets are present about the central atom — three sets of electron pairs involved in single bonds and one electron pair that is nonbonding.

Step 3. The arrangement of four electron sets about a central atom is always tetrahedral. It does not matter whether the sets are bonding or nonbonding. Arrangement depends only on the number of sets and not on how the sets function.

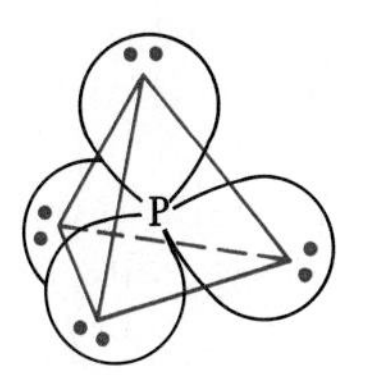

Step 4. The molecular shape will be different from the electron-set geometry since a nonbonding electron pair is present. The molecular shape is trigonal pyramidal.

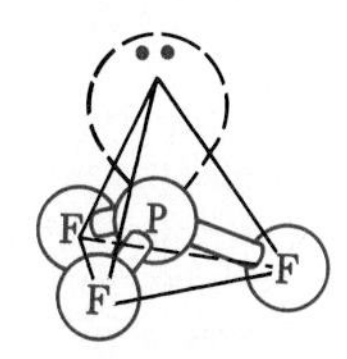

Note that the vacant corner of the tetrahedron (the corner that has a nonbonding

electron pair instead of an atom) is not considered when coining a word to describe the molecular geometry. Molecular geometry describes the arrangement of *atoms only*.

(b) ***Step 1.*** The electron-dot structure for SO_2 is

$$:\ddot{\underset{\cdot\cdot}{O}}:\ddot{S}::\underset{\cdot}{\dot{O}}\!\cdot \longleftrightarrow \cdot\dot{\underset{\cdot}{O}}::\ddot{S}:\ddot{\underset{\cdot\cdot}{O}}:$$

The two resonance structures are equivalent except for the location of the double bond. Because of their equivalency, either one may be used to determine molecular shape; they will both give the same answer.

Step 2. The number of electron sets present about the central S atom is three — a nonbonding pair, two electrons involved in a single bond, and four electrons involved in a double bond.

Step 3. The arrangement of three electron sets about the central atoms will be trigonal planar.

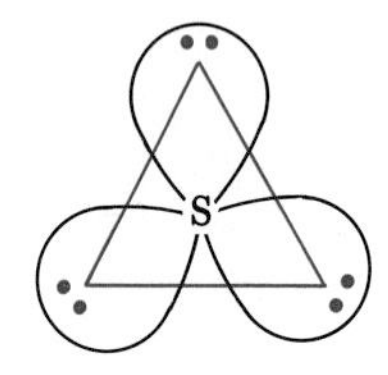

Step 4. The molecular shape is angular.

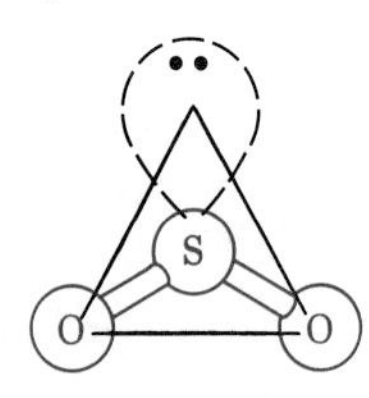

Even though the nonbonding electron pair is ignored when coining a word to describe the molecular shape its presence is very important. It is the nonbonding pair that causes the molecule to be angular; without it the molecule would be linear. Nonbonding electron pairs are just as important as bonding electron pairs in determining the geometry of the molecule.

Learning Objectives

After completing this chapter you should be able to

- Determine the number of valence electrons a representative element has (given its location in the periodic table or its electron configuration) and write an electron-dot structure for the element (Sec. 10.2).
- State the octet rule and the basis for the rule (Sec. 10.3).
- Identify by name and symbol monoatomic ions that are isoelectronic with a given rare gas and write the electron configuration of these ions (Sec. 10.4).
- Use electron-dot structures to describe the bonding in simple ionic compounds (Sec. 10.4).
- Use electron-dot structures to describe the bonding in simple covalent bonds (Secs. 10.5 and 10.7).
- Understand the relationship between the magnitude of

an element's electronegativity and its position in the periodic table (Sec. 10.6).

- Classify, with the help of a table of electronegativities, a given bond as nonpolar covalent, polar covalent, or ionic (Sec. 10.6).
- Understand what is meant by the terms single-, double-, triple-, multiple-, and coordinate-covalent bond (Secs. 10.7 and 10.8).
- Determine when resonance structures are needed to describe the bonding in a molecule (Sec. 10.9).
- Write an electron-dot structure for a complicated molecule (many electrons) or a polyatomic ion using a systematic approach (Sec. 10.10).
- Predict, given its geometry, whether a molecule is polar or nonpolar (Sec. 10.11).
- Determine the geometry of a molecule by VSEPR theory (Sec. 10.12) [optional].

Terms and Concepts for Review

The new terms or concepts defined in this chapter are

chemical bond (Sec. 10.1)
ionic bond (Sec. 10.1)
covalent bond (Sec. 10.1)
valence electron (Sec. 10.2)
electron-dot structure (Sec. 10.2)
octet rule (Sec. 10.3)
isoelectronic (Sec. 10.4)
electronegativity (Sec. 10.6)
polarity (Sec. 10.6)
nonpolar covalent bond (Sec. 10.6)
polar covalent bond (Sec. 10.6)
single covalent bond (Sec. 10.7)
multiple covalent bond (Sec. 10.7)
double covalent bond (Sec. 10.7)
triple covalent bond (Sec. 10.7)
coordinate covalent bond (Sec. 10.8)
bond strength (Sec. 10.9)
bond length (Sec. 10.9)
resonance structure (Sec. 10.9)
molecular geometry (Sec. 10.11)
VSEPR theory (Sec. 10.12)

Questions and Problems

Valence Electrons

10-1 How many valence electrons do atoms with the following electron configurations have?
a. $1s^2 2s^2 2p^1$
b. $1s^2 2s^2 2p^6 3s^2$
c. $1s^2 2s^2 2p^6 3s^2 3p^6 4s^2 3d^{10} 4p^5$
d. $1s^2 2s^2 2p^6 3s^2 3p^6 4s^2 3d^{10}$
e. $1s^2 2s^1$
f. $1s^2 2s^2 2p^6 3s^2 3p^6 4s^2 3d^{10} 4p^6 5s^2 4d^{10} 5p^6 6s^2 4f^{14} 5d^{10} 6p^1$

10-2 How many valence electrons are there in atoms of the following elements?
a. $_{20}Ca$ **b.** $_{16}S$ **c.** $_{53}I$ **d.** $_{15}P$ **e.** $_{6}C$ **f.** $_{31}Ga$

10-3 For a representative element, what is the relationship between its position in the periodic table (group number) and the number of valence electrons it contains?

Electron-dot Structures for Atoms

10-4 Draw electron-dot structures for atoms of the following elements.
a. $_{17}Cl$ **b.** $_{7}N$ **c.** $_{56}Ba$ **d.** $_{37}Rb$ **e.** $_{34}Se$ **f.** $_{13}Al$

10-5 Write the elemental symbols that might be represented by X if the electron-dot structure is
a. ·Ẋ· **b.** ·Ẍ: (with two dots below) **c.** X·

The Octet Rule

10-6 State the octet rule.

10-7 What is unique about the electron configurations of the rare gases?

Ions and Ionic Charge

10-8 Predict the general kind of behavior (that is, loss or gain of electrons) you would expect from atoms with the following electron configurations.
a. $1s^2 2s^2 2p^3$
b. $1s^2 2s^2 2p^6 3s^2$
c. $1s^2 2s^2 2p^6 3s^2 3p^6 4s^2 3d^{10} 4p^1$
d. $1s^2 2s^2 2p^6 3s^2 3p^5$

10-9 Predict the charge on the monoatomic ion that each of the following elements would form.
a. $_{15}P$ **b.** $_{19}K$ **c.** $_{52}Te$ **d.** $_{35}Br$ **e.** $_{4}Be$ **f.** $_{13}Al$

10-10 Write the elemental symbols for representative element ions that might be represented by the following notations.
a. X^{3-} **b.** X^{2+} **c.** X^-

10-11 Write the electron configuration for the following ions.
a. $_{9}F^-$ **b.** $_{19}K^+$ **c.** $_{16}S^{2-}$ **d.** $_{53}I^-$ **e.** $_{38}Sr^{2+}$ **f.** $_{13}Al^{3+}$

10-12 Identify two negatively charged monoatomic ions and two positively charged monoatomic ions that are isoelectronic with each of the following.
a. $_{10}Ne$ **b.** $_{36}Kr$ **c.** $_{17}Cl^-$ **d.** $_{11}Na^+$

10-13 In each of the following sets indicate those ions that are isoelectronic with each other.
a. Al^{3+}, F^-, Cl^-, Li^+ **b.** I^-, Ba^{2+}, Se^{2-}, Cs^+
c. S^{2-}, P^{3-}, Rb^+, Sr^{2+} **d.** F^-, S^{2-}, Ca^{2+}, Cs^+

Ionic Bonds

10-14 Show the formation of the following ionic compounds using electron-dot structures.
a. K_2S **b.** CaO **c.** Na_3P **d.** $AlCl_3$

10-15 Using electron-dot structures, show how ionic compounds are formed by atoms of
a. Na and Cl **b.** Mg and Br **c.** Al and N
d. Ca and P

Covalent Single Bonds

10-16 Draw electron-dot structures to illustrate the covalent bonding found in each of the following molecules, all of which contain single bonds. (The skeletal arrangement of atoms in each molecule is given.)

a. I_2: I I

b. NCl_3:
```
Cl  N  Cl
    Cl
```

c. PH_3:
```
H  P  H
   H
```

d. CCl_4:
```
    Cl
Cl  C  Cl
    Cl
```

e. CH_2Cl_2:
```
    H
Cl  C  H
    Cl
```

f. N_2H_4:
```
H       H
   N  N
H       H
```

10-17 Write electron-dot structures for the compounds must likely to be formed between these pairs of elements.
a. Se and F **b.** H and Br **c.** Cl and F
d. H and Te

Polarity and Electronegativity

10-18 Arrange each of the following sets of atoms in order of increasing electronegativity.
a. Na, Al, P, Mg **b.** F, Cl, S, Se
c. P, As, N, Bi **d.** F, I, Rb, Sr
e. B, Ga, K, Co

10-19 Arrange each of the following sets of bonds in order of increasing polarity.
a. H—Cl, H—O, H—Br
b. O—F, P—O, Al—O
c. H—Cl, Br—Br, B—N
d. Al—Cl, C—N, H—F
e. P—N, S—O, Br—F

10-20 Place a $\delta+$ above the atom that is relatively positive and a $\delta-$ above the atom that is relatively negative in each of the following bonds.
a. B—N **b.** F—O **c.** C—Cl **d.** Al—S
e. N—C **f.** Br—P

10-21 Classify the bonding in each of the following compounds as nonpolar covalent, polar covalent, or ionic.
a. $BeCl_2$ **b.** MgS **c.** NO **d.** NaBr
e. Rb_2S **f.** H_2O

Multiple Covalent Bonds

10-22 What is the difference between a single covalent bond, a double covalent bond, and a triple covalent bond?

10-23 Draw the electron-dot structures for the following molecules, each of which contains at least one double bond. (The skeletal arrangement of atoms in each molecule is given.)

a. CS_2: S C S

b. C_3H_4:
```
H       H
C   C   C
H       H
```

c. C_2H_3Cl:
```
H       H
   C  C
H       Cl
```

d. N_2F_2: F N N F

10-24 Draw the electron-dot structures for the following molecules, each of which contains at least one triple covalent bond. (The skeletal arrangement of atoms in each molecule is given.)

a. C_2H_3N:
```
   H
H  C  C  N
   H
```

b. C_3H_4:
```
            H
H  C  C  C  H
            H
```

c. C_2N_2: N C C N

Coordinate Covalent Bonds

10-25 What is a coordinate covalent bond?

10-26 Draw electron-dot structures for the following molecules. Each of the bond schemes will require the use of the coordinate covalency concept.

a. H_3PO_4:
```
      O
H  O  P  O  H
      O
      H
```

b. $SOCl_2$:
```
Cl  S  O
    Cl
```

c. S_2O: S S O

Resonance Structures

10-27 Write all resonance structures, if any, for each of the following molecules.
a. O_3: O O O **b.** HN_3: H N N N

c. N_2O_4:

```
O       O

   N  N

O       O
```

d. H_2SO_4:

```
   O

O  S  O  H

   O

   H
```

e. $[SCN]^-$: S C N

10-28 How would you expect the C—O bond length in CO_3^{2-} to compare with that in CO_2?

Electron-Dot Structures — General Considerations

10-29 Which of each pair of arrangements of atoms is more likely?

a.

```
O—C—O—O   or   O     O
                \   /
                  C
                  |
                  O
```

b.

```
    O                  O
    |                  |
O—P—O      or     O—O—P
    |                  |
    O                  O
```

c. H—H—O—O or H—O—O—H

d.

```
                          H
                          |
O—S—O—H     or     O—S—O—O
  |                       |
  O                       H
  |
  H
```

10-30 How many electron dots should appear in the electron-dot structures for each of the following molecules or ions?

a. HNO_3 **b.** PF_3 **c.** O_2F_2 **d.** NF_4^+
e. PO_4^{3-} **f.** S_2^{2-}

10-31 Draw electron-dot structures for the following polyatomic ions.

a. NO_3^- **b.** SO_4^{2-} **c.** PO_4^{3-} **d.** OH^-
e. ClO_3^- **f.** CN^-

10-32 What is wrong with each of the following electron-dot structures?

a. :Br::Cl: **b.** $[:H::O:]^-$

c. H:Ö:::Ö:H **d.**

```
[O:N::O:]⁻
   :
   O
```

10-33 Draw electron-dot structures for the following molecules or polyatomic ions.

a. H_3PO_4
b. H_2O_2 (HOOH)
c. S_2^{2-}
d. ClO_4^-
e. $SOCl_2$ (S is the central atom)
f. $POBr_3$ (P is the central atom)
g. $(CN)_2$ (NCCN)
h. $SCNBr_3$ (all Br atoms are bonded to the N atom)

10-34 Discuss, using electron-dot structures, the bonding in the following compounds.

a. Cs_2SO_4 **b.** NH_4ClO_3

Molecular Polarity

10-35 For each of the following hypothetical triatomic molecules indicate whether the *bonds* are polar or nonpolar and whether the *molecule* is polar or nonpolar. Assume A, X, and Y have different electronegativities.

a. X—A—X **b.** A—X—X **c.**

```
   X
  / \
 X   X
```

d.

```
   A
  / \
 X   X
```

e. Y—A—X

10-36 Determine the polarity of each of the following molecules.

a. NCl_3 (trigonal pyramid)
b. $CClH_3$ (tetrahedral)
c. $SOCl_2$ (trigonal pyramid)
d. H_2S (angular)
e. CS_2 (linear)

Molecular Geometry (optional)

10-37 Using VSEPR theory, determine the geometry of the following molecules.

a. NCl_3 **b.** CCl_3H **c.** OF_2 **d.** H_2S
e. H_2CO **f.** CS_2 **g.** HCN **h.** CH_4

10-38 Using VSEPR theory, determine the geometry of the following polyatomic ions.

a. SO_3^{2-} **b.** PO_4^{3-} **c.** $AlCl_4^-$
d. CO_3^{2-} **e.** NF_2^+ **f.** NO_2^+

11 States of Matter

11.1 Physical States of Matter

In this chapter we consider the physical states of matter, that is, the solid state, the liquid state, and the gaseous state.

From everyday experience we know that the physical state of a substance is determined both by what it is, that is, its chemical identity, and by the temperature and pressure it is under. At room temperature and pressure some substances are solids (gold, sodium chloride, etc.), others liquids (water, mercury, etc.), and still others gases (oxygen, carbon dioxide, etc.). Thus chemical identity must be a determining factor for physical state, since all three states are observed at room temperature and pressure. On the other hand, when the physical state of a single substance is considered temperature and pressure are determining variables. Liquid water may be changed to a solid by lowering the temperature or to a gas by raising the temperature.

We tend to characterize a substance almost exclusively in terms of its most common physical state, that is, the state in which it is found at room temperature and pressure. Oxygen is almost always thought of as a gas, its most common state; gold is almost always thought of as a solid, its most common state. A major reason for such "single state" characterization is the narrow range of temperatures encountered on this planet. Most substances are never encountered in more than one state under "natural" conditions. We must be careful not to fall into the error of assuming that the commonly observed state of a substance is the *only* state in which it can exist. Under laboratory conditions states other than the "natural" one may be obtained for almost all substances. Figure 11-1 shows the solid-, liquid-, and gaseous-state temperature ranges for selected elements and compounds. As can be seen in this figure, the length and location of the physical-state ranges vary widely among chemical substances. Extremely high temperatures are required to obtain some substances in the gaseous

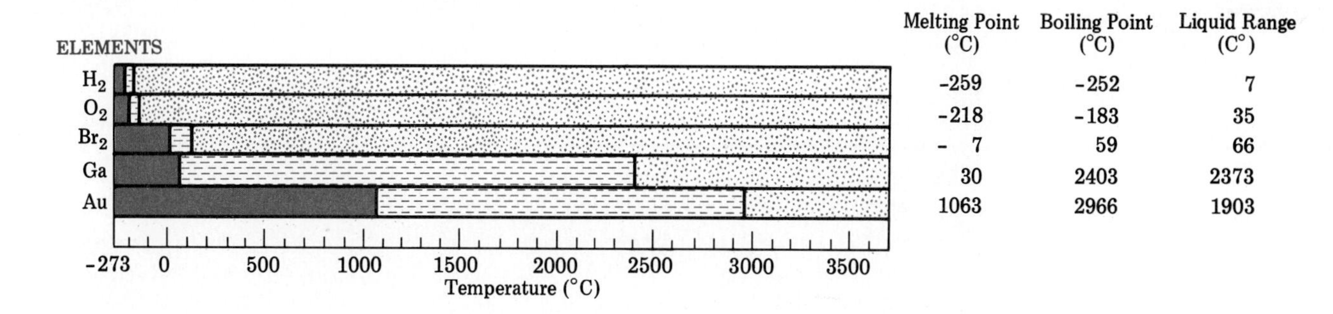

ELEMENTS	Melting Point (°C)	Boiling Point (°C)	Liquid Range (C°)
H_2	-259	-252	7
O_2	-218	-183	35
Br_2	- 7	59	66
Ga	30	2403	2373
Au	1063	2966	1903

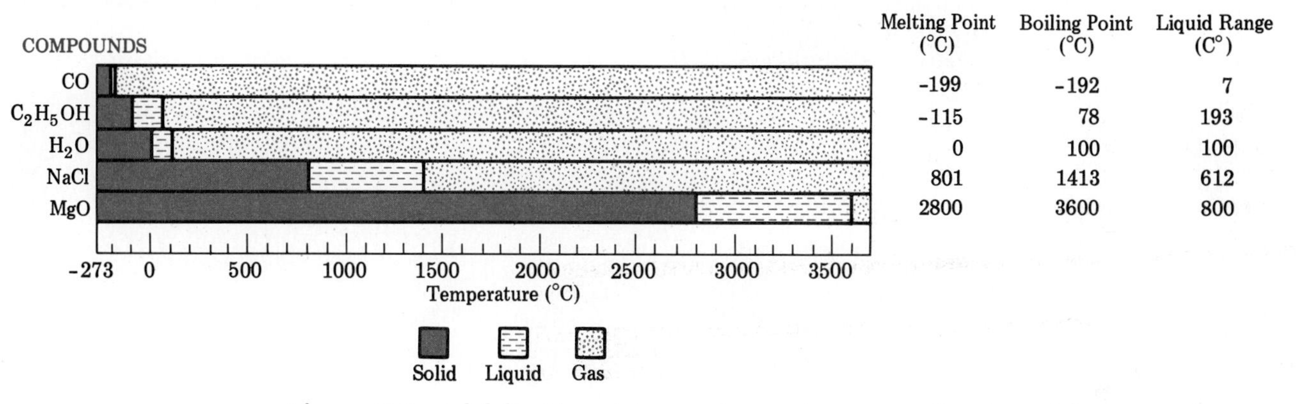

COMPOUNDS	Melting Point (°C)	Boiling Point (°C)	Liquid Range (C°)
CO	-199	-192	7
C_2H_5OH	-115	78	193
H_2O	0	100	100
NaCl	801	1413	612
MgO	2800	3600	800

Figure 11-1. Solid, liquid, and gaseous temperature ranges for selected elements and compounds.

states; other substances are gases at temperatures below room temperature. The length of a given physical state range also varies dramatically. For example, the elements H_2 and O_2 are liquids over a very short temperature range, whereas the elements Ga and Au remain liquids over ranges of hundreds of degrees.

Among the elements, at room temperature and pressure the solid state is the dominant physical state. Ninety-three of the 106 elements are solids at room temperature, as shown in Figure 11-2. Only two elements (bromine and mercury) are liquids, and eleven are gases.

Explanations for the experimentally observed physical-state variations among substances may be derived from the concepts of atomic structure and bonding we have considered in previous chapters. Such explanations are found in later parts of this chapter.

11.2 Property Differences Between Physical States

The differences among solids, liquids, and gases are so great that only a few gross distinguishing features need to be mentioned in order to clearly differentiate them. Certain obvious differences among the three states of matter are apparent to even the

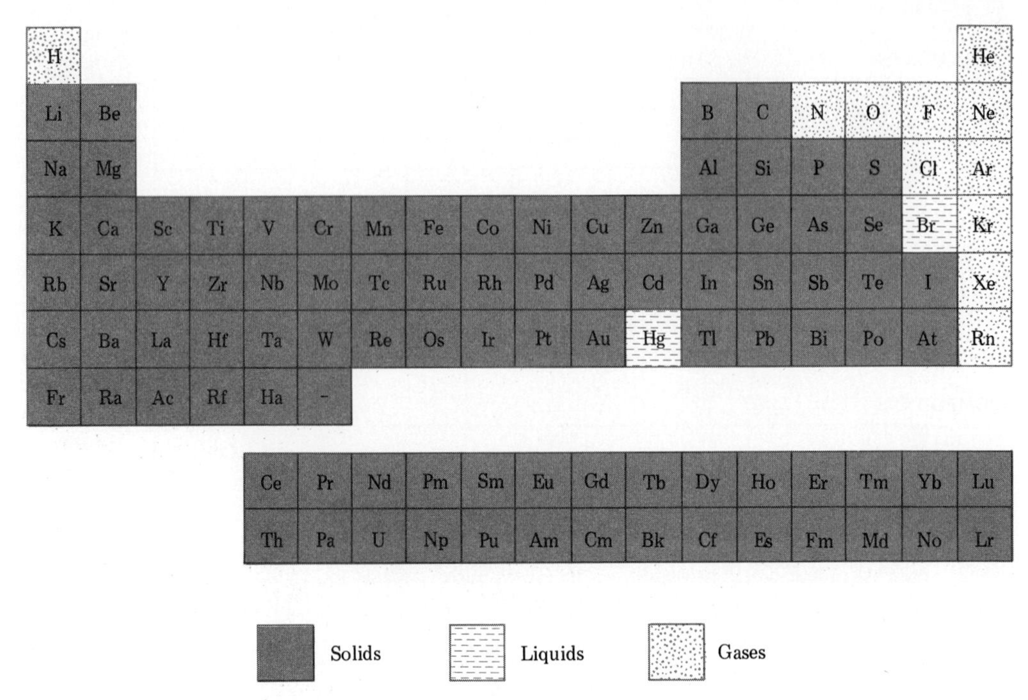

Figure 11-2. Physical states of the elements at room temperature and pressure.

most casual observer—differences related to (1) volume and shape, (2) density, (3) compressibility, and (4) thermal expansion. These distinguishing properties are compared in Table 11-1 for the three states of matter. The properties of volume and density have been discussed in detail previously (Secs. 3.1 and 3.3, respectively).

compressibility

Compressibility *measures the change in volume resulting from a pressure change.*

Table 11-1. Distinguishing Properties of Solids, Liquids, and Gases

Property	Solid State	Liquid State	Gaseous State
Volume and shape	definite volume and definite shape	definite volume and indefinite shape; takes the shape of container to the extent it is filled	indefinite volume and indefinite shape; takes the volume and shape of container that it fills
Density	high	high, but usually lower than corresponding solid	low
Compressibility	small	small, but usually greater than corresponding solid	large
Thermal expansion	very small: about 0.01% per °C	small: about 0.10% per °C	moderate: about 0.30% per °C

thermal expansion

Thermal expansion *measures the volume change resulting from a temperature change.*

The contents of Table 11-1 should be studied in detail; they will serve as the starting point for further discussions about the states of matter.

11.3 The Kinetic Molecular Theory of Matter

The physical characteristics of the solid, liquid, and gaseous states can be explained by *kinetic molecular theory,* one of the fundamental theories of chemistry. A basic idea of this theory is that the particles (atoms or molecules) present in a substance, independent of the physical state of the substance, have *motion* associated with them. The word *kinetic* comes from the Greek word *kinesis,* which means movement; hence, the name kinetic molecular theory.

kinetic molecular theory

The specific postulates of **kinetic molecular theory** are

1. Matter is ultimately composed of tiny particles (atoms or molecules) with definite and characteristic sizes that never change.
2. The particles are in constant random motion and therefore possess kinetic energy.
3. The particles interact with each other through attractions and repulsions and therefore possess potential energy.
4. The velocity of the particles increases as the temperature is increased. The average kinetic energy of all particles in a system depends on the temperature, increasing as the temperature increases.
5. The particles in a system transfer energy from one to another during collisions in which no net energy is lost from the system. The energy of any given particle is thus continually changing.

kinetic energy

The above postulates contain two new terms related to energy: kinetic energy and potential energy. **Kinetic energy** *is the energy an object (particle) has because of its motion.* The amount of kinetic energy a particle possesses depends on both its mass and its velocity. The exact mathematical relationship between the kinetic energy and the mass and velocity of a particle is

$$\text{Kinetic energy} = \tfrac{1}{2}mv^2$$

where m is the mass of the particle and v is its velocity. From this expression we see that any differences in kinetic energy between particles of the same mass must be caused by differences in their velocities. Similarly, differences in kinetic energy between particles moving at the same velocity must be due to mass differences.

potential energy

Potential energy *is the energy an object (particle) possesses because of attractions and repulsions between it and other objects (particles).* Attractive and repulsive interactions are operative in all systems, from macroscopic to submicroscopic. Familiar examples include gravitational attraction, magnetic attractions and repulsions, and electrostatic attractions and repulsions. Electrostatic interactions are the potential energy interactions of most importance in atomic-sized particles. **Electrostatic interactions** *occur between charged particles; objects of opposite electrical charge (one positive and one negative) attract each other and objects of identical charge (both positive or both*

electrostatic interactions

negative) *repel each other*. The magnitude of an electrostatic interaction depends on the sizes of the charges associated with the particles and their separation distance. Potential energy of attraction increases as the separation between particles increases, while that of repulsion increases with decreasing separation, as shown graphically in Figure 11-3. An analogy to the dependence of electrostatic potential energy on distance is found in two particles connected by a spring. When the spring is either stretched or compressed, the system has more potential energy than when the spring is in a nonextended position. The stretched spring represents attractive potential energy (the particles will come together if the spring is released), and the compressed spring represents repulsive potential energy (the particles will move apart when the spring is released). The greater the stretching of the spring (the greater the distance between particles), the greater the potential energy of attraction. The greater the compression of the spring (the smaller the distance between particles), the greater the potential energy of repulsion.

The relative influence of kinetic energy and potential energy in a chemical system is the major consideration in uses of kinetic molecular theory to explain the general properties of the solid, liquid, and gaseous states of matter. The important question is whether the kinetic energy or the potential energy dominates the energetics of the chemical system under study.

Kinetic energy may be considered a *disruptive force* within the chemical system tending to make the particles of the system increasingly independent of each other. As the result of energy of motion the particles will tend to move away from each other. Potential energy may be considered a *cohesive force* tending to cause order and stability among the particles of the system.

The role that temperature plays in determining the state of a system is related to kinetic energy magnitude. Kinetic energy increases as temperature increases (postulate 4 of the kinetic molecular theory of matter). Thus, the higher the temperature the greater the magnitude of disruptive influences within a chemical system. Potential energy magnitude is essentially independent of temperature change. Neither charge or separation distance, the two factors on which potential energy magnitude depends, is affected by temperature change.

Sections 11.4, 11.5, and 11.6 deal, respectively, with kinetic molecular theory explanations for the general properties of the solid, liquid, and gaseous states.

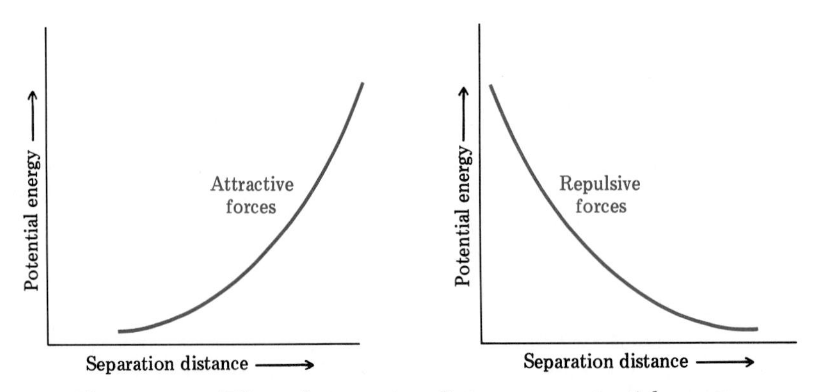

Figure 11-3. Effect of separation distance on potential energy.

11.4 The Solid State

The solid state is characterized by a dominance of potential energy (cohesive forces) over kinetic energy (disruptive forces). The particles in a solid are drawn close together in a regular pattern by the strong cohesive forces present. Each particle occupies a fixed position about which it vibrates, owing to disruptive kinetic energy. An explanation of the characteristic properties of solids is obtained from this model.

1. *Definite volume and definite shape.* The strong cohesive forces hold the particles in essentially fixed positions, resulting in definite volume and definite shape.

2. *High density.* The constituent particles of solids are located as close together as possible. Therefore, large numbers of particles are contained in a unit volume, resulting in a high density.

3. *Small compressibility.* Since there is very little space between particles, increased pressure cannot push them any closer together and therefore has little effect on the solid's volume.

4. *Very small thermal expansion.* An increased temperature increases the kinetic energy (disruptive forces) thereby causing more vibrational motion of the particles. Each particle "occupies" a slightly larger volume. The result is a slight expansion of the solid. The strong cohesive forces prevent this effect from becoming very large.

11.5 The Liquid State

The liquid state consists of particles randomly packed relatively near to each other. The molecules are in constant random motion, freely sliding over one another, but without sufficient energy to separate from each other. *The liquid state is, thus, a situation in which neither potential energy (cohesive forces) nor kinetic energy (disruptive forces) dominates.* The fact that the particles freely slide over each other indicates the influence of disruptive forces, but the fact that the particles do not separate indicates a fairly strong influence from cohesive forces. The characteristic properties of liquids are explained by this model.

1. *Definite volume and indefinite shape.* Attractive forces are strong enough to restrict particles to movement within a definite volume. They are not strong enough, however, to prevent the particles from moving over each other in a random manner, limited only by the container walls. Thus liquids have no definite shape, with the exception that they maintain a horizontal upper surface in containers that are not completely filled.

2. *High density.* The particles in a liquid are not widely separated; they essentially touch each other. Therefore, there will be a large number of particles per unit volume and a resultant high density.

3. *Small compressibility.* Since the particles in a liquid essentially touch each other, there is very little empty space. Therefore, a pressure increase cannot squeeze the particles much closer together.

4. *Small thermal expansion.* Most of the particle movement in a liquid is vibrational because a particle can move only a short distance before colliding with a neighbor. The increased particle velocity that accompanies a temperature increase results only in increased vibrational amplitudes. The net effect is an increase in the effective volume a particle "occupies," which causes a slight volume increase in the liquid.

11.6 The Gaseous State

Kinetic energy (disruptive forces) completely dominates potential energy (cohesive forces) in the gaseous state. As a result, the particles of a gas move essentially independently of one another in a totally random manner. Under ordinary pressure, the particles are relatively far apart except, of course, when they collide with each other. Between collisions with each other or with the container walls, gas particles travel in straight lines. The particle velocities and resultant collision frequencies are extremely high; at room temperature the collisions experienced by one molecule in one second at one atmosphere pressure is of the order of 10^{10}.

The kinetic theory explanation of gaseous-state properties follows the same pattern we saw earlier for solids and liquids.

1. *Indefinite volume and indefinite shape.* The attractive (cohesive) forces between particles have been overcome by kinetic energy, and the particles are free to travel in all directions. Therefore, the particles completely fill the container the gas is in and assume its shape.

2. *Low density.* The particles of a gas are widely separated. There are relatively few of them in a given volume, which means little mass per unit volume.

3. *Large compressibility.* Particles in a gas are widely separated; a gas essentially is mostly empty space. When pressure is applied, the particles are easily pushed closer together, decreasing the amount of empty space and the volume of the gas (see Fig. 11-4).

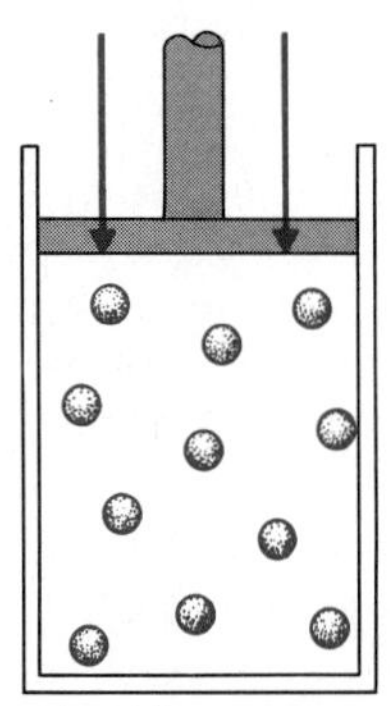

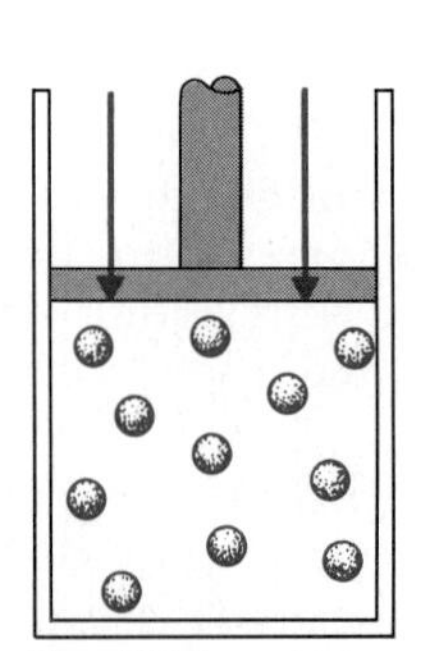

Figure 11-4. The compression of a gas—decreasing the amount of empty space in the container. [From *General, Organic and Biochemistry: A Brief Introduction* by H. Stephen Stoker and Michael R. Slabaugh. Glenview, Ill.: Scott, Foresman and Company, 1981.]

4. *Moderate thermal expansion.* An increase in temperature means an increase in particle velocity. The increased kinetic energy of the particles enables them to push back whatever barrier is confining them into a given volume. Hence, the volume increases.

It must be understood that the size of the particles is not changed during expansion or compression of gases, solids, or liquids. The particles merely move farther apart or closer together; the space between them is what changes.

11.7 A Comparison of Solids, Liquids, and Gases

Two obvious conclusions about the similarities and differences between the various states of matter may be drawn from comparing the descriptive materials in Sections 11.4, 11.5, and 11.6.

1. One of the states of matter, the gaseous state, is markedly different from the other two states, and
2. Two of the states of matter, the solid and the liquid states, have many similar characteristics.

These two conclusions are illustrated diagrammatically in Figure 11-5.

The average distance between particles is only slightly different in the solid and liquid states, but markedly different in the gaseous state. Roughly speaking, at ordinary temperatures and pressures, particles in a liquid are about 10% and particles in a gas about 1000% farther apart than those in the solid state. The distance ratio between particles in the three states (solid to liquid to gas) is thus 1 to 1.1 to 10.

11.8 Physical Changes of State

Physical changes of state were discussed in Section 4.4. The terminology associated with such changes — evaporation, condensation, sublimation, etc. — was introduced at that time (Fig. 4-2).

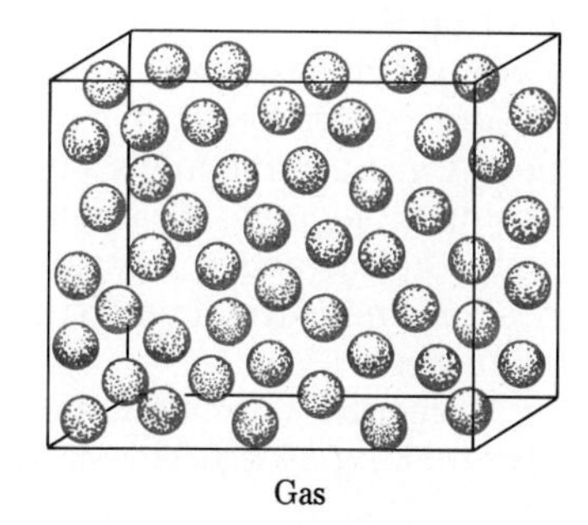

Gas

Molecules far apart and disordered
Negligible interactions between molecules

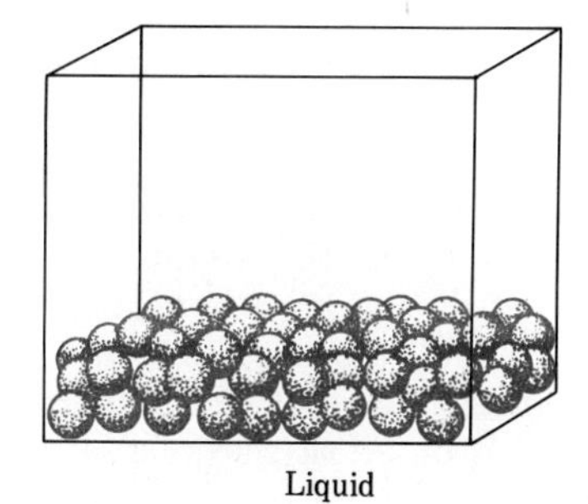

Liquid

Intermediate situation

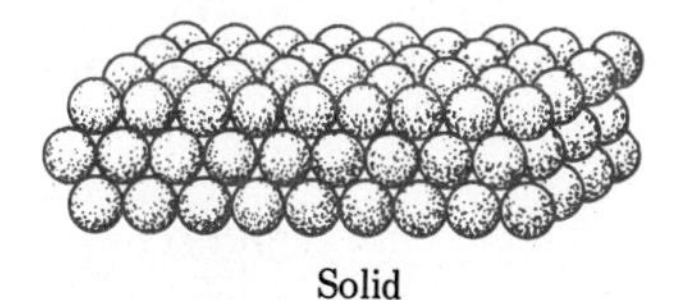

Solid

Molecules close together and ordered
Strong interactions between molecules

Figure 11-5. Similarities and differences in the states of matter. [From *General Chemistry* by Jerry March and Stanley Windwer. New York: Macmillan Publishing Co., Inc., 1979.]

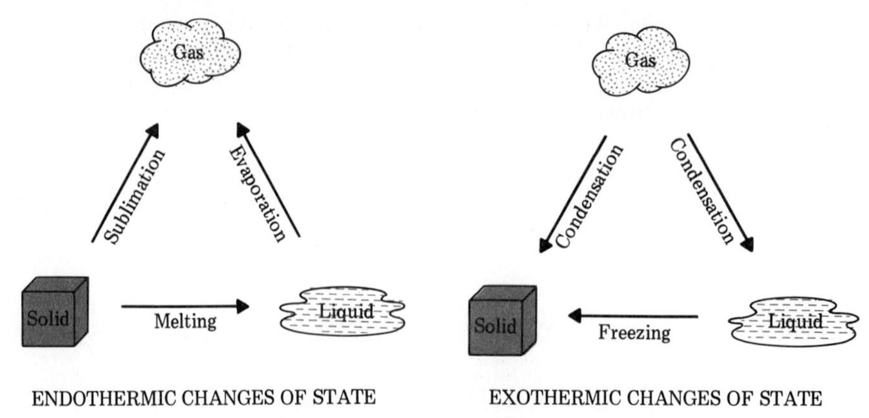

Figure 11-6. Endothermic and exothermic changes of state.

Changes of state are usually accomplished through heating or cooling a substance. (Pressure change is also a factor in some systems.) Changes of state may be classified into two categories based on whether heat (thermal energy) is given up or absorbed. An **endothermic change of state** *is a change that requires the input (absorption) of heat.* The endothermic changes of state are melting, sublimation, and evaporation. An **exothermic change of state** *is a change that requires heat to be given up (released).* Exothermic changes of state are the reverse of endothermic changes of state and include condensation and freezing. Figure 11-6 summarizes the classification of changes of state as endothermic or exothermic.

endothermic change of state

exothermic change of state

11.9 Evaporation of Liquids

evaporation

Evaporation *is the process by which molecules escape from a liquid phase to the gas phase.* It is a familiar process to us. We are all aware that water left in an open container, at room temperature, will slowly disappear by evaporation.

The phenomenon of evaporation can readily be explained using kinetic molecular theory. Postulate 5 of this theory (Sec. 11.3) indicates that the molecules in a liquid (or solid or gas) do not all possess the same kinetic energy. At any given instant some molecules will have above average kinetic energies and others below average kinetic energies as a result of collisions between molecules. A given molecule's energy constantly changes as a result of collisions with neighboring molecules. Molecules considerably above average in kinetic energy can overcome the attractive forces (potential energy) holding them in the liquid and escape if they are near the liquid surface and are moving in a favorable direction relative to the surface.

Note that evaporation is a surface phenomenon. Molecules within the interior of a liquid are surrounded on all sides by other molecules making escape very improbable. Surface molecules are subject to fewer attractive forces since they are not completely surrounded by other molecules; escape is much more probable. Liquid surface area is an important factor in determining the rate at which evaporation occurs. Increased surface area results in an increased evaporation rate; a greater fraction of molecules occupy "surface" locations.

Water evaporates faster from a glass of hot water than from a glass of cold water. Why is this so? A certain minimum kinetic energy is required for molecules to escape from the attractions of its neighboring molecules. As the temperature of a liquid increases, a larger fraction of the molecules present acquire this needed minimum kinetic energy. Consequently, the rate of evaporation always increases as liquid temperature increases. Figure 11-7 contrasts the fraction of molecules possessing the needed minimum kinetic energy for escape at two temperatures. Note that at both the lower and higher temperatures a broad distribution of kinetic energies is present and that at each temperature some molecules possess the needed minimum kinetic energy. However, at the higher temperature a larger fraction of molecules present have the requisite kinetic energy.

The escape of high-energy molecules from a liquid during evaporation affects the liquid in two ways: the amount of liquid decreases, and the liquid temperature is lowered. The temperature lowering reflects the fact that the average kinetic energy of the remaining molecules is lower than the pre-evaporation value due to the loss of the most energetic molecules. (Analogously, if all the tall people are removed from a classroom of students, the average height of the remaining students decreases.) A lower average kinetic energy corresponds to a lower temperature (postulate 4 of kinetic molecular theory); hence a cooling effect is produced.

Evaporative cooling is important in many processes. Our own bodies use evaporation to maintain a constant temperature. We perspire in hot weather because evaporation of the perspiration cools our skin. The cooling effect of evaporation is quite noticeable when someone first comes out of a swimming pool on a hot day (especially if a breeze is blowing). A canvas water bag keeps water cool because some of the water seeps through the canvas and evaporates, a process that removes heat from the remaining water. In medicine, the local skin anesthetic ethyl chloride (C_2H_5Cl) exerts its effect through evaporative cooling (freezing). The minimum kinetic energy molecules of this substance need to acquire to escape (evaporate) is very low resulting in a very rapid evaporation rate. Evaporation is so fast that the cooling "freezes" tissue near the surface of the skin with temporary loss of feeling in the region of application. Certain alcohols also evaporate quite rapidly. An alcohol "rub" is sometimes used to reduce body temperature when a high fever is present.

For a liquid in a container the decrease in temperature that occurs as a result of evaporation can be actually measured only if the container is an *insulated* one. When a

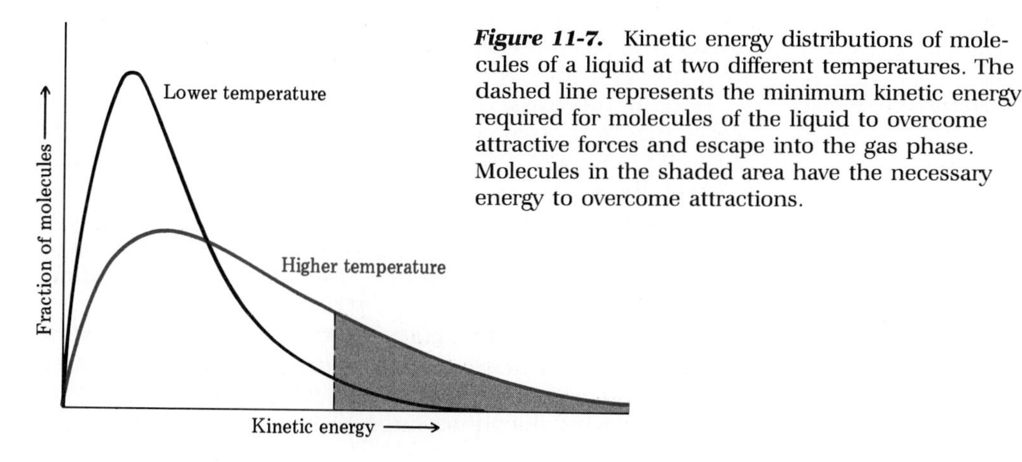

Figure 11-7. Kinetic energy distributions of molecules of a liquid at two different temperatures. The dashed line represents the minimum kinetic energy required for molecules of the liquid to overcome attractive forces and escape into the gas phase. Molecules in the shaded area have the necessary energy to overcome attractions.

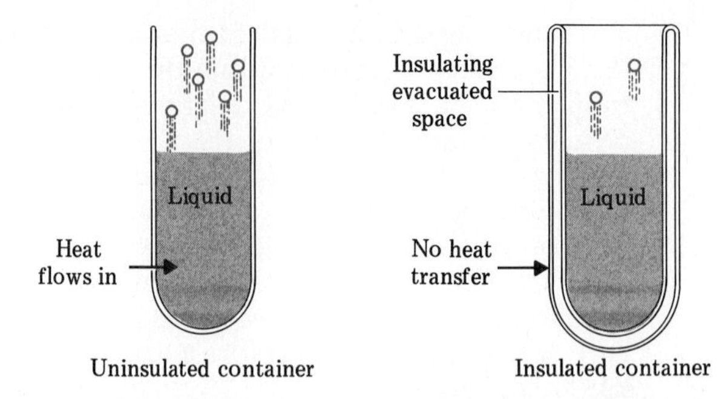

Figure 11-8. Evaporation from insulated and uninsulated containers. [From *Chemistry: A Science for Today* by Spencer L. Seager and H. Stephen Stoker. Glenview, Ill.: Scott, Foresman and Company, 1973.]

liquid evaporates from a noninsulated container, there is sufficient heat flow from the surroundings into the container to counterbalance the loss of energy in the escaping molecules and thus prevent any cooling effect. The Thermos bottle is an insulated container that minimizes heat transfer, hence its use in maintaining liquids at a cool temperature for a short period of time. The evaporations of a liquid from a Thermos bottle and from an uninsulated container are compared in Figure 11-8.

Collectively, the molecules that escape from an evaporating liquid are often referred to as *vapor* rather than gas. The term *vapor* describes gaseous molecules of a substance at a temperature and pressure at which we ordinarily would think of the substance as a liquid or solid. For example, at room temperature and atmospheric pressure the normal state for water is the liquid state. Molecules that escape (evaporate) from liquid water at these conditions are frequently called water vapor.

11.10 Vapor Pressure of Liquids

The evaporative behavior of a liquid in a closed container is quite different from that in an open container. In a closed container we observe that some liquid evaporation occurs, as indicated by a drop in liquid level. However, unlike the open container system, the liquid level, with time, ceases to drop (becomes constant), an indication that not all of the liquid will evaporate.

Kinetic molecular theory explains these observations in the following way. The molecules that do evaporate are unable to move completely away from the liquid as they did in the open container. They find themselves confined in a fixed space immediately above the liquid (see Fig. 11-9a). These "trapped" vapor molecules undergo many random collisions with the container walls, other vapor molecules, and the liquid surface. Molecules colliding with the liquid surface are recaptured by the liquid. Thus, two processes, evaporation (escape) and condensation (recapture), take place in the closed container.

In a closed container, for a short time, the rate of evaporation exceeds the rate of condensation and the liquid level drops. However, as more and more of the liquid evaporates, the number of vapor molecules increases and the chance of their recapture through striking the liquid surface also increases. Eventually the rate of condensation

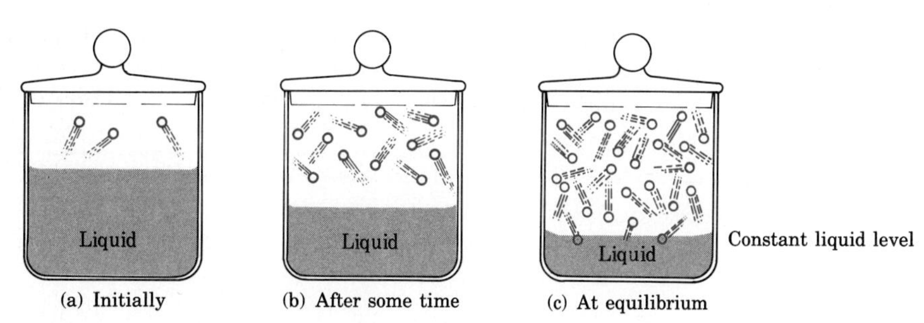

Figure 11-9. Evaporation of a liquid in a closed container. [From *Chemistry: A Science for Today* by Spencer L. Seager and H. Stephen Stoker. Glenview, Ill.: Scott, Foresman and Company, 1973.]

becomes equal to the rate of evaporation and the liquid level stops dropping (see Fig. 11-9c). At this point the number of molecules that escape in a given time is the same as the number recaptured; a steady-state situation has been reached. The amount of liquid and vapor in the container is not changing, even though both evaporation and condensation are still occurring.

state of equilibrium

This steady-state situation, which will continue as long as the temperature of the system remains constant, is an example of a *state of equilibrium*. A **state of equilibrium** *is a situation in which two opposite processes take place at equal rates.* For systems in a state of equilibrium no net macroscopic changes may be detected. However, the system is dynamic; both forward and reverse processes are still occurring, but in a manner such that they balance each other.

vapor pressure

For a liquid–vapor equilibrium in a closed container the vapor in the fixed space immediately above the liquid exerts a constant pressure on the liquid surface and the walls of the container. This pressure is called the liquid's vapor pressure. **Vapor pressure** *is the pressure exerted by a vapor above a liquid when the liquid and vapor are in equilibrium.*

The magnitude of a vapor pressure depends on the nature and temperature of a liquid. Liquids with strong attractive forces between molecules will have lower vapor pressures than liquids in which only weak attractive forces exist between particles. Substances with high vapor pressures at room temperature are said to be *volatile*.

The vapor pressure of all liquids increases with temperature. Why? An increase in temperature results in more molecules having the minimum energy required for evaporation. Hence, at equilibrium the pressure of the vapor is greater. Table 11-2 shows the variation in vapor pressure with increasing temperature for water.

When the temperature of a liquid–vapor system in equilibrium is changed, the equilibrium is upset. The system immediately begins the process of establishing a new equilibrium. Let us consider a case where the temperature is increased. The higher temperature signifies that energy has been added to the system. More molecules will have the minimum energy needed to escape. Thus, immediately after the temperature is increased molecules begin escaping at a rate greater than that at which they are recaptured. With time, however, the rates of escape and recapture will again become equal. The new rates will, however, be different from those for the previous equilibrium. Since energy has been added to the system, the rates will be higher, resulting in a higher vapor pressure.

The size (volume) of the space that the vapor occupies does not affect the magni-

*Table 11-2. **Vapor Pressure of Water at Various Temperatures***

Temperature (°C)	Vapor Pressure (mm Hg)[a]	Temperature (°C)	Vapor Pressure (mm Hg)[a]
0	4.6	60	149.4
10	9.2	70	233.7
20	17.5	80	355.1
30	31.8	90	525.8
40	55.3	100	760.0
50	92.5		

[a]The units used to specify vapor pressure in this table will be discussed in detail in Section 12.1.

tude of the vapor pressure. A larger fixed space will enable more molecules to be present in the vapor at equilibrium. However, the larger number of molecules spread over a larger volume results in the same pressure as a small number of molecules in a small volume.

11.11 Boiling and the Boiling Point

boiling

Usually, for a molecule to escape from the liquid state, it must be on the surface of the liquid. **Boiling** *is a special form of evaporation in which conversion from the liquid to the vapor state occurs within the body of a liquid through bubble formation.* This phenomenon begins to occur when the vapor pressure of a liquid, which is steadily increasing as a liquid is heated, reaches a value equal to that of the prevailing external pressure on the liquid; for liquids in open containers this value is atmospheric pressure. When these two pressures become equal, bubbles of vapor form around any speck of dust or on any rough surface of the container. Being less dense than the liquid itself, these vapor bubbles quickly rise to the surface and escape. The quick ascent of the bubbles causes the agitation associated with a boiling liquid.

Bubbles formed during the boiling process, which are gaseous liquid, should not be confused with the small stationary bubbles often present in liquid at temperatures significantly below their boiling points. These latter bubbles are dissolved gases, usually oxygen.

Like evaporation, boiling is actually a cooling process. When heat is taken away from a boiling liquid, it almost immediately ceases to boil. It is the highest energy molecules which are escaping. Quickly the temperature of the remaining molecules drops below the boiling point of the liquid.

boiling point

The **boiling point** of a liquid is defined as *the temperature at which the vapor pressure of the liquid becomes equal to the external* (*atmospheric*) *pressure exerted on it.* Since atmospheric pressure fluctuates from day to day, so does the boiling point of a liquid. Thus, in order to compare the boiling points of different liquids the external pressure must be the same. The boiling point of a liquid most often used for comparison and tabulation purposes (reference books, for example) is the *normal* boiling point.

Table 11-3. ***Variation of the Boiling Point of Water with Elevation***

Location	Elevation (feet above sea level)	Boiling Point of Water (°C)
San Francisco, Calif.	sea level	100.0
Salt Lake City, Utah	4,390	95.6
Denver, Colo.	5,280	95.0
La Paz, Bolivia	12,795	91.4
Mount Everest	28,028	76.5

normal boiling point

A liquid's **normal boiling point** *is the temperature at which a liquid boils under a pressure of 760 torr.**

At any given location the changes in the boiling point of liquids due to *natural variation* in atmospheric pressure seldom exceeds a few degrees; in the case of water the maximum is about 2°C. However, variations in boiling points *between* locations at different elevations can be quite striking as shown by the data in Table 11-3.

The boiling point of a liquid can be increased by increasing the external pressure. Use is made of this principle in the operation of a pressure cooker. Foods cook faster in pressure cookers because the elevated pressure causes water to boil above 100°C. An increase in temperature of only 10°C will cause food to cook in approximately one-half the normal time. (Cooking food involves chemical reactions and the rate of a chemical reaction generally doubles with every 10°C increase in temperature.) Table 11-4 gives the boiling temperatures reached by water under normal household pressure-cooker conditions. Hospitals use the same principle involved in the operation of a pressure cooker in sterilizing instruments and laundry in autoclaves. Sufficiently high temperatures are reached to destroy bacteria.

Liquids that have high normal boiling points or that undergo undesirable chemical reactions at boiling temperatures can be made to boil at low temperatures by reducing the external pressure. This principle is used in the preparation of numerous food products including frozen fruit juice concentrates. At a reduced pressure some of the water in a fruit juice is boiled away concentrating the juice without having to heat the juice to a high temperature. Heating to a high temperature would cause changes that would spoil the taste of the juice and/or reduce its nutritional value.

Table 11-4. ***Boiling Point of Water in a Pressure Cooker***

Pressure above Atmospheric		Boiling Point of Water (°C)
lb/in.^{2}	torr	
5	259	108
10	517	116
15	776	121

*Torr, a unit for specifying pressure, will be discussed in detail in Sec. 12.1.

11.12 Intermolecular Forces in Liquids

In order for a liquid in an open container to boil its vapor pressure must reach atmospheric pressure. For some substances this occurs at temperatures well below zero; that is, oxygen has a boiling point of −183°C. Other substances do not boil until the temperature is much higher. Mercury, for example, has a boiling point of 357°C, which is 540 C° higher than that of oxygen. An explanation for this variation involves a consideration of the nature of the *intermolecular forces* that must be overcome in order for molecules to escape from the liquid state into the vapor state. **Intermolecular forces** *are forces that act* BETWEEN *a molecule or ion and another molecule.*

intermolecular forces

Intermolecular forces are similar in one way to the previously discussed *intra*molecular forces (*within* molecules) involved in covalent and ionic bonding (Sec. 10.4 and 10.5). They are electrostatic in origin. A major difference between inter- and intramolecular forces is their magnitude; the former are much weaker. However, intermolecular forces, despite their relative weakness, are sufficiently strong to influence the behavior of liquids, often in a very dramatic way. There are three principal types of intermolecular forces: dipole–dipole interactions, hydrogen bonds, and London forces.

dipole–dipole interactions

Dipole–dipole interactions *are electrostatic attractions between polar molecules.* Polar molecules (which are often called dipoles), it should be recalled, are electrically unsymmetrical (Sec. 10.12). Therefore, when polar molecules approach each other, they tend to line up so that the relatively positive end of one molecule is directed to the relatively negative end of the other molecule. As a result, there is an electrostatic attraction between the molecules. The greater the polarity of the molecules, the greater the strength of the dipole–dipole interactions.

hydrogen bond

A **hydrogen bond** *is a special type of dipole–dipole interaction that occurs between polar molecules when one molecule contains hydrogen bonded to a very electronegative element.* These interactions are sufficiently unique to be given the special name of hydrogen bonds. When hydrogen is bonded to a very electronegative element (F, O, and N are the main elements involved), the electronegativity difference is sufficient essentially to strip the H atom of its electron, leaving a nearly exposed nucleus. (Remember, hydrogen has only one electron.) The small size of the hydrogen nucleus allows it to approach the F, N, or O atoms of other molecules very closely, resulting in a much stronger than normal dipole–dipole interaction. It is significant that hydrogen bonding appears to be limited to compounds containing the three elements F, O, and N, all of which are very small in addition to being very electronegative. The larger Cl and S atoms, even though Cl has an electronegativity similar to that of N, show little tendency to hydrogen bond. The strongest hydrogen bond is about one-tenth as strong as a covalent bond and is the strongest of all intermolecular forces. Many polar molecules of biological importance contain O—H and N—H bonds, and hydrogen bonding plays a role in the behavior of these substances. Figure 11-10 shows hydrogen bonding in the specific case of the molecule H_2O.

The third type of intermolecular force, and the weakest, is the London force, named after the German physicist Fritz London (1900–1954), who first postulated its existence. **London forces** *are "instantaneous dipole–dipole interactions" found between all atoms and molecules, nonpolar as well as polar.* The origin of London forces is more difficult to visualize than that of dipole–dipole interactions.

London forces

London forces result from momentary (temporary) uneven electron distributions in

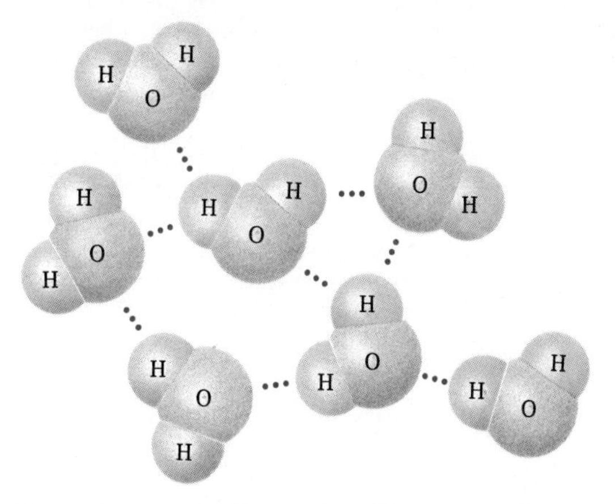

Figure 11-10. Hydrogen bonding in liquid water.

molecules. Most of the time the electrons can be visualized as being distributed in a molecule in a definite pattern determined by their energies and the electronegativities of the atoms. However, there is a small statistical chance (probability) that the electrons will deviate from their normal pattern. For example, in the case of a nonpolar diatomic molecule more electron density may temporarily be located on one side of the molecule than the other. This condition causes the molecule to become polar for an instant. The negative side of this instantaneously polar molecule will tend to repel electrons of adjoining molecules and cause these molecules to also become polar (an induced polarity). The original (statistical) polar molecule and all of the molecules with induced polarity are then attracted to each other. This happens many, many times per second throughout the liquid resulting in a net attractive force. Figure 11-11 depicts the situation present when London forces exist.

The strength of London forces depends on the ease with which an electron distribution in a molecule can be distorted (polarized) by the statistical polarity present in another molecule. In large molecules the outermost electrons necessarily are located farther from the nucleus than are the outermost electrons in small molecules. The farther electrons are located from the nucleus, the more freedom they have and the more susceptible they are to polarization. This leads to the observation that for *related* molecules boiling points increase with molecular weight, which usually parallels size. This trend is reflected in the boiling points given in Table 11-5 for two series of related molecules — the rare gases and the elements of group VIIA.

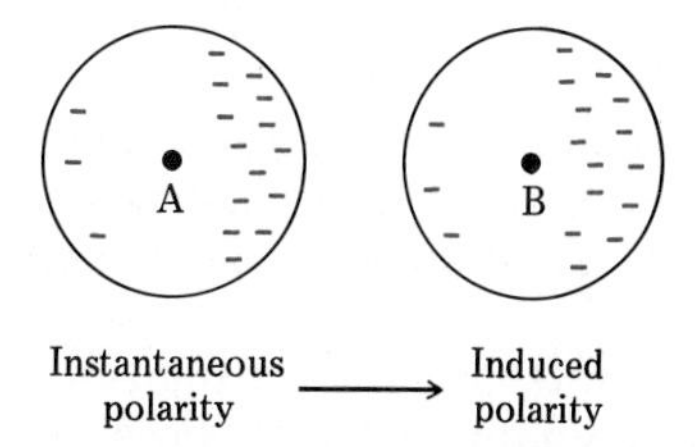

Figure 11-11. A London force. The instantaneous (temporary) polarity present in atom A results in induced polarity in atom B.

Table 11-5. Boiling Point Trends for Related Series of Nonpolar Molecules

Substance	Molecular Weight (amu)	Boiling Point (°C)
He	4.0	−269
Ne	20.2	−246
Ar	39.9	−186
Kr	83.7	−153
Xe	131.1	−107
Rn	222	−62
F_2	39.0	−187
Cl_2	70.9	−35
Br_2	159.8	+59
I_2	253.8	+184

If two molecules have approximately the same molecular weight and polarity but are of different size, the smaller will have the lower boiling point, since the electrons in the smaller molecules are less susceptible to polarization. The nonpolar SF_6 and $C_{10}H_{22}$ molecules (molecular weights 146 and 142 amu, respectively) reflect this trend with boiling points of −64°C and 174°C, respectively.

Of the three types of intermolecular forces (dipole–dipole interactions, hydrogen bonds, and London forces) the London force is the most common and, in the majority of cases, the most important. They are the only attractive forces present between *non-polar* molecules. It is estimated that London forces contribute 85% of the total intermolecular force in the *polar* HCl molecule. Only where hydrogen bonding is involved do London forces play a minor role. In water, for example, about 80% of the intermolecular attraction is attributed to hydrogen bonding and only 20% to London forces.

In this section it has been assumed that the particles making up the liquids are molecules or atoms. This is a valid assumption for liquids as they are normally encountered. Only at extremely high temperatures are liquids encountered where this is not the case. Liquids obtained by melting ionic compounds, which always requires an extremely high temperature, are ionic in the liquid state. The attractive forces in such liquids are those that result from the attraction of positive and negative ions.

11.13 Types of Solids

crystalline solids

Solids may be classified into two categories: *crystalline solids* and *amorphous solids.* Most solids are crystalline solids. **Crystalline solids** *are solids characterized by a regular three-dimensional arrangement of the atoms, ions, or molecules present.* This regular arrangement of particles is manifested in the outward appearance of the solid; crystalline features are discernible. Many times these features are very obvious; other times, however, the crystals may be so tiny as to be visible only with the aid of a microscope. Many times the crystals are imperfect making it difficult to discern them, but they are there.

amorphous solids

Amorphous solids *are solids characterized by a random, nonrepetitive three-dimensional arrangement of the atoms or molecules present.* Only a few such solids exist. The word *amorphous* comes from the Greek word *amorphos* meaning "without form." Examples of amorphous solids include glass, tar, rubber, and many plastics. In some textbooks amorphous solids are referred to as supercooled liquids, since they have the molecular disorder associated with liquids. However, they differ from liquids and resemble solids in being rigid and immobile.

Most amorphous solids consist of very long chainlike molecules that are randomly intertwined. To form a crystalline material from such a solid, the intertwined molecules would have to become regularly arranged in the melted state. In cooling such a liquid to get a solid, molecular motion decreases to the point where "solidification" takes place before the untwining can occur; hence, the disordered structure.

The remainder of this section deals with the more commonly encountered crystalline solid. Indeed, in this text the word solid without a modifier will always mean a crystalline solid.

The highly ordered pattern of particles found in a crystalline solid is called a *crystal lattice.* The positions occupied by the particles in the lattice are called *crystal lattice sites.* Crystalline solids may be classified into five groups based on the type of particles at the crystal lattice sites and the forces that hold the particles together. A consideration of these classifications provides some insight into the reasons why some solids have very low and others very high melting points and why some solids are more volatile than others. The five classes of crystalline solids are: ionic, polar molecular, nonpolar molecular, macromolecular, and metallic.

ionic solids

Ionic solids *consist of positive and negative ions arranged in such a way that each ion is surrounded by nearest neighbors of the opposite charge* (as was previously discussed in Sec. 5.2). Any given ion is bonded by electrostatic attractions to all the ions of opposite charge immediately surrounding it. Since these interionic attractions are relatively strong, ionic solids have high melting points and negligible vapor pressures. Figure 11-12 shows a two-dimensional cross section and also a three-dimensional view of the arrangement of ions for the compound NaCl.

polar molecular solids

Polar molecular solids *have polar molecules at the crystal lattice sites.* The molecules are oriented so that the positive end of each molecule is near the negative end of an adjacent molecule in the lattice. Dipole–dipole interactions and London forces (Sec. 11.12) hold the molecules in the lattice positions. Since these forces are relatively weak compared to other types of electrostatic forces (ionic and covalent bonds), the resulting solids have moderate melting points and moderate vapor pressures.

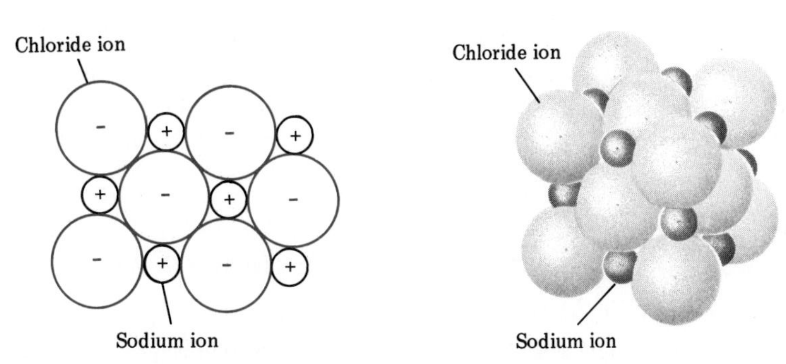

Figure 11-12. Two-dimensional cross section and three-dimensional view of the ionic compound NaCl.

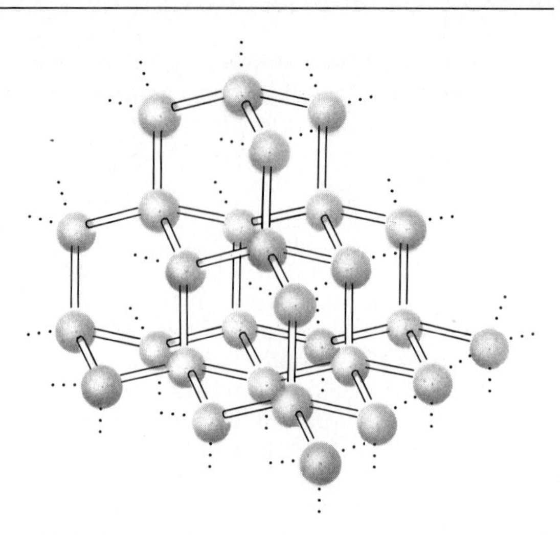

Figure 11-13. The diamond crystal lattice — a macromolecular molecule.

nonpolar molecular solids

Nonpolar molecular solids *have nonpolar molecules (or atoms in the case of the rare gases) at the crystal lattice sites.* They are held together solely by London forces, the weakest of electrostatic forces. Consequently, these solids have low melting points and are volatile.

macromolecular solids

Macromolecular solids *have atoms that are bonded to their nearest neighbors by covalent bonds at the crystal lattice sites.* A crystal of such a solid may be visualized as a gigantic molecule. Such a molecule, for the form of carbon called diamond, is shown in Figure 11-13. Each lattice site in diamond is occupied by a carbon atom. Macromolecular solids are not common, but are noteworthy because of their extremely high melting points and their low volatility. Covalent bonds must be broken in order to cause physical change in such solids. The atoms at the lattice sites in a macromolecular solid need not all be the same, as was the case for diamond. Examples of macromolecular solids where different kinds of atoms are found at the lattice sites include quartz (SiO_2, the main component of sand) and silicon carbide (SiC, an abrasive used in sandpaper and other like materials).

metallic solids

Metallic solids *have metal atoms occupying the crystal lattice sites.* The nature of bonding in such solids is not completely understood. However, the solids are often considered to have a lattice of metal ions with the outer electrons of each metal atom free to move about in the lattice. The bonding results from the interaction of the electrons with the various nuclei. The movement of the free electrons also accounts for the electrical conductivity of metals. The melting points of metallic solids show a wide range. Their volatility is generally low.

Table 11-6 gives a summary of the various types of solids and their distinguishing characteristics along with actual examples of each type.

11.14 Energy and the States of Matter

For a given pure substance, molecules in the gaseous state contain more energy than molecules in the liquid state, which in turn contain more energy than molecules in the solid state. This fact is obvious; we know that it takes energy (heat) to melt a solid and still more energy (heat) to change the resulting liquid to a gas. Additional information

Table 11-6. Characteristics of and Examples of the Various Types of Crystalline Solids

Type	Particles Occupying Lattice Sites	Forces Between Particles	Properties of Solids	Examples
Ionic	positive and negative ions	electrostatic attraction between oppositely charged ions	high melting points; nonvolatile	NaCl $MgSO_4$ KBr
Polar molecular	polar molecules	dipole–dipole attractions and London forces	moderate melting points; moderate volatility	H_2O NH_3 SO_2
Nonpolar molecular	nonpolar molecules (or atoms)	London forces	low melting points; volatile	CO_2 CH_4 I_2 O_2 Ar
Covalent network	atoms	covalent bonds between atoms	extremely high melting points; nonvolatile	C (diamond) SiO_2 (sand) SiC
Metallic	metal atoms	attraction between outer electrons and positive atomic centers	variable melting points; low volatility	Cu Ag Au Fe Al

concerning the relationship of energy to the states of matter may be obtained by a closer examination of what happens, step by step, to a solid (which is below its melting point) as heat is continuously supplied, causing it to melt and ultimately to change to a gas. The heating curve shown in Figure 11-14 gives the steps involved in changing a solid to a gas. As the solid is heated (region I of the graph) its temperature rises until the melting point is reached. The temperature increase indicates that the added heat causes an increase in the kinetic energy of the particles (recall Sec 11.3, kinetic molecular theory). Once the melting point is reached, the temperature remains constant while the solid melts (region II of the graph). The constant temperature during melting indicates that the added heat has increased the potential energy of the particles without increasing their kinetic energy — the interparticle attractions are being weakened by the increase in potential energy. The addition of more heat to the system, which is now in the liquid state, increases the temperature until the boiling point is reached (region III of the graph). During this stage the molecules in the system are again gaining kinetic energy. At the boiling point another state change occurs as heat is added with the temperature remaining constant (region IV of the graph). The constant temperature again indicates an increase in potential energy. Once the system is completely vaporized, the temperature of the gas will again increase with further heating (region V of the graph).

The actual amount of energy that must be added to a system in order to cause it to undergo the series of changes just described depends on values of three properties of the substance. These properties are (1) specific heat, (2) heat of fusion, and (3) heat of vaporization.

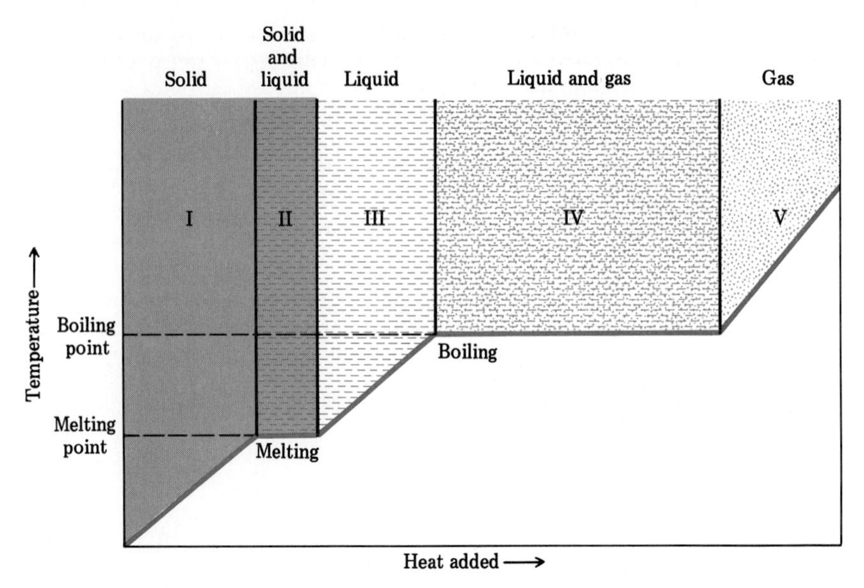

Figure 11-14. A heating curve depicts the addition of heat to a substance — a solid at a temperature below its melting point — until it becomes a gas at a temperature above its boiling point.

specific heat

Specific heat *is the amount of heat required to change the temperature of a specified amount of a substance in a specific physical state by 1°C.* Units most commonly used for specific heat, in scientific work, are calories per gram per degree Celsius (cal/g °C), the number of calories of heat required to raise the temperature of one gram of material by one degree Celsius. Specific heats for a number of substances in various states are given in Table 11-7.

Table 11-7. ***Specific Heats of Selected Elements and Compounds***

Substance and State	Specific Heat (cal/g °C)
Aluminum (solid)	0.24
Copper (solid)	0.093
Ethylene glycol (liquid)	0.57
Helium (gas)	1.25
Hydrogen (gas)	3.39
Lead (solid)	0.031
Mercury (liquid)	0.033
Nitrogen (gas)	0.25
Oxygen (gas)	0.22
Sodium (solid)	0.29
Sodium (liquid)	0.32
Water (solid)	0.50
Water (liquid)	1.00
Water (gas)	0.48

The higher the specific heat of a substance, the less its temperature will change when it absorbs a given amount of heat. For liquids, water has a relatively high specific heat; it is thus a very effective coolant. The moderate climates of geographical areas where large amounts of water are present, for example, the Hawaiian Islands, are related to water's ability to absorb large amounts of heat without undergoing drastic temperature changes. Desert areas, areas that lack water, are the areas where the extremes of high temperature are encountered on this earth. The temperature of a living organism remains remains relatively constant because of the large amounts of water present in it.

The amount of heat energy needed to cause a fixed amount of a substance to undergo a specific temperature change (within a range which causes no change of state) can easily be calculated if the substance's specific heat is known. The specific heat (in cal/g °C) is multiplied by the mass (in grams) and by the temperature change (in °C) to eliminate the units of g and °C and obtain the units of cal.

$$\text{Heat absorbed} = \text{specific heat} \times \text{mass} \times \text{temperature change}$$
$$= \left(\frac{\text{cal}}{\cancel{\text{g}}\ {}^\circ\cancel{\text{C}}}\right) \times \cancel{\text{grams}} \times {}^\circ\cancel{\text{C}}$$
$$= \text{calories}$$

EXAMPLE 11.1

Calculate the amount of heat (in cal) needed to increase the temperature of each of the following substances from 81°C to 93°C.

(a) 3.4 g of ethylene glycol (automobile antifreeze)
(b) 3.4 g of water (also functional as a coolant in automobile radiators)

SOLUTION

(a) From Table 11-7 we determine that the specific heat of liquid ethylene glycol is 0.57 cal/g °C. Using this value, the given mass in g and the temperature change in °C, which is 12°C, we calculate the heat needed as follows.

$$\text{Heat absorbed} = \left(\frac{0.57\ \text{cal}}{\cancel{\text{g}}\ {}^\circ\cancel{\text{C}}}\right) \times 3.4\ \cancel{\text{g}} \times 12{}^\circ\cancel{\text{C}}$$
$$= 23.256\ \text{cal} \quad \text{(calculator answer)}$$
$$= 23\ \text{cal} \quad \text{(correct answer)}$$

(b) From Table 11-7 we determine that the specific heat of liquid water is 1.00 cal/g °C. Note that three specific heats for water are given in Table 11-7—one for each of the three physical states. In general, as is the case for water, a substance's specific heat as a solid is different from that as a liquid, which in turn is different from that as a gas.

$$\text{Heat absorbed} = \left(\frac{1.00 \text{ cal}}{\cancel{g}\ ^\circ\cancel{C}}\right) \times 3.4 \cancel{g} \times 12^\circ\cancel{C}$$

$= 40.8$ cal (calculator answer)

$= 41$ cal (correct answer)

Note that the heat-carrying capacity of water is almost twice that of ethylene glycol when both are liquids.

heat of fusion

heat of vaporization

Heat energy is absorbed or evolved in changes of state. The **heat of fusion** *is the amount of energy required to convert one gram of a solid to a liquid at its melting point.* This same amount of energy is released to the surroundings when 1 gram of the liquid is changed back to a solid during a freezing process. The **heat of vaporization** *is the amount of energy required to convert 1 gram of a liquid to a gas at its boiling point.* This same amount of energy is released to the surroundings when 1 gram of the gas condenses to a liquid.

Numerical values of the heats of fusion and vaporization for water are 79.8 and $54\bar{0}$ (5.40×10^2) cal/g, respectively. Note that the units in both cases are cal/g. No temperature units are involved, since temperature remains constant during a change of state. The use of heat of fusion and heat of vaporization values in problem solving is illustrated in Examples 11.2 and 11.3.

EXAMPLE 11.2

Calculate the heat released when 32.0 g of steam at 122°C is condensed to water at 100°C.

SOLUTION

We may consider this process as occurring in two steps.

1. Cooling the steam from 122°C to 100°C (the condensation temperature for steam).
2. Condensing the steam to liquid water at 100°C.

We will calculate the amount of heat released in each of the steps and then add these amounts together to get our final answer.

Step 1. $\text{Gas}_{122^\circ\text{C}} \longrightarrow \text{gas}_{100^\circ\text{C}}$

Heat released = specific heat × mass × temperature change

$$= \left(\frac{0.48 \text{ cal}}{\cancel{g}\ ^\circ\cancel{C}}\right) \times 32.0 \cancel{g} \times 22^\circ\cancel{C}$$

$= 337.92$ cal (calculator answer)

$= 340$ cal (correct answer)

(The answer may contain only two significant figures.)

Step 2. $Gas_{100°C} \longrightarrow liquid_{100°C}$

Heat released = heat of vaporization × mass

$$= \left(\frac{54\bar{0}\ \text{cal}}{\cancel{g}}\right) \times 32.0\ \cancel{g} = 17{,}280\ \text{cal} \quad \text{(calculator answer)}$$

$$= 17{,}300\ \text{cal} \quad \text{(correct answer)}$$

(The correct answer may contain only three significant figures.)

The total heat released is the sum of that released in each step.

From step 1:	338 cal
From step 2:	17,300 cal
	17,638 cal (calculator answer)

The **correct answer** is **17,600 cal,** obtained by noting that the last digit position where both numbers of the sum have a significant number is the hundreds position.

Last digit position where all numbers are significant

```
   338
17,300
------
17,638
```

Note that the heat of vaporization was used in step 2 of the calculation even though the water was changing from the vapor to the liquid state. The only difference between vaporization and condensation is the direction of heat flow; the amount of heat remains the same for a specific quantity of material.

Our results make it clear that most of the transported heat (98%) was carried in the form of potential energy, which was released when the steam condensed (step 2). A calculation such as this also explains why a burn caused by steam at 100°C is more severe than one caused by water at 100°C. The steam will release $54\bar{0}$ cal of heat for each gram that condenses on the skin. The condensed water is then still at 100°C.

EXAMPLE 11.3

Calculate the number of calories of heat needed to change 12 g of ice at −18°C to 12 g of steam at 121°C.

SOLUTION

We may consider the process as occurring in five steps.

1. Heating the ice from −18°C to 0°C (its melting point).
2. Melting the ice (while the temperature remains at 0°C).
3. Heating the liquid from 0°C to 100°C (its boiling point).
4. Evaporating the water at 100°C.
5. Heating the steam from 100°C to 121°C.

We will calculate the amount of heat required in each of the steps and then add these amounts together to get our desired answer.

Step 1. $\text{Solid}_{-18°C} \longrightarrow \text{solid}_{0°C}$

Heat required = specific heat × mass × temperature change

$$= \left(\frac{0.50 \text{ cal}}{\not{g}\ °\not{C}}\right) \times 12\ \not{g} \times 18°\not{C} = 108 \text{ cal} \quad \text{(calculator answer)}$$

$$= 110 \text{ cal} \quad \text{(correct answer)}$$

Step 2. $\text{Solid}_{0°C} \longrightarrow \text{liquid}_{0°C}$

Heat required = heat of fusion × mass

$$= \left(\frac{8\bar{0} \text{ cal}}{\not{g}}\right) \times 12\ \not{g} = 960 \text{ cal} \quad \text{(calculator answer)}$$

$$= 960 \text{ cal} \quad \text{(correct answer)}$$

Step 3. $\text{Liquid}_{0°C} \longrightarrow \text{liquid}_{100°C}$

Heat required = specific heat × mass × temperature change

$$= \left(\frac{1.00 \text{ cal}}{\not{g}\ °\not{C}}\right) \times 12\ \not{g} \times 100°\not{C} = 1200 \text{ cal} \quad \text{(calculator answer)}$$

$$= 1200 \text{ cal} \quad \text{(correct answer)}$$

Step 4. $\text{Liquid}_{100°C} \longrightarrow \text{gas}_{100°C}$

Heat required = heat of vaporization × mass

$$= \left(\frac{54\bar{0} \text{ cal}}{\not{g}}\right) \times 12\ \not{g} = 6480 \text{ cal} \quad \text{(calculator answer)}$$

$$= 6500 \text{ cal} \quad \text{(correct answer)}$$

Step 5. $\text{Gas}_{100°C} \longrightarrow \text{gas}_{121°C}$

Heat required = specific heat × mass × temperature change

$$= \left(\frac{0.48 \text{ cal}}{\not{g}\ °\text{C}}\right) \times 12\ \not{g} \times 21°\not{C} = 120.96 \text{ cal} \quad \text{(calculator answer)}$$

$$= 120 \text{ cal} \quad \text{(correct answer)}$$

The total heat required is the sum of that added in the five steps.

$$(110 + 960 + 1200 + 6500 + 120) \text{ cal} = 8890 \text{ cal} \quad \text{(calculator answer)}$$

$$= 8900 \text{ cal} \quad \text{(correct answer)}$$

(The answer has only two significant figures. In adding, the numbers 1200 and 6500 are the limiting numbers, with uncertainty in the hundreds place.)

Learning Objectives

After completing this chapter you should be able to

- Compare and contrast the gross distinguishing properties of gases, liquids, and solids (Sec. 11.2).
- Understand the roles that kinetic energy (disruptive forces) and potential energy (cohesive forces) play in determining the physical state of a system (Sec. 11.3).
- Use kinetic molecular theory to explain the characteristic properties of solids, liquids, and gases (Secs. 11.3 through 11.6).
- Contrast, on a molecular level, the differences between solids, liquids, and gases (Sec. 11.7).
- Understand the difference between exothermic and endothermic changes of state (Sec. 11.8).
- Explain, on a molecular basis, what happens during the process of evaporation, and also list the factors that affect evaporation rates (Sec. 11.9).
- Understand the relationship between vapor pressure and an equilibrium state, and list the factors that affect the magnitude of the vapor pressure of a given liquid (Sec. 11.10).
- Understand the process of boiling from a molecular viewpoint, understand what is meant by the term normal boiling point, and know the relationship between the boiling point of a liquid and the external pressure (Sec. 11.11).
- Understand the difference between intramolecular and intermolecular forces, and describe and distinguish the three types of intermolecular forces that may be operative in a liquid (Sec. 11.12).
- Distinguish between crystalline and amorphous solids, and know the nature of the crystal lattice structure for each of the five types of crystalline solids (Sec. 11.13).
- Calculate the heat released or absorbed when a substance is heated, cooled, or undergoes a change in state given the specific heat, heat of vaporization, and heat of fusion of the substance (Sec. 11.14).

Terms and Concepts for Review

The new terms or concepts defined in this chapter are

compressibility (Sec. 11.2)
thermal expansion (Sec. 11.2)
kinetic molecular theory (Sec. 11.3)
kinetic energy (Sec. 11.3)
potential energy (Sec. 11.3)
electrostatic interaction (Sec. 11.3)
endothermic change (Sec. 11.8)
exothermic change (Sec. 11.8)
evaporation (Sec. 11.9)
state of equilibrium (Sec. 11.10)
vapor pressure (Sec. 11.10)
boiling (Sec. 11.11)
boiling point (Sec. 11.11)
normal boiling point (Sec. 11.11)
intermolecular force (Sec. 11.12)
dipole–dipole interaction (Sec. 11.12)
hydrogen bond (Sec. 11.12)
London force (Sec. 11.12)
crystalline solid (Sec. 11.13)
amorphous solid (Sec. 11.13)
ionic solid (Sec. 11.13)
polar molecular solid (Sec. 11.13)
nonpolar molecular solid (Sec. 11.13)
macromolecular solid (Sec. 11.13)
metallic solid (Sec. 11.13)
specific heat (Sec. 11.14)
heat of fusion (Sec. 11.14)
heat of vaporization (Sec. 11.14)

Questions and Problems

States of Matter

11-1 The following statements relate to the terms *solid state, liquid state,* and *gaseous state.* Match the terms to the appropriate statements.
- **a.** This state is characterized by the lowest density of the three.
- **b.** This state is characterized by an indefinite shape and a high density.
- **c.** Temperature changes influence the volume of this state significantly.
- **d.** Pressure changes influence the volume of this state more than that of the other two.
- **e.** In this state constituent particles are less free to move about than in other states.

Kinetic Molecular Theory of Matter

11-2 According to the postulates of the kinetic molecular theory of matter:
- **a.** What two types of energy do particles possess?
- **b.** How do molecules transfer energy from one to another?
- **c.** What is the relationship between temperature and the average velocity with which the particles move?

11-3 Contrast the effects of cohesive and disruptive forces on a system of particles. What type of energy is related to cohesive forces and to disruptive forces?

11-4 What effect does temperature have on the magnitude of disruptive forces and cohesive forces?

11-5 Distinguish between the gaseous, liquid, and solid states of a substance from the point of view of the magnitude of the kinetic and potential energies of the constituent particles.

11-6 Explain each of the following observations using the kinetic molecular theory of matter.
- **a.** Gases have a low density.
- **b.** Solids maintain characteristic shapes.

c. Liquids show little change in volume with changes in temperature.
d. Both liquids and solids are practically incompressible.
e. A gas always exerts a pressure on the object or container with which it is in contact.

11-7 Using kinetic molecular theory, explain why it is dangerous to heat an aerosol can in an open fire.

Changes of State

11-8 Indicate whether each of the following is an exothermic or endothermic change.
a. sublimation b. melting
c. freezing d. condensation
e. evaporation

11-9 Match the following statements to the appropriate term: *evaporation, vapor pressure,* and *boiling point.*
a. This is a temperature at which the liquid vapor pressure is equal to the external pressure on a liquid.
b. The process takes place when a liquid changes into a vapor.
c. This can be measured by allowing a liquid to evaporate in a closed container.
d. At this temperature bubbles of vapor form within a liquid.
e. This temperature changes appreciably with changes in atmospheric pressure.

11-10 Offer a clear explanation for each of the following observations.
a. Evaporation lowers the temperature of a liquid.
b. Changing the volume of a container in which there is a liquid–vapor equilibrium does not change the magnitude of the vapor pressure.
c. Changing the temperature in a container in which there is a liquid–vapor equilibrium causes a change in the magnitude of the vapor pressure.
d. Foods cook faster in a pressure cooker than in an open pan.
e. When your skin is cleansed with alcohol it feels cold, even though the bottle of alcohol is at room temperature.
f. It is possible to boil water at room temperature.
g. Different liquids may have different vapor pressures at the same temperatures.

11-11 Distinguish between the terms *boiling point* and *normal boiling point.*

11-12 What two factors affect the rate at which a substance evaporates?

11-13 What factors affect the magnitude of the vapor pressure a liquid exerts?

Intermolecular Forces in Liquids

11-14 What is the difference in meaning of the terms *intermolecular force* and *intramolecular force?*

11-15 What are the three major types of intermolecular forces, between which types of molecules do they occur, and what are their origins?

11-16 Which gas in each of the following pairs of gases should have the larger intermolecular attractive forces and why?
a. F_2 and Cl_2 (both are in period VIIA)
b. CO_2 (molecular weight = 44 amu) and F_2 (molecular weight = 38 amu) (both are nonpolar molecules).

Types of Solids

11-17 Name the five types of crystalline solids. For each type identify (a) the particles that occupy the crystal lattice sites and (b) the kinds of bonding forces that hold the particles in their positions.

11-18 What is the difference between a *crystalline solid* and an *amorphous solid?*

11-19 Carbon and silicon are in the same period of the periodic table and would be expected to form compounds of a similar nature. Yet CO_2 is a soft low melting solid (dry ice), and SiO_2 is a hard very high melting solid (sand). Explain this apparent anomaly using the concepts of interparticle forces and the nature of the particles that occupy lattice sites in the solids.

Energy Calculations and the States of Matter

11-20 Define the terms *specific heat, heat of fusion,* and *heat of vaporization.*

11-21 Use the specific heat data of Table 11-7 to calculate the amount of heat needed to effect the following changes.
a. Raise the temperature of a 48 g Al bar from 25°C to 55°C.
b. Raise the temperature of 2500 g of liquid ethylene glycol from 80°C to 85°C.
c. Raise the temperature of 321 g of steam from 110°C to 120°C.

11-22 In order to use solar energy effectively, collected heat must be stored for use during periods of decreased sunshine. One proposal is that heat can be stored by melting solids, which, on solidification, would release the heat. Calculate the heat that could be stored by melting 1000 kg (four significant figures) of calcium chloride; melting point = 30.2°C, and heat of fusion = 40.7 cal/g.

11-23 Liquid freon, CCl_2F_2, is used as a refrigerant. It is circulated inside the cooling coils of a refrigerator or freezer. As it vaporizes, it absorbs heat. How much heat could be removed by 2.21 kg of freon as it vaporizes inside the coils of a refrigerator? The heat of vaporization of freon is 38.6 cal/g.

11-24 Calculate the total amount of heat needed to change 483 g of ice at −20°C into 483 g of steam at 110°C.

11-25 How many grams of copper can be heated from 20°C to 30°C when 245 g of aluminum cools from 80°C to 50°C?

11-26 Steam at 100°C causes more severe burns than does boiling water at 100°C. Why is this so?

12 Gas Laws

12.1 Gas-Law Variables

Our discussion of solids, liquids, and gases in the last chapter was very qualitative. The general properties of the three states of matter were discussed in essentially non-numerical terms. In this chapter we consider the gaseous state again, this time in quantitative (numerical) terms. The behavior of a gas can be described reasonably well by *simple* quantitative relationships called *gas laws*. **Gas laws** *are generalizations that describe in mathematical terms the relationships among the* PRESSURE, TEMPERATURE, *and* VOLUME *of a specific quantity of a gas.*

gas laws

It is only the gaseous state that is describable by simple mathematical relationships. Laws describing liquid- and solid-state behavior are mathematically extremely complex. Consequently, quantitative treatments of liquid-state and solid-state behavior will not be given in this text.

Prior to discussing the mathematical form of the various gas laws, some comments concerning the major variables involved in gas-law calculations — volume, temperature, and pressure — are in order. Two of these three variables, volume and temperature, have been discussed previously (Secs. 3.1 and 3.5, respectively). The units of *liter* or *milliliter* are usually used in specifying gas volume. Only one of the three temperature scales discussed in Section 3.5, *the Kelvin scale,* may be used in gas law calculations if the results are to be valid. Therefore, you should be thoroughly familiar with the conversion of Celsius and Fahrenheit scale readings to a Kelvin scale (Sec. 3.5). We have not yet discussed pressure, the third variable. Comments concerning pressure will occupy the remainder of this section.

pressure

Pressure *is defined as the force applied per unit area, that is, the total force on a surface divided by the area of that surface.*

$$P\text{ (pressure)} = \frac{F\text{ (force)}}{A\text{ (area)}}$$

Note that pressure and force are not one and the same. Identical forces give rise to different pressures if they are acting on areas of different size. For areas of the same size, the larger the force, the greater the pressure.

For a gas, the force involved in pressure is that exerted by the gas molecules or atoms as they constantly collide with the walls of their container. Barometers, manometers, and gauges are the instruments commonly used by chemists to measure gas pressures. Barometers and manometers measure pressure in terms of the height of a column of mercury, whereas gauges usually are calibrated in terms of force per area, for example, in pounds per square inch.

barometer

The air that surrounds the earth exerts a pressure on all objects with which it has contact. The **barometer** *is the most commonly used device for measuring atmospheric pressure.* It was invented by the Italian physicist Evangelista Torricelli (1608–1647) in 1643. The essential components of a barometer are shown in Figure 12-1. A barometer may be constructed by filling a long glass tube, sealed at one end, all the way to the top with mercury and then inverting the tube (without letting any air in) into a dish of mercury. The mercury in the tube falls until the pressure from the mass of the mercury in the tube is just balanced by the pressure of the atmosphere on the mercury in the dish. The pressure of the atmosphere is then expressed in terms of the height of the supported column of mercury.

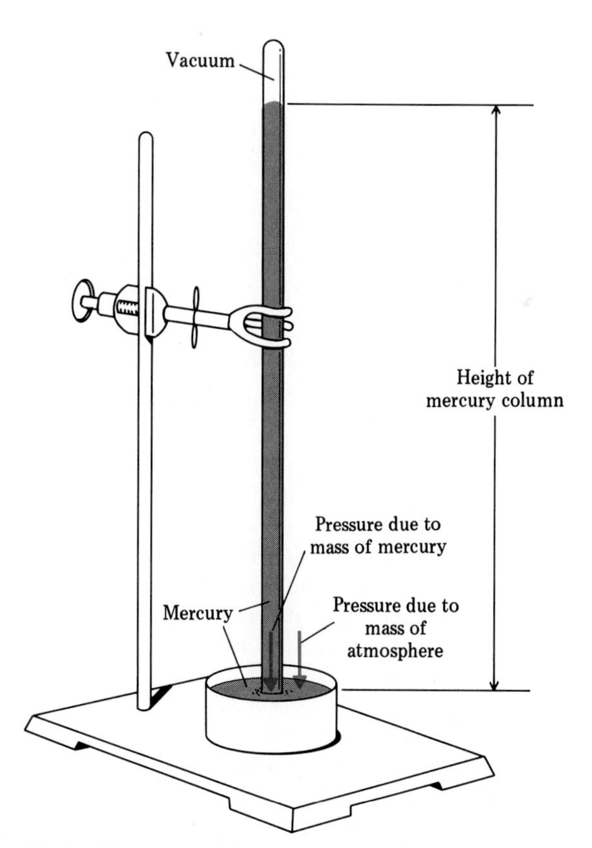

Figure 12-1. The essential components of a mercury barometer.

Mercury is the liquid of choice in a barometer for two reasons. (1) It is a very dense liquid, hence the length of the glass tube needed is short; and (2) it has a very low vapor pressure, hence the pressure reading does not have to be corrected for vapor pressure.

The height of the mercury column is most often expressed in millimeter or inch units. Millimeters of mercury (mm Hg) are used in laboratory work. The most common use of inches of mercury (in Hg) is in weather reporting. A pressure corresponding to 1 mm Hg is also called a *torr,* in recognition of Torricelli's invention of the barometer; that is,

$$1 \text{ torr} = 1 \text{ mm Hg}$$

Pressures expressed in terms of torr and mm Hg will always be numerically equal.

The pressure of the atmosphere varies with altitude, decreasing at the rate of approximately 25 torr per 1000 feet increase in altitude. It also fluctuates with weather conditions. Recall the terminology used in a weather report — high pressure front, low pressure front, and so on. At sea level the height of a column of Hg in a barometer fluctuates with weather conditions between about 740 and 770 torr and averages about 760 torr. Another pressure unit, the atmosphere (atm), is defined in terms of this average sea-level pressure.

$$1 \text{ atm} = 760 \text{ torr} = 760 \text{ mm Hg}$$

Because of its size, 760 times larger than the torr, the atmosphere unit is frequently used to express the high pressures encountered in many industrial processes and in some experimental work.

manometer

It is impractical to measure the pressure of gases other than air with a barometer. One cannot usually introduce a mercury barometer directly into a container of a gas. A **manometer** *is the device used to measure gas pressures in a laboratory.* It is a U-tube filled with mercury. One side of the U-tube is connected to the container in which the pressure is to be measured, and the other side is connected to a region of known pressure. Such an arrangement is depicted in Figure 12-2. The gas in the container exerts a pressure on the Hg on that side of the tube, while atmospheric pressure pushes on the open end. A difference in the heights of the Hg columns in the two arms of the manometer indicates a pressure difference between the gas and the atmosphere. In Figure 12-2 the gas pressure in the container exceeds atmospheric pressure by an amount, ΔP, which is equal to the difference in the heights of the Hg columns in the two arms.

$$P_{\text{container}} = P_{\text{atmosphere}} + \Delta P$$

Pressures of contained gas samples may also be measured by special gas pressure gauges attached to their containers. Such gauges are commonly found on tanks (cylinders) of gas purchased commercially. These gauges are most often calibrated in terms of pounds per square inch (lb/in.^{2} or psi). The relationship between psi and atmospheres is

$$14.68 \text{ psi} = 1 \text{ atm}$$

Table 12-1 summarizes the relationships among the various pressure units we have discussed and also lists one additional pressure unit, the *pascal.* The pascal is the SI unit (Sec. 3.1) of pressure.

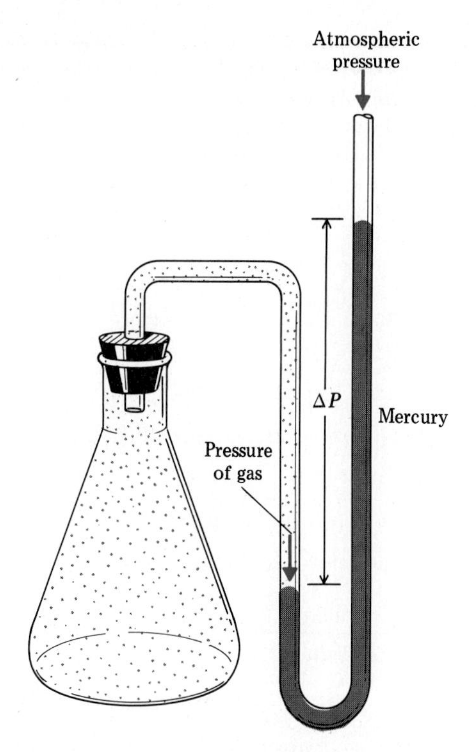

Figure 12-2. Pressure measurement using a manometer.

The values 1 atm and 760 torr (or 760 mm Hg) are exact numbers since they arise from a definition. Consequently, these two values have an infinite number of significant figures. All other values in Table 12-1 are not exact numbers; they are given to four significant figures in the table.

The equalities of Table 12-1 can be used to construct conversion factors for problem solving in the usual way (Sec. 3.2). Example 12.1 illustrates the use of such conversion factors in the context of change in pressure unit problems.

Table 12-1. Units of Pressure and Their Relationship to the Unit 1 Atmosphere

Unit	Relationship to Atmosphere	Area of Use
Atmosphere	—	gas-law calculations
Millimeter of Hg	760 mm Hg = 1 atm	gas-law calculations
Torr	760 torr = 1 atm	gas-law calculations
Inches of Hg	29.92 in. Hg = 1 atm	weather reports
Pounds per square inch	14.68 psi = 1 atm	stored or bottled gases
Pascal	1.013×10^5 Pa = 1 atm	calculations requiring SI units

EXAMPLE 12.1

A weather reporter gives the day's barometric pressure as 30.12 in. of mercury. Express this pressure in the units of

(a) atmospheres **(b)** torr **(c)** millimeters of mercury
(d) pounds per square inch

SOLUTION

(a) The given quantity is 30.12 in. Hg and the unit of the desired quantity is atmospheres.

$$30.12 \text{ in. Hg} = ? \text{ atm}$$

The conversion factor relating these two units, found in Table 12-1, is

$$\left(\frac{1 \text{ atm}}{29.92 \text{ in. Hg}}\right)$$

The result of using this conversion factor is

$$30.12 \cancel{\text{in. Hg}} \times \left(\frac{1 \text{ atm}}{29.92 \cancel{\text{in. Hg}}}\right) = 1.0066845 \text{ atm} \quad \text{(calculator answer)}$$

$$= 1.007 \text{ atm} \quad \text{(correct answer)}$$

(b) This time, the unit for the desired quantity is torr.

$$30.12 \text{ in. Hg} = ? \text{ torr}$$

A direct conversion factor between in. Hg and torr is not given in Table 12-1. However, the table does give the relationships of both in. Hg and torr to atmospheres. Hence, we can use atmospheres as an intermediate step in the unit conversion sequence.

$$\text{in. Hg} \longrightarrow \text{atm} \longrightarrow \text{torr}$$

The dimensional-analysis setup is

$$30.12 \cancel{\text{in. Hg}} \times \left(\frac{1 \cancel{\text{atm}}}{29.92 \cancel{\text{in. Hg}}}\right) \times \left(\frac{760 \text{ torr}}{1 \cancel{\text{atm}}}\right) = 765.08021 \text{ torr} \quad \text{(calculator answer)}$$

$$= 765.1 \text{ torr} \quad \text{(correct answer)}$$

(c) Since the torr and mm Hg are two names for the same unit, we work this problem as we did part (b) with torr being replaced with mm Hg. The numerical answer is the same.

$$30.12 \cancel{\text{in. Hg}} \times \left(\frac{1 \cancel{\text{atm}}}{29.92 \cancel{\text{in. Hg}}}\right) \times \left(\frac{760 \text{ mm Hg}}{1 \cancel{\text{atm}}}\right) = 765.08021 \text{ mm Hg} \quad \text{(calculator answer)}$$

$$= 765.1 \text{ mm Hg} \quad \text{(correct answer)}$$

(d) The problem to be solved here is

$$30.12 \text{ in. Hg} = ? \text{ psi}$$

The unit conversion sequence will be

$$\text{in. Hg} \longrightarrow \text{atm} \longrightarrow \text{psi}$$

The dimensional-analysis sequence of conversion factors, by the above pathway, is

$$30.12\ \cancel{\text{in. Hg}} \times \left(\frac{1\ \cancel{\text{atm}}}{29.92\ \cancel{\text{in. Hg}}}\right) \times \left(\frac{14.68\ \text{psi}}{1\ \cancel{\text{atm}}}\right) = 14.778128\ \text{psi} \quad \text{(calculator answer)}$$

$$= \mathbf{14.78\ psi} \quad \text{(correct answer)}$$

12.2 Boyle's Law: A Pressure–Volume Relationship

Boyle's law

Of the several relationships that exist between gas-law variables, the first to be discovered was the one that relates gas pressure to gas volume. It was formulated over 300 years ago, in 1662, by the British chemist and physicist Robert Boyle (1627–1691) and is known as Boyle's law. **Boyle's law** states that *the volume of a sample of gas is* INVERSELY PROPORTIONAL *to the pressure applied to the gas if the temperature is kept constant.* This means that if the pressure on the gas increases the volume decreases proportionally, and conversely if the pressure is decreased the volume will increase. Doubling the pressure cuts the volume in half; tripling the pressure cuts the volume to one-third its original value; quadrupling the pressure cuts the volume to one-fourth; and so on. Any time two quantities are *inversely proportional,* which pressure and volume are (Boyle's law), one increases as the other decreases.

Boyle's law can be illustrated quite simply with the J-tube apparatus shown in Figure 12-3. The pressure on the trapped gas is increased by adding mercury to the J-tube. The volume of the trapped gas decreases as the pressure is raised.

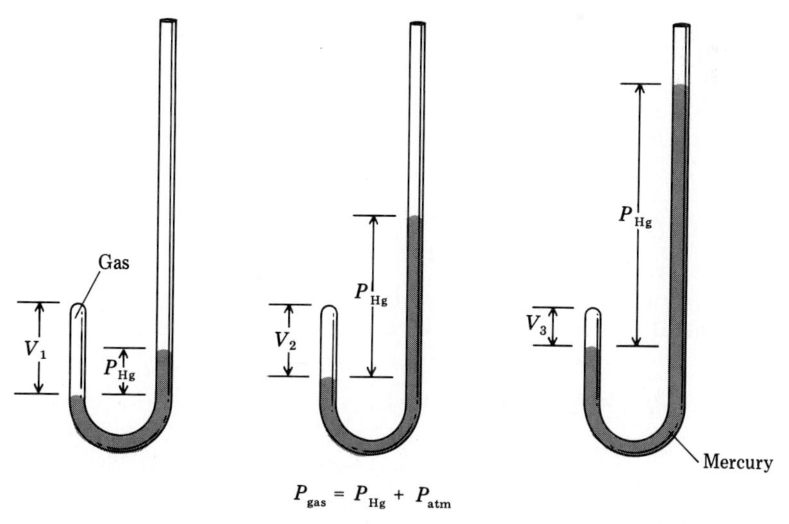

Figure 12-3. Boyle's law apparatus. The volume of the trapped gas in the closed end of the tube (V_1, V_2, and V_3) decreases as mercury is added through the open end of the tube.

Boyle's law may be stated mathematically as

$$P \times V = \text{constant}$$

In this expression V is the volume of the gas at a given temperature and P is the pressure. This expression, thus, indicates that at constant temperature the product of the pressure times the volume is always the same (or constant). (Note that Boyle's law is valid only if the temperature of the gas does not change.)

An alternate, and more useful, mathematical form of Boyle's law can be derived by considering the following situation. Suppose a gas is at an initial pressure, P_1, and has a volume V_1. (We will use the subscript 1 to indicate the initial conditions.) Now imagine that the pressure is changed to some final pressure, P_2. The volume will also change, and we will call the final volume V_2. (We will use the subscript 2 to indicate the final conditions.) According to Boyle's law

$$P_1 \times V_1 = \text{constant}$$

After the change in pressure and volume, we have

$$P_2 \times V_2 = \text{constant}$$

The constant is the same in both cases; we are dealing with the same sample of gas. Thus, we may combine the two *PV* products to give the equation

$$P_1 \times V_1 = P_2 \times V_2$$

When we know any three of the four quantities in this equation we can calculate the fourth, which will usually be the final pressure, P_2, or the final volume, V_2.

EXAMPLE 12.2

A sample of O_2 gas occupies a volume of 1.51 L at a pressure of 1.82 atm and a temperature of 25°C. What volume will it occupy if the pressure is reduced to 1.00 atm with no change in temperature?

SOLUTION

A suggested first step in working all gas-law problems involving two sets of conditions is to analyze the given data in terms of initial and final conditions. Doing this, we find that

$$P_1 = 1.82 \text{ atm} \qquad P_2 = 1.00 \text{ atm}$$

$$V_1 = 1.51 \text{ L} \qquad V_2 = ?\text{ L}$$

Next, we arrange Boyle's law to isolate V_2 (the quantity to be calculated) on one side of the equation. This is accomplished by dividing both sides of the Boyle's law equation by P_2.

$$P_1V_1 = P_2V_2 \qquad \text{(Boyle's law)}$$

$$\frac{P_1V_1}{P_2} = \frac{\cancel{P_2}V_2}{\cancel{P_2}} \qquad \text{(division of each side of the equation by } P_2\text{)}$$

$$V_2 = V_1 \times \frac{P_1}{P_2}$$

Substituting the given data in the rearranged equation and doing the arithmetic gives

$$V_2 = 1.51\ \text{L} \times \left(\frac{1.82\ \cancel{\text{atm}}}{1.00\ \cancel{\text{atm}}}\right)$$

$$= 2.7482\ \text{L} \quad \text{(calculator answer)}$$

$$= 2.75\ \text{L} \quad \text{(correct answer)}$$

In gas-law problems, when possible, it is always a good idea to check qualitatively the reasonableness of your final answer. This is a double check against two types of errors frequently made: improper substitution of variables into the equation and improper rearrangement of the equation. Decreasing the pressure on a fixed amount of gas at constant pressure, which is the case in this problem, should result in a volume increase for the gas. Our answer is consistent with this conclusion.

EXAMPLE 12.3

A sample of F_2 occupies a volume of 1.43 L at 17°C and 735 torr pressure. What volume will the gas occupy at 17°C and 1.13 atm pressure?

SOLUTION

Analyzing the given data in terms of initial and final conditions gives

$$P_1 = 735\ \text{torr} \qquad P_2 = 1.33\ \text{atm}$$

$$V_1 = 1.43\ \text{L} \qquad V_2 = ?\ \text{L}$$

$$T_1 = 17°\text{C} \qquad T_2 = 17°\text{C}$$

The temperature is constant; it will not enter into the calculation.

A slight complication exists with the pressure values; they are not given in the same units. We must make the units the same before proceeding with the calculation. It does not matter whether they are both torr or both atmospheres. What is important is that they are the same. The same answer is obtained with either unit. Let us arbitrarily decide to change atmospheres to torr.

$$1.13\ \cancel{\text{atm}} \times \left(\frac{760\ \text{torr}}{1\ \cancel{\text{atm}}}\right) = 858.8\ \text{torr} \quad \text{(calculator answer)}$$

$$= 859\ \text{torr} \quad \text{(correct answer)}$$

Our given conditions are now

$$P_1 = 735\ \text{torr} \qquad P_2 = 859\ \text{torr}$$

$$V_1 = 1.43\ \text{L} \qquad V_2 = ?\ \text{L}$$

Boyle's law with V_2 isolated on the left side has the form

$$V_2 = V_1 \times \frac{P_1}{P_2}$$

Plugging the given quantities into this equation and doing the arithmetic gives

$$V_2 = 1.43 \text{ L} \times \left(\frac{735 \text{ torr}}{859 \text{ torr}}\right) = 1.2235739 \text{ L} \quad \text{(calculator answer)}$$

$$= 1.22 \text{ L} \quad \text{(correct answer)}$$

The answer is reasonable. Increased pressure, at constant temperature, should produce a volume decrease.

In solving this problem we arbitrarily chose to use torr pressure units. If we had used atmosphere pressure units instead, the answer obtained would still be the same, as we now illustrate.

$$735 \text{ torr} \times \left(\frac{1 \text{ atm}}{760 \text{ torr}}\right) = 0.96710526 \text{ atm} \quad \text{(calculator answer)}$$

$$= 0.967 \text{ atm} \quad \text{(correct answer)}$$

Using $P_1 = 0.967$ atm and $P_2 = 1.13$ atm, we get

$$V_2 = 1.43 \text{ L} \times \left(\frac{0.967 \text{ atm}}{1.13 \text{ atm}}\right) = 1.2237257 \text{ L} \quad \text{(calculator answer)}$$

$$= 1.22 \text{ L} \quad \text{(correct answer)}$$

Boyle's law is consistent with kinetic molecular theory (Sec. 11.3). The theory states that the pressure the gas exerts results from collisions of the gas molecules with the sides of the container. The number of collisions within a given area on the container wall in a given time is proportional to the pressure of the gas at a given temperature. If the volume of a container holding a specific number of gas molecules is increased, the total wall area of the container will also increase and the number of collisions in a given area (the pressure) will decrease due to the greater wall area. Conversely, if the volume of the container is decreased the wall area will be smaller and there will be more collisions in a given wall area. This means an increase in pressure. Figure 12-4 illustrates this idea.

The phenomenon described by Boyle's law has practical importance. Helium-filled research balloons, used to study the upper atmosphere, are only half-filled with helium when launched. As the balloon ascends, it encounters lower and lower pressures. As the pressure decreases, the balloon expands until it reaches full inflation. If the balloon were launched at full inflation, it would burst in the upper atmosphere due to the reduced external pressure.

Breathing is an example of Boyle's law in action as is the operation of a respirator, a machine designed to help patients with respiration difficulties breathe. A respirator contains a movable diaphragm that works in opposition to the patient's lungs. When the diaphragm is moved out so that the volume inside the respirator increases, the lower pressure in the respirator allows air to expand out of the patient's lungs. When the diaphragm is moved in the opposite direction, the higher pressure inside the respirator compresses the air into the lungs and causes them to increase in volume.

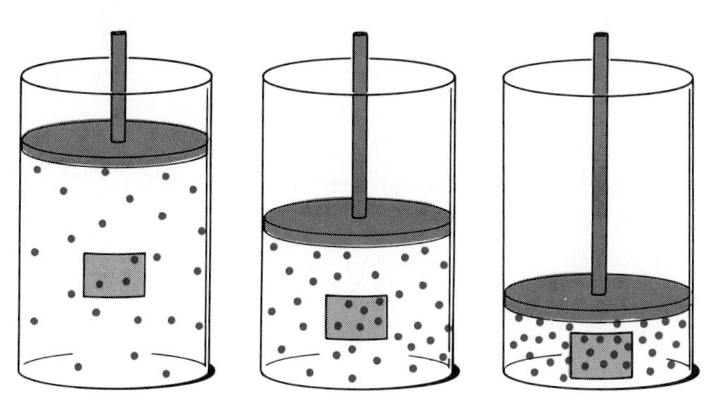

Figure 12-4. The number of gas molecule–container wall collisions per unit area of wall per unit time, for a given quantity of gas at constant temperature, increases as the volume of the container decreases. (Regard each dot as the same number of gas molecules capable of hitting the wall of the container per unit time.)

12.3 Charles' Law: A Temperature–Volume Relationship

How does the volume of a fixed quantity of gas respond to a temperature change at constant pressure? In 1787 the French scientist Jacques Alexander Cesar Charles (1746–1823) showed that a simple mathematical relationship exists between the volume and temperature of a gas, at constant pressure, *provided the temperature is expressed in degrees Kelvin.* **Charles' law** states that *the volume of a sample of gas is* DIRECTLY PROPORTIONAL *to its Kelvin temperature if the pressure is kept constant.* Contained within the wording of this law is the phrase *directly proportional;* this contrasts with Boyle's law, which contains the phrase *indirectly proportional.* Any time a *direct* proportion exists between two quantities, one increases when the other increases, and one decreases when the other decreases. Thus a direct proportion and an indirect proportion portray "opposite" behaviors. The direct proportion relationship of Charles' law means that if the temperature increases the volume will also increase and if the temperature decreases the volume will also decrease.

Charles' law

Charles' law can be qualitatively illustrated by a balloon filled with air. If the balloon is placed near a heat source, for example, a light bulb that has been on for some time, the heat will cause the balloon to increase in size (volume). The change in volume is visually noticeable. Putting the same balloon in the refrigerator will cause it to shrink. Quantitative Charles' law data can be obtained with an apparatus like that shown in Figure 12-5.

Charles' law may be stated mathematically as

$$\frac{V}{T} = \text{constant}$$

In this expression V is the volume of a gas at a given pressure and T is the temperature, expressed on the Kelvin temperature scale. A consideration of two sets of temperature–volume conditions for a gas, in a manner similar to that done for Boyle's law, leads to the following useful form of the law.

$$\frac{V_1}{T_1} = \frac{V_2}{T_2}$$

Again, note that in any of the mathematical expressions of Charles' law, or any other gas law, the symbol T is understood to mean the Kelvin temperature.

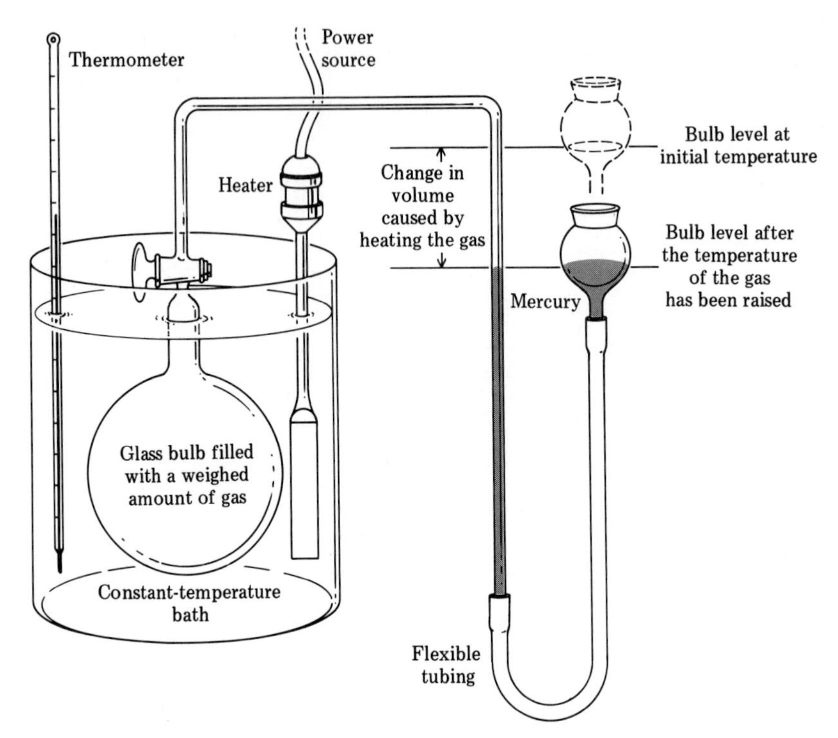

Figure 12-5. An apparatus for studying Charles' law behavior of a gas. [From *Fundamentals of Chemistry* by James E. Brady and John R. Holum. New York: John Wiley & Sons, 1981.]

EXAMPLE 12.4

At constant pressure, a 675 mL sample of N_2 gas is warmed from 23°C (a comfortable room temperature) to 38°C (the temperature outdoors on a hot summer day). What is the new volume of the N_2 gas?

SOLUTION

Writing all the given data in the form of initial and final conditions, we have

$$V_1 = 675 \text{ mL} \qquad V_2 = ?\text{ mL}$$

$$T_1 = 23°\text{C} = 296 \text{ K} \qquad T_2 = 38°\text{C} = 311 \text{ K}$$

Note that both of the given temperatures have been converted to Kelvin scale readings.

Rearranging Charles' law to isolate V_2, the desired quantity, on one side of the equation is accomplished by multiplying each side of the equation by T_2.

$$\frac{V_1}{T_1} = \frac{V_2}{T_2} \qquad \text{(Charles' law)}$$

$$\frac{V_1 T_2}{T_1} = \frac{V_2 \cancel{T_2}}{\cancel{T_2}} \quad \text{(multiplication of each side by } T_2\text{)}$$

$$V_2 = V_1 \times \frac{T_2}{T_1}$$

Substituting the given data into the equation and doing the arithmetic gives

$$V_2 = 675 \text{ mL} \times \left(\frac{311 \cancel{K}}{296 \cancel{K}}\right) = 709.20608 \text{ mL} \quad \text{(calculator answer)}$$

$$= 709 \text{ mL} \quad \text{(correct answer)}$$

Our answer is consistent with what reasoning says it should be. Increasing the temperature of a gas, at constant pressure, should result in a volume increase.

Charles' law is readily understood in terms of kinetic molecular theory. The theory states that when the temperature of a gas increases, the velocity of the gas particles increases. The speedier particles hit the container walls harder and more often. In order for the pressure of the gas to remain constant, it is necessary for the container volume to increase. Moving in a larger volume, the particles will hit the container walls less often and thus the pressure can remain the same. A similar argument applies if the temperature of the gas is lowered, only this time the velocity of the molecules decreases and the wall area (volume) must decrease in order to increase the number of collisions in a given area in a given time.

Charles' law is the principle used in the operation of an incubator. The air in contact with the heating element expands (its density becomes less). The hot less-dense air rises, causing continuous circulation of warm air. This same principle has ramifications in closed rooms in which there is not effective air circulation. The warmer and less-dense air stays near the top of the room. This is desirable in the summer, but not in the winter.

12.4 Gay-Lussac's Law: A Temperature–Pressure Relationship

If a gas is placed in a rigid container, one that cannot expand, the volume of the gas must remain constant. What is the relationship between the pressure and temperature of a fixed amount of gas in such a situation? This question was answered in 1802 by the French scientist Joseph Gay-Lussac (1778–1850). Performing a number of experiments similar to those performed by Charles, he discovered the relationship between pressure and the temperature of a gas when it is held at constant volume. **Gay-Lussac's law** states that *the pressure of a sample of gas, at constant volume, is* DIRECTLY PROPORTIONAL *to its Kelvin temperature.* Thus, as the temperature of the gas increases, the pressure increases and conversely as temperature is decreased pressure decreases.

Gay-Lussac's law

The kinetic-molecular-theory explanation for Gay-Lussac's law is very simple. As the temperature increases, the velocity of the gas molecules increases. This increases the number of times the molecules hit the container walls in a given time, which translates into an increase in pressure.

Gay-Lussac's law explains the observed pressure increase inside an automobile tire after a car has been driven for a period of time on a hot day. It also explains why aerosol cans should not be disposed of in a fire. The aerosol can itself keeps the gas at a constant volume. The elevated temperatures encountered in the fire increase the pressure of the confined gas to the point that the can may explode.

Mathematical forms of Gay-Lussac's law are

$$\frac{P}{T} = \text{constant} \qquad \text{and} \qquad \frac{P_1}{T_1} = \frac{P_2}{T_2}$$

Again, note that the symbol T is understood to mean the Kelvin temperature. Any pressure unit is acceptable; P_1 and P_2 must, however, be in the same units.

EXAMPLE 12.5

Early in the morning with the temperature at 17°C, prior to leaving on a trip, automobile tire pressure is found to be 30.1 psi. After traveling 8 hr, the combination of tire friction and hot pavement has raised the temperature of the air inside the tires to 110°C. (A temperature this high, which is equivalent to 230°F, is definitely attainable on a hot summer day.) What will be the tire pressure (gauge pressure) at this higher temperature?

SOLUTION

In working this problem we are assuming a negligible change in tire volume—a reasonable assumption.

We also note that the pressure inside a tire is the pressure shown by the gauge plus atmospheric pressure. We will assume atmospheric pressure for the day of 14.5 psi. The total pressure in the tire initially will be

$$30.1 \text{ psi} + 14.5 \text{ psi} = 44.6 \text{ psi}$$

Writing all the given data in the form of initial and final conditions, we have

$$P_1 = 44.6 \text{ psi} \qquad P_2 = ? \text{ psi}$$

$$T_1 = 17°\text{C} = 290 \text{ K} \qquad T_2 = 110°\text{C} = 383 \text{ K}$$

Rearrangement of Gay-Lussac's law to isolate P_2, the quantity we desire, on the left side of the equation gives

$$P_2 = P_1 \times \frac{T_2}{T_1}$$

Substituting the given data into the equation and doing the arithmetic gives

$$P_2 = 44.6 \text{ psi} \times \left(\frac{383 \not{\text{K}}}{290 \not{\text{K}}}\right) = 58.902759 \text{ psi} \quad \text{(calculator answer)}$$

$$= 58.9 \text{ psi} \quad \text{(correct answer)}$$

The gauge pressure observed for the "hot" tire will be

$$58.9 \text{ psi} - 14.5 \text{ psi} = 44.4 \text{ psi}$$

The answer is reasonable. Increased temperature, at constant volume, should produce a pressure increase.

12.5 The Combined Gas Law

combined gas law

The **combined gas law** *is a single expression obtained from mathematically combining Boyle's, Charles', and Gay-Lussac's laws.*

$$\frac{P_1V_1}{T_1} = \frac{P_2V_2}{T_2}$$

This combined gas law is a much more versatile equation than the individual gas laws. With it, a change in any one of the three gas-law variables, brought about by changes in *both* of the other two variables, can be calculated. Each of the individual gas laws requires that one of the three variables be held constant.

The combined gas law reduces (simplifies) to each of the equations for the individual gas laws when the constancy provision is invoked for the appropriate variable. These reduction relationships are given in Table 12-2. Because of the ease with which they may be derived from the combined gas law, the mathematical forms for the individual gas laws need not be memorized if the mathematical form for the combined gas law is known.

Table 12-2. Relationship of the Individual Gas Laws to the Combined Gas Law

Law	Variable Constancy Requirement	Mathematical Form of the Law
Combined gas law	none	$\frac{P_1V_1}{T_1} = \frac{P_2V_2}{T_2}$
Boyle's law	$T_1 = T_2$	Since T_1 and T_2 are equal, substitute T_1 for T_2 in the combined gas law and cancel. $\frac{P_1V_1}{\cancel{T_1}} = \frac{P_2V_2}{\cancel{T_1}}$ or $P_1V_1 = P_2V_2$
Charles' law	$P_1 = P_2$	Since P_1 and P_2 are equal, substitute P_1 for P_2 in the combined gas law and cancel. $\frac{\cancel{P_1}V_1}{T_1} = \frac{\cancel{P_1}V_2}{T_2}$ or $\frac{V_1}{T_1} = \frac{V_2}{T_2}$
Gay-Lussac's law	$V_1 = V_2$	Since V_1 and V_2 are equal, substitute V_1 for V_2 in the combined gas law and cancel. $\frac{P_1\cancel{V_1}}{T_1} = \frac{P_2\cancel{V_1}}{T_2}$ or $\frac{P_1}{T_1} = \frac{P_2}{T_2}$

The three most used forms of the combined gas law are those that have V_2, P_2, and T_2, respectively, isolated on the left side of the equation.

$$V_2 = V_1 \times \frac{P_1}{P_2} \times \frac{T_2}{T_1}$$

$$P_2 = P_1 \times \frac{V_1}{V_2} \times \frac{T_2}{T_1}$$

$$T_2 = T_1 \times \frac{P_2}{P_1} \times \frac{V_2}{V_1}$$

Students frequently have questions about the algebra involved in accomplishing the above rearrangements. Example 12.6 should clear up any such questions. Examples 12.7 and 12.8 illustrate the use of the combined gas law equation.

EXAMPLE 12.6

Rearrange the standard form of the combined gas law equation such that the variable T_2 is by itself on the left side of the equation.

SOLUTION

In rearranging the standard form of the combined gas law into various formats the following rule from algebra is useful.

If two fractions are equal,

$$\frac{a}{b} = \frac{c}{d},$$

then the numerator of the first fraction (a) times the denominator of the second fraction (d) is equal to the numerator of the second fraction (c) times the denominator of the first fraction (b).

$$\text{If} \quad \frac{a}{b} = \frac{c}{d}, \quad \text{then} \quad a \times d = c \times b.$$

Applying this rule to the standard form of the combined gas law gives

$$\frac{P_2V_2}{T_2} = \frac{P_1V_1}{T_1} \longrightarrow P_2V_2T_1 = P_1V_1T_2$$

With the combined gas law in the form $P_1V_1T_2 = P_2V_2T_1$ any of the six variables can be isolated by a simple division. To isolate T_2, we divide both sides of the equation by P_1V_1 (the other quantities on the same side of the equation as T_2).

$$\frac{\cancel{P_1V_1}T_2}{\cancel{P_1V_1}} = \frac{P_2V_2T_1}{P_1V_1}$$

$$T_2 = \frac{P_2V_2T_1}{P_1V_1} \quad \text{or} \quad T_2 = T_1 \times \frac{P_2}{P_1} \times \frac{V_2}{V_1}$$

EXAMPLE 12.7

A sample of SO_2 gas occupies a volume of 1.23 L at 755 torr and 0°C. What volume will this same gas sample occupy at 735 torr and 50°C?

SOLUTION

Writing all the given data in the form of initial and final conditions, we have

$$P_1 = 755 \text{ torr} \qquad P_2 = 735 \text{ torr}$$
$$V_1 = 1.23 \text{ L} \qquad V_2 = ?\text{ L}$$
$$T_1 = 0°\text{C} = 273 \text{ K} \qquad T_2 = 50°\text{C} = 323 \text{ K}$$

Rearrangement of the combined gas law expression to isolate V_2 on the left side gives

$$V_2 = V_1 \times \frac{P_1}{P_2} \times \frac{T_2}{T_1}$$

Substituting numerical values into this equation and doing the arithmetic gives

$$V_2 = 1.23 \text{ L} \times \left(\frac{755 \text{ torr}}{735 \text{ torr}}\right) \times \left(\frac{323 \text{ K}}{273 \text{ K}}\right)$$
$$= 1.494874 \text{ L} \quad \text{(calculator answer)}$$
$$= 1.49 \text{ L} \quad \text{(correct answer)}$$

In this problem both the pressure correction factor (P_1/P_2) and the temperature correction factor (T_2/T_1) contribute to the increase in volume — both of the factors have values greater than unity. The reasonableness of each correction factor contributing to the volume increase can be seen by considering each factor independent of the other. Decreasing the pressure (at constant temperature) should increase the volume. Increasing the temperature (at constant pressure) should also increase the volume.

EXAMPLE 12.8

A sample of argon gas (the gas used in electric light bulbs) occupies a volume of 80.0 mL at a pressure of 1.10 atm and a temperature of 29°C. What will be the temperature of the gas if the volume of the container is decreased to 40.0 mL and the pressure is increased to 2.20 atm?

SOLUTION

Writing all the given data in the form of initial and final conditions we have

$$P_1 = 1.10 \text{ atm} \qquad P_2 = 2.20 \text{ atm}$$
$$V_1 = 80.0 \text{ mL} \qquad V_2 = 40.0 \text{ mL}$$
$$T_1 = 29°\text{C} = 302 \text{ K} \qquad T_2 = ?\,°\text{C}$$

Rearrangement of the combined gas law to isolate T_2 on the left side of the equation gives

$$T_2 = T_1 \times \frac{P_2}{P_1} \times \frac{V_2}{V_1}$$

Substituting the given data into this equation and doing the arithmetic gives

$$T_2 = 302\ \text{K} \times \left(\frac{2.20\ \cancel{\text{atm}}}{1.10\ \cancel{\text{atm}}}\right) \times \left(\frac{40.0\ \cancel{\text{mL}}}{80.0\ \cancel{\text{mL}}}\right)$$

$$= 302\ \text{K} \quad \text{(calculator and correct answer in degrees Kelvin)}$$

Converting the temperature to the Celsius scale by subtracting 273 gives 29°C as the final answer.

$$302\ \text{K} - 273 = 29^\circ\text{C}$$

The temperature did not change! The pressure correction factor, considered by itself, would cause the temperature to increase by a factor of 2; the pressure was doubled. The volume correction factor, considered by itself, would cause the temperature to decrease by a factor of 2; the volume was halved. Considered together, the effects of the two factors cancel each other and result in the temperature not changing.

12.6 Standard Conditions of Temperature and Pressure

The volumes of liquids and solids change only slightly with temperature and pressure changes (Sec. 11.2). This is not the case for volumes of gases. As the gas-law discussions of previous sections pointedly show, the volume of a gas is very dependent on temperature and pressure.

standard temperature
standard pressure
STP conditions

Comparisons of gas volumes may be made only if the gases are at the same temperature and pressure. It is convenient to specify a particular temperature and pressure as standards for comparison purposes. **Standard temperature** *is* 0°C (273 K). **Standard pressure** *is* 1 *atmosphere* (760 *torr*). **STP conditions** *are those of standard temperature and standard pressure.*

Example 12.9 is a problem involving STP conditions.

EXAMPLE 12.9

A sample of H_2 gas occupies a volume of 1.37 L at STP. What volume will it occupy at a pressure of 1.00 atm and a temperature of 340°C?

SOLUTION

STP conditions denote a temperature of 0°C and a pressure of 1.00 atm.

Writing all the given data in the form of initial and final conditions, we have

$$P_1 = 1.00 \text{ atm} \qquad P_2 = 1.00 \text{ atm}$$

$$V_1 = 1.37 \text{ L} \qquad V_2 = ?\text{ L}$$

$$T_1 = 0°\text{C} = 273 \text{ K} \qquad T_2 = 340°\text{C} = 613 \text{ K}$$

Rearrangement of the combined gas law expression to isolate V_2 on the left side gives

$$V_2 = V_1 \times \frac{P_1}{P_2} \times \frac{T_2}{T_1}$$

This equation may be simplified since $P_1 = P_2$ for this particular problem; the pressure factor cancels from the equation.

$$V_2 = V_1 \times \frac{T_2}{T_1}$$

This simplified equation is a form of Boyle's law.

Substituting numerical values into the equation gives

$$V_2 = 1.37 \text{ L} \times \left(\frac{613 \cancel{\text{K}}}{273 \cancel{\text{K}}}\right) = 3.0762271 \text{ L} \quad \text{(calculator answer)}$$

$$= 3.08 \text{ L} \quad \text{(correct answer)}$$

Our answer is consistent with what reasoning says it should be. A significant increase in the temperature of a gas (at constant pressure) should produce a significant increase in the volume of the gas, which is the case.

12.7 Avogadro's Law: A Volume–Quantity Relationship

In the year 1811, Amedeo Avogadro, the Avogadro involved in Avogadro's number (Sec. 7.3), published work in which he proposed a relationship between gaseous volumes and numbers of molecules present. His proposal, a hypothesis at the time it was first published, is now known as Avogadro's law. Its validity has been demonstrated in a number of different ways.

Avogadro's law

Avogadro's law states that *equal volumes of different gases, when measured at the same temperature and pressure, contain equal numbers of molecules* (*or moles of molecules*). Although Avogadro's law seems very simple in terms of today's scientific knowledge, it was a very astute conclusion at the time it was first proposed. At that time scientists were still struggling with the differences between atoms and molecules, and Avogadro's proposal played an important role in resolving this problem.

Avogadro's law may also be stated in an alternate form, which uses terminology similar to that used in the previously discussed gas laws. This alternate form deals with two samples of the same gas, rather than two samples of different gases. The alternate form is: "The volume of a gas, at constant temperature and pressure, is directly proportional to the number of moles of gas present." When the original number of moles of gas present is doubled, the volume of the gas increases twofold. Halving the number of moles present halves the volume.

Experimentally, the direct relationship between volume and amount of gas present, at constant temperature and pressure, is easy to show. All that is needed is an inflated balloon. Adding more gas to the balloon increases the volume of the balloon. The more gas added to the balloon the greater the volume increase. Conversely, letting gas out of the balloon decreases its volume.

Mathematical statements of Avogadro's law, where n represents the number of moles, are

$$\frac{V}{n} = \text{constant} \quad \text{and} \quad \frac{V_1}{n_1} = \frac{V_2}{n_2}$$

Note the similarity in mathematical form between this law and the laws of Charles and Gay-Lussac. The similarity results from all three laws being direct proportionality relationships between two variables.

Avogadro's law and the combined gas law may be combined to give the expression

$$\frac{P_1V_1}{n_1T_1} = \frac{P_2V_2}{n_2T_2}$$

The equation covers the situation where none of the four variables, P, T, V, and n, is constant. With it, a change in any one of the four variables, brought about by changes in the other three variables can be calculated.

EXAMPLE 12.10

A balloon containing 1.83 moles of He has a volume of 0.673 L at a given temperature and pressure. If an additional 0.50 mol of He gas is introduced into the balloon, without changing temperature and pressure conditions, what will be the new volume of the balloon?

SOLUTION

We will use the equation

$$\frac{V_1}{n_1} = \frac{V_2}{n_2}$$

in solving the problem.

Writing the given data in terms of initial and final conditions, we have

$V_1 = 0.673 \text{ L}$ $\qquad V_2 = ?\text{ L}$

$n_1 = 1.83 \text{ moles}$ $\qquad n_2 = (1.83 + 0.50) \text{ moles} = 2.33 \text{ moles}$

Rearrangement of the Avogadro's law expression to isolate V_2 on the left side gives

$$V_2 = V_1 \times \frac{n_2}{n_1}$$

Substituting numerical values into the equation gives

$$V_2 = 0.673 \text{ L} \times \left(\frac{2.33 \text{ \sout{moles}}}{1.83 \text{ \sout{moles}}}\right) = 0.85687978 \text{ L} \quad \text{(calculator answer)}$$

$$= 0.857 \text{ L} \quad \text{(correct answer)}$$

Our answer is consistent with reasoning. Increasing the number of moles of gas present in the balloon, at constant temperature and pressure, should increase the volume of the balloon.

12.8 Molar Volume of a Gas

molar volume

The **molar volume** *of a gas is the volume occupied by one mole of the gas at STP conditions.* It follows from Avogadro's law (Sec. 12.7) that all gases will have the same molar volume. *Experimentally, it is found that the molar volume of a gas is 22.4 liters.* To visualize a volume this size, 22.4 liters, think of standard-sized basketballs. The volume occupied by three standard-sized basketballs is almost exactly equal to 22.4 liters.

The fact that all gases have the same molar volume at STP (or any other temperature–pressure combination) is a property unique to the gaseous state. Similar statements cannot be made about the liquid and gaseous states. The reason for the difference can be understood by considering the relationship between the volume the molecules occupy in the given physical state and the volume of the molecules themselves. In the gaseous state, because the molecules are so far apart — a gas is mostly empty space — the volume of the gas molecules themselves is negligible compared to the total volume. This is not the case in the liquid and solid states where the molecules are in close contact with each other.

The molar volume concept is very useful in a variety of types of calculations. When used in calculations, the concept "translates" into a conversion factor having either the form

$$\left(\frac{1 \text{ mole gas}}{22.4 \text{ L gas}}\right) \quad \text{or} \quad \left(\frac{22.4 \text{ L gas}}{1 \text{ mole gas}}\right)$$

A most common type of problem involving these conversion factors is one where the volume of a gas at STP is known and you are asked to calculate from it either the moles, grams, or particles of gas present, or vice versa. Figure 12-6 summarizes the relationships needed in performing calculations of this general type.

Perhaps you recognize Figure 12-6 as being very similar to a diagram you have previously encountered many times in problem solving, Figure 7-1. This "new diagram" differs from Figure 7-1 in only one way — a volume "box" has been added. This diagram is used in an analogous manner to the previous one. The given and desired quantities are determined and the arrows of the diagram are used to "map out" the pathway to be used in going from the given to the desired quantity. Examples 12.11 through 12.13 illustrate the usefulness of Figure 12-6.

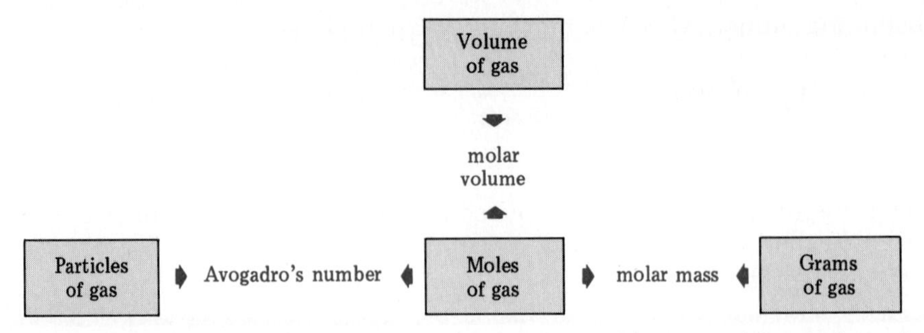

Figure 12-6. Quantitative relationships needed to solve problems involving molar volume, grams, moles, and particles.

EXAMPLE 12.11

What is the volume at STP, in L, occupied by 10.3 g of N_2 gas?

SOLUTION

Step 1. The given quantity is 10.3 g of N_2 and the unit of the desired quantity is liters of N_2.

$$10.3 \text{ g } N_2 = ?\text{ L } N_2 \text{ at STP}$$

Step 2. Using Figure 12-6 as a guide in setting up the problem, we see that the pathway is

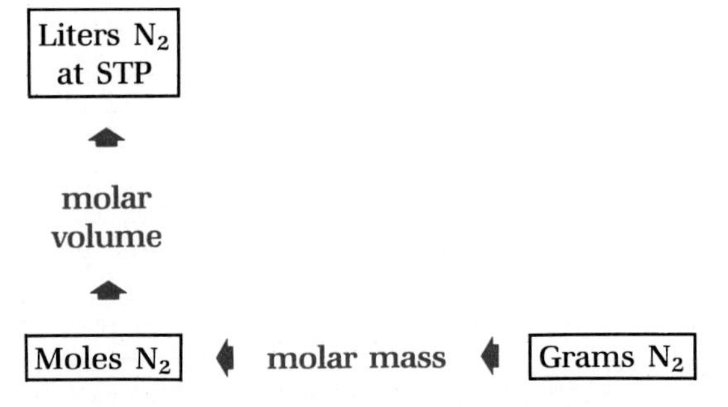

Step 3. The dimensional-analysis setup for this sequence of conversion steps is

$$10.3 \cancel{\text{g } N_2} \times \left(\frac{1 \cancel{\text{mole } N_2}}{28.0 \cancel{\text{g } N_2}}\right) \times \left(\frac{22.4 \text{ L } N_2}{1 \cancel{\text{mole } N_2}}\right)$$

Step 4. The solution, obtained by doing the arithmetic, is

$$\left(\frac{10.3 \times 1 \times 22.4}{28.0 \times 1}\right) \text{L } N_2 = 8.24 \text{ L } N_2 \quad \text{(calculator answer and correct answer)}$$

EXAMPLE 12.12

What volume, at STP, will 1 trillion (1.00×10^{12}) O_2 molecules occupy?

SOLUTION

Step 1. The given quantity is particles of O_2, 1.00×10^{12} O_2 molecules, and the desired quantity is liters of O_2 at STP. In the "jargon" of Figure 12-6 this is a particle-to-volume problem.

Step 2. The sequence of conversion factors needed for this problem, with Figure 12-6 as a guide, is

Particles O_2 → Avogadro's number → Moles O_2 → molar volume → Liters O_2 at STP

Step 3. Translating this pathway into a dimensional-analysis setup of conversion factors gives

$$(1.00 \times 10^{12} \text{ molecules } O_2) \times \left(\frac{1 \text{ mole } O_2}{6.02 \times 10^{23} \text{ molecules } O_2}\right) \times \left(\frac{22.4 \text{ L } O_2}{1 \text{ mole } O_2}\right)$$

Step 4. Collecting numerical terms, after cancellation of units, and doing the arithmetic gives

$$\left(\frac{1.00 \times 10^{12} \times 1 \times 22.4}{6.02 \times 10^{23} \times 1}\right) \text{L } O_2 = 3.7209302 \times 10^{-11} \text{ L } O_2 \quad \text{(calculator answer)}$$

$$= 3.72 \times 10^{-11} \text{ L } O_2 \quad \text{(correct answer)}$$

Even in the gaseous state, where molecules are relatively far apart, the volume of a trillion molecules at STP is so small it cannot be measured. The message, again, is that atoms and molecules are very, very small.

Equal volumes of different gases at STP do not have equal masses. This is because different molecules have different masses. Example 12.13 illustrates this point.

EXAMPLE 12.13

What is the mass, in grams, at STP of 4.78 L of each of the following gases?
(a) O_2 **(b)** F_2

SOLUTION

(a) ***Step 1.*** The given quantity is 4.78 L of O_2 at STP, and the desired quantity is grams of O_2.

$$4.78 \text{ L } O_2 = ? \text{ g } O_2$$

Step 2. This is a liter-to-gram problem. The pathway appropriate for solving this problem, as indicated in Figure 12-6, is

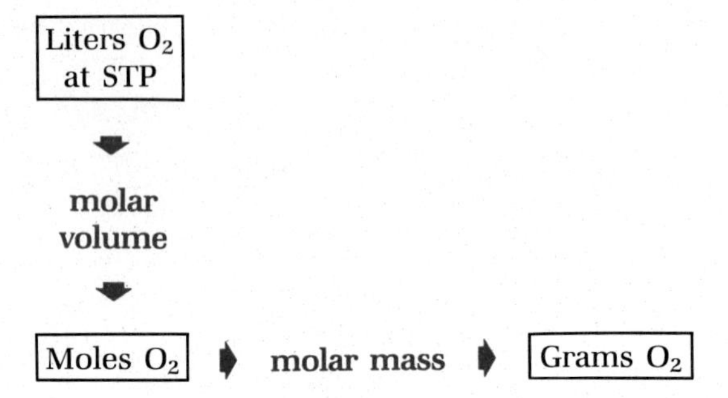

Step 3. The setup, from dimensional analysis, is

$$4.78 \text{ L } O_2 \times \left(\frac{1 \text{ mole } O_2}{22.4 \text{ L } O_2}\right) \times \left(\frac{32.0 \text{ g } O_2}{1 \text{ mole } O_2}\right)$$

Step 4. Collecting the numerical terms, after cancellation of units, and doing the arithmetic gives

$$\left(\frac{4.78 \times 1 \times 32.0}{22.4 \times 1}\right) \text{g } O_2 = 6.8285714 \text{ g } O_2 \quad \text{(calculator answer)}$$

$$= \mathbf{6.83 \text{ g } O_2} \quad \text{(correct answer)}$$

(b) The analysis and setup for this part are identical to those in the previous part except F_2 and its molecular weight replace O_2 and its molecular weight. Thus, the setup is

$$4.78 \text{ L } F_2 \times \left(\frac{1 \text{ mole } F_2}{22.4 \text{ L } F_2}\right) \times \left(\frac{38.0 \text{ g } F_2}{1 \text{ mole } F_2}\right) = 8.1089286 \text{ g } F_2 \quad \text{(calculator answer)}$$

$$= \mathbf{8.11 \text{ g } F_2} \quad \text{(correct answer)}$$

Equal volumes of different gases at STP do not have equal masses. 4.78 L of O_2 has a mass of 6.83 g and 4.78 L of F_2 has a mass of 8.11 g.

The density of a gas at STP conditions may be calculated by the molar volume relationship for gases and the molecular weight (molar mass relationship) for the particular gas. This type of calculation is illustrated in Example 12.14.

EXAMPLE 12.14

Calculate the density at STP conditions, in g/L, of CO gas.

SOLUTION

Density is mass divided by volume (Sec. 3.3). For gases, usual density units are g/L. The two relationships needed to calculate the density of CO are

$$1 \text{ mole CO} = 28.0 \text{ g CO} \qquad \text{(molar mass)}$$

$$1 \text{ mole CO} = 22.4 \text{ L CO at STP} \qquad \text{(molar volume)}$$

The setup for the density calculation is

$$\left(\frac{28.0 \text{ g CO}}{1 \cancel{\text{mole CO}}}\right) \times \left(\frac{1 \cancel{\text{mole CO}}}{22.4 \text{ L CO}}\right)$$

The moles units cancel, leaving the units g/L, which are density units. Doing the arithmetic, we find that CO has a density of 1.25 g/L at STP.

$$\frac{(28.0 \times 1) \text{ g}}{(1 \times 22.4) \text{ L}} = 1.25 \text{ g/L} \quad \text{(calculator answer and correct answer)}$$

12.9 The Ideal Gas Law

ideal gas law

The **ideal gas law** *describes the relationships among the four variables temperature* (T), *pressure* (P), *volume* (V), *and moles of gas* (n) *for gaseous substances.*

In previous sections we have discussed three independent relationships dealing with the volume of a gas. They are

$$\text{Boyle's law:} \quad V = k\left(\frac{1}{P}\right) \quad (n \text{ and } T \text{ constant})$$

$$\text{Charles' law:} \quad V = k(T) \quad (n \text{ and } P \text{ constant})$$

$$\text{Avogadro's law:} \quad V = k(n) \quad (P \text{ and } T \text{ constant})$$

We may combine these three equations into a single expression

$$V = \frac{k(T)(n)}{(P)}$$

since we know (from mathematics) that if a quantity is independently proportional to two or more quantities, this quantity (volume in our case) is also proportional to the product of these quantities. This combined equation is a mathematical statement of the ideal gas law. Any gas that obeys the individual laws of Boyle, Charles, and Avogadro will also obey the ideal gas law.

The usual form in which the ideal gas law is written is

$$PV = nRT$$

This form of the ideal gas law, besides being rearranged, differs from the previous form in that the proportionality constant k has been given the symbol R. The constant, R, is called the ideal gas constant. In order to use the ideal gas equation we must know the value of R. This can be determined by substituting STP values for P, V, T, and n into the ideal gas equation and solving for R. For 1 mole of gas at STP, $P = 1$ atm, $V = 22.4$ L, $T = 273$ K, and $n = 1$ mole. Substituting these values into the ideal gas equation arranged to have R isolated on the left side gives

$$R = \frac{PV}{nT} = \frac{(1\ \text{atm})(22.4\ \text{L})}{(1\ \text{mole})(273\ \text{K})} = 0.0821\,\frac{\text{atm L}}{\text{mole K}}$$

Notice the complex units associated with R—the four variables temperature, pressure, volume, and moles are all involved. This will always be the case. The value of R is dependent on the units used to express pressure and volume. If we use 760 torr, instead of 1 atm, as standard pressure, then R would have the value 62.4 and the units (torr)(L)/(mole)(K).

$$R = \frac{PV}{nT} = \frac{(760\ \text{torr})(22.4\ \text{L})}{(1\ \text{mole})(273\ \text{K})} = 62.4\,\frac{\text{torr L}}{\text{mole K}}$$

The most used values of R are the two values just encountered. When pressure units other than torr or atmosphere and volume units other than liter are encountered, convert them to these units and then use the appropriate known R value.

If three of the four variables in the ideal gas equation are known then the fourth can be calculated by the equation. The ideal gas equation is used in calculations when *one* set of conditions is given with one missing variable. The combined gas law (Sec. 12.5) is used when *two* sets of conditions are given with one missing variable. Examples 12.15 and 12.16 illustrate the use of the ideal gas equation.

EXAMPLE 12.15

Calculate the volume, in liters, occupied by 1.73 moles of N_2 gas at 0.992 atm pressure and a temperature of 75°C.

SOLUTION

This problem deals with only one set of conditions, a situation where the ideal gas equation is applicable. Three of the four variables in the ideal gas equation (n, P, and T) are given, and the fourth (V) is to be calculated.

$$P = 0.992\ \text{atm} \qquad n = 1.73\ \text{moles}$$

$$V = ?\ \text{L} \qquad T = 75°\text{C} = 348\ \text{K}$$

Rearranging the ideal gas equation to isolate V on the left side of the equation gives

$$V = \frac{nRT}{P}$$

Since the pressure is given in atmospheres and the volume unit is liters, the appropriate R value is

$$R = 0.0821\,\frac{\text{atm L}}{\text{mole K}}$$

Substituting the given numerical values into the equation and cancelling units gives

$$V = \frac{(1.73 \text{ moles})\left(0.0821 \dfrac{\text{atm L}}{\text{mole K}}\right)(348 \text{ K})}{(0.992 \text{ atm})}$$

Note how all of the parts of the ideal gas constant unit cancel except for one, the volume part.

Doing the arithmetic, we get as an answer 49.8 L.

$$V = \left(\frac{1.73 \times 0.0821 \times 348}{0.992}\right) \text{L N}_2$$

$$= 49.826093 \text{ L} \quad \text{(calculator answer)}$$

$$= 49.8 \text{ L} \quad \text{(correct answer)}$$

EXAMPLE 12.16

What is the temperature, in degrees Celsius, of a 1.23 mole sample of O_2 gas under a pressure of 4.00 atm in a 9.00 L container?

SOLUTION

Three of the four variables in the ideal gas equation (P, V, and n) are given and the fourth (T) is to be calculated.

$$P = 4.00 \text{ atm} \qquad n = 1.23 \text{ moles}$$

$$V = 9.00 \text{ L} \qquad T = ?\text{ K}$$

Rearranging the ideal gas equation to isolate T on the left side gives $T = PV/nR$.

Since the pressure is given in atmospheres and the volume in liters, the value of R to be used is

$$R = 0.0821 \frac{\text{atm L}}{\text{mole K}}$$

Substituting numerical values into the equation gives

$$T = \frac{(4.00 \text{ atm})(9.00 \text{ L})}{(1.23 \text{ moles})\left(0.0821 \dfrac{\text{atm L}}{\text{mole K}}\right)}$$

Notice again how the gas constant units, except for K, cancel. After cancellation, the expression 1/(1/K) remains. This expression is equivalent to K. That this is the case can be easily shown. All we need to do is multiply both the numerator and denominator of the fraction by K.

$$\frac{1 \times \text{K}}{\dfrac{1}{\text{K}} \times \text{K}} = \text{K}$$

Doing the arithmetic, we get as an answer 356 K for the temperature of the O_2 gas.

$$T = \frac{(4.00)(9.00)}{(1.23)(0.0821)}\text{ K} = 356.49565\text{ K} \quad \text{(calculator answer)}$$

$$= 356\text{ K} \quad \text{(correct answer)}$$

The calculated temperature is in degrees Kelvin. To convert to degrees Celsius, the units specified in the problem statement, we subtract 273 degrees from the Kelvin temperature.

$$T\,(°\text{C}) = 356\text{ K} - 273 = 83°\text{C}$$

ideal gas

A word about the phrase *ideal gas* in the name ideal gas law is in order. An **ideal gas** *is a gas that obeys exactly all of the postulates of kinetic molecular theory* (Sec. 11.3) *and obeys exactly the ideal gas law.* Real gases are not ideal gases; that is, real gases do not obey exactly the ideal gas equation. Nonetheless, for real gases under ordinary conditions of temperature and pressure, deviations from ideal gas behavior are small, and the ideal gas law (as well as the other laws discussed in this chapter) gives accurate information about gas behavior. It is only at low temperatures and/or high pressures (near liquefaction conditions) that ideal gas behavior breaks down. At low temperatures, because of the slower motion of the molecules, attractive forces between molecules begin to become important. At high pressures, where the molecules are forced closer together, again attractive forces begin to cause deviation from gas law predicted behavior.

12.10 Modified Forms of the Ideal Gas Equation

Some of the most useful calculations involving the ideal gas equation are those in which the mass, molecular weight, or density of a gas is determined. Such calculations are performed by using modified forms of the ideal gas equation.

The number of moles of any substance is equal to the number of grams of the substance divided by the substance's molecular weight.

$$n = \frac{\text{g}}{\text{MW}}$$

Replacing n in the ideal gas equation with this equivalent expression gives

$$PV = \frac{\text{g}}{\text{MW}}RT$$

This equation, in the rearranged form

$$\text{g} = \frac{PV(\text{MW})}{RT}$$

is used to calculate the mass, in grams, of a gas. This same equation, in the rearranged form

$$\mathrm{MW} = \frac{gRT}{PV}$$

is used to calculate the molecular weight of a gas.

The density, d, of a gas has the units of mass (grams) per unit volume (liters).

$$d = g/V$$

Solving the modified ideal gas equation for g/V gives

$$\frac{g}{V} = \frac{P(\mathrm{MW})}{RT} \quad \text{or} \quad d = \frac{P(\mathrm{MW})}{RT}$$

Examples 12.17 and 12.18 illustrate the use of these new ideal-gas-equation relationships.

EXAMPLE 12.17

A 0.276 g sample of oxygen gas (O_2) occupies a volume of 0.270 L at 739 torr and 98°C. Calculate, from these data, the molecular weight of gaseous O_2.

SOLUTION

The molecular weight of a gas is calculated by the ideal gas equation in the modified form

$$\mathrm{MW} = \frac{gRT}{PV}$$

All of the quantities on the right side of the equation are known.

$$g = 0.276\ \mathrm{g} \qquad T = 98°\mathrm{C} = 371\ \mathrm{K}$$

$$R = 62.4\ \frac{\mathrm{torr\ L}}{\mathrm{mole\ K}} \qquad P = 739\ \mathrm{torr}$$

$$V = 0.270\ \mathrm{L}$$

Substitution of these values into the equation gives

$$\mathrm{MW} = \frac{(0.276\ \mathrm{g})\left(62.4\ \frac{\cancel{\mathrm{torr}}\ \cancel{\mathrm{L}}}{\mathrm{mole}\ \cancel{\mathrm{K}}}\right)(371\ \cancel{\mathrm{K}})}{(739\ \cancel{\mathrm{torr}})(0.270\ \cancel{\mathrm{L}})}$$

All units cancel except for g/mole, the units of molecular weight. Recall that the molecular weight of a substance is the mass in g of 1 mole of the substance.

Doing the arithmetic, we obtain a value of 32.0 g/mole for the molecular weight of oxygen (O_2).

$$\text{MW} = \frac{(0.276)(62.4)(371)}{(739)(0.270)} \frac{\text{g}}{\text{mole}}$$

$$= 32.022806 \frac{\text{g}}{\text{mole}} \quad \text{(calculator answer)}$$

$$= 32.0 \frac{\text{g}}{\text{mole}} \quad \text{(correct answer)}$$

This experimental molecular-weight value is in agreement with the molecular-weight value obtained with a table of atomic weights, 32.0 g/mole.

EXAMPLE 12.18

Calculate the density of CO_2 gas, in g/L, at 1.21 atm pressure and a temperature of 35°C.

SOLUTION

The ideal gas equation in the modified form

$$d = \frac{P(\text{MW})}{RT}$$

is used to calculate the density of a gas.

All of the quantities on the right side of this equation are known.

$$P = 1.21 \text{ atm} \qquad R = 0.0821 \frac{\text{atm L}}{\text{mole K}}$$

$$\text{MW} = 44.0 \frac{\text{g}}{\text{mole}} \quad \text{(calculated from a table of atomic weights)} \qquad T = 35°\text{C} = 308 \text{ K}$$

Substitution of these values into the equation gives

$$d = \frac{(1.21 \cancel{\text{atm}})\left(44.0 \frac{\text{g}}{\cancel{\text{mole}}}\right)}{\left(0.0821 \frac{\cancel{\text{atm}} \text{ L}}{\cancel{\text{mole}} \cancel{\text{K}}}\right)(308 \cancel{\text{K}})}$$

All units cancel except for the desired ones, g/L.

Doing the arithmetic, we obtain a value of 2.11 g/L for the density of CO_2 gas at the specified temperature and pressure.

$$d = \frac{(1.21)(44.0)}{(0.0821)(308)} \frac{\text{g}}{\text{L}}$$

$$= 2.1054463 \text{ g/L} \quad \text{(calculator answer)}$$

$$= 2.11 \text{ g/L} \quad \text{(correct answer)}$$

12.11 Gas Laws and Chemical Equations

Gases are involved as reactants or products in many chemical reactions. In such reactions it is usually easier to determine the volumes rather than the masses of the gases involved.

A very simple relationship exists between the volumes of different gases consumed or produced in chemical reactions, provided the volumes are all determined at the same temperature and pressure. *The volumes of different gases involved in a reaction, if measured at the same temperature and pressure, are in the same ratio as the coefficients for these gases in the balanced equation for the reaction.*

This coefficient–volume relationship follows directly from Avogadro's law (Sec. 12.7), which says that moles and volume of a gas are directly proportional at constant temperature and pressure. In Section 8.5 we learned that coefficients in a balanced equation may be interpreted in terms of moles. For example, for the equation

$$4\ NH_3(g) + 3\ O_2(g) \longrightarrow 2\ N_2(g) + 6\ H_2O(g)$$

it is correct to say

$$4\ \textbf{moles}\ NH_3(g) + 3\ \textbf{moles}\ O_2(g) \longrightarrow 2\ \textbf{moles}\ N_2(g) + 6\ \textbf{moles}\ H_2O(g)$$

As a result of the direct proportionality between moles and volume (Avogadro's law) all of the mole designations in this equation may be replaced with volume designations.

$$4\ \textbf{volumes}\ NH_3(g) + 3\ \textbf{volumes}\ O_2(g) \longrightarrow 2\ \textbf{volumes}\ N_2(g) + 6\ \textbf{volumes}\ H_2O(g)$$

Again, all of the volumes must be measured at the same temperature and pressure for the above relationship to be valid.

In calculations involving chemical reactions, where two or more gases are participants, this volume interpretation of coefficients may be used to generate conversion factors useful in problem solving. Consider the balanced equation

$$2\ CO(g) + O_2(g) \longrightarrow 2\ CO_2(g)$$

At constant temperature and pressure, three volume–volume relationships are obtainable from this equation.

2 volumes CO produces 2 volumes CO_2

1 volume O_2 produces 2 volumes CO_2

2 volumes CO react with 1 volume O_2

From these three relationships, six conversion factors can be written.

From the first relationship:

$$\left(\frac{2\ \text{volumes CO}}{2\ \text{volumes CO}_2}\right) \quad \text{and} \quad \left(\frac{2\ \text{volumes CO}_2}{2\ \text{volumes CO}}\right)$$

From the second relationship:

$$\left(\frac{1\ \text{volume O}_2}{2\ \text{volumes CO}_2}\right) \quad \text{and} \quad \left(\frac{2\ \text{volumes CO}_2}{1\ \text{volume O}_2}\right)$$

From the third relationship:

$$\left(\frac{2 \text{ volumes CO}}{1 \text{ volume } O_2}\right) \quad \text{and} \quad \left(\frac{1 \text{ volume } O_2}{2 \text{ volumes CO}}\right)$$

The more gaseous reactants and products there are in a chemical reaction, the greater the number of volume–volume conversion factors obtainable from the equation for the chemical reaction. The use of volume–volume conversion factors is illustrated in Example 12.19.

EXAMPLE 12.19

Nitrogen reacts with hydrogen to produce ammonia as shown by the equation

$$N_2(g) + 3\ H_2(g) \longrightarrow 2\ NH_3(g)$$

What volume of H_2 gas at 750 torr and 25°C is required to produce 1.75 L of NH_3 gas at the same temperature and pressure?

SOLUTION

Step 1. The given quantity is 1.75 L of H_2 and the desired quantity is liters of NH_3.

$$1.75 \text{ L } H_2 = ?\text{ L } NH_3$$

Step 2. This is a volume-to-volume problem. The conversion factor needed for this one-step problem is derived from the coefficients of H_2 and NH_3 in the equation for the chemical reaction. The equation tells us that at constant temperature and pressure three volumes of H_2 produces two volumes of NH_3. From this relationship the conversion factor

$$\left(\frac{2 \text{ L } NH_3}{3 \text{ L } H_2}\right)$$

is obtained.

Step 3. The dimensional-analysis setup for this problem is

$$1.75 \cancel{\text{L } H_2} \times \left(\frac{2 \text{ L } NH_3}{3 \cancel{\text{L } H_2}}\right)$$

Note that it makes no difference what the temperature and pressure are, so long as they are the same for the two gases involved in the calculation.

Step 4. Doing the arithmetic, after cancellation of units, gives

$$\left(\frac{1.75 \times 2}{3}\right) \text{L } NH_3 = 1.1666667 \text{ L } NH_3 \quad \text{(calculator answer)}$$

$$= 1.17 \text{ L } NH_3 \quad \text{(correct answer)}$$

In Section 8.6 we learned to calculate the mass of any component (reactant or product) of a chemical reaction given the balanced equation for the reaction and the mass of any other component — a mass–mass (or gram-to-gram) problem. A simple

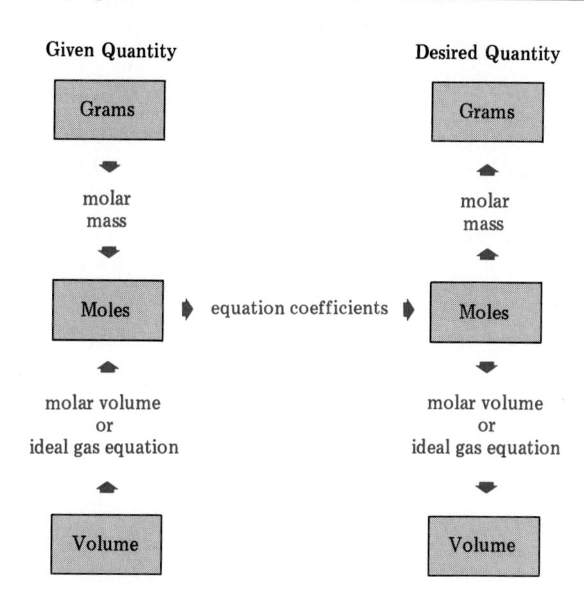

Figure 12-7. Conversion factor relationships needed for solving mass–volume and volume–mass chemical-equation-based problems.

extension of the procedures involved in solving such problems will enable us to do mass–volume and volume–mass calculations for reactions where at least one gas is involved.

If you are given the number of liters of any gaseous components of a reaction and asked to calculate the mass of any other component (gaseous or otherwise) — *a volume–mass problem* — first convert the given volume of gas to moles of gas, using the principles in this chapter. If the reaction is at STP conditions, use molar volume as the conversion factor to go from volume to moles. At other temperatures and pressures, the ideal gas law (or the combined gas law and molar volume) is used to make the conversion. Once moles of gas is obtained, the rest of the calculation is a standard mole-to-gram conversion (Sec. 8.5).

If you are given the number of grams of any component (gaseous or otherwise) in a chemical reaction and asked to calculate the volume of any gaseous component (reactant or product) — *a mass–volume problem* — use standard procedures to calculate the moles of gas and then the new procedures of this chapter to go from moles to volume.

Figure 12-7 summarizes the relationships between the conversion factors needed to solve mass–volume and volume–mass type chemical-equation based problems.

EXAMPLE 12.20

Aluminum and oxygen react to produce aluminum oxide as shown by the following equation.

$$4\,Al(s) + 3\,O_2(g) \longrightarrow 2\,Al_2O_3(s)$$

What volume, in L, of O_2 at STP conditions will completely react with 75.0 g of Al?

SOLUTION

Step 1. The given quantity is 75.0 g of Al and the desired quantity is L of O_2 at STP conditions.

$$75.0 \text{ g Al} = ? \text{ L } O_2$$

Step 2. This problem is a "gram-to-volume" type problem. The pathway used in solving it, in terms of Figure 12-7, is

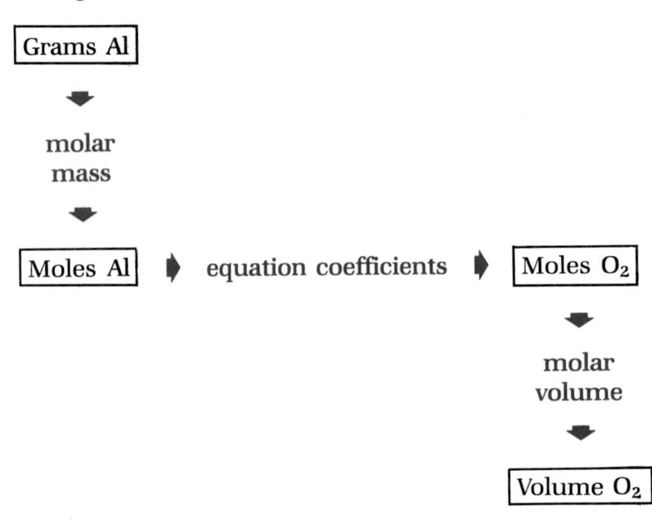

Step 3. The dimensional-analysis setup for the calculation is

$$75.0 \text{ g Al} \times \left(\frac{1 \text{ mole Al}}{27.0 \text{ g Al}}\right) \times \left(\frac{3 \text{ moles } O_2}{4 \text{ moles Al}}\right) \times \left(\frac{22.4 \text{ L } O_2}{1 \text{ mole } O_2}\right)$$

Step 4. The solution, obtained from combining all of the numerical factors, is

$$\left(\frac{75.0 \times 1 \times 3 \times 22.4}{27.0 \times 4 \times 1}\right) \text{L } O_2 = 46.666667 \text{ L } O_2 \quad \text{(calculator answer)}$$

$$= 46.7 \text{ L } O_2 \quad \text{(correct answer)}$$

EXAMPLE 12.21

Lead(IV) chloride, $PbCl_4$, may be produced from its constituent elements as shown by the equation

$$Pb(s) + 2\,Cl_2(g) \longrightarrow PbCl_4(s)$$

How many grams of $PbCl_4$ can be produced from the reaction of 5.00 L of Cl_2 at 1.33 atm pressure and a temperature of 7°C with an excess of Pb metal?

SOLUTION

Step 1. The given quantity is 5.00 L of Cl_2 and the desired quantity is grams of $PbCl_4$.

$$5.00 \text{ L } Cl_2 = ? \text{ g } PbCl_4$$

Step 2. This is a volume-to-gram problem. The pathway, in terms in Figure 12-7, is

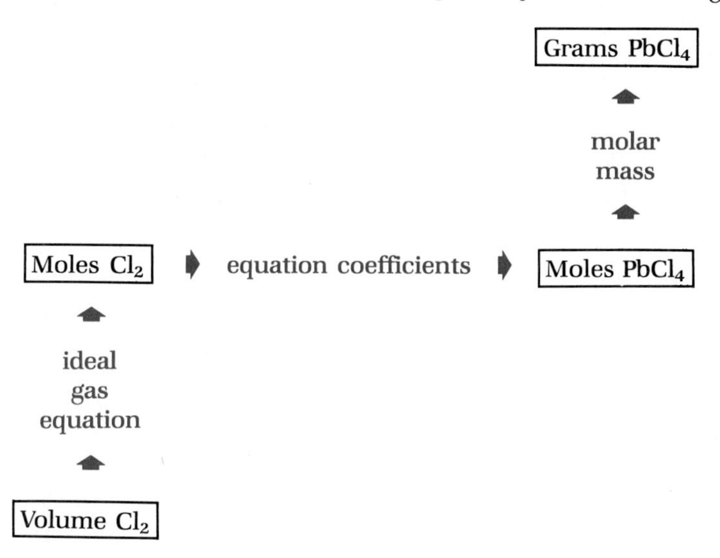

Step 3. The first conversion step, from volume (Cl_2) to moles (Cl_2), can be made by applying the ideal gas law.

$$n_{Cl_2} = \frac{PV}{RT} = \frac{(1.33 \text{ atm})(5.00 \text{ L})}{\left(0.0821 \dfrac{\text{atm L}}{\text{mole K}}\right)(280 \text{ K})}$$

$$= 0.28928136 \text{ mole} \quad \text{(calculator answer)}$$

$$= 0.289 \text{ mole } Cl_2 \quad \text{(correct answer)}$$

The dimensional analysis setup for the remaining conversion steps is

$$(0.289 \text{ mole } Cl_2) \times \left(\frac{1 \text{ mole } PbCl_4}{2 \text{ moles } Cl_2}\right) \times \left(\frac{349.2 \text{ g } PbCl_4}{1 \text{ mole } PbCl_4}\right)$$

Step 4. The solution, obtained from combining all of the numerical factors, is

$$\left(\frac{0.289 \times 1 \times 349.2}{2 \times 1}\right) \text{g } PbCl_4 = 50.4594 \text{ g } PbCl_4 \quad \text{(calculator answer)}$$

$$= 50.5 \text{ g } PbCl_4 \quad \text{(correct answer)}$$

An alternate method exists for converting the liters of Cl_2 to moles of Cl_2. The combined gas law can be used to calculate the STP volume of chlorine, and then the molar volume relationship can be used to calculate the moles of chlorine.

The combined gas law, rearranged to isolate V_2 on the left side, is

$$V_2 = V_1 \times \left(\frac{P_1}{P_2}\right) \times \left(\frac{T_2}{T_1}\right)$$

The given values for the variables are

$P_1 = 1.33$ atm $\quad P_2 = 1.00$ atm (STP)

$V_1 = 5.00$ L $\quad V_2 = ?$ L

$T_1 = 7°C = 280$ K $\quad T_2 = 0°C = 273$ K (STP)

Substituting these values into the combined gas law equation we get

$$V_2 = 5.00 \text{ L} \times \left(\frac{1.33 \text{ atm}}{1.00 \text{ atm}}\right) \times \left(\frac{273 \text{ K}}{280 \text{ K}}\right)$$

$= 6.48375$ L (calculator answer)

$= 6.48$ L (correct answer)

Now, changing the STP volume to moles by the molar volume relationship, we get

$$6.48 \text{ L} \times \left(\frac{1 \text{ mole } Cl_2}{22.4 \text{ L } Cl_2}\right) = 0.28928571 \text{ mole } Cl_2$$ (calculator answer)

$= 0.289$ mole Cl_2 (correct answer)

This is the same value we previously obtained with the ideal gas law.

12.12 Dalton's Law of Partial Pressures

In a mixture of gases that do not react with each other, each type of molecule moves about in the container as if the other kinds were not there. This type of behavior is possible because attractions between molecules in the gaseous state are negligible at most temperatures and pressures and because a gas is mostly empty space (Sec. 11.6). Each gas in the mixture occupies the entire volume of the container, that is, it distributes itself uniformly throughout the container. The molecules of each type strike the walls of the container as frequently and with the same energy as if they were the only gas in the mixture. Consequently, the pressure exerted by a gas in a mixture is the same as it would be if the gas were alone in the same container under the same conditions.

John Dalton (1766–1844), an English scientist, was the first to notice this independent behavior of gases when in mixtures. In 1803 he published a summary statement concerning such behavior now known as **Dalton's law of partial pressures:** *The total pressure exerted by a mixture of gases is the sum of the partial pressures of the individual gases.* A new term, *partial pressure,* is used in stating Dalton's law. A **partial pressure** *is the pressure that a gas in a mixture would exert if it were present alone under the same conditions.*

Dalton's law of partial pressures

partial pressure

Expressed mathematically, Dalton's law states that

$$P_T = P_1 + P_2 + P_3 + \cdots$$

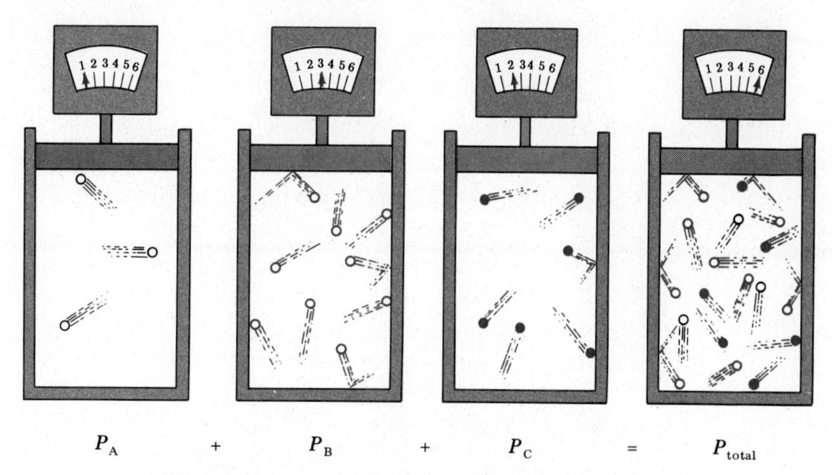

Figure 12-8. Dalton's law of partial pressures.

where P_T is the total pressure of a gaseous mixture and P_1, P_2, P_3, and so on are the partial pressures of the individual gaseous components of the mixture. (When the identity of a gas is known, its molecular formula is used as a subscript in the partial-pressure notation; for example, P_{CO_2} is the partial pressure of carbon dioxide in a mixture.)

To illustrate Dalton's law, consider the four identical gas containers shown in Figure 12-8. Suppose we place amounts of three different gases (represented by A, B, and C) into three of the containers and measure the pressure exerted by each sample. We then place all three samples in the fourth container and measure the pressure exerted by this mixture of gases. It is found that

$$P_{total} = P_A + P_B + P_C$$

Using the pressures given in Figure 12-8, we see that

$$P_{total} = 1 + 3 + 2 = 6$$

EXAMPLE 12.22

Assume that a sample of air, collected when the atmospheric pressure is 742 torr, contains only three gases: nitrogen, oxygen, and water vapor. The partial pressures of nitrogen and oxygen in the sample are found to be 581 torr and 143 torr, respectively. What is the partial pressure of the water vapor present in the air?

SOLUTION

Dalton's law says that

$$P_{total} = P_{N_2} + P_{O_2} + P_{H_2O}$$

The known values for variables in this equation are

$$P_{total} = 742 \text{ torr (atmospheric pressure)}$$

$$P_{N_2} = 581 \text{ torr}$$

$$P_{O_2} = 143 \text{ torr}$$

Rearranging Dalton's law to isolate P_{H_2O} on the left side of the equation gives

$$P_{H_2O} = P_{total} - P_{N_2} - P_{O_2}$$

Substituting the known numerical values into this equation and doing the arithmetic gives

$$P_{H_2O} = 742 \text{ torr} - 581 \text{ torr} - 143 \text{ torr} = \mathbf{18 \text{ torr}}$$

A common application of Dalton's law of partial pressures is encountered in a laboratory. Gases prepared in the laboratory are quite often collected by displacement of water. Figure 12-9 shows O_2, prepared from the decomposition of $KClO_3$, being collected by water displacement. A gas collected by water displacement is never pure. It always contains some water vapor. The total pressure exerted by the gaseous mixture is the sum of the partial pressures of the gas being collected and the water vapor.

$$P_{total} = P_{gas} + P_{H_2O}$$

The pressure exerted by the water vapor in the mixture will be constant at any given temperature if sufficient time has been allowed to permit equilibrium conditions to be established.

When a gas is collected by water displacement, by an apparatus similar to that in Figure 12-9, the total pressure of the mixture is atmospheric pressure. The partial pressure of the water vapor, the vapor pressure of water at the specific temperature of the bulk water, can be obtained from a table showing the variation of water vapor pressure with temperature (see Table 12-3). Thus the partial pressure of the collected gas is easily determined.

$$P_{gas} = P_{atm} - P_{H_2O}$$

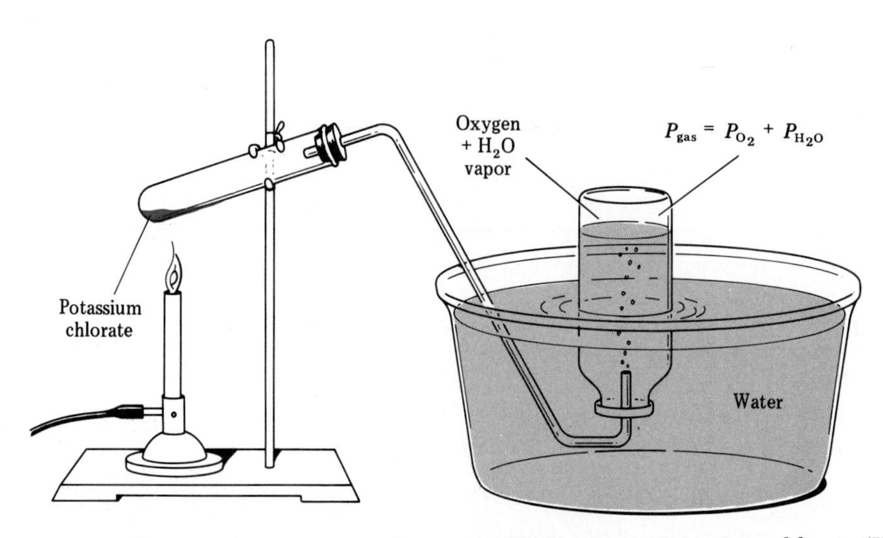

Figure 12-9. Collection of oxygen gas using water displacement. Potassium chlorate ($KClO_3$) decomposes to form oxygen (O_2), which is collected over water.

Table 12-3. Vapor Pressure of Water at Various Temperatures

T (°C)	Vapor Pressure (torr)	*T* (°C)	Vapor Pressure (torr)	*T* (°C)	Vapor Pressure (torr)
15	12.8	22	19.8	29	30.0
16	13.6	23	21.1	30	31.8
17	14.5	24	22.4	31	33.7
18	15.5	25	23.8	32	35.7
19	16.5	26	25.2	33	37.7
20	17.5	27	26.7	34	39.9
21	18.7	28	28.3	35	42.2

EXAMPLE 12.23

What is the partial pressure of oxygen collected over water at 24°C when the atmospheric pressure is 752 torr?

SOLUTION

From Table 12-3 we determine that water has a vapor pressure of 22.4 torr at 24°C. Using the equation

$$P_{O_2} = P_{atm} - P_{H_2O}$$

we find the partial pressure of the oxygen to be $73\bar{0}$ torr.

$$P_{O_2} = 752 \text{ torr} - 22.4 \text{ torr} = 729.6 \text{ torr} \quad \text{(calculator answer)}$$
$$= 73\bar{0} \text{ torr} \quad \text{(correct answer)}$$

Learning Objectives

After completing this chapter you should be able to

- Understand the restriction on temperature-scale use required by the gas laws (Sec. 12.1).
- State the definition of pressure, know the commonly used pressure units, and be able to convert from one pressure unit to another (Sec. 12.1).
- Describe the principles of operation of barometers and manometers (Sec. 12.1).
- State Boyle's, Charles', and Gay-Lussac's laws and use them to solve problems involving changes in the condition of a gas; be able to explain how one variable is affected by a change in another (Secs. 12.2–12.4).
- Use the combined gas law to calculate, for a fixed amount of gas, the new value for a gas law variable brought about by changes in the other two gas law variables (Sec. 12.5).
- Define STP conditions (Sec. 12.6).
- State Avogadro's law in words and mathematically, and use it in problem solving (Sec. 12.7).
- Define molar volume and use this relationship in problem-solving situations (Sec. 12.8).
- Give the units and value of the ideal gas constant, *R*, for commonly encountered unit combinations (Sec. 12.9).
- Use the ideal gas law to calculate a value for *P*, *V*, *T*, or *n* given values for the other three (Sec. 12.9).
- Use the ideal gas law in modified form to calculate the mass of a gas, its molecular weight, or its density from appropriate data (Sec. 12.10).
- Understand the relationship, at constant temperature and pressure, between the volumes of different gases

consumed or produced in a chemical reaction (Sec. 12.11).

- Calculate the volume or mass of a gas, under a given set of conditions, that is required as a reactant or formed as a product in a chemical reaction (Sec. 12.11).
- Use Dalton's law of partial pressures to calculate the partial or total pressure from appropriate data on mixtures of gas and also to correct for the effects of water vapor pressure in gases collected by water displacement (Sec. 12.12).

Terms and Concepts for Review

The new terms or concepts defined in this chapter are

gas laws (Sec. 12.1)
pressure (Sec. 12.1)
barometer (Sec. 12.1)
manometer (Sec. 12.1)
Boyle's law (Sec. 12.2)
Charles' law (Sec. 12.3)
Gay-Lussac's law (Sec. 12.4)
combined gas law (Sec. 12.5)
standard temperature (Sec. 12.6)
standard pressure (Sec. 12.6)
STP conditions (Sec. 12.6)
Avogadro's law (Sec. 12.7)
molar volume (Sec. 12.8)
ideal gas law (Sec. 12.9)
ideal gas (Sec. 12.9)
Dalton's law of partial pressures (Sec. 12.12)
partial pressure (Sec. 12.12)

Questions and Problems

Measurement of Pressure

12-1 Express each of the following pressures in torr.
a. 735 mm Hg **b.** 0.371 atm **c.** 21.6 in. Hg
d. 14.3 psi

12-2 Express each of the following pressures in atmospheres.
a. 775 torr **b.** 775 psi **c.** 775 mm Hg
d. 775 in. Hg

12-3 Describe what would happen to the level of the mercury column in the barometer of Figure 12-1 in the following situations.
a. The atmospheric pressure decreases.
b. A small leak develops in the sealed end of the glass tube.
c. More mercury is added to the circular container.

12-4 The mercury level in the arm of an open-end manometer (see Fig. 12-2) open to the atmosphere is found to be 237 mm higher than the Hg level in the arm of the manometer connected to the container of gas. Measured barometric pressure is 762 mm Hg. What is the pressure, in mm Hg, of the gas in the container?

Boyle's Law

12-5 A sample of cyclopropane (C_3H_6), a colorless gas with a pleasant odor that finds use as a general anesthetic, occupies a volume of 2.00 L at 27°C and 1.00 atm pressure. What volume will this sample occupy at the same temperature, but at each of the following pressures?
a. 2.73 atm **b.** 0.283 atm **c.** $78\bar{0}$ torr
d. $1\overline{000}$ psi

12-6 A sample of carbon monoxide (CO) gas, in an expandable container, has a volume of 3.73 L at a temperature of 23°C and a pressure of 742 torr. What pressure, in torr, will the gas be under if the volume of the container is changed, at constant temperature, to the following values?
a. 1.75 L **b.** 10.00 L **c.** $65\bar{0}$ mL
d. 232 mL

12-7 A balloon is inflated to a volume of 12.6 L on a day when the atmospheric pressure is 674 torr. The next day, as a storm front arrives, the atmospheric pressure drops to 651 torr. Assuming the temperature remains constant, what is the change, in liters, in the volume of the balloon?

Charles' Law

12-8 A sample of H_2 gas has a volume of 2.73 L at 27°C. What volume will the H_2 gas occupy at each of the following temperatures if the pressure is held constant?
a. 227°C **b.** −28°C **c.** $12\overline{00}$°C
d. 95°F

12-9 A sample of N_2 gas occupies a volume of $35\bar{0}$ mL at 25°C and a pressure of 1.00 atm. At the same pressure, at what temperature, in °C, would the volume of the gas be equal to each of the following?
a. 385 mL **b.** 12.0 mL **c.** $3\overline{000}$ mL
d. 1.23 L

12-10 It is desired to increase the volume of 75 mL of Ne gas by $1\bar{0}$% while holding the pressure constant. To what temperature must the gas be heated if the initial temperature is 28°C?

Gay-Lussac's Law

12-11 At room temperature, 27°C, the pressure exerted by some oxygen gas stored in a 20.0 L steel cylinder is 1.50 atm. What will be the pressure of the oxygen in the cylinder if the temperature is allowed to reach the following levels?
a. $42\bar{0}$°C **b.** 527°C **c.** $2\overline{000}$°C
d. $2\overline{000}$°F

12-12 A sample of F_2 gas at 55°C in a constant volume container exerts a pressure of 1.31 atm. To what temperatures would the gas have to be heated for it to exert each of the following pressures?

a. 2.00 atm b. 10.0 atm c. 1.35 atm
d. 100.0 psi

12-13 An aerosol spray can contains gas under a pressure of 2.00 atm at 27°C. The can itself can only withstand a pressure of 3.00 atm. To what temperature may the can and its contents be heated before the can explodes?

The Combined Gas Law

12-14 Rearrange the standard form of the combined gas law equation such that each of the following variables is by itself on the left side of the equation.
a. T_1 **b.** V_1 **c.** P_2 **d.** T_2

12-15 A sample of Cl_2 gas has a volume of 17.5 L at a pressure of 4.00 atm and a temperature of 27°C. Determine the volume this gas sample will occupy after undergoing temperature and pressure changes to the following values.
a. 35°C and 2.43 atm **b.** 382°C and 25.0 atm
c. −78°C and 0.532 atm **d.** 600°C and 2.00 atm

12-16 A student collects 20.0 mL of O_2 gas at a temperature of 21°C and a pressure of 645 mm Hg. What is the pressure of the O_2 gas when the volume is halved and the temperature is increased by 20 Celsius degrees?

12-17 A helium-filled weather balloon, when released, has a volume of 10.0 L at 27.0°C and a pressure of 0.833 atm. Calculate the volume of the balloon after it encounters the following upper atmospheric conditions: temperature of −30.0°C and pressure of 250 mm Hg.

STP Conditions

12-18 Change the following volumes of neon gas to STP conditions.
a. 5.73 L at 30°C and 2.00 atm
b. 25.1 mL at 0°C and 5.35 atm
c. 0.431 L at 235°C and 150.3 atm
d. 275 mL at −198°C and 285 torr

12-19 A quantity of NO_2 gas has a volume of 0.0835 L at STP. What is the pressure if the gas undergoes the following volume and temperature changes?
a. 1.03 L and 21°C **b.** 21.0 mL and 325°C
c. 2.10 L and −125°C **d.** 0.167 L and 546°C

Avogadro's Law

12-20 A 0.527 mole sample of O_2 gas at 1.00 atm and 32°C occupies a volume of 1.37 L. What volume would a 0.527 mole sample of N_2 gas occupy at the same temperature and pressure?

12-21 A 1.20 mole quantity of argon gas has a volume of 2.00 L at a certain temperature and pressure. What is the volume of a 1.50 mole quantity of the same gas at the same temperature and pressure?

12-22 A balloon containing 1.83 moles of He has a volume of 0.673 L at a certain temperature and pressure. How many *grams* of He would have to be added to the balloon in order for the volume to increase to 0.811 L at the same temperature and pressure?

Molar Volume

12-23 What is the volume, in L, of 18.0 g of each of the following gases at STP?
a. N_2 **b.** O_2 c. CO_2 **d.** He

12-24 How many moles of gas are present in each of the following volumes of gas at STP?
a. 26.3 L H_2 **b.** 888 mL of He
c. 3.04 L O_2 **d.** 7.00 L CO_2

12-25 Calculate the density at STP conditions, in g/L, of each of the following gases.
a. H_2 **b.** Cl_2 c. CO_2 **d.** Ne

The Ideal Gas Law

12-26 Using the ideal gas law, calculate the volume of 0.100 mole of O_2 gas at each of the following sets of conditions.
a. STP **b.** 98°C and 1.21 atm
c. 500°C and 20.0 atm **d.** −7°C and 755 torr

12-27 A 5.00 L gas cylinder contains 2.00 moles of N_2 gas at 27°C. What is the pressure, in atmospheres, of this gas?

12-28 3.00 moles of He gas is introduced into a 6.00 L fixed-volume container. The container is heated until the pressure of the gas becomes 27.5 atm. What is the temperature of the Ne gas?

12-29 How many moles of SO_2 will be present in a 3.00 L cylinder if the temperature is 150°C and the pressure is 13.3 atm?

12-30 A sample of CO gas in a 6.00 L container is at a pressure of 3.00 atm and a temperature of 27°C. One-tenth (0.100) mole of CO gas is removed from the container. The remaining CO gas is then transferred to a new 4.00 L container and heated to a temperature of 127°C. What is the pressure in the new container?

Mass, Molecular Weight, and Density Calculations

12-31 Calculate the mass, in grams, of each of the following quantities of gas.
a. 30.0 L of CH_4 at 1.25 atm and 31°C
b. 1.11 L of H_2 at 546 torr and 123°C
c. 4.00 L of O_2 at STP
d. 6.75 L of N_2 at 100 torr and −100°C

12-32 If 3.00 g of a gas occupies a volume of 6.00 L at 85°C and 1.11 atm pressure what is its molecular weight?

12-33 Calculate the density of CO_2 gas, in g/L, at each of the following temperature–pressure conditions.
a. 27°C and 0.889 atm **b.** 127°C and 0.889 atm
c. 27°C and 8.89 atm **d.** 500°C and 1.00 atm

12-34 Calculate the mass, in grams, of 3.50 L of NO gas measured at 35°C and 835 torr.

12-35 Calculate the molecular weight of a gas having a density of 2.20 g/L at 31°C and 745 torr.

12-36 The elemental analysis of a certain compound is 24.3% C, 4.1% H, and 71.6% Cl by mass. If 0.132 g of compound occupies 41.4 mL at 741 torr and 96°C, what are the molecular weight and molecular formula of the compound?

Gases in Chemical Reactions

12-37 Consider the reaction

$$4\ NH_3(g) + 5\ O_2(g) \longrightarrow 4\ NO(g) + 6\ H_2O(g)$$

a. How many liters of NH_3 must be consumed to produce 2.80 L of NO, if all volumes are measured at STP?

b. How many liters of NH_3 are needed to react with 0.78 L of O_2, if all volumes are measured at STP?

c. How many liters of NH_3 are needed to react with 0.78 L of O_2 if all volumes are measured at 2.00 atm and 56°C?

d. How many liters of O_2, measured at STP, are consumed in the production of 6.73 L of NO, measured at 5.00 atm and 27°C?

12-38 A mixture of 50.0 g of nitrogen and an excess of oxygen react according to the balanced equation

$$N_2(g) + O_2(g) \longrightarrow 2\ NO(g)$$

How many liters of nitric oxide (NO) at STP are produced?

12-39 A common laboratory preparation for oxygen gas involves the thermal decomposition of potassium nitrate.

$$2\ KNO_3(s) \longrightarrow 2\ KNO_2(s) + O_2(g)$$

How many grams of KNO_3 must be decomposed in order to produce $20\bar{0}$ L of O_2 gas at 27°C and 741 torr?

12-40 In the reaction

$$3\ NO_2(g) + H_2O(l) \longrightarrow 2\ HNO_3(l) + NO(g)$$

it is desired to produce 75.0 L of NO at 38°C and 645 torr. How many liters of NO_2 gas at 21°C and 2.31 atm must be consumed, assuming complete reaction?

Dalton's Law of Partial Pressure

12-41 A steel cylinder contains a mixture of nitrogen, oxygen, and carbon dioxide gases. The total pressure in the cylinder is 2235 torr. The pressure exerted by the nitrogen and oxygen is 545 and 685 torr, respectively. What is the partial pressure of the CO_2? (Assume all gases are at the same constant temperature.)

12-42 A 2.00 L mixture of gases is produced from 2.00 L of N_2 at $20\bar{0}$ torr, 2.00 L of O_2 at $50\bar{0}$ torr, and 2.00 L of H_2 at $70\bar{0}$ torr. What is the pressure of the mixture? (Assume all gases are at the same constant temperature.)

12-43 A 2.00 L mixture of gases is produced from 1.00 L of N_2 at $35\bar{0}$ torr, 6.00 L of O_2 at $30\bar{0}$ torr, and 1.00 L of H_2 at $20\bar{0}$ torr. What is the pressure of the mixture? (Assume all gases are at the same constant temperature.)

12-44 Suppose 30.0 mL of nitrogen (N_2) at 27°C and 645 torr is added to a 40.0 mL container that already contains helium at 37°C and 765 torr. If the temperature of the mixture is brought to 32°C, what is the total pressure in torr?

12-45 What would be the partial pressure of O_2 collected over water at the following conditions of temperature and atmospheric pressure?

a. 19°C and 743 torr **b.** 28°C and 645 torr
c. 34°C and $76\bar{0}$ torr **d.** 21°C and 0.933 atm

12-46 At a temperature of 27°C, 50.0 mL of O_2 gas is collected by water displacement. Barometric pressure is 638 torr. What would be the volume of O_2 gas (dry) at STP?

13 Solutions: Terminology and Concentrations

13.1 Types of Solutions

solution

A **solution** *is a homogeneous (uniform) mixture of two or more substances.* To achieve a homogeneous mixture the intermingling of components must be on the molecular level; that is, the particles present must be of atomic and molecular size.

solvent

solute

In discussing solutions it is often convenient to call one component the *solvent* and the others *solutes*. The **solvent** *is the component of the solution present in the greatest amount.* The solvent may be thought of as the medium in which the other substances present are *dissolved.* A **solute** *is a solution component present in a small amount relative to that of solvent.* More than one solute may be present in the same solution. For example, both sugar and salt (two solutes) may be dissolved in a container of water.

In most situations we will encounter, the solutes present in a solution will be of more interest to us than the solvent. The solutes are the "active ingredients" in the solution. They are the substances that undergo reaction when solutions are mixed.

Solutions used in the laboratory are usually liquids, and the solvent is almost always water. However, as we shall see shortly, gaseous solutions and solid solutions of numerous types do exist.

A solution, since it is homogeneous, will have the same properties throughout. No matter where we take a sample from a solution we will obtain material with the same composition as that of any other sample from the same solution. The composition of a solution can be varied, usually within certain limits, by changing the relative amounts of solvent and solute present. (If the composition limits are transgressed, a heterogeneous mixture is formed.)

Nine types of two-component solutions can exist according to a classification scheme based on the physical states of the solvent and solute before mixing. These types, along with an example of each, are listed in Table 13-1. Solutions in which the

Table 13-1. Examples of Various Types of Solutions

Solution Type (solute listed first)	Example
Gaseous Solutions	
Gas dissolved in gas	Dry air (oxygen and other gases dissolved in nitrogen)
Liquid dissolved in gas[a]	Wet air (water vapor in air)
Solid dissolved in gas[a]	Moth repellent (or moth balls) sublimed into air
Liquid Solutions	
Gas dissolved in liquid	Carbonated beverage (CO_2 in water)
Liquid dissolved in liquid	Vinegar (acetic acid dissolved in water)
Solid dissolved in liquid	Salt water
Solid Solutions	
Gas dissolved in solid	Hydrogen in platinum
Liquid dissolved in solid	Dental filling (mercury dissolved in silver)
Solid dissolved in solid	Sterling silver (copper dissolved in silver)

[a] An alternate viewpoint is that liquid-in-gas and solid-in-gas solutions do not actually exist as true solutions. From this viewpoint water vapor or moth repellent in air is considered to be a gas-in-gas solution since the water or moth repellent must evaporate or sublime first in order to enter the air.

final state of the solution components is a liquid are the most common and are the type that will be emphasized in this book.

The physical state of a solute becomes that of the solvent when a solution is formed. For example, solid naphthalene (moth repellent) must be sublimed (Sec. 11.8) in order for it to dissolve in air. Merely finely pulverizing a solid and dispersing it in air does not produce a solution. (Dust particles in air would be an example of this.) The particles of the solid must be subdivided to the molecular level; the solid must be sublimed. Similarly, fog is a suspension of water droplets in air; the droplets are large enough to reflect light, a fact that becomes evident when we drive an automobile on a foggy night. Thus, fog is not a solution. Water vapor, however, is present in solution form in air. When hydrogen gas dissolves in platinum metal (a gas in solid solution), the gas molecules take up fixed positions in the metal lattice. The gas is "solidified" as a result.

13.2 Terminology Used in Describing Solutions

In addition to *solvent* and *solute,* several other terms are useful in describing characteristics of solutions.

solubility

The **solubility** *of a solute is the amount of solute that will dissolve in a given amount of solvent.* Numerous factors affect the numerical value of a solute's solubility in a given solvent including the nature of the solvent itself, the temperature, and in some cases the pressure and the presence of other solutes.

Common units for expressing solubility are grams of solute per 100 g of solvent. The temperature of the solvent must also be specified. Table 13-2 gives the solubilities of selected solutes in the solvent water at three different temperatures.

Table 13-2. Solubilities of Various Compounds in Water at 0°C, 50°C, and 100°C

	Solubility (g solute/100 g H_2O)		
Solute	**0°C**	**50°C**	**100°C**
Lead(II) bromide ($PbBr_2$)	0.455	1.94	4.75
Silver sulfate (Ag_2SO_4)	0.573	1.08	1.41
Copper(II) sulfate ($CuSO_4$)	14.3	33.3	75.4
Sodium chloride (NaCl)	35.7	37.0	39.8
Silver nitrate ($AgNO_3$)	122	455	952
Cesium chloride (CsCl)	161.4	218.5	270.5

The use of specific units for specifying solubility, as in Table 13-2, allows us to compare solubilities quite precisely. Such precision is often unnecessary and instead *qualitative* statements about solubilities are made by using terms such as *very soluble, slightly soluble,* and so forth. The guidelines for the use of such terms are given in Table 13-3.

Table 13-3. Qualitative Solubility Terms

Solute Solubility (g solute/100 g solvent)	Qualitative Solubility Description
Less than 0.1	insoluble
0.1–1	slightly soluble
1–10	soluble
Greater than 10	very soluble

saturated solution

A **saturated solution** *is a solution that contains the maximum amount of solute that can be dissolved under the conditions at which the solution exists.* A saturated solution containing excess undissolved solute is an equilibrium situation where the rate of dissolution of undissolved solute is equal to the rate of crystallization of dissolved solute. Consider the process of adding table sugar (sucrose) to a container of water. Initially the added sugar dissolves as the solution is stirred. Finally, as we add more sugar, a point is reached where no amount of stirring will cause the added sugar to dissolve. The last-added sugar remains as a solid on the bottom of the container; the solution is saturated. Although it appears to the eye that nothing is happening once the saturation point is reached, on the molecular level this is not the case. Solid sugar from the bottom of the container is continuously dissolving in the water and an equal amount of sugar is coming out of solution. Accordingly, the net number of sugar molecules in the liquid remains the same, and outwardly it appears that the dissolution process has stopped. This equilibrium situation in the saturated solution is somewhat similar to the previously discussed evaporation of a liquid in a closed container (Sec. 11.10). Figure 13-1 illustrates the dynamic equilibrium process occurring in a saturated solution containing undissolved excess solute.

unsaturated solution

An **unsaturated solution** *is a solution where less solute than the maximum amount possible is dissolved in the solution.*

dilute solution

concentrated solution

The terms *dilute* and *concentrated* are also used to convey qualitative information about the degree of saturation of a solution. A **dilute solution** *is a solution where a small amount of solute is present in solution relative to the amount that could dissolve.* On the other hand, a **concentrated solution** *is a solution that contains a relatively large amount of solute relative to the amount that could dissolve.* A concentrated solution need not be a saturated solution.

miscible

partially miscible

immiscible

In dealing with liquid in liquid solutions the terms *miscible, partially miscible,* and *immiscible* are frequently used to describe solubility characteristics associated with the liquids. **Miscible** *substances dissolve in any amount in each other.* For example, methyl alcohol (CH_3OH) and water are miscible, that is, they completely mix with each other in any and all proportions. Always after mixing these two liquids only one phase is present. **Partially miscible** *substances have limited solubility in each other.* Benzene (C_6H_6) and water are partially miscible. If benzene is added slowly to water a small amount of benzene initially dissolves; a single phase results. However, as soon as the benzene solubility limit is reached the excess benzene forms a separate layer on top of the water (on top because it is less dense). **Immiscible** *substances do not dissolve in each other.* When such substances are mixed, two layers (phases) immediately form. Very few liquids are totally immiscible in each other; toluene (C_7H_8) and water approach this limiting case. Figure 13-2 illustrates the results obtained by mixing liquids of various miscibilities with each other.

aqueous solution

Another term commonly encountered in solution discussions is *aqueous solution.* An **aqueous solution** *is simply a solution in which water is the solvent.*

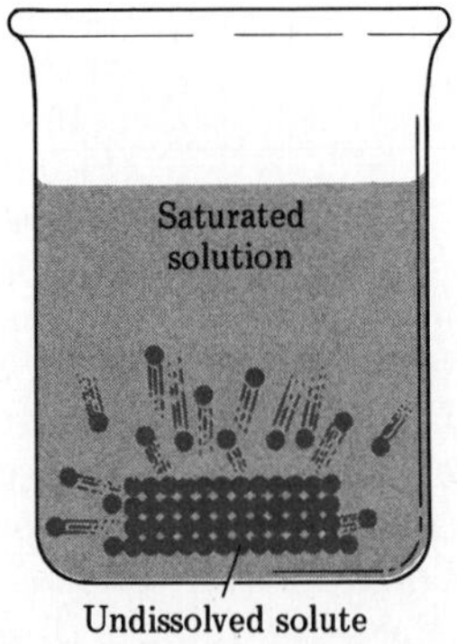

Figure 13-1. The dynamic equilibrium process occurring in a saturated solution containing undissolved excess solute.

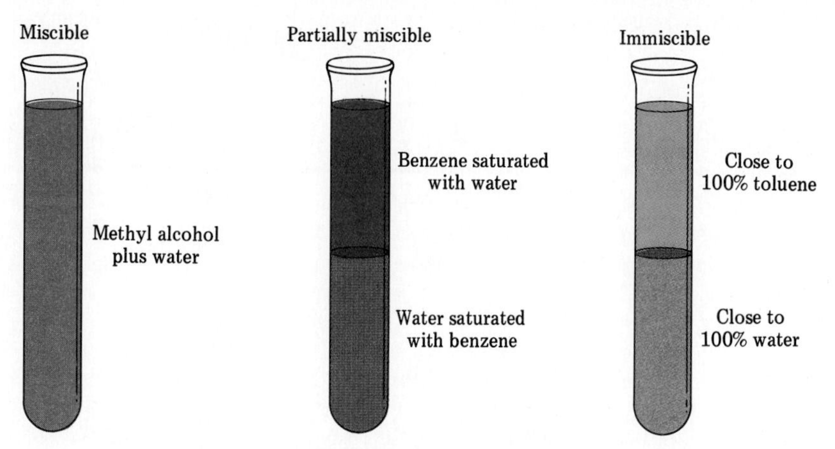

Figure 13-2. Miscibility of selected liquids with each other.

13.3 Solution Formation

In a solution, solute particles are uniformly dispersed throughout the solvent. Considering what happens at the molecular level during the solution process will help us to understand how this is achieved.

For a solute to dissolve in a solvent two types of interparticle attractions must be overcome (1) attractions between solute particles (solute–solute attractions) and (2) attractions between solvent particles (solvent–solvent attractions). Only when these attractions are overcome can particles in both pure solute and pure solvent separate from each other and begin to intermingle. A new type of interaction, which does not exist prior to solution formation, arises as the result of the mixing of solute and solvent. This new interaction is the attraction between solute and solvent particles (solute–solvent attractions). These new attractions are the primary driving force for solution formation. The extent to which a substance dissolves depends on the degree to which the newly formed solute–solvent attractions are able to compensate for the energy needed to overcome the solute–solute and solvent–solvent interactions. A solute will not dissolve in a solvent if either solute–solute or solvent–solvent interactions are too strong to be compensated for by the formation of the new solute–solvent interactions.

A most important type of solution process is that where an ionic solid dissolves in water. Let us consider in detail the process of dissolving sodium chloride, a typical ionic solid, in water. We will consider the process to occur in steps. The fact that water molecules are polar (Sec. 10.12) will become very important in our considerations.

Figure 13-3 shows what is thought to happen when sodium chloride is placed in water. The polar water molecules become oriented such that the negative oxygen portion points toward positive sodium ions, and the positive hydrogen portion points toward negative chloride ions. As the polar water molecules begin to surround ions on the crystal surface they exert sufficient attraction to cause these ions to break away from the crystal surface. After leaving the crystal, the ion retains its surrounding group of water molecules; it has become a *hydrated ion.* As each hydrated ion leaves the surface, other ions are exposed to the water, and the crystal is picked apart ion by ion.

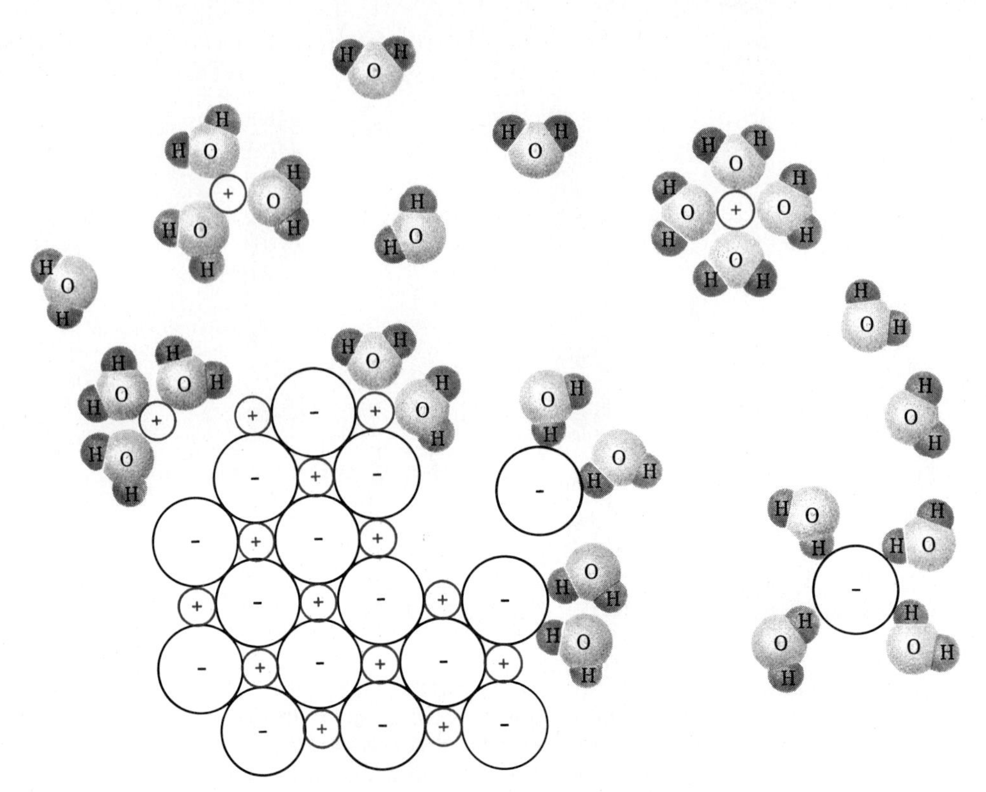

Figure 13-3. The solution process for an ionic solid in water.

Once in solution, the hydrated ions are uniformly distributed by stirring or random collisions with other molecules or ions.

The random motion of solute ions in solution causes them to collide with each other, with solvent molecules, and occasionally with the surface of the undissolved solute. Ions undergoing this latter type of collision occasionally stick to the solid surface and thus leave the solution. When the number of ions in solution is low, the chances for collision with the undissolved solute are low. However, as the number of ions in solution increases, so do the chances for such collisions, and more ions are recaptured by the undissolved solute. Eventually, the number of ions in solution reaches a level such that ions return to the undissolved solute at the same rate as other ions leave. At this point, the solution is saturated, and the equilibrium process discussed in the last section is in operation.

13.4 Solubility Rules

In this section we present some rules for qualitatively predicting solute solubilities. These rules summarize in a concise form the results of thousands of experimental solute–solvent solubility determinations.

A very useful generalization that relates polarity to solubility is *"Substances of like polarity tend to be more soluble in each other than substances that differ in polarity."* This conclusion is often expressed as the simple phrase "*likes dissolve likes.*" Polar substances, in general, are good solvents for other polar substances, but not for nonpolar substances. Similarly, nonpolar substances exhibit greater solubility in nonpolar solvents than they do in polar solvents.

The generalization "likes dissolve likes" is a useful tool for predicting solubility behavior in many, but not all, solute–solvent situations. Results that agree with the generalization are almost always obtained in the cases of gas-in-liquid and liquid-in-liquid solutions and for solid-in-liquid solutions in which the solute is not an ionic compound. For example, NH_3 gas (a polar gas) is much more soluble in H_2O (a polar liquid) than is O_2 gas (a nonpolar gas). (The actual solubilities of NH_3 and O_2 in water at 20°C are, respectively, 51.8 g/100 g H_2O and 0.0043 g/100 g H_2O.)

In the case of those solid-in-liquid solutions in which the solute is an ionic compound — a very common situation — the rule "likes dissolve likes" is not adequate. Because of their polar nature, one would predict that all ionic compounds are soluble in polar solvents such as water. This is not the case. The failure of the generalization here is related to the complexity of the factors involved in determining the magnitude of the solute–solute (ion–ion) and solvent–solute (ion–polar solvent molecule) interactions. Among other things, both the charge on and size of the ions in the solute must be considered. Changes in these factors affect both types of interactions, but not to the same extent.

Some guidelines concerning the solubility of ionic compounds in water, which should be used in place of "likes dissolve likes," are given in Table 13-4.

Table 13-4. Solubility Guidelines for Ionic Compounds in Water

Ion Contained in the Compound	Solubility	Exceptions
Group IA (Li^+, Na^+, K^+, etc.)	soluble	
Ammonium (NH_4^+)	soluble	
Acetates ($C_2H_3O_2^-$)	soluble	
Nitrates (NO_3^-)	soluble	
Chlorides (Cl^-), bromides (Br^-), and iodides (I^-)	soluble	Ag^+, Pb^{2+}, Hg_2^{2+}
Sulfates (SO_4^{2-})	soluble	Ca^{2+}, Sr^{2+}, Ba^{2+}, Pb^{2+}
Carbonates (CO_3^{2-})	insoluble	group IA and NH_4^+
Phosphates (PO_4^{3-})	insoluble	group IA and NH_4^+
Sulfides (S^{2-})	insoluble	groups IA and IIA and NH_4^+
Hydroxides (OH^-)	insoluble	group IA, Ba^{2+}, Sr^{2+}

EXAMPLE 13.1

Predict the solubility of the following solutes in the solvent indicated.

(a) SO_2 (a polar gas) in water
(b) paraffin wax (nonpolar) in CCl_4 (nonpolar)
(c) $AgNO_3$ in water

(d) $CaSO_4$ in water
(e) NH_4Cl in water

SOLUTION

(a) Soluble. SO_2 is polar as is water—likes dissolve likes.
(b) Soluble. Since both substances are nonpolar, they should be relatively soluble in each other—likes dissolve likes.
(c) Soluble. Table 13-4 indicates that all nitrates are soluble.
(d) Insoluble. Table 13-4 indicates that calcium is an exception to the rule that all sulfates are soluble.
(e) Soluble. Table 13-4 indicates that all compounds containing the ammonium ion are soluble. Alternately, we could use the rule that all chlorides are soluble, except for Ag^+, Pb^{2+}, and Hg_2^{2+}, to predict that NH_4Cl would be soluble.

All ionic compounds, even the most insoluble ones, dissolve to a slight extent in water. The insoluble classification used in Table 13-4 thus really means ionic compounds that have very limited solubility in water.

You should become thoroughly familiar with the rules in Table 13-4. They find extensive use in chemical discussions. We will next encounter them in Section 14.6 when the topic of net ionic equations is presented.

13.5 Solution Concentrations

In Section 13.2 we learned that, in general, there is a limit to the amount of solute that can be dissolved in a specified amount of solvent and also that a solution is said to be saturated when this maximum amount of solute has been dissolved. We can determine the amount of dissolved solute in a saturated solution by the solute's solubility.

Most solutions chemists deal with are *unsaturated* rather than saturated solutions. The amount of solute present in an unsaturated solution is specified by stating the *concentration* of the solution. The **concentration** *of a solution is the amount of solute present in a specified amount of solvent or a specified amount of solution.* Thus, concentration is a ratio of two quantities, being either the ratio

concentration

$$\frac{\text{Amount of solute}}{\text{Amount of solvent}} \quad \text{or} \quad \frac{\text{Amount of solute}}{\text{Amount of solution}}$$

In specifying a concentration what are the units used to indicate the amounts of solute and solvent or solution present? In practice, a number of different unit combinations are used, with the choice of units depending on the use to be made of the solution. In each of the next four sections we shall discuss a commonly encountered set of units used to express solution concentration. The concentration expressions to be discussed are (1) percentage of solute, Section 13.6, (2) molarity, Section 13.7, (3) molality, Section 13.8, and (4) normality, Section 13.9.

13.6 Concentration: Percentage of Solute

The concentration of a solution is often specified in terms of the percentage of solute in the total amount of solution. Since the amounts of solute and solution present may be stated in terms of either mass or volume, different types of percentage units exist. The three most common are

1. Weight–weight percent (or percent by weight).
2. Volume–volume percent (or percent by volume).
3. Weight–volume percent.

percent by weight

Weight–weight percent (or percent by weight) is the percentage unit most frequently used by chemists. **Percent by weight** *is equal to the mass of solute divided by the total mass of solution multiplied by 100 (to put the value in terms of percentage).* (Percentage is always part of the whole divided by the whole times 100, Sec. 3.4.)

$$\text{Percent by weight} = \frac{\text{mass solute}}{\text{mass solution}} \times 100$$

The solute and solution masses must be the same units. The mass of solution is equal to the mass of *solute* plus the mass of *solvent.*

$$\text{Percent by weight} = \frac{\text{mass solute}}{\text{mass solute} + \text{mass solvent}} \times 100$$

A solution of 5.0% by weight concentration would contain 5.0 g of solute in 100.0 g of solution (5.0 g of solute and 95.0 g of solvent). Thus, percent by weight gives directly the number of g of solute in 100 g of solution.

As was the case with atomic weight (Sec. 5.7), the term *percent by weight* (or weight–weight percent) is a misnomer. This concentration unit should really be called *percent by mass.* However, the term percent by weight has become so firmly entrenched in the working vocabulary of scientists that we will continue to use it. Remember, however, that masses are used in calculating weight percent.

EXAMPLE 13.2

What is the percent by weight concentration of Na_2SO_4 in a solution made by dissolving 7.6 g of Na_2SO_4 in enough water to give 87.3 g of solution?

SOLUTION

Both the mass of solute and mass of solution are directly given in the problem statement. Substituting these numbers into the equation

$$\text{Percent by weight} = \left(\frac{\text{mass solute}}{\text{mass solution}}\right) \times 100$$

gives

$$\text{Percent by weight} = \left(\frac{7.6\ \cancel{g}\ Na_2SO_4}{87.3\ \cancel{g}\ \text{solution}}\right) \times 100 = 8.7056128\% \quad \text{(calculator answer)}$$

$$= 8.7\% \quad \text{(correct answer)}$$

EXAMPLE 13.3

How many grams of NaCl must be added to 375 g of water to prepare a 2.75% by weight solution of NaCl?

SOLUTION

Often, when a solution concentration is given as part of a problem statement, the concentration information is used in the form of a conversion factor in solving the problem. That will be the case in this problem.

The given quantity is 375 g of H_2O (grams of solvent) and the desired quantity is g of NaCl (grams of solute).

$$375 \text{ g } H_2O = ? \text{ g NaCl}$$

The conversion factor relating these two quantities (solvent and solute) is obtained from the given concentration. In a 2.75% by weight NaCl solution there are 2.75 g of NaCl per every 97.25 g of H_2O.

$$100.00 \text{ g solution} - 2.75 \text{ g NaCl} = 97.25 \text{ g } H_2O$$

This relationship between grams of solute and grams of solvent (2.75 to 97.25) gives us the needed conversion factor.

$$\frac{2.75 \text{ g NaCl}}{97.25 \text{ g } H_2O}$$

Dimensional analysis gives the problem setup that is solved in the following manner.

$$375 \cancel{\text{g } H_2O} \times \left(\frac{2.75 \text{ g NaCl}}{97.25 \cancel{\text{g } H_2O}}\right) = 10.604113 \text{ g NaCl} \quad \text{(calculator answer)}$$

$$= 10.6 \text{ g NaCl} \quad \text{(correct answer)}$$

percent by volume

Volume–volume percent (or percent by volume) finds use as a concentration unit in situations where both the solute and solvent are both liquids or gases. In such cases it is often more convenient to measure volumes than masses. **Percent by volume** *is equal to the volume of solute divided by the total volume of solution multiplied by 100.*

$$\text{Percent by volume} = \frac{\text{volume solute}}{\text{volume solution}} \times 100$$

Solute and solution volumes must always be expressed in the same units when using this expression.

The numerical value of a concentration expressed as a percent by volume gives directly the number of milliliters of solute in 100 mL of solution. Thus, a 100 mL sample of a 5.0% alcohol in water solution contains 5.0 mL of alcohol dissolved in enough water to give 100 mL of solution. Note that such a 5.0% by volume solution could not be made by adding 5 mL of alcohol to 95 mL of water since volumes of liquids are not usually additive. Differences in the way molecules are packed as well as in the distances between molecules almost always result in the volume of a solution being different from the sum of the volumes of solute and solvent. For example, the final volume resulting from the addition of 50.0 mL of ethyl alcohol to 50.0 mL of water is 96.5 mL of solution.

EXAMPLE 13.4

80.0 mL of methyl alcohol and 80.0 mL of water are mixed to give a solution that has a final volume of 154 mL. What is the concentration of the solution expressed as percent by volume methyl alcohol?

SOLUTION

To calculate a percent by volume, the volumes of solute and solution are needed. Both are given in this problem.

$$\text{Solute volume} = 80.0 \text{ mL}$$

$$\text{Solution volume} = 154 \text{ mL}$$

Note that the solution volume is not the sum of the solute and solvent volumes. As previously mentioned, generally liquid volumes are not additive.

Substituting the given quantities in the equation

$$\text{Percent by volume} = \frac{\text{volume solute}}{\text{volume solution}} \times 100$$

gives

$$\text{Percent by volume} = \left(\frac{80.0 \cancel{\text{mL}}}{154 \cancel{\text{mL}}}\right) \times 100 = 51.948052\% \quad \text{(calculator answer)}$$

$$= \mathbf{51.9\%} \quad \text{(correct answer)}$$

The third type of percentage unit in common use is weight–volume percentage. This unit, which is often encountered in hospital and industrial settings, is particularly convenient to use when working with a solid solute (which is easily weighed) and a liquid solvent. Concentrations are specified using this unit when dealing with physiological fluids such as blood and urine. **Weight–volume percent** *is equal to the mass of solute (in grams) divided by the total volume of solution (milliliters) multiplied by 100.*

weight–volume percent

$$\text{Weight–volume percent} = \frac{\text{mass of solute (g)}}{\text{volume of solution (mL)}} \times 100$$

Note that in defining weight–volume percent specific mass and volume units are given. This is necessary because the units do not cancel as was the case with mass percent and volume percent. Note also that this percentage unit should really be called mass–volume percent but is not for the reasons previously cited.

EXAMPLE 13.5

In the treatment of certain illnesses of the human body, a 0.92 weight–volume percent sodium chloride (NaCl) solution is administered intravenously. How many grams of sodium chloride are required to prepare 345 mL of this solution?

SOLUTION

The given quantity is 345 mL of solution and the desired quantity is grams of NaCl.

$$345 \text{ mL solution} = ? \text{ g NaCl}$$

The given concentration of 0.92 weight–volume percent, which means 0.92 g NaCl per $1\bar{0}0$ mL solution, can be used as a conversion factor to go from milliliters of solution to g of NaCl. The setup for the conversion is

$$345 \cancel{\text{mL solution}} \times \left(\frac{0.92 \text{ g NaCl}}{1\bar{0}0 \cancel{\text{mL solution}}}\right)$$

Doing the arithmetic, after cancellation of units, gives

$$\left(\frac{345 \times 0.92}{1\bar{0}0}\right) \text{ g NaCl} = 3.174 \text{ g NaCl} \quad \text{(calculator answer)}$$

$$= 3.2 \text{ g NaCl} \quad \text{(correct answer)}$$

13.7 Concentration: Molarity

molarity

The **molarity** *of a solution, abbreviated M, is a ratio giving the number of moles of solute per liter of solution.*

$$\text{Molarity } (M) = \frac{\text{moles of solute}}{\text{liters of solution}}$$

A 1.00 M (molar) solution of KBr would contain 1 mole of KBr in 1 L of solution.

Molarity is the concentration expression most often used in a chemical laboratory. A major reason for this is the fact that the amount of solute is expressed in moles, a most convenient unit for dealing with chemical reactions. (Most laboratory solutions are liquids prepared for use in chemical reactions.) Because chemical reactions occur between molecules and atoms, a unit that counts particles, which moles does, is desirable.

To find the molarity of a solution we need to know the solution volume in liters and the number of moles of solute present. An alternative to knowing the number of moles of solute is knowledge about the grams of solute present and the solute's formula or molecular weight.

EXAMPLE 13.6

Determine the molarities of the following solutions.
(a) 2.37 moles of KNO_3 are dissolved in enough water to give 650 mL of solution.
(b) 25.0 g of NaOH are dissolved in enough water to give 2.50 L of solution.

SOLUTION

(a) The number of moles of solute is given in the problem statement.

$$\text{Moles of solute} = 2.37 \text{ moles } KNO_3$$

The volume of the solution is also given in the problem statement, but not in the right units. Molarity requires liters for volume units. Making the unit change gives

$$65\bar{0} \text{ mL} \times \left(\frac{1 \text{ L}}{1000 \text{ mL}}\right) = 0.65 \text{ L} \quad \text{(calculator answer)}$$

$$= 0.650 \text{ L} \quad \text{(correct answer)}$$

The molarity of the solution is obtained by substituting the known quantities into the equation

$$M = \frac{\text{moles of solute}}{\text{L of solution}}$$

which gives

$$M = \frac{2.37 \text{ moles } KNO_3}{0.650 \text{ L solution}} = 3.6461538 \, \frac{\text{moles } KNO_3}{\text{L solution}} \quad \text{(calculator answer)}$$

$$= 3.65 \, \frac{\text{moles } KNO_3}{\text{L solution}} \quad \text{(correct answer)}$$

Note that the units of molarity are always moles/L.
(b) This time the volume of solution is given in the right units, liters.

$$\text{L of solution} = 2.50 \text{ L}$$

The moles of solute must be calculated from the grams of solute (given) and the solute's formula weight, which is 40.0 amu (calculated from a table of atomic weights).

$$25.0 \text{ g NaOH} \times \left(\frac{1 \text{ mole NaOH}}{40.0 \text{ g NaOH}}\right) = 0.625 \text{ mole NaOH} \quad \text{(calculator and correct answer)}$$

Substituting the known quantities into the defining equation for molarity gives

$$M = \frac{0.625 \text{ mole NaOH}}{2.50 \text{ L solution}} = 0.25 \, \frac{\text{mole NaOH}}{\text{L solution}} \quad \text{(calculator answer)}$$

$$= 0.250 \, \frac{\text{mole NaOH}}{\text{L solution}} \quad \text{(correct answer)}$$

The mass of solute present in a known volume of solution is an easily calculable quantity if the molarity of the solution is known. In doing such a calculation, molarity serves as a conversion factor relating liters of solution to moles of solute.

EXAMPLE 13.7

How many grams of sucrose (table sugar, $C_{12}H_{22}O_{11}$) are present in 125 mL of a 1.07 M sucrose solution?

SOLUTION

The given quantity is 125 mL of solution and the desired quantity is grams of $C_{12}H_{22}O_{11}$.

$$125 \text{ mL solution} = ? \text{ g } C_{12}H_{22}O_{11}$$

The pathway to be used in solving this problem is

$$\text{mL solution} \longrightarrow \text{L solution} \longrightarrow \text{moles } C_{12}H_{22}O_{11} \longrightarrow \text{g } C_{12}H_{22}O_{11}$$

The given molarity (1.07 M) will serve as the conversion factor for the second unit change; the molecular weight of sucrose (which must be calculated as it is not given) is used in accomplishing the third unit change.

The dimensional-analysis setup from this pathway is

$$125 \cancel{\text{ mL solution}} \times \left(\frac{1 \cancel{\text{ L solution}}}{1000 \cancel{\text{ mL solution}}}\right) \times \left(\frac{1.07 \cancel{\text{ moles } C_{12}H_{22}O_{11}}}{1 \cancel{\text{ L solution}}}\right) \times \left(\frac{342 \text{ g } C_{12}H_{22}O_{11}}{1 \cancel{\text{ mole } C_{12}H_{22}O_{11}}}\right)$$

Cancelling units and doing the arithmetic gives

$$\frac{125 \times 1 \times 1.07 \times 342}{1000 \times 1 \times 1} \text{ g } C_{12}H_{22}O_{11} = 45.7425 \text{ g } C_{12}H_{22}O_{11} \quad \text{(calculator answer)}$$

$$= 45.7 \text{ g } C_{12}H_{22}O_{11} \quad \text{(correct answer)}$$

EXAMPLE 13.8

How many liters of 0.750 M $CaSO_4$ solution can be prepared from 50.0 g of $CaSO_4$?

SOLUTION

The given quantity is 50.0 g of $CaSO_4$ and the desired quantity is liters of $CaSO_4$ solution.

$$50.0 \text{ g } CaSO_4 = ? \text{ L } CaSO_4 \text{ solution}$$

The pathway to be used in solving this problem will involve the following steps

$$\text{g } CaSO_4 \longrightarrow \text{moles } CaSO_4 \longrightarrow \text{L } CaSO_4 \text{ solution}$$

The first unit conversion will be accomplished by using the molecular weight of $CaSO_4$ (which must be calculated as it is not given) as a conversion factor. The second unit conversion involves the use of the given molarity as a conversion factor.

$$50.0 \cancel{\text{ g NaCl}} \times \left(\frac{1 \cancel{\text{ mole NaCl}}}{58.5 \cancel{\text{ g NaCl}}}\right) \times \left(\frac{1 \text{ L solution}}{0.750 \cancel{\text{ mole NaCl}}}\right)$$

Cancelling units and doing the arithmetic gives

$$\frac{50.0 \times 1 \times 1}{58.5 \times 0.750} \text{ L solution} = 1.1396011 \text{ L solution} \quad \text{(calculator answer)}$$

$$= 1.14 \text{ L solution} \quad \text{(correct answer)}$$

Molar concentrations do not give information about the amount of *solvent* present. All that is known is that enough solvent is present to give a specific volume of *solution.* The amount of solvent present in a solution of a known molarity can be calculated if the density of the solution is known. Without the density it cannot be calculated.

EXAMPLE 13.9

A 1.350 *M* NaCl solution has a density of 1.054 g/mL at 20°C. How many grams of solvent are present in $65\bar{0}$ mL of this solution?

SOLUTION

To find the grams of solvent present we must first find the grams of solute (NaCl) and the grams of solution. The grams of solvent present is then obtained by calculating the difference.

$$\text{g solvent} = \text{g solution} - \text{g solute}$$

Grams of Solution. The volume of solution is given. Density, used as a conversion factor, will enable us to convert this volume to grams of solution.

$$65\bar{0} \cancel{\text{mL solution}} \times \left(\frac{1.054 \text{ g solution}}{1 \cancel{\text{mL solution}}}\right) = 685.1 \text{ g solution} \quad \text{(calculator answer)}$$

$$= 685 \text{ g solution} \quad \text{(correct answer)}$$

Grams of Solute. We will use the molarity of the solution as a conversion factor in obtaining the grams of solute. The setup for this calculation is similar to that in Example 13.7.

$$650 \cancel{\text{mL solution}} \times \left(\frac{1 \cancel{\text{L solution}}}{1000 \cancel{\text{mL solution}}}\right) \times \left(\frac{1.350 \cancel{\text{moles NaCl}}}{1 \cancel{\text{L solution}}}\right) \times \left(\frac{58.5 \text{ g NaCl}}{1 \cancel{\text{mole NaCl}}}\right)$$

$$= 51.33375 \text{ g NaCl} \quad \text{(calculator answer)}$$

$$= 51.3 \text{ g NaCl} \quad \text{(correct answer)}$$

Grams of Solvent. The grams of solvent will be the difference in mass between the grams of solution and the grams of solute.

$$685 \text{ g solution} - 51.3 \text{ g solute} = 633.7 \text{ g solvent} \quad \text{(calculator answer)}$$

$$= 634 \text{ g solvent} \quad \text{(correct answer)}$$

13.8 Concentration: Molality

molality

Molality is a concentration unit based on a fixed amount of *solvent* and is used in areas where this is a concern. Despite this unit having a name very similar to molarity, molality differs distinctly from molarity; molarity is a unit based on a fixed amount of *solution* rather than a fixed amount of *solvent*. The **molality** *of a solution, abbreviated m, is a ratio giving the number of moles of solute per kilogram of solvent.*

$$\text{Molality } (m) = \frac{\text{moles of solute}}{\text{kg of solvent}}$$

Molality also finds use, in preference to molarity, in experimental situations where changes in temperature are of concern. Molality is a temperature-independent concentration unit, molarity is not. To be temperature independent, a concentration unit cannot involve a volume measurement. Volumes of solutions change (expand or contract) with changes in temperature. A change in temperature thus means a change in concentration, even though the amount of solute remains constant, if a concentration unit has a volume dependency. Volume changes caused by temperature change are usually very, very small; consequently, temperature independence or dependence is a factor in only the most precise experimental measurements.

Careful note should be taken of the fact that the same letter of the alphabet is used as an abbreviation for both molality and molarity—a lowercase m for molality (m) and a capitalized M for molarity (M).

In dilute aqueous solutions molarity and molality are practically identical in numerical value. This results from water having a density of 1.0 g/L. Molarity and molality have significantly different values when the solvent has a density that is not equal to unity or when the solution is concentrated.

EXAMPLE 13.10

Calculate the molality of a solution made by dissolving 10.00 g of $AgNO_3$ (molecular weight = 169.9 amu) in 275 g of H_2O.

SOLUTION

To calculate molality the number of moles of solute and the solvent mass in kg must be known.

In this problem the solvent mass is given, but in g rather than kg. We, thus, need to change the g unit to kg.

$$275 \cancel{\text{g H}_2\text{O}} \times \left(\frac{1 \text{ kg H}_2\text{O}}{1000 \cancel{\text{g H}_2\text{O}}}\right) = 0.275 \text{ kg H}_2\text{O} \quad \text{(calculator and correct answer)}$$

Information about the solute is given in terms of g. We can calculate moles of solute from the given information by using molecular weight as a conversion factor.

$$10.00 \cancel{\text{g AgNO}_3} \times \left(\frac{1 \text{ mole AgNO}_3}{169.9 \cancel{\text{g AgNO}_3}}\right) = 0.05885815 \text{ mole AgNO}_3 \quad \text{(calculator answer)}$$

$$= 0.05886 \text{ mole AgNO}_3 \quad \text{(correct answer)}$$

Substituting moles of solute and kilograms of solvent into the defining equation for molality gives

$$m = \frac{0.05886 \text{ mole AgNO}_3}{0.275 \text{ kg H}_2\text{O}} \quad \begin{array}{l}\leftarrow \text{Moles solute} \\ \leftarrow \text{Kilograms solvent}\end{array}$$

$$= 0.21403636 \frac{\text{mole AgNO}_3}{\text{kg H}_2\text{O}} \quad \text{(calculator answer)}$$

$$= 0.214 \frac{\text{mole AgNO}_3}{\text{kg H}_2\text{O}} \quad \text{(correct answer)}$$

EXAMPLE 13.11

Calculate the number of grams of KOH which must be added to 25.0 g of water to prepare a 0.0100 m solution.

SOLUTION

The given quantity is 25.0 g of H_2O and the desired quantity is grams of KOH.

$$25.0 \text{ g H}_2\text{O} = ? \text{ g KOH}$$

The pathway to be used in solving this problem is

$$\text{g solvent} \longrightarrow \text{kg solvent} \longrightarrow \text{moles solute} \longrightarrow \text{g solute}$$

The molality of the solution, which is given, will serve as a conversion factor to effect the change from kilograms of solvent to moles of solute.

The dimensional-analysis set up for the problem is

$$25.0 \cancel{\text{g H}_2\text{O}} \times \left(\frac{1 \cancel{\text{kg H}_2\text{O}}}{1000 \cancel{\text{g H}_2\text{O}}}\right) \times \left(\frac{0.0100 \cancel{\text{mole KOH}}}{1 \cancel{\text{kg H}_2\text{O}}}\right) \times \left(\frac{56.1 \text{ g KOH}}{1 \cancel{\text{mole KOH}}}\right)$$

Cancelling units and then doing the arithmetic gives

$$\left(\frac{25.0 \times 1 \times 0.0100 \times 56.1}{1000 \times 1 \times 1}\right) \text{g KOH} = 0.014025 \text{ g KOH} \quad \text{(calculator answer)}$$

$$= 0.0140 \text{ g KOH} \quad \text{(correct answer)}$$

Molal concentrations do not give information about the volume of solution present. All that is known is that a specific amount of solute has been dissolved in a definite mass of solvent. The volume of solution can be calculated if the density of the solution is known. Without the density it cannot be calculated.

EXAMPLE 13.12

Calculate the total mass (in g) and the total volume (in mL) of a 1.20 *m* aqueous solution of $Pb(NO_3)_2$ containing 20.0 g of $Pb(NO_3)_2$. The density of the solution is 1.33 g/mL.

SOLUTION

Solution Mass. The total mass of the solution is the sum of the solute and solvent masses.

The solute mass is given as 20.0 g $Pb(NO_3)_2$.

The solvent mass can be calculated from the solute mass by using molality as a conversion factor to convert from moles of $Pb(NO_3)_2$ to kilograms of solvent. The pathway is

$$\text{g Pb(NO}_3)_2 \longrightarrow \text{moles Pb(NO}_3)_2 \longrightarrow \text{kg H}_2\text{O} \longrightarrow \text{g H}_2\text{O}$$

The dimensional-analysis setup for the calculation of solvent mass is

$$20.0\ \cancel{\text{g Pb(NO}_3)_2} \times \left(\frac{1\ \cancel{\text{mole Pb(NO}_3)_2}}{331.2\ \cancel{\text{g Pb(NO}_3)_2}}\right) \times \left(\frac{1\ \cancel{\text{kg H}_2\text{O}}}{1.20\ \cancel{\text{moles Pb(NO}_3)_2}}\right) \times \left(\frac{1000\ \text{g H}_2\text{O}}{1\ \cancel{\text{kg H}_2\text{O}}}\right)$$

Cancelling the units and doing the arithmetic gives

$$\left(\frac{20.0 \times 1 \times 1 \times 1000}{331.2 \times 1.20 \times 1}\right) \text{g H}_2\text{O} = 50.322061\ \text{g H}_2\text{O} \quad \text{(calculator answer)}$$

$$= 50.3\ \text{g H}_2\text{O} \quad \text{(correct answer)}$$

The total solution mass is the sum of the solute and solvent masses.

$$\underset{\text{Solute}}{20.0\ \text{g}} + \underset{\text{Solvent}}{50.3\ \text{g}} = \underset{\text{Solution}}{70.3\ \text{g}} \quad \text{(calculator and correct answer)}$$

Solution Volume. The solution volume can be obtained from the solution mass by using density as a conversion factor.

$$70.3\ \cancel{\text{g solution}} \times \left(\frac{1\ \text{mL solution}}{1.33\ \cancel{\text{g solution}}}\right) = 52.857143\ \text{mL solution} \quad \text{(calculator answer)}$$

$$= 52.9\ \text{mL solution} \quad \text{(correct answer)}$$

13.9 Concentration: Normality

The normality concentration unit is most often encountered in situations that involve the reaction of acids with bases (Sec. 14.7). Even though we will not discuss such reactions until later, normality will be introduced here (in a limited form) so that its relationship to other concentration units may be seen. Normality is closely related to molarity; both are concerned with the amount of solute in a quantity of solution whose volume is expressed in liters. The difference lies in the way the quantity of solute is

normality

expressed. The **normality** *of a solution, abbreviated N, is a ratio giving the number of equivalents of solute per liter of solution.*

$$\text{Normality } (N) = \frac{\text{equivalents of solute}}{\text{liters of solution}}$$

Thus, normality deals with the number of equivalents of solute present while molarity deals with the number of moles of solute present.

What is an equivalent? An equivalent, like a mole, corresponds to a specific quantity. It can be defined in several ways. We will define it for *ionic solutes* here and for acids and bases in Section 14.13.

For ionic solutes, the definition of an equivalent is tied to the behavior of such solutes as they dissolve in a solvent. Upon dissolving in water, ionic solutes dissociate (break up) into their constituent ions. This dissociation process may be represented by equations. For the ionic solutes KNO_3 and $Ca(NO_3)_2$, the dissociation equations are as follows.

$$KNO_3 \longrightarrow K^+ + NO_3^-$$

$$Ca(NO_3)_2 \longrightarrow Ca^{2+} + 2\,NO_3^-$$

Consideration of each of these equations in further detail will give us the insights we need to understand what an equivalent is.

Interpreting the KNO_3 equation in terms of moles leads to the following statement.

$$1 \text{ mole of } KNO_3 \longrightarrow 1 \text{ mole of } K^+ \text{ ions} + 1 \text{ mole of } NO_3^- \text{ ions}$$

Since each K^+ ion carries one positive charge, and each NO_3^- ion one negative charge, we can also write

$$1 \text{ mole of } KNO_3 \longrightarrow 1 \text{ mole of positive charge} + 1 \text{ mole of negative charge}$$

equivalent

We are now ready to define an equivalent. An **equivalent** *of ionic solute is the quantity of the substance that will supply one mole of positive charge or one mole of negative charge upon complete dissociation.* For KNO_3, one mole of compound produces one mole of positive or negative charge. Therefore, for this compound one mole is equal to one equivalent.

$$1 \text{ mole } KNO_3 = 1 \text{ equiv } KNO_3$$

For KNO_3 an equivalent and a mole are equal. Such equality is not the general case, as we will see when $Ca(NO_3)_2$ is the solute.

Interpreting the $Ca(NO_3)_2$ equation in terms of moles leads to the following statement.

$$1 \text{ mole } Ca(NO_3)_2 \longrightarrow 1 \text{ mole of } Ca^{2+} \text{ ions} + 2 \text{ moles of } NO_3^- \text{ ions}$$

The one mole of Ca^{2+} ions produced from the dissociation is equivalent to 2 moles of positive charge, since each ion carries a +2 charge. Two moles of negative charge are also produced; although each NO_3^- carries only a −1 charge, 2 moles of nitrate ions are produced. Thus, we can write

$$1 \text{ mole of } Ca(NO_3)_2 \longrightarrow 2 \text{ moles of positive charge} + 2 \text{ moles of negative charge}$$

Therefore, since an equivalent is associated with one mole of charge, we can write

$$1 \text{ mole } Ca(NO_3)_2 = 2 \text{ equiv } Ca(NO_3)_2$$

or, dividing each side of this equation by two,

$$0.5 \text{ mole } Ca(NO_3)_2 = 1 \text{ equiv } Ca(NO_3)_2$$

equivalent weight

The **equivalent weight** *of a substance is the mass, in grams, of one equivalent of the substance.* If the molecular weight of the substance is known and also the number of equivalents of the substance present in 1 mole, then equivalent weight can be easily determined. Example 13.13 illustrates this type of calculation.

EXAMPLE 13.13

Calculate the equivalent weight of each of the following ionic solutes.

(a) Na_2SO_4 (molecular weight = 142.1 amu)
(b) K_3PO_4 (molecular weight = 212.3 amu)
(c) LiCl (molecular weight = 42.4 amu)

SOLUTION

(a) The dissociation reaction in solution for Na_2SO_4 is

$$Na_2SO_4 \longrightarrow 2\,Na^+ + SO_4^{2-}$$

Thus, 1 mole of Na_2SO_4, upon dissociation, produces 2 moles of positive charge and 2 moles of negative charge. Therefore, since an equivalent is a mole of positive or negative charge, we can write

$$\textbf{1 mole of } Na_2SO_4 = \textbf{2 equiv of } Na_2SO_4$$

With this relationship known, by dimensional analysis, we calculate the equivalent weight as follows.

$$\left(\frac{142.1 \text{ g } Na_2SO_4}{1 \cancel{\text{ mole } Na_2SO_4}}\right) \times \left(\frac{1 \cancel{\text{ mole } Na_2SO_4}}{2 \text{ equiv } Na_2SO_4}\right) = \textbf{71.05 g/equiv} \quad \text{(calculator and correct answer)}$$

Notice how the mole–equivalent relationship was used as a conversion factor in calculating the equivalent weight.

(b) The dissociation reaction in solution for K_3PO_4 is

$$K_3PO_4 \longrightarrow 3\,K^+ + PO_4^{3-}$$

We note that 1 mole of K_3PO_4, upon dissociation, produces 3 moles of positive charge and 3 moles of negative charge. Thus,

$$\textbf{1 mole of } K_3PO_4 = \textbf{3 equiv of } K_3PO_4$$

By dimensional analysis, the equivalent-weight calculation becomes

$$\left(\frac{212.3 \text{ g } K_3PO_4}{1 \cancel{\text{ mole } K_3PO_4}}\right) \times \left(\frac{1 \cancel{\text{ mole } K_3PO_4}}{3 \text{ equiv } K_3PO_4}\right) = 70.766667 \text{ g/equiv} \quad \text{(calculator answer)}$$

$$= \textbf{70.77 g/equiv} \quad \text{(correct answer)}$$

(c) The dissociation reaction in solution for LiCl is

$$LiCl = Li^+ + Cl^-$$

Since the dissociation of one mole of compound produces only one mole of positive and one mole of negative charge, an equivalent and a mole will be the same.

$$1 \text{ mole LiCl} = 1 \text{ equiv LiCl}$$

Consequently, the equivalent weight and molecular weight will be the same. The molecular weight was given as 42.4 amu. Therefore, LiCl has an **equivalent weight** of 42.4 g/equiv.

Once the concept of an equivalent is understood, normality is a very simple concentration unit to work with. Calculational manipulations involving it are identical to those for molarity, except that equivalents are used in place of moles.

EXAMPLE 13.14

Calculate the normality of the solution that results when 4.00 g of $Al(NO_3)_3$ (molecular weight = 213.0 amu) is dissolved in enough water to give 250.0 mL of solution.

SOLUTION

To calculate normality we need to know the number of equivalents of solute and the solution volume in liters.

To obtain equivalents of solute, we must first consider how the solute dissociates in solution. The dissociation reaction is

$$Al(NO_3)_3 \longrightarrow Al^{3+} + 3\ NO_3^-$$

Thus, 3 moles of positive charge and 3 moles of negative charge result from the dissociation of 1 mole of $Al(NO_3)_3$. This means that

$$1 \text{ mole } Al(NO_3)_3 = 3 \text{ equiv of } Al(NO_3)_3$$

The number of equivalents in 4.00 g of solute is obtained by converting the gram units to moles and then using our mole–equivalent relationship as a conversion factor to obtain equivalents. The dimensional-analysis setup for this sequence of operations is

$$4.00\ \cancel{\text{g Al(NO}_3)_3} \times \left(\frac{1\ \cancel{\text{mole Al(NO}_3)_3}}{213.0\ \cancel{\text{g Al(NO}_3)_3}}\right) \times \left(\frac{3 \text{ equiv Al(NO}_3)_3}{1\ \cancel{\text{mole Al(NO}_3)_3}}\right)$$

$$= 0.05633803 \text{ equiv Al(NO}_3)_3 \quad \text{(calculator answer)}$$

$$= 0.0563 \text{ equiv Al(NO}_3)_3 \quad \text{(correct answer)}$$

The solution volume is given, but in the wrong units; mL must be converted to L.

$$250.0\ \cancel{\text{mL solution}} \times \left(\frac{1 \text{ L solution}}{1000\ \cancel{\text{mL solution}}}\right) = 0.25 \text{ L solution} \quad \text{(calculator answer)}$$

$$= 0.2500 \text{ L solution} \quad \text{(correct answer)}$$

Both quantities called for in the defining equation for normality

$$N = \frac{\text{equiv solute}}{\text{L solution}}$$

are now known. Substituting these quantities into the equation gives

$$N = \frac{0.0563 \text{ equiv Al(NO}_3)_3}{0.2500 \text{ L solution}} = 0.2252 \frac{\text{equiv Al(NO}_3)_3}{\text{L solution}} \quad \text{(calculator answer)}$$

$$= \mathbf{0.225 \text{ equiv/L}} \quad \textbf{(correct answer)}$$

13.10 Dilution

A common problem encountered when working with solutions in the laboratory is that of diluting a solution of known concentration (usually called a stock solution) to a lower concentration. **Dilution** *is the process in which more solvent is added to a solution in order to lower its concentration.* Dilution always lowers the concentration of a solution. The same amount of solute is present but it is now distributed in a larger amount of solvent (the original solvent plus the added solvent).

dilution

Since laboratory solutions are almost always liquids, dilution is normally a volumetric procedure. Most often, a solution of a specific molarity must be prepared by adding a predetermined volume of solvent to a specific volume of stock solution.

With molar concentration units, a very simple mathematical relationship exists between the volumes and molarities of the diluted and stock solutions. This relationship is derived from the fact that the same amount of solute is present in both solutions; only solvent is added in a dilution procedure.

$$\text{Moles solute}_{\text{stock solution}} = \text{moles solute}_{\text{diluted solution}}$$

The number of moles of solute in either solution is given by the expression

$$\text{Moles solute} = \text{molarity } (M) \times \text{liters of solution } (V)$$

(This equation is just a rearrangement of the defining equation for molarity to isolate moles of solute on the left side.) Substitution of this second expression into the first one gives the equation

$$M_s \times V_s = M_d \times V_d$$

In this equation M_s and V_s are the molarity and volume of the stock solution (the solution to be diluted) and M_d and V_d the molarity and volume of the solution resulting from the dilution. Because volume appears on both sides of the equation, any volume unit, not just liters, may be used as long as it is the same on both sides of the equation. Again, the validity of this equation is based on there being no change in the amount of solute present.

For dilution problems in which the concentration unit is normality, the similar equation

$$N_s \times V_s = N_d \times V_d$$

applies.

EXAMPLE 13.15

What is the molarity of the solution prepared by diluting 12.0 mL of 0.405 M NaCl to a final volume of 80.0 mL?

SOLUTION

Three of the four variables in the equation

$$M_s \times V_s = M_d \times V_d$$

are known.

$$M_s = 0.405\,M \qquad M_d = ?\,M$$

$$V_s = 12.0 \text{ mL} \qquad V_d = 80.0 \text{ mL}$$

Rearranging the equation to isolate M_d on the left side and substituting the known variables into it gives

$$M_d = M_s \times \frac{V_s}{V_d}$$

$$= 0.405\,M \times \left(\frac{12.0 \cancel{\text{mL}}}{80.0 \cancel{\text{mL}}}\right) = 0.06075\,M \quad \text{(calculator answer)}$$

$$= 0.0608\,M \quad \text{(correct answer)}$$

Thus, the diluted solution's concentration is 0.0608 M.

EXAMPLE 13.16

How much solvent must be added to 100.0 mL of 1.50 M $AgNO_3$ solution in order to decrease its concentration to 0.350 M?

SOLUTION

The volume of solvent added is equal to the difference between the final and initial volumes. The initial volume is known. The final volume can be calculated using the equation

$$M_s \times V_s = M_d \times V_d$$

Once the final volume is known, the difference between the two volumes may be obtained.

Substituting the known quantities into the dilution equation, rearranged to isolate V_d on the left side, gives

$$V_d = V_s \times \frac{M_s}{M_d}$$

$$V_d = 100.0 \text{ mL} \times \left(\frac{1.50 \cancel{M}}{0.350 \cancel{M}}\right) = 428.57143 \text{ mL} \quad \text{(calculator answer)}$$

$$= 429 \text{ mL} \quad \text{(correct answer)}$$

The solvent added is

$$V_d - V_s = (429 - 100.0) \text{ mL} = 329 \text{ mL} \quad \text{(calculator and correct answer)}$$

When two "like" solutions, that is, solutions that contain the same solute and the same solvent, of differing known molarities and volumes are mixed together, the molarity of the newly formed solution can be calculated by the same principles that apply in a simple dilution problem.

Again, the key concept involves the amount of solute present; it is constant. The sum of the amount of solute present in each of the individual solutions prior to mixing is the same as the amount of solute present in the solution obtained from the mixing. No solute is lost or gained in the mixing process. Thus, we can write

$$\text{moles solute}_{\text{first solution}} + \text{moles solute}_{\text{second solution}} = \text{moles solute}_{\text{combined solution}}$$

Substituting the expression $(M \times V)$ for moles solute in this equation gives

$$(M_1 \times V_1) + (M_2 \times V_2) = (M_3 \times V_3)$$

where the subscripts 1 and 2 denote the solutions to be mixed and the subscript 3 is the solution resulting from the mixing. Again, this expression is valid only when the solutions that are mixed are "like" solutions.

EXAMPLE 13.17

What is the molarity of the solution obtained by mixing 50.0 mL of 3.75 *M* NaCl solution with 160.0 mL of 1.75 *M* NaCl solution?

SOLUTION

Five of the six variables in the equation

$$(M_1 \times V_1) + (M_2 \times V_2) = (M_3 \times V_3)$$

are known.

$$M_1 = 3.75\ M \qquad V_1 = 50.0\ \text{mL}$$

$$M_2 = 1.75\ M \qquad V_2 = 160.0\ \text{mL}$$

$$M_3 = ?\ M \qquad V_3 = 210.0\ \text{mL}$$

Note that in the mixing process we consider the volumes of the solutions to be additive; that is,

$$V_3 = V_1 + V_2$$

This is a valid assumption for "like" solutions.

Solving our equation for M_3 and then substituting the known quantities into it gives

$$M_3 = \frac{(M_1 \times V_1) + (M_2 \times V_2)}{V_3}$$

$$= \frac{(3.75\ M \times 50.0\ \cancel{\text{mL}}) + (1.75\ M \times 160.0\ \cancel{\text{mL}})}{(210.0\ \cancel{\text{mL}})}$$

$$= 2.2261905\ M \quad \text{(calculator answer)}$$

$$= 2.23\ M \quad \text{(correct answer)}$$

13.11 Molarity and Chemical Equations

In Section 8.6 we were introduced to a general problem-solving procedure useful in setting up problems that involve chemical equations. With this procedure, if information is given about one reactant or product in a chemical reaction (number of grams, moles, or particles), similar information can easily be obtained for any other reactant or product.

In Section 12.11 this procedure was refined to allow us to do mass–volume or volume–mass calculations for reactions where at least one reactant or product is a gas.

In this section we further refine our problem-solving procedure in order to deal efficiently with reactions that occur in aqueous solution. Of primary importance to us in this new area of problem solving will be *solution volume.* In most situations, solution volume is more conveniently determined than solution mass.

When solution concentrations are expressed in terms of molarity, a direct relationship exists between solution volume (in liters) and moles of solute present. The definition of molarity itself gives the relationship; molarity is the ratio of moles of solute to volume (in liters) of solution. Thus, molarity is the connection that links volume of solution to the other common problem-solving parameters, such as moles and grams. Figure 13-4 shows diagrammatically the place that volume of solution occupies, relative to other parameters, in the overall scheme of chemical equation-based problem solving. This diagram is a simple expansion of Figure 12-7 and is used in the same way as was that diagram.

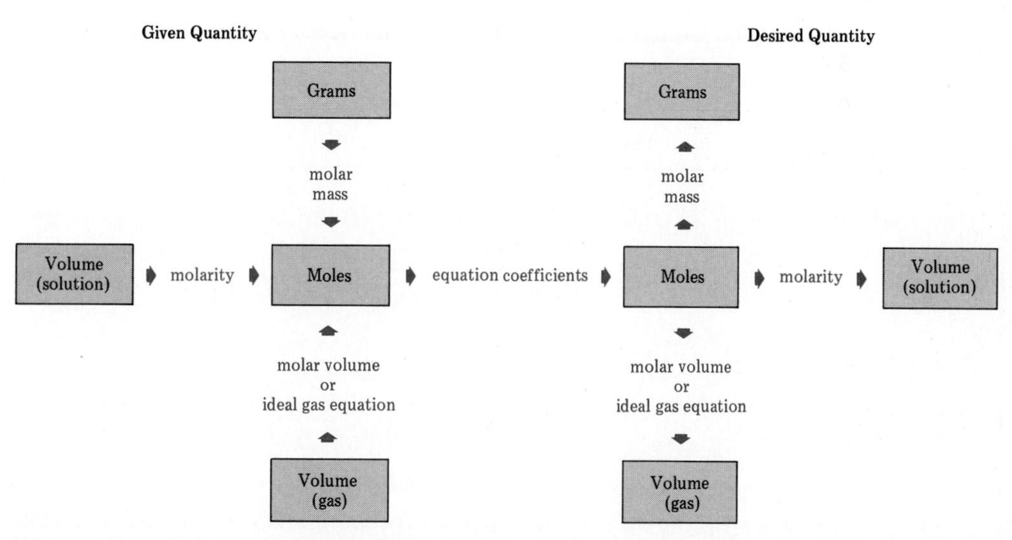

Figure 13-4. Conversion factor relationships needed to solve problems involving chemical reactions occurring in aqueous solution.

EXAMPLE 13.18

What volume, in liters, of a 3.40 *M* KOH solution is needed to react completely with 0.100 L of a 6.72 *M* H_2SO_4 solution according to the equation

$$2\ KOH(aq) + H_2SO_4(aq) \longrightarrow K_2SO_4(aq) + 2\ H_2O(l)$$

SOLUTION

Step 1. The given quantity is 0.100 L of H_2SO_4 solution and the desired quantity is L of KOH solution.

$$0.100\ L\ H_2SO_4 = ?\ L\ KOH$$

Step 2. This problem is a (volume of solution)-to-(volume of solution) type problem. The pathway used in solving it, in terms of Figure 13-4, is

Given → **Desired**

Volume H_2SO_4 → molarity → Moles H_2SO_4 → equation coefficients → Moles KOH → molarity → Volume KOH

Step 3. The dimensional analysis setup for the calculation is

$$0.100\ \cancel{L\ H_2SO_4} \times \left(\frac{6.72\ \cancel{moles\ H_2SO_4}}{1\ \cancel{L\ H_2SO_4}}\right) \times \left(\frac{2\ \cancel{moles\ KOH}}{1\ \cancel{mole\ H_2SO_4}}\right) \times \left(\frac{1\ L\ KOH}{3.40\ \cancel{moles\ KOH}}\right)$$

Step 4. The solution, obtained from combining all of the numerical factors, is

$$\left(\frac{0.100 \times 6.72 \times 2 \times 1}{1 \times 1 \times 3.40}\right) L\ KOH = 0.39529412\ L\ KOH \quad \text{(calculator answer)}$$

$$= 0.395\ L\ KOH \quad \text{(correct answer)}$$

EXAMPLE 13.19

How many grams of AgCl can be produced from the reaction of 1.00 L of 0.375 *M* $AgNO_3$ solution with an excess of 1.00 *M* NaCl solution according to the equation

$$AgNO_3(aq) + NaCl(aq) \longrightarrow AgCl(s) + NaNO_3(aq)$$

SOLUTION

Step 1. The given quantity is 1.00 L of $AgNO_3$ solution and the desired quantity is grams of AgCl.

$$1.00\ L\ AgNO_3 = ?\ g\ AgCl$$

Step 2. This is a (volume of solution)-to-gram problem. The pathway, in terms of Figure 13-4, is

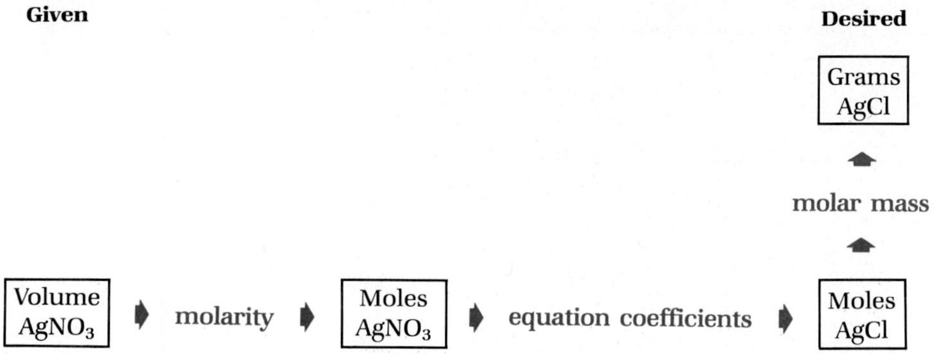

Given **Desired**

Grams AgCl
↑
molar mass
↑
Volume $AgNO_3$ → molarity → Moles $AgNO_3$ → equation coefficients → Moles AgCl

Step 3. The dimensional-analysis setup for the calculation is

$$1.00\ \cancel{\text{L } AgNO_3} \times \left(\frac{0.375\ \cancel{\text{mole } AgNO_3}}{1\ \cancel{\text{L } AgNO_3}}\right) \times \left(\frac{1\ \cancel{\text{mole AgCl}}}{1\ \cancel{\text{mole } AgNO_3}}\right) \times \left(\frac{143\text{ g AgCl}}{1\ \cancel{\text{mole AgCl}}}\right)$$

Step 4. The solution, obtained from combining all of the numerical factors, is

$$\left(\frac{1.00 \times 0.375 \times 1 \times 143}{1 \times 1 \times 1}\right)\text{ g AgCl} = 53.625\text{ g AgCl} \quad \text{(calculator answer)}$$

$$= 53.6\text{ g AgCl} \quad \text{(correct answer)}$$

Note that the concentration of NaCl solution, given as 1.00 *M* in the problem statement, did not enter into the calculation. This is because the NaCl solution is present in excess; we know that we have enough of it. If a specific volume of NaCl solution had been given in the problem statement, we would have had to determine the limiting reactant (NaCl or $AgNO_3$) as the first step in working the problem. The concept of a limiting reactant was discussed in Section 8.7.

EXAMPLE 13.20

What volume, in liters, of NO gas measured at STP can be produced from 1.75 L of 0.550 *M* HNO_3 solution according to the reaction

$$2\ HNO_3(aq) + 3\ H_2S(aq) \longrightarrow 2\ NO(g) + 3\ S(s) + 4\ H_2O(l)$$

SOLUTION

Step 1. The given quantity is 1.75 L of HNO_3 solution and the desired quantity is liters of NO gas at STP.

$$1.75\text{ L } HNO_3 = ?\text{ L NO (at STP)}$$

Step 2. This is a (volume of solution)-to-(volume of gas) problem. The pathway, in terms of Figure 13-4, for solving this problem is

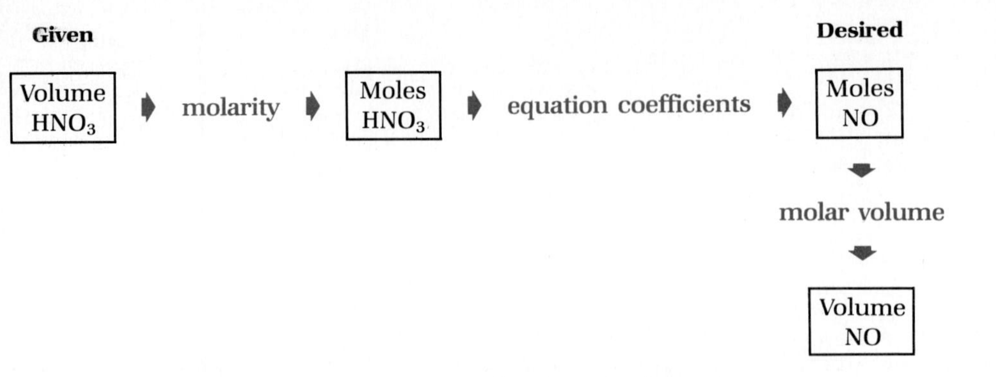

Step 3. The dimensional-analysis setup for the calculation is

$$1.75\ \text{L HNO}_3 \times \left(\frac{0.550\ \text{mole HNO}_3}{1\ \text{L HNO}_3}\right) \times \left(\frac{2\ \text{moles NO}}{2\ \text{moles HNO}_3}\right) \times \left(\frac{22.4\ \text{L NO}}{1\ \text{mole NO}}\right)$$

Step 4. The solution, obtained by combining all of the numerical factors, is

$$\left(\frac{1.75 \times 0.550 \times 2 \times 22.4}{1 \times 2 \times 1}\right)\ \text{L NO} = 21.56\ \text{L NO} \quad \text{(calculator answer)}$$

$$= 21.6\ \text{L NO} \quad \text{(correct answer)}$$

Learning Objectives

After completing this chapter you should be able to

- Define the terms solution, solute, and solvent (Sec. 13.1).
- List the nine types of two-component solutions (Sec. 13.1).
- Define and distinguish between the terms saturated and unsaturated, dilute and concentrated, and miscible and immiscible (Sec. 13.2).
- Define the term solubility and know what is qualitatively meant by the terms insoluble, slightly soluble, soluble, and very soluble (Sec. 13.2).
- Describe the solution process, at a molecular level, as an ionic solute dissolves in water (Sec. 13.3).
- Apply the solubility rule "likes dissolve likes" and use the solubility guidelines for ionic solutes in water (Sec. 13.4).
- Define the term concentration and distinguish between it and the term solubility (Sec. 13.5).
- Give the defining equation for each of the concentration units percentage by weight, percentage by volume, weight–volume percent, molarity, molality, and normality, and use these defining equations to calculate any one of the three quantities within an equation given the other two (Secs. 13.6 through 13.9).
- Calculate the molarity of a solution obtained by diluting a stock solution of known molar concentration (Sec. 13.10).
- Calculate, given a balanced equation, the volume and molarity of one reactant, and the volume (or molarity) of a second reactant, the molarity (or volume) of the second reactant (Sec. 13.11).
- Calculate, given a balanced chemical equation and the volume and molarity of one reactant combining with an excess of other reactants, the mass or volume (for gases) of any product that is produced (Sec. 13.11).

Terms and Concepts for Review

The new terms or concepts defined in this chapter are

solution (Sec. 13.1)
solvent (Sec. 13.1)
solute (Sec. 13.1)
solubility (Sec. 13.2)
saturated solution (Sec. 13.2)
unsaturated solution (Sec. 13.2)
dilute solution (Sec. 13.2)
concentrated solution (Sec. 13.2)
miscible (Sec. 13.2)
partially miscible (Sec. 13.2)
immiscible (Sec. 13.2)
aqueous solution (Sec. 13.2)
concentration (Sec. 13.5)

percent by weight (Sec. 13.6)
percent by volume (Sec. 13.6)
weight–volume percent (Sec. 13.6)
molarity (Sec. 13.7)
molality (Sec. 13.8)
normality (Sec. 13.9)
equivalent (Sec. 13.9)
equivalent weight (Sec. 13.9)
dilution (Sec. 13.10)

Questions and Problems

Solution Terminology

13-1 Indicate which substance is the *solvent* in each of the following solutions.
- **a.** a solution containing 10.0 g of NaCl and 100.0 g of water
- **b.** a solution containing 20.0 mL of ethyl alcohol and 15.0 mL of water
- **c.** a solution containing 0.50 g of $AgNO_3$ and 10 mL of water

13-2 Use the terms *soluble, insoluble,* or *immiscible* to describe the behavior of the following pairs of substances. More than one term may apply to a given case.
- **a.** 75 mL of water and 1 g of table sugar are shaken together. The resulting solution is clear and colorless.
- **b.** 10 mL of water and 0.1 g of fine white sand are shaken together. The resulting mixture is very cloudy and the sand rapidly settles to the bottom of the container.
- **c.** 25 mL of water and 10 mL of mineral oil are shaken together. The mixture is cloudy and gradually separates into two layers.
- **d.** 100 mL of water and 100 mL of ethyl alcohol are shaken together. A clear colorless solution results that with time shows no separation into layers.

13-3 Use Table 13-2 to determine whether each of the following solutions is *saturated* or *unsaturated.*
- **a.** 1.94 g of $PbBr_2$ in 100 g of H_2O at 50°C
- **b.** 34.0 g of NaCl in 100 g of H_2O at 0°C
- **c.** 75.4 g of $CuSO_4$ in 200 g of H_2O at 100°C
- **d.** 0.540 g of Ag_2SO_4 in 50 g of H_2O at 50°C

13-4 Using the solubilities given in Table 13-2, characterize each of the following solids as being *insoluble, slightly soluble, soluble,* or *very soluble* in water at the indicated temperature.
- **a.** lead(II) bromide (at 50°C)
- **b.** cesium chloride (at 0°C)
- **c.** silver nitrate (at 100°C)
- **d.** silver sulfate (at 0°C)

13-5 Based on the solubilities in Table 13-2, characterize each of the following solutions as being a *dilute* or a *concentrated* solution.
- **a.** 0.20 g of $CuSO_4$ dissolved in 100 g of H_2O at 100°C
- **b.** 1.50 g of $PbBr_2$ dissolved in 100 g of H_2O at 50°C
- **c.** 60 g of $AgNO_3$ dissolved in 100 g of H_2O at 50°C
- **d.** 0.50 g of Ag_2SO_4 dissolved in 100 g of H_2O at 0°C

Solubility Rules

13-6 Predict whether the following solutes are very soluble or slightly soluble in water.
- **a.** O_2 (a nonpolar gas)
- **b.** CH_3OH (a polar liquid)
- **c.** CBr_4 (a nonpolar liquid)
- **d.** NaCl (an ionic solid)
- **e.** AgCl (an ionic solid)

13-7 Using Table 13-4, predict whether each of the following ionic compounds is soluble or insoluble in water.
- **a.** KBr
- **b.** $PbCl_2$
- **c.** $Ca(NO_3)_2$
- **d.** $Be(C_2H_3O_2)_2$
- **e.** $AlPO_4$
- **f.** Ag_2CO_3
- **g.** $BaSO_4$
- **h.** NH_4I

Weight Percent

13-8 Calculate the weight percent of solute in the following solutions.
- **a.** 7.37 g of NaCl in 95.0 g of H_2O
- **b.** 3.73 g of KCl in 53.0 g of solution
- **c.** 10.3 g of KBr in 131 g of H_2O
- **d.** 0.0513 mole of Na_2SO_4 in 100.0 g of H_2O

13-9 Calculate the mass, in grams, of solute needed to prepare each of the following amounts of solution.
- **a.** 300 g of a 5.00% by weight NaI solution
- **b.** 37.5 g of a 10.0% by weight glucose ($C_6H_{12}O_6$) solution
- **c.** a 7.0% by weight KNO_3 solution containing 150 g of H_2O
- **d.** 50.0 mL of a 20.0% by weight NaCl solution whose density is 1.15 g/mL

13-10 How many grams of water must be added to 50.0 g of glucose ($C_6H_{12}O_6$) in order to prepare a 5.5% by weight glucose solution?

13-11 Using Table 13-2, calculate the weight percent of solute in each of the following saturated solutions.
- **a.** NaCl at 0°C
- **b.** NaCl at 100°C
- **c.** $AgNO_3$ at 0°C
- **d.** $AgNO_3$ at 100°C

Volume Percent

13-12 Calculate the volume percent of solute in the following solutions.
- **a.** 30.0 mL of methyl alcohol (CH_3OH) in enough water to give 500.0 mL of solution
- **b.** 3.75 mL of bromine in enough carbon tetrachloride (CCl_4) to produce 96.0 mL of solution
- **c.** 60.0 mL of ethylene glycol ($C_2H_6O_2$) in enough water to produce 354 mL of solution

d. 450 mL of ethyl alcohol (C_2H_5OH) in enough water to produce 3.75 L of solution

13-13 The final volume of a solution resulting from the addition of 50.0 mL of ethyl alcohol (C_2H_5OH) to 50.0 mL of water is 96.5 mL. What is the volume percent ethyl alcohol in this solution?

13-14 How much methyl alcohol (CH_3OH), in mL, is needed to make 450 mL of a 25.0% by volume solution of methyl alcohol in water?

13-15 What volume of ethylene glycol ($C_2H_6O_2$), in quarts, is contained in 5.00 qt of a 45.0% by volume solution of ethylene glycol in water?

Weight–Volume Percent

13-16 Calculate the weight–volume percent of solute in each of the following solutions.

a. 5.0 g of magnesium chloride ($MgCl_2$) added to enough water to give 250 mL of solution

b. 85.0 g of magnesium chloride ($MgCl_2$) added to enough water to give 1.00 L of solution

c. 1.00 mole of magnesium chloride ($MgCl_2$) added to enough water to give 100 mL of solution

d. 30.0 g of magnesium chloride ($MgCl_2$) added to 70.0 g of water to give a solution with a density of 1.27 g/mL

13-17 How many grams of glucose ($C_6H_{12}O_6$) are there in 50.0 mL of a 5.0% by weight–volume glucose solution?

Molarity

13-18 Calculate the molarity of the following solutions.

a. 2.0 moles of sodium chloride (NaCl) in 0.50 L of solution

b. 13.7 g of potassium nitrate (KNO_3) in 80.0 mL of solution

c. 53.0 g of sucrose ($C_{12}H_{22}O_{11}$) in 1250 mL of solution

d. 0.0010 mole of baking soda ($NaHCO_3$) in 3.03 mL of solution

13-19 Calculate the number of moles of solute in each of the following solutions.

a. 3.00 L of 6.0 *M* HNO_3 solution

b. 50.0 mL of 0.350 *M* NaCl solution

c. 375 mL of 1.75 *M* $BaCl_2$ solution

d. 150.0 g of a 6.00 *M* H_2SO_4 solution with a density of 1.3 g/mL.

13-20 Calculate the number of milliliters of solution required to provide the following.

a. 1.00 g of sodium chloride (NaCl) from a 0.100 *M* sodium chloride solution

b. 200.0 g of glucose ($C_6H_{12}O_6$) from a 5.6 *M* glucose solution

c. 3.67 moles of silver nitrate ($AgNO_3$) from a 0.0030 *M* silver nitrate solution

d. 0.0010 mole of sucrose ($C_{12}H_{22}O_{11}$) from a 10.0 *M* sucrose solution

13-21 How many liters of 0.300 *M* potassium bromide (KBr) solution can be prepared from 100.0 g of KBr?

13-22 If the sodium chloride obtained from the evaporation to dryness of 900 mL of 0.50 *M* sodium chloride solution is added to enough water to give 5.00 L of solution, what is the molarity of the new solution?

13-23 A 0.92 weight–volume percent solution of NaCl is administered intravenously in the treatment of certain illnesses. What is the molarity of such a solution?

13-24 Calculate the molarity of a 36.0% by weight solution of HCl that has a density of 1.18 g/mL.

13-25 A 2.999 *M* $NaNO_3$ solution has a density of 1.1609 g/mL at 20°C. How many grams of solvent are present in 1.375 L of this solution?

Molality

13-26 Calculate the molality of each of the following solutions.

a. 165 g of ethylene glycol ($C_2H_6O_2$) in 1.35 kg of water

b. 3.75 moles of sodium sulfate (Na_2SO_4) in 500 g of water

c. 0.0350 g of ammonium chloride (NH_4Cl) in 12.0 g of water

d. 50.0 g of calcium chloride ($CaCl_2$) in enough water to give 543 mL of solution with a density of 1.10 g/mL

13-27 Calculate the number of grams of solute necessary to prepare the following solutions.

a. 100.0 g of 0.353 *m* aqueous potassium nitrate (KNO_3) solution

b. 50.0 g of 1.50 *m* aqueous sodium chloride (NaCl) solution

c. 10.0 g of 0.0034 *m* aqueous potassium iodide (KI) solution

d. 46.78 g of 3.444 *m* aqueous iron(III) chloride ($FeCl_3$) solution

13-28 How many grams of water must be added to 100.0 g of sucrose ($C_{12}H_{22}O_{11}$) to prepare a 0.350 *m* solution?

13-29 Calculate the total mass (in grams) and the total volume (in mL) of a 2.9 *m* aqueous solution of potassium chromate (K_2CrO_4) containing 35.0 g of K_2CrO_4. The density of the solution is 1.40 g/mL.

13-30 Calculate the molality of a 14.0% by weight HCl solution.

13-31 Calculate the molality of a 2.00 *M* H_3PO_4 whose density is 1.120 g/mL.

Normality

13-32 Calculate the equivalent weight of the following ionic compounds.

a. $CaCl_2$ **b.** KBr **c.** $(NH_4)_2SO_4$
d. NH_4NO_3 **e.** Na_2S **f.** AlP
g. $Fe_3(PO_4)_2$ **h.** Be_2C

13-33 Calculate the normality of each of the following solutions.
- **a.** 1.85 equiv of magnesium chloride ($MgCl_2$) in 1.00 L of solution
- **b.** 1.85 moles of magnesium chloride ($MgCl_2$) in 1.00 L of solution
- **c.** 1.85 g of magnesium chloride ($MgCl_2$) in 1.00 L of solution
- **d.** 1.85 g of magnesium chloride ($MgCl_2$) in 0.674 L of solution

13-34 Calculate the normality of the following solutions.
- **a.** 0.333 *M* NaCl **b.** 1.45 *M* Li_2SO_4
- **c.** 1.56 *M* K_3PO_4 **d.** 0.033 *M* $Al(NO_3)_3$

13-35 How many grams of $Ca(C_2H_3O_2)_2$ are there in $8\overline{0}0$ mL of 2.00 *N* $Ca(C_2H_3O_2)_2$ solution?

13-36 How many mL of 0.0400 *N* sodium sulfate (Na_2SO_4) solution are required to provide 1.00 g of Na_2SO_4?

Dilution

13-37 Determine the volume to which each of the following solutions should be diluted to produce the solution indicated.
- **a.** a 5.0 *M* HCl solution from 60.0 mL of 8.0 *M* HCl
- **b.** a 0.57 *M* NaCl solution from 80.0 mL of 0.60 *M* NaCl
- **c.** a 0.00500 *M* $CaCl_2$ solution from $5\overline{0}0$ mL of 12.0 *M* $CaCl_2$
- **d.** a 1.5 *M* KI solution from 1.3 L of 6.0 *M* KI

13-38 In each of the following, how many mL of water should be added to obtain a solution that has a concentration of 0.100 *M*?
- **a.** 50.0 mL of 3.00 *M* NaCl
- **b.** 2.00 mL of 1.00 *M* NaCl
- **c.** 1.45 L of 6.00 *M* NaCl
- **d.** 75.0 mL of 3.76*M* NaCl

13-39 What will be the final concentration of each of the following if the solution is diluted with 20.0 mL of water?
- **a.** 25.0 mL of 6.0 *M* Na_2SO_4
- **b.** $1\overline{0}0$ mL of 3.0 *M* Na_2SO_4
- **c.** 0.155 L of 10.0 *M* Na_2SO_4
- **d.** 2.00 mL of 0.100 *M* Na_2SO_4

13-40 What would be the molarity of a solution made by mixing 235 mL of 0.350 *M* KBr with 351 mL of 6.53 *M* KBr?

Molarity and Chemical Equations

13-41 What volume, in L, of 1.00 *M* H_2S is needed to react completely with 0.500 L of 4.00 *M* HNO_3 solution according to the equation

$$2\ HNO_3(aq) + 3\ H_2S(aq) = 2\ NO(g) + 3\ S(s) + 4\ H_2O(l)$$

13-42 What volume of 0.50 *M* HBr is required to react with 18.0 g of zinc according to the equation

$$2\ HBr(aq) + Zn(s) \longrightarrow ZnBr_2(aq) + H_2(g)$$

13-43 How many grams of $Ca_3(PO_4)_2$ can be produced from the reaction of 2.50 L of 0.200 *M* $CaCl_2$ solution with an excess of 0.750 *M* H_3PO_4 according to the equation

$$3\ CaCl_2(aq) + 2\ H_3PO_4(aq) \longrightarrow Ca_3(PO_4)_2(s) + 6\ HCl(aq)$$

13-44 How many grams of AgCl can be produced from the reaction of 30.0 mL of 12.0 *M* NaCl solution with an excess of 10.0 *M* $AgNO_3$ solution according to the equation

$$AgNO_3(aq) + NaCl(aq) \longrightarrow AgCl(s) + NaNO_3(aq)$$

13-45 What is the molarity of a 37.5 mL sample of H_2SO_4 solution that will completely react with 23.7 mL of 0.100 *M* NaOH solution according to the following equation

$$H_2SO_4(aq) + 2\ NaOH(aq) = Na_2SO_4(aq) + 2\ H_2O(l)$$

13-46 What volume, in liters, of H_2 gas measured at STP can be produced from 50.0 mL of 6.0 *M* H_2SO_4 solution according to the reaction

$$Ni(s) + H_2SO_4(aq) \longrightarrow NiSO_4(aq) + H_2(g)$$

13-47 What is the molarity of a 1.75 L solution of $Ca(OH)_2$ that would completely react with 2.00 L of CO_2 gas measured at STP according to the reaction

$$CO_2(g) + Ca(OH)_2(aq) \longrightarrow CaCO_3(s) + H_2O(l)$$

13-48 What mass, in grams, of $BaCrO_4$ would be produced by mixing $35\overline{0}$ mL of 3.25 *M* $BaCl_2$ solution with $45\overline{0}$ mL of 4.50 *M* K_2CrO_4 solution. The two solutions react according to the following equation

$$BaCl_2(aq) + K_2CrO_4(aq) \longrightarrow BaCrO_4(s) + 2\ KCl(aq)$$

14 Acids, Bases, and Salts

14.1 Acid–Base Definitions

Acids and bases are among the most common and important compounds known. Aqueous solutions of such compounds are key materials in both biological systems and chemical industrial processes.

Historically, as early as the seventeenth century, acids and bases were recognized as important groups of compounds. Such early recognition was based on what the substances did rather than on their chemical composition. It was not until 1884 that acids and bases were defined in terms of chemical composition.

Early known facts about acids include

1. Acids, when dissolved in water, have a sour taste. (The name acid comes from the Latin word *acidus,* which means "sour.")
2. Acids cause the dye litmus to change from a blue to a red color. (Litmus is a naturally occurring vegetable dye obtained from lichens.)
3. When certain metals, such as zinc and iron, are placed in acids, they dissolve with the liberation of hydrogen gas.

Early known characteristics of bases include

1. Water solutions of bases feel slippery or soapy to the touch and have a bitter taste.
2. Bases cause the dye litmus to change from a red to a blue color.
3. When certain greases are placed in a base solution, they dissolve.

Arrhenius acid

Arrhenius base

In 1884, the Swedish chemist Svante August Arrhenius (1859–1927) proposed that acids and bases be defined in terms of the species they form upon dissolution in water. His definitions, now called the Arrhenius definitions, are still in use today. An **Arrhenius acid** *is a substance that releases hydrogen ions* (H^+) *in aqueous solution.* An **Arrhenius base** *is a substance that releases hydroxide ions* (OH^-) *in aqueous solution.*

Some common examples of acids, according to the Arrhenius definitions, are the substances HNO_3 and HCl.

$$HNO_3(l) \xrightarrow{H_2O} H^+(aq) + NO_3^-(aq)$$

$$HCl(g) \xrightarrow{H_2O} H^+(aq) + Cl^-(aq)$$

Arrhenius acids when in the pure state (not in solution) are covalent compounds, that is, they do not contain H^+ ions. These ions are formed, through a chemical reaction, when the acid is mixed with water. This chemical reaction, between water and the acid molecules, results in removal of H^+ ions from acid molecules.

Two common examples of Arrhenius bases are NaOH and KOH.

$$NaOH(s) \xrightarrow{H_2O} Na^+(aq) + OH^-(aq)$$

$$KOH(s) \xrightarrow{H_2O} K^+(aq) + OH^-(aq)$$

Arrhenius bases are usually ionic compounds in the pure state, in direct contrast to acids. When such compounds dissolve in water, the ions separate to yield the OH^- ions.

The Arrhenius definitions for acids and bases adequately explain, at an introductory level, the behaviors, in aqueous solution, of acids and bases. These definitions are the only definitions found in many elementary textbooks. However, the definitions do have two drawbacks: (1) They are adequate only in the case of aqueous solution, and (2) the identity of the acidic species present in aqueous solutions is oversimplified. Concerning this latter point, unknown to Arrhenius, *free* hydrogen ions (H^+) cannot exist in water. The attraction between a hydrogen ion and polar water molecules is sufficiently strong to bond the hydrogen ion with a water molecule to form a hydronium ion (H_3O^+).*

$$H^+ + :\ddot{O}\!-\!H \longrightarrow \left[H:\ddot{O}\!-\!H\right]^+$$
(each oxygen also bonded to an H below)

Hydronium ion

The bond holding the hydrogen ion to the water molecule is a coordinate covalent bond (Sec. 10.8) since both electrons are furnished by the oxygen atom.

Because of the limitations inherent in the Arrhenius acid–base definitions other more general definitions for these substances exist. These alternate definitions are not contradictory to the Arrhenius ones but rather are complimentary; they expand upon them. The most used of the newer acid–base definitions is the Brønsted–Lowry definition, proposed in 1923, independently and almost simultaneously by the Danish chemist Johannes Nicholas Brønsted (1879–1947) and the English chemist Thomas Martin Lowry (1874–1936). Occasionally it will be advantageous to discuss acid–base chemistry using the concepts of Brønsted and Lowry; however, the Arrhenius acid–base definitions will be adequate for most of our discussions.

Brønsted–Lowry acid

Brønsted–Lowry base

A **Brønsted–Lowry acid** *is any substance that can donate a proton (H^+) to some other substance.* (Remember, a H^+ ion is a hydrogen atom (proton plus electron) that has lost its electron; hence it is a bare proton.) A **Brønsted–Lowry base** *is any sub-*

*To keep our ions straight, let us summarize: a hydrogen ion is H^+; a hydroxide ion is OH^-; and a hydronium ion is H_3O^+.

stance that can accept a proton from some other substance. In short, a Brønsted–Lowry acid is a *proton donor* and a Brønsted–Lowry base is a *proton acceptor.* The Brønsted–Lowry definitions of acid and base change the concept of acids and bases from that of the production of specific chemical species (H^+ and OH^-) to one dealing with chemical reactions involving proton exchange. This focus on proton exchange (1) extends the number of substances that may be considered to be acids and bases, (2) allows us to discuss acid–base chemistry in nonaqueous solutions, and (3) accounts for the fact that the acidic species in aqueous solution is the hydronium ion.

The Brønsted–Lowry acid–base definitions can best be illustrated by example. Let us consider the formation reaction for hydrochloric acid, which involves the dissolving of hydrogen chloride gas in water.

$$HCl(g) + H_2O(l) \longrightarrow H_3O^+(aq) + Cl^-(aq)$$

The HCl behaves as a Brønsted–Lowry acid by donating a proton to a water molecule. Note that a hydronium ion is formed as a result. The base in this reaction is water since it has accepted a proton; no hydroxide ions are involved. The Brønsted–Lowry definition of a base includes any species capable of accepting a proton; hydroxide ions can do this, but so can many other substances.

It is not necessary that a water molecule be one of the reactants in a Brønsted–Lowry acid–base reaction, nor that the reaction take place in the liquid state. An important application of Brønsted–Lowry acid–base theory is to gas-phase reactions. The white solid haze that often covers glassware in chemistry laboratory results from the gas-phase reaction between HCl and NH_3.

$$HCl(g) + NH_3(g) \longrightarrow NH_4^+(g) + Cl^-(g) \longrightarrow NH_4Cl(s)$$

This is a Brønsted–Lowry acid–base reaction, since the HCl molecules donate protons to the NH_3 forming NH_4^+ and Cl^- ions. These ions instantaneously combine to form the white solid NH_4Cl.

All Arrhenius acids are Brønsted–Lowry acids, and all Arrhenius bases are Brønsted–Lowry bases. However, the converse of this statement is not true. Brønsted–Lowry theory includes Arrhenius theory but also much more.

14.2 Strengths of Acids and Bases

A 0.1 *M* solution of nitric acid (HNO_3) or sulfuric acid (H_2SO_4) when spilled on your clothes and not immediately washed off will "eat" holes in your clothing. If 0.10 *M* solutions of either acetic acid ($HC_2H_3O_2$) or carbonic acid (H_2CO_3) were spilled on your clothes, the previously noted "corrosive" effects would not be observed. Why? All four acid solutions are of equal concentration; all are 0.1 *M* solutions. The difference in behavior relates to the *strength* of the acids; nitric and sulfuric acids are *strong* acids, whereas acetic and carbonic acids are *weak* acids.

strong acid

weak acid

Acids may be classified as strong or weak depending on the number of H^+ ions (or H_3O^+ ions) they produce in aqueous solution. A **strong acid** *dissociates 100%, or very nearly 100%, in solution; that is, all, or almost all, of the acid molecules present dissociate into ions.* Because of this extensive dissociation, many, many hydrogen ions are present in the solution of a strong acid. A **weak acid** *dissociates only slightly in solution; that is, most of the acid molecules are present in solution in undissociated form.*

Table 14-1. Commonly Encountered Strong Acids

Name[a]	Molecular Formula	Molecular Structure
Nitric acid	HNO_3	H—O—N(=O)—O
Sulfuric acid	H_2SO_4	H—O—S(O)(O)—O—H
Perchloric acid	$HClO_4$	H—O—Cl(O)(O)—O
Hydrochloric acid	HCl	H—Cl
Hydrobromic acid	HBr	H—Br
Hydroiodic acid	HI	H—I

[a]Nomenclature for acids was discussed in Sec. 6.12.

Generally, less than 5% of acid molecules are found in dissociated form in a solution of a weak acid. Relatively speaking, then, there are only a few H^+ ions present in a solution containing a weak acid.

The extent to which an acid dissociates in solution, that is, whether it is strong or weak, depends on the molecular structure of the acid; molecular polarity and strength and polarity of individual bonds are particularly important factors.

There are very few strong acids; the formulas and structures of the six most commonly encountered strong acids are given in Table 14-1. You should know the identity of these six strong acids; such knowledge is necessary in order to write net ionic equations, the topic of Section 14.6.

The vast majority of acids that exist are weak acids. Familiar weak acids include acetic acid ($HC_2H_3O_2$), the acidic component of vinegar; boric acid (H_3BO_3), a common ingredient for eye washes; and carbonic acid (H_2CO_3), found in carbonated beverages.

Just as there are strong acids and weak acids, there are also strong bases and weak bases. As with acids, there are only a few strong bases. Strong bases are limited to the hydroxides of groups IA and IIA of the periodic table listed in Table 14-2. Of the strong

Table 14-2. Common Strong Bases

Group IA Hydroxides	Group IIA Hydroxides
LiOH	
NaOH	
KOH	
RbOH	$Sr(OH)_2$
CsOH	$Ba(OH)_2$

bases, only NaOH and KOH are commonly used in the chemical laboratory. The limited solubility of the group IIA hydroxides is water limits their use. However, despite this limited solubility, these hydroxides are still considered to be strong bases; that which dissolves dissociates into ions 100%.

Only one common weak base exists — ammonia. It furnishes small amounts of OH^- ions through reaction with water molecules.

$$NH_3(g) + H_2O(l) \longrightarrow NH_4^+(aq) + OH^-(aq)$$

A solution of ammonia in water is most properly called *aqueous ammonia* although it is also, somewhat erroneously, called ammonium hydroxide. Aqueous ammonia is the preferred designation, since most of the NH_3 present is in molecular form. Only a very few NH_3 molecules have reacted with the water to give ammonium (NH_4^+) and hydroxide (OH^-) ions.

It is important to remember that the terms *strong* and *weak* apply to the extent of dissociation, and not to the concentrations of acid or base. For example, stomach acid (gastric juice) is a dilute (not weak) solution of a strong acid; it is a 5% by weight solution of hydrochloric acid. On the other hand, a 35% by weight solution of hydrochloric acid would be considered to be a concentrated (not strong) solution of a strong acid.

14.3 Polyprotic Acids

monoprotic acid

Acids may be classified according to the number of hydrogen ions they produce per molecule on dissociation in solution. A **monoprotic acid** *is an acid that yields one H^+ ion (proton) per molecule on dissociation.* Hydrochloric acid (HCl) and nitric acid (HNO_3) are both monoprotic acids.

diprotic acid

A **diprotic acid** *is an acid that yields two H^+ ions (two protons) per molecule on complete dissociation.* Sulfuric acid (H_2SO_4) is a diprotic acid. The dissociation process for a diprotic acid occurs in steps. For H_2SO_4, the two steps are

$$H_2SO_4 \longrightarrow H^+ + HSO_4^-$$
$$HSO_4^- \longrightarrow H^+ + SO_4^{2-}$$

The second proton is not as easily removed as the first, because it must be pulled away from a negatively charged particle, HSO_4^-. Accordingly, HSO_4^- is a weaker acid than H_2SO_4. In general, each successive step in the dissociation of an acid occurs to a lesser extent than the previous step. Note that four sulfuric-acid-related species are present in a solution of sulfuric acid: H_2SO_4, HSO_4^-, SO_4^{2-}, and H^+.

triprotic acid

A few triprotic acids exist. A **triprotic acid** *is an acid that yields three H^+ ions (three protons) per molecule on complete dissociation.* Phosphoric acid (H_3PO_4) is the most common triprotic acid. The three dissociation steps involved in the removal of H^+ ions from this molecule are

$$H_3PO_4 \longrightarrow H^+ + H_2PO_4^-$$
$$H_2PO_4^- \longrightarrow H^+ + HPO_4^{2-}$$
$$HPO_4^{2-} \longrightarrow H^+ + PO_4^{3-}$$

polyprotic acid

The general term **polyprotic acid** *describes acids capable of producing two or more H^+ ions (protons) per molecule on complete dissociation.*

The formula of an acid, which gives the number of hydrogen atoms present in a molecule of the acid, cannot always be used to classify an acid as mono-, di-, or triprotic. For example, a molecule of acetic acid ($C_2H_4O_2$) contains four hydrogen atoms and yet it is a monoprotic acid. Only one of the hydrogen atoms in this molecule is *acidic;* that is, only one of the hydrogen atoms leaves the molecule when it is in solution. The dissociation of acetic acid in solution can be represented by the equation

```
    H  O                      ┌    H  O    ┐-
    |  ‖                      │    |  ‖    │
 H—C—C—O—H  ⟶  H+ +  │ H—C—C—O │
    |                         │    |       │
    H                         └    H       ┘
```

Structural formulas are used to emphasize that all of the hydrogens in the molecule are not equivalent. One hydrogen atom is bonded to an oxygen atom; the other three hydrogen atoms are each bonded to a carbon atom. Only the hydrogen bound to the oxygen is acidic. Those bound to carbon atoms are too tightly held to be removed. The reason why the hydrogen leaves the oxygen but not the carbon atoms is related to electronegativity differences and bond polarity (Sec. 10.6). Carbon–hydrogen bonds are essentially nonpolar, while oxygen–hydrogen bonds are very polar. Water, a polar molecule, can exert a greater influence on the polar oxygen–hydrogen bond.

Table 14-3. Selected Common Mono-, Di-, and Triprotic Acids

Name	Formula	Classification	Common Occurrence
Acetic	$HC_2H_3O_2$	monoprotic; weak	vinegar
Lactic	$HC_3H_5O_3$	monoprotic; weak	sour milk, cheese; produced during muscle contraction
Salicylic	$HC_7H_5O_3$	monoprotic; weak	present in chemically combined form in aspirin
Hydrochloric	HCl	monoprotic; strong	constituent of gastric juice; industrial cleaning agent
Nitric	HNO_3	monoprotic; strong	used in urinalysis test for protein; used in manufacture of dyes and explosives
Tartaric	$H_2C_4H_4O_6$	diprotic; weak	grapes
Carbonic	H_2CO_3	diprotic; weak	carbonated beverages; produced in the body from carbon dioxide
Sulfuric	H_2SO_4	diprotic; strong	storage batteries; manufacture of fertilizer
Citric	$H_3C_6H_5O_7$	triprotic; weak	citrus fruits
Boric	H_3BO_3	triprotic; weak	antiseptic eyewash
Phosphoric	H_3PO_4	triprotic; weak	found in dissociated form (HPO_4^{-2}, $H_2PO_4^-$) in intracellular fluid; fertilizer manufacture

We now see why the formula for acetic acid is usually written as $HC_2H_3O_2$ rather than as $C_2H_4O_2$. In the situation where some hydrogens are easily removed (acidic) and others are not (nonacidic), it is accepted procedure to write the acidic hydrogen first, separated from the other hydrogens in the formula. Citric acid, the principal acid in citrus fruits, is another example of an acid that contains both acidic and nonacidic hydrogens. Its formula, $H_3C_6H_5O_6$, indicates that three of the eight hydrogen atoms present in a molecule are acidic. Table 14-3 contains information about the use and occurrence of selected common mono-, di-, and triprotic acids, many of which contain nonacidic hydrogen atoms.

We have focused our attention on acids in the preceding discussion. It should be noted that in a similar manner molecules of Arrhenius bases may be the source of more than one hydroxide ion. For example, calcium hydroxide, $Ca(OH)_2$, is a base that yields two hydroxides per molecule when it dissociates in solution.

14.4 Acid and Base Stock Solutions

Acids and bases are used so often in most laboratories that stock solutions (Sec. 13.10) of the most common ones are made readily available at each work space. The concentrations of such solutions are traditionally the same from laboratory to laboratory and are given in Table 14-4. Ordinarily the concentrations of the stock solutions are not given on their containers. Only the name of the acid or base and the term dilute (dil) or concentrated (con) are found. It is assumed that the student or researcher knows what

Table 14-4. Concentrations of Common Laboratory Stock Solutions of Acids and Bases

	Concentration	
Label Designation	**Molarity**	**Normality**
Acids		
Dilute hydrochloric acid	6	6
Concentrated hydrochloric acid	12	12
Dilute nitric acid	6	6
Concentrated nitric acid	16	16
Dilute sulfuric acid	3	6
Concentrated sulfuric acid	18	36
Glacial acetic acid	6	6
Dilute acetic acid	18	18
Bases		
Dilute aqueous ammonia[b]	6	6
Concentrated aqueous ammonia[b]	15	15
Dilute sodium hydroxide[a]	6	6

[a] Often labeled simply "sodium hydroxide."
[b] Often labeled "ammonium hydroxide."

is implied by the designations dil and con in each specific case, that is, that they know the information found in Table 14-4. Note from Table 14-4 that the designations dil and con do not have a constant meaning in terms of molarity. For example, concentrated solutions of sulfuric acid, nitric acid, and hydrochloric acid have molarities of 18, 16, and 12, respectively. There is not as much variance in the meaning of the term dilute; except for sulfuric acid, all of the listed dilute solutions are 6 *M*. Note that all the listed dilute solutions, including sulfuric acid, are 6 *N*.

14.5 Salts

To a nonscientist the term salt connotes a white granular substance used as a seasoning for food. To the chemist the term salt has a much, much broader meaning; sodium chloride, table salt, is only one of thousands of salts known to a chemist. "Pass the salt" is a very ambiguous request to a chemist.

salts

Salts *are ionic compounds that contain any negative ion except hydroxide ion and any positive ion except hydrogen ion.* Salts, as a class of compounds, differ from acids and bases in that they contain no common ion or species that characterizes the class. The relationship between an acid, a base, and a salt is that a salt is one of the products resulting from the reaction of an acid with a base (Sec. 14.7).

Many salts occur in nature and numerous others have been prepared in the laboratory. The wide variety of uses found for salts can be seen from Table 14-5, a listing of selected salts and their uses.

*Table 14-5. **Some Common Salts and Their Uses***

Name	Formula	Uses
Ammonium nitrate	NH_4NO_3	fertilizer; explosive manufacture
Barium sulfate	$BaSO_4$	X-rays of gastrointestinal tract
Calcium carbonate	$CaCO_3$	chalk; limestone
Calcium chloride	$CaCl_2$	drying agent for removal of small amounts of water
Iron(II) sulfate	$FeSO_4$	treatment for anemia
Potassium chloride	KCl	"salt" substitute for low sodium diets
Sodium chloride	NaCl	table salt; used as a deicer (to melt ice)
Sodium bicarbonate	$NaHCO_3$	ingredient in baking powder; stomach antiacid
Sodium hypochlorite	NaClO	bleaching agent
Silver bromide	AgBr	light-sensitive material in photographic film
Tin(II) fluoride	SnF_2	toothpaste additive

Much information concerning salts has been presented in previous chapters, although the term *salt* was not explicitly used in these discussions. Formula writing and nomenclature for binary ionic compounds (salts) was covered in Sections 6.5 and 6.6. Many salts, as shown in Table 14-5, contain polyatomic ions such as nitrate and sulfate. Such ions were discussed in Sections 6.7 through 6.9. The solubility of ionic compounds (salts) in water was the topic of Section 13.4.

All common salts in solution are dissociated into ions (Sec. 13.3). Even if a salt is only slightly soluble, the small amount that does dissolve completely dissociates. Thus, the terms weak and strong, used to denote qualitatively the percent dissociation of acids and bases, are not applicable to common salts. We do not use the term weak salt.

14.6 Ionic and Net Ionic Equations

Soluble strong acids, soluble strong bases, and soluble salts all produce ions in aqueous solution. It is extremely useful to discuss the reactions of such acids, bases, and salts in aqueous solution in terms of the ions present. This is most easily done using a new type of equation — a *net ionic equation.*

ionic equation

net ionic equation

Up to this point in the text, all the equations we have used have been *molecular equations,* equations where the complete formulas of all reactants and products are shown. From molecular equations, *ionic equations* may be written. An **ionic equation** *is an equation in which the formulas of the predominant form of each compound in aqueous solution are used; dissociated compounds are written as ions, undissociated compounds are written in molecular form. Net ionic equations* are derived from ionic equations. A **net ionic equation** *is an ionic equation from which nonparticipating (spectator) species have been eliminated.*

The differences between molecular, ionic, and net ionic equations can best be illustrated by examples. Let us consider the chemical reaction that results when a solution of potassium chloride (KCl) is added to a solution of silver nitrate ($AgNO_3$). An insoluble salt, silver chloride (AgCl) is produced as a result of the mixing. The *molecular equation* for this reaction is

$$AgNO_3(aq) + KCl(aq) \longrightarrow KNO_3(aq) + AgCl(s)$$

Three of the four substances involved in this reaction — $AgNO_3$, KCl, and KNO_3 — are soluble salts and thus exist in solution in ionic form. This is shown by writing the *ionic equation* for the reaction.

$$\underbrace{Ag^+(aq) + NO_3^-(aq)}_{AgNO_3\text{ in ionic form}} + \underbrace{K^+(aq) + Cl^-(aq)}_{KCl\text{ in ionic form}} \longrightarrow \underbrace{K^+(aq) + NO_3^-(aq)}_{KNO_3\text{ in ionic form}} + AgCl(s)$$

In this equation each of the three soluble salts is shown in dissociated (ionic) form rather than in undissociated form. A close look at this ionic equation shows that the potassium ions (K^+) and nitrate ions (NO_3^-) appear on both sides of the equation, indicating that they did not undergo any chemical change. In other words, they are *spectator ions;* they did not participate in the reaction. The *net ionic equation* for this reaction is written by dropping (cancelling) all spectator ions from the ionic equation. In our case, the net ionic equation becomes

$$Ag^+(aq) + Cl^-(aq) \longrightarrow AgCl(s)$$

This net ionic equation indicates that the product AgCl was formed by the reaction of silver ions (Ag^+) with chloride ions (Cl^-). It totally ignores the presence of those ions that are not taking part in the reaction. Thus, a net ionic equation focuses on only those species in a solution actually involved in a chemical reaction. It does not give all species present in the solution.

If a student can write equations in molecular form, the conversion of these equations to net ionic form is a straightforward process. Three steps are involved.

1. Check the given molecular equation to make sure that it is balanced.
2. Expand the molecular equation into an ionic equation.
3. Convert the ionic equation into a net ionic equation by eliminating spectator ions.

In expanding a molecular equation into an ionic equation (step 2) the decision to write each reactant and product in dissociated (ionic) form or undissociated (molecular) form must be made. The following rules serve as guidelines in making such decisions.

1. Soluble compounds that completely dissociate in aqueous solution are written in ionic form. They include
   (a) all soluble salts, see Section 13.4 for solubility rules.
   (b) all strong acids, see Table 14-1.
   (c) all strong bases, see Table 14-2.
2. Soluble weak acids and weak bases are written in molecular form, since they are incompletely ionized in solution and thus exist predominantly in the undissociated form. The following are considered to be weak acids or weak bases.
   (a) All acids not listed in Table 14-1 as strong acids. Common examples are HNO_2, H_2SO_3, HF, H_2S, $HC_2H_3O_2$, H_2CO_3, and H_3PO_4.
   (b) All bases not listed in rule 1 as strong bases. Aqueous ammonia (NH_3) is the most common weak base.
3. All insoluble substances (solids, liquids, and gases), whether ionic or covalent, exist as molecules or neutral ionic units and are written as such.
4. All soluble covalent substances, for example, carbon dioxide (CO_2) or sucrose ($C_{12}H_{22}O_{11}$), are written in molecular form.
5. Water, the solvent, if it appears in the equation, is written in molecular form.

Now, let's apply these guidelines by writing some net ionic equations.

EXAMPLE 14.1

Write the net ionic equation for the following aqueous solution reaction.

$$K_2CO_3 + CaCl_2 \longrightarrow CaCO_3 + KCl \quad \text{(unbalanced equation)}$$

SOLUTION

Step 1. To balance the given molecular equation, the coefficient 2 must be placed in front of KCl.

$$K_2CO_3 + CaCl_2 \longrightarrow CaCO_3 + 2\ KCl$$

Step 2. A decision must be made to write each reactant and product in ionic or molecular form. Let us consider them one by one.

K_2CO_3: This compound is a salt. The solubility rules indicate that it is soluble. Thus, K_2CO_3 will be written in ionic form—$2\,K^+ + CO_3^{2-}$. Note that three ions (two K^+ ions and one CO_3^{2-} ion) are produced from the dissociation of one K_2CO_3 unit.

$CaCl_2$: This compound is also a soluble salt; it will be written in ionic form. Again, three ions are produced upon dissociation of one $CaCl_2$ unit—Ca^{2+} and $2\,Cl^-$.

$CaCO_3$: The solubility rules indicate that this compound is an insoluble salt. Thus, it will be written in molecular form in the ionic equation.

KCl: All potassium salts are soluble. Thus, KCl will be written in ionic form—$K^+ + Cl^-$.

The ionic equation will have the form

$$2\,K^+ + CO_3^{2-} + Ca^{2+} + 2\,Cl^- \longrightarrow CaCO_3 + 2\,K^+ + 2\,Cl^-$$

Note how the coefficient 2 in front of KCl in the molecular equation affects the ionic equation. The dissociation of two KCl units produces two K^+ ions and two Cl^- ions.

Step 3. Inspection of the ionic equation shows that K^+ ions (two of them) and Cl^- ions (two of them) are spectator ions. Cancellation of these ions from the equation will give the net ionic equation.

$$\cancel{2\,K^+} + CO_3^{2-} + Ca^{2+} + \cancel{2\,Cl^-} \longrightarrow CaCO_3 + \cancel{2\,K^+} + \cancel{2\,Cl^-}$$

$$Ca^{2+} + CO_3^{2-} \longrightarrow CaCO_3$$

EXAMPLE 14.2

Write the net ionic equation for the following aqueous solution reaction.

$$H_2S + AlCl_3 \longrightarrow Al_2S_3 + HCl \qquad \text{(unbalanced equation)}$$

SOLUTION

Step 1. Balancing the molecular equation, we get

$$3\,H_2S + 2\,AlCl_3 \longrightarrow Al_2S_3 + 6\,HCl$$

Step 2. The expansion of the molecular equation into an ionic equation is accomplished by the following analysis.

H_2S: This is a weak acid. All weak acids are written in molecular form in ionic equations.

$AlCl_3$: This is a soluble salt. All chloride salts are soluble, with three exceptions; this is not one of the exceptions. Soluble salts are written in ionic form in ionic equations.

Al_2S_3: This is an insoluble salt. All sulfides are insoluble except for groups IA and IIA and NH_4^+. Thus, Al_2S_3 will remain in molecular form in the ionic equation.

HCl: This compound is an acid. It is one of the six strong acids listed in Table 14-1. Strong acids are written in ionic form.

The ionic equation for the reaction is

$$3\,H_2S + 2\,Al^{3+} + 6\,Cl^- \longrightarrow Al_2S_3 + 6\,H^+ + 6\,Cl^-$$

Note again that the coefficient present in the balanced molecular equation must be taken into consideration when determining the total number of ions produced from dissociation. On dissociation an $AlCl_3$ unit produces four ions — an Al^{3+} ion and three Cl^- ions. This number must be doubled for the ionic equation since $AlCl_3$ carries the coefficient 2 in the balanced molecular equation. Similar considerations apply to HCl in this equation.

Step 3. Inspection of the ionic equation shows that only Cl^- ions (six of them) are spectator ions. Cancellation of these ions from the equation will give the net ionic equation.

$$3\,H_2S + 2\,Al^{3+} + \cancel{6\,Cl^-} \longrightarrow Al_2S_3 + 6\,H^+ + \cancel{6\,Cl^-}$$

$$\mathbf{3\,H_2S + 2\,Al^{3+} \longrightarrow Al_2S_3 + 6\,H^+}$$

EXAMPLE 14.3

Write the net ionic equation for the following aqueous solution reaction.

$$H_2SO_4 + NaOH \longrightarrow Na_2SO_4 + H_2O \qquad \text{(unbalanced equation)}$$

SOLUTION

Step 1. Balancing the given molecular equation, we get

$$H_2SO_4 + 2\,NaOH \longrightarrow Na_2SO_4 + 2\,H_2O$$

Step 2. The expansion of the molecular equation into an ionic equation is based on the following analysis.

H_2SO_4: This compound is an acid. It is one of the six strong acids listed in Table 14-1. Strong acids are written in **ionic form** in ionic equations.

NaOH: This compound is a base. It is one of the strong bases listed in Table 14-2. Strong bases are written in **ionic form** in ionic equations.

Na_2SO_4: This compound is a soluble salt. All sodium salts are soluble. Thus, Na_2SO_4 is written in **ionic form.**

H_2O: This compound is a covalent compound; two nonmetals are present. Covalent compounds are always written in **molecular form.**

The ionic equation for the reaction, from the above information, is

$$2\,H^+ + SO_4^{2-} + 2\,Na^+ + 2\,OH^- \longrightarrow 2\,Na^+ + SO_4^{2-} + 2\,H_2O$$

Step 3. Inspection of the ionic equation shows that SO_4^{2-} ions and Na^+ ions are spectator ions. Cancellation of these ions from the equation gives the net ionic equation.

$$2\,H^+ + \cancel{SO_4^{2-}} + \cancel{2\,Na^+} + 2\,OH^- \longrightarrow \cancel{2\,Na^+} + \cancel{SO_4^{2-}} + 2\,H_2O$$

$$\mathbf{2\,H^+ + 2\,OH^- \longrightarrow 2\,H_2O}$$

The coefficients in the net ionic equation should be the smallest set of numbers that correctly balance the equation. In this case, all the coefficients are divisible by two.

Dividing by two, we get

$$H^+ + OH^- \longrightarrow H_2O$$

The final net equation should always be checked to see that the coefficients are in the lowest possible integral ratio.

14.7 Reactions of Acids

All acids have some unique properties that adapt them for use in specific situations. In addition, all acids have certain chemical properties in common, properties related to the presence of H^+ ions in aqueous solution. In this section we consider three types of chemical reactions that acids characteristically undergo. The three reaction types are

1. Acids react with active metals to produce hydrogen gas and a salt.
2. Acids react with bases to produce a salt and water.
3. Acids react with carbonates and bicarbonates to produce carbon dioxide, a salt, and water.

Reaction with Metals

Acids react with many, but not all, metals. When they do react, the metal dissolves and hydrogen gas (H_2) is liberated. In the reaction the metal atoms lose electrons and become metal ions. The lost electrons are taken up by the hydrogen ions (protons) of the acid; the hydrogen ions become electrically neutral, combine into molecules, and emerge from the reaction mixture as hydrogen gas. Illustrative of the reaction of an acid and a metal is the reaction between zinc and sulfuric acid.

$$\text{Molecular equation:} \quad Zn + H_2SO_4 \longrightarrow ZnSO_4 + H_2$$

$$\text{Net ionic equation:} \quad Zn + 2\,H^+ \longrightarrow Zn^{2+} + H_2$$

Metals can be arranged in a reactivity order based on their ability to react with acids. Such an ordering for the more common metals is given in Table 14-6. Any metal above hydrogen in the activity series will dissolve in an acid solution and form H_2. The closer a metal is to the top of the series the more rapid the reaction. Those metals below hydrogen in the series do not dissolve in an acid to form H_2.

The right side of Table 14-6 also indicates that the most active metals (those nearest the top in the activity series) also react with water. Again, hydrogen gas production is the result of the reaction. In the cases of potassium and sodium, the reaction is violent enough sometimes to cause explosions as the result of H_2 ignition. The equation for the reaction of potassium with water is

$$2\,K + 2\,H_2O \longrightarrow 2\,KOH + H_2$$

Note that the resulting solution is basic when a metal reacts with water; hydroxide ions are present. In the previous equation KOH is the source of the hydroxide ions.

Table 14-6. Activity Series for Common Metals

Tendency	Reaction with H^+	Metal	Symbol	Remarks
INCREASING TENDENCY TO REACT (↑)	React with H^+ ions to liberate hydrogen gas	Potassium	K	react violently with cold water
		Sodium	Na	
		Calcium	Ca	reacts slowly with cold water
		Magnesium	Mg	react slowly with hot water (steam)
		Aluminum	Al	
		Zinc	Zn	
		Chromium	Cr	
		Iron	Fe	
		Nickel	Ni	
		Tin	Sn	
		Lead	Pb	
		HYDROGEN	H	
	Do not react with H^+ ions	Copper	Cu	
		Mercury	Hg	
		Silver	Ag	
		Platinum	Pt	
		Gold	Au	

Reaction with Bases

neutralization

When acids and bases are mixed, they react with each other; their acid and basic properties disappear, and we say that they have *neutralized* each other. **Neutralization** *is the reaction between an acid and a base to form a salt and water.* The hydrogen ions from the acid combine with the hydroxide ions from the base to form water. The salt formed contains the negative ion from the acid and the positive ion from the base.

Any time an acid is completely reacted with a base, neutralization occurs. It does not matter whether the acid and base are strong or weak. Sodium hydroxide (a strong base) and nitric acid (a strong acid) react as follows.

$$\text{Molecular equation:} \quad HNO_3 + NaOH \longrightarrow NaNO_3 + H_2O$$

$$\text{Net ionic equation:} \quad H^+ + OH^- \longrightarrow H_2O$$

The equations for the reaction of potassium hydroxide (a strong base) with hydrocyanic acid (a weak acid) are

$$\text{Molecular equation:} \quad HCN + KOH \longrightarrow KCN + H_2O$$

$$\text{Net ionic equation:} \quad HCN + OH^- \longrightarrow CN^- + H_2O$$

Note that in each case the products are a salt ($NaNO_3$ in the first reaction, KCN in the second) and water. Note also that the net ionic equations for the two previous neutralization reactions are different. In the second set of reactions the acid must remain written in molecular form, since it is a weak acid.

Reaction with Carbonates and Bicarbonates

Carbon dioxide gas (CO_2), water, and a salt are always the products of the reaction of acids with carbonates or bicarbonates, as illustrated by the following equations.

$$2\ HCl + Na_2CO_3 \longrightarrow 2\ NaCl + CO_2 + H_2O$$

$$HCl + NaHCO_3 \longrightarrow NaCl + CO_2 + H_2O$$

Baking powder is a mixture of a bicarbonate and an acid-forming solid. The addition of water to this mixture generates the acid that then reacts with the bicarbonates to release carbon dioxide into the batter. It is the generated carbon dioxide that causes the batter to rise. Baking soda is pure $NaHCO_3$. To cause it to release carbon dioxide, an acid-containing substance, such as buttermilk, must be added to it.

14.8 Reactions of Bases

The most important characteristic reaction of bases — their reaction with acids (neutralization) — was discussed in the previous section. Another characteristic reaction, that of bases with certain salts, will be discussed in Section 14.9.

Bases react with fats and oils and convert them into smaller, soluble molecules. For this reason most household cleaning products contain basic substances. Lye (impure NaOH) is an active ingredient in numerous drain cleaners. Also, many advertisements for liquid household cleaners emphasize the fact that aqueous ammonia is present in the product.

14.9 Reactions of Salts

Dissolved salts will react with metals, acids, bases, and other salts under specific conditions. Characteristic reactions that are possible include

1. Salts react with some metals to form another salt and another metal.
2. Salts react with some acid solutions to form other acids and salts.
3. Salts react with some base solutions to form other bases and salts.
4. Salts react with some solutions of other salts to form new salts.

The tendency for salts to react with metals is related to the relative positions of the two involved metals in the activity series (Table 14-6). In order for salts to react with acids, bases, or other salts, one of the reaction products must be (1) an insoluble salt, (2) a gas that is evolved from the solution, or (3) an undissociated soluble species, such as a weak acid or a weak base. The formation of any of these products serves as the driving force to cause the reaction to occur.

Reaction with Metals

If an iron nail is dipped into a solution of copper sulfate ($CuSO_4$), metallic copper will be deposited on the nail and some of the iron will dissolve.

Molecular equation: $Fe(s) + CuSO_4(aq) \longrightarrow Cu(s) + FeSO_4(aq)$

Net ionic equation: $Fe + Cu^{2+} \longrightarrow Cu + Fe^{2+}$

One metal has replaced the other. This type of reaction will occur only if the metal going into solution is above the replaced metal in the activity series. Iron is above copper in the activity series; hence, it can replace it. If a strip of copper were dipped into a solution of $FeSO_4$—just the opposite situation to what we have been discussing—no reaction would occur since copper is below iron in the activity series.

Reaction with Acids

In order for a salt to react with an acid either a new weaker acid, a new insoluble salt, or a gaseous compound must be one of the products.

An example of a reaction in which the formation of an *insoluble salt* is the driving force for the reaction to occur is

$$AgNO_3(aq) + HCl(aq) \longrightarrow AgCl(s) + HNO_3(aq)$$

The conclusion that this reaction will occur comes from a consideration of the possible recombinations of the reacting species. In a solution made by mixing silver nitrate ($AgNO_3$) and hydrochloric acid (HCl), four kinds of ions are present initially (before any reaction occurs): Ag^+ and NO_3^- (since $AgNO_3$ is a soluble salt) and H^+ and Cl^- (since HCl is a strong acid). The question is whether these ions can get together in new appropriate combinations. The possible new combinations of oppositely charged ions are $H^+NO_3^-$ and Ag^+Cl^-. The first of these combinations would result in the formation of the strong acid HNO_3. Strong acids in solution exist in dissociated form; therefore, these ions will not combine. The second combination does occur, because AgCl is an insoluble salt. Thus, the overall reaction takes place as a result of the formation of this insoluble salt. The net result of the reaction is that the original ions exchange partners.

The reaction of sodium fluoride (a soluble salt) and hydrochloric acid (a strong acid) illustrates the case where a new weaker acid is the driving force for the occurrence of the reaction.

$$NaF(aq) + HCl(aq) \longrightarrow NaCl(aq) + HF(aq)$$

Using an analysis pattern similar to that in the previous example, we find that four types of ions are present initially: Na^+ and F^- (from the soluble salt) and H^+ and Cl^- (from the strong acid). Possible new combinations are Na^+Cl^- and H^+F^-. Sodium chloride, the result of the first combination, will not form because this salt is soluble. The combination of H^+ ion with F^- ion does occur because it results in the formation of the weak acid HF. In solution weak acids exist predominantly in molecular form. In all reactions of this general type, the acid formed in the reaction must be weaker than the reactant acid. If the reactant acid is strong, as in this example, such a determination is obvious. If both the reactant and product acids are weak, more information than you have been given in this chapter would be needed to predict which of the two acids was the weakest. You will not be expected to distinguish between the strengths of two weak acids.

The most common type of reaction in which the driving force is the *evolution of a gas* is that where the salt is a carbonate or bicarbonate. This type of reaction was previously discussed in Section 14.7.

Note that a reaction does not always occur when acid and salt solutions are mixed.

Consider the possible reaction of NaCl and HNO_3 solutions. Initially four types of ions are present: Na^+ and Cl^- (from the soluble salt) and H^+ and NO_3^- (from the strong acid). The new combinations, if a reaction did occur, would be $Na^+NO_3^-$ (a soluble salt) and H^+Cl^- (a strong acid). Since both of the products would exist in dissociated form in solution, no recombination of ions occurs; hence, no reaction occurs.

Reaction with Bases

The criteria for the reaction of bases with salts are similar to those for acid–salt reactions, except that weaker base formation replaces weaker acid formation as one of the three driving forces. An example of a base–salt reaction involving the formation of an insoluble salt is

$$Ba(OH)_2(aq) + Na_2SO_4(aq) \longrightarrow BaSO_4(s) + 2\ NaOH(aq)$$

The most common situation in which gas evolution is the driving force for base–salt reactions is where ammonium salts are involved. In such cases, ammonia gas is given off, as illustrated by the reaction of NH_4Cl and KOH.

$$NH_4Cl(aq) + KOH(aq) \longrightarrow KCl(aq) + NH_3(g) + H_2O(l)$$

Reaction of Salts with Each Other

Two different salt solutions will react when mixed only if an insoluble salt can be formed. Consider the following possible reactions.

$$AgNO_3(aq) + NaCl(aq) \longrightarrow AgCl(s) + NaNO_3(aq)$$

$$KNO_3(aq) + NaCl(aq) \longrightarrow KCl(aq) + NaNO_3(aq)$$

The first reaction occurs because AgCl is an insoluble salt. The second reaction does not occur since both of the possible products are soluble salts, which means there is no driving force for the reaction.

EXAMPLE 14.4

Write molecular, ionic, and net ionic equations for the reaction that occurs, if any, when 0.1 *M* solutions of the following substances are mixed.

(a) $Ni(NO_3)_2$ and Na_2S (b) $BaCl_2$ and H_2SO_4 (c) HNO_3 and $KC_2H_3O_2$

SOLUTION

(a) Both of the reactants are salts. Two different salt solutions will react when mixed only if an insoluble salt can be formed.

In a solution made by mixing $Ni(NO_3)_2$ and Na_2S, four kinds of ions are present initially (before any reaction occurs): Ni^{2+} and NO_3^- (from the $Ni(NO_3)_2$) and Na^+ and S^{2-} (from the Na_2S). The possible new combinations of oppositely charged ions are Ni^{2+} with S^{2-} and Na^+ with NO_3^-.

Original Ion Combinations **Possible New Combinations**

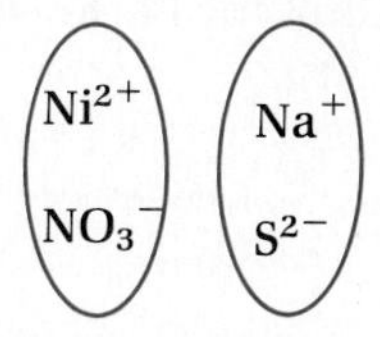

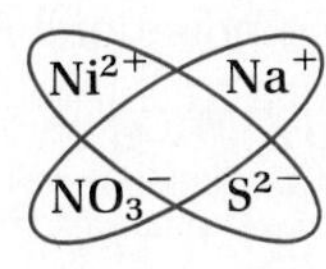

The first of these new combinations, the formation of NiS, is the one that will be the driving force for the reaction to occur; NiS is an insoluble salt. The second new combination, the formation of $NaNO_3$, does not occur since $NaNO_3$ is a soluble salt and soluble salts exist in dissociated form in solution. The equations for the reaction are

Molecular: $Ni(NO_3)_2 + Na_2S \longrightarrow NiS + 2\ NaNO_3$

Ionic: $Ni^{2+} + \cancel{2\ NO_3^-} + \cancel{2\ Na^+} + S^{2-} \longrightarrow NiS + \cancel{2\ Na^+} + \cancel{2\ NO_3^-}$

Net ionic: $Ni^{2+} + S^{2-} \longrightarrow NiS$

(b) One of the reactants, $BaCl_2$, is a soluble salt and the other reactant, H_2SO_4, is a strong acid. Both reactants exist in solution in dissociated form; thus, four types of ion are present in the mixed solution (before any reaction occurs): Ba^{2+}, Cl^-, H^+, and SO_4^-. The conclusion that a reaction will occur comes from a consideration of the possible new combinations of the reacting species.

Original Ion Combinations **Possible New Combinations**

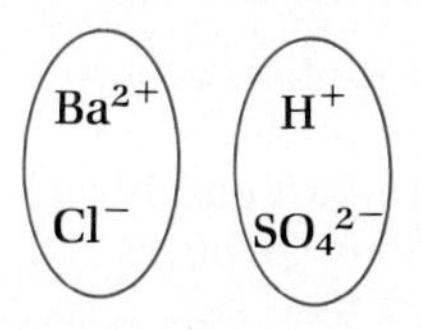

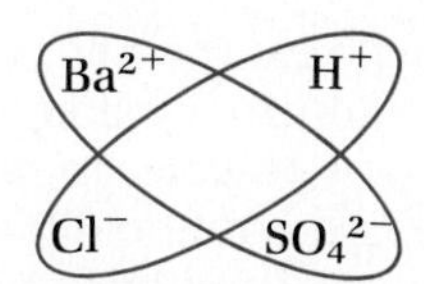

The first of these new combinations, the formation of $BaSO_4$, is the driving force for the reaction to occur; $BaSO_4$ is an insoluble salt. The second recombination, the formation of HCl, does not occur since HCl is a strong acid and will exist in solution in dissociated form. The equations for the reaction are

Molecular: $BaCl_2 + H_2SO_4 \longrightarrow BaSO_4 + 2\ HCl$

Ionic: $Ba^{2+} + \cancel{2\ Cl^-} + \cancel{2\ H^+} + SO_4^{2-} \longrightarrow BaSO_4 + \cancel{2\ H^+} + \cancel{2\ Cl^-}$

Net ionic: $Ba^{2+} + SO_4^{2-} \longrightarrow BaSO_4$

(c) The reactants are a strong acid (HNO_3) and a soluble salt ($KC_2H_3O_2$). Both are dissociated in solution; hence, H^+, NO_3^-, K^+, and $C_2H_3O_2^-$ ions are present in the reaction mixture (before any reaction occurs). Possible new combinations of the reacting species are

Original Ion Combinations **Possible New Combinations**

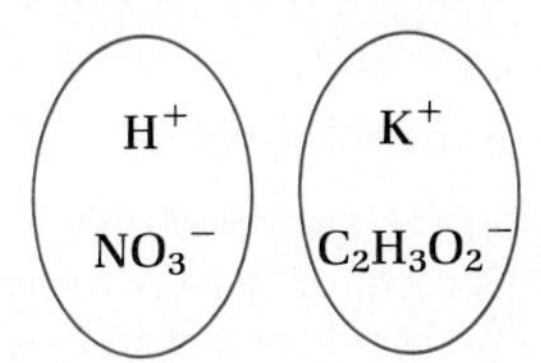

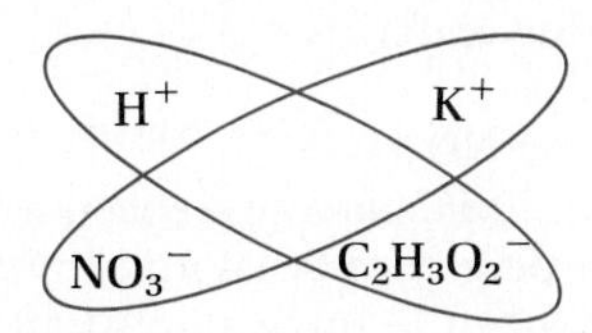

The driving force for the reaction will be the formation of acetic acid ($HC_2H_3O_2$) from the combination of H^+ and $C_2H_3O_2^-$ ions. The K^+ and NO_3^- will not combine because a soluble salt would be the product. The equations for the reaction are

Molecular: $HNO_3 + KC_2H_3O_2 \longrightarrow HC_2H_3O_2 + KNO_3$

Ionic: $H^+ + \cancel{NO_3^-} + \cancel{K^+} + C_2H_3O_2^- \longrightarrow HC_2H_3O_2 + \cancel{K^+} + \cancel{NO_3^-}$

Net ionic: $H^+ + C_2H_3O_2^- \longrightarrow HC_2H_3O_2$

14.10 Dissociation of Water

Normally, we think of water as a covalent, nondissociating substance. Experiments show, however, that in a sample of pure water a very small percentage of the water molecules have undergone dissociation to produce ions. The dissociation reaction may be thought of as involving the transfer of a proton from one water molecule to another (Brønsted–Lowry theory, Sec. 14.1)

$$H_2O + H_2O \longrightarrow H_3O^+ + OH^-$$

or simply as the dissociation of a single water molecule (Arrhenius theory, Sec. 14.1)

$$H_2O\ (HOH) \longrightarrow H^+ + OH^-$$

From either viewpoint, the net result is the formation of *equal amounts* of hydrogen (hydronium) ion and hydroxide ion.

The dissociation of water molecules is part of an equilibrium situation. Individual water molecules are continually dissociating. This process is balanced by hydroxide and hydrogen ions recombining to form water at the same rate. At equilibrium, at 25°C, the H^+ and OH^- ion concentrations are each $1.00 \times 10^{-7}\,M$ (0.000000100 M). This very, very small concentration is equivalent to there being one H^+ and one OH^- ion present for every 550,000,000 undissociated water molecules. Even though the H^+ and OH^- ion concentrations are very minute, they are important as we shall shortly see.

Experimentally it is found that, at any given temperature, the product of the concentrations of H^+ ion and OH^- ion in water is a constant. We can calculate the value of this constant, at 25°C, since we know the concentration of each ion is $1.00 \times 10^{-7}\,M$ at this temperature.

$$[H^+] \times [OH^-] = \text{constant}$$

$$(1.00 \times 10^{-7}) \times (1.00 \times 10^{-7}) = 1.00 \times 10^{-14}$$

ion product for water

Ion product for water *is the name given to the numerical value* (1.00×10^{-14}) *associated with the product of the* H^+ *ion and* OH^- *ion molar concentrations in water.* Note that the ion concentrations must be expressed in moles/L (molarity) in order to obtain the value 1.00×10^{-14} for the ion product for water. In the general expression for the ion product for water

$$[H^+] \times [OH^-] = \text{ion product for water}$$

the brackets [ ] specifically denote ion concentrations in moles/L.

The ion-product expression for water is valid not only in pure water but also when solutes are present in the water. At all times, the product of the hydrogen and hydroxide ion molarities in an aqueous solution, at 25°C, must equal 1.00×10^{-14}. Thus if the $[H^+]$ is increased by the addition of an acidic solute, the $[OH^-]$ must decrease until the expression

$$[H^+] \times [OH^-] = 1.00 \times 10^{-14}$$

is satisfied. Similarly, if additional OH^- ions are added to the water, the $[H^+]$ must correspondingly decrease. The extent of the decrease in $[H^+]$ or $[OH^-]$, as the result of the addition of quantities of the other ion, is easily calculated by the ion-product expression.

EXAMPLE 14.5

Sufficient acidic solute is added to a quantity of water to produce a $[H^+] = 2.50 \times 10^{-2}$ What is the $[OH^-]$ in this solution?

SOLUTION

The $[OH^-]$ can be calculated by the ion-product expression for water. Solving this expression for $[OH^-]$ gives

$$[OH^-] = \frac{1.00 \times 10^{-14}}{[H^+]}$$

Substituting into this expression the known $[H^+]$ and doing the arithmetic gives

$$[OH^-] = \frac{1.00 \times 10^{-14}}{2.50 \times 10^{-2}} = 4 \times 10^{-13} \quad \text{(calculator answer)}$$

$$= \mathbf{4.00 \times 10^{-13}} \quad \text{(correct answer)}$$

In Section 14.1 we defined an Arrhenius acid as a substance that produces H^+ ions in solution. Now that we have noted that a small amount of H^+ ions are present in all aqueous solutions, even basic ones, a slight modification of our definition of an Arrhenius acid is in order. An acid is a substance that increases the H^+ ion concentration in water. All **acidic solutions** *have a higher* $[H^+]$ *than* $[OH^-]$. In a similar manner, a base is a substance that increases the OH^- ion concentration in water. All **basic solutions** *have a higher* $[OH^-]$ *than* $[H^+]$. In a **neutral solution,** *the concentrations of both* H^+ *ions and* OH^- *ions are equal.* Table 14-7 summarizes the relationships between $[H^+]$ and $[OH^-]$ that we have just considered.

acidic solution

basic solution

neutral solution

Table 14-7. Relationships Between $[H^+]$ and $[OH^-]$ in Aqueous Solutions at 25°C

Neutral Solution:	$[H^+] = [OH^-] = 1.00 \times 10^{-7}$
Acidic Solution:	$[H^+]$ is greater than 1.00×10^{-7} $[OH^-]$ is less than 1.00×10^{-7}
Basic Solution:	$[H^+]$ is less than 1.00×10^{-7} $[OH^-]$ is greater than 1.00×10^{-7}

14.11 The pH Scale

Hydrogen ion concentrations in aqueous solution range from relatively high values (10 M) to extremely small ones (10^{-14} M). It is somewhat inconvenient to work with numbers that extend over such a wide range; a hydrogen ion concentration of 10 M is 1000 trillion times larger than a hydrogen ion concentration of 10^{-14} M. The pH scale, proposed by the Danish chemist Sören Sörensen (1868–1939) in 1909, was devised to express such a wide range of acidities in a more convenient way. **pH** *is a mathematical definition of* $[H^+]$ *that involves a numerical scale that runs from* 0 *to* 14.

pH

If the coefficient of the hydrogen ion concentration (expressed in exponential notation), is 1.0, for example, $\underline{1.0} \times 10^{-4}$ or $\underline{1.0} \times 10^{-7}$, the pH–$[H^+]$ relationship assumes a very simple form. The pH is given directly by the exponent of the power of ten without its minus sign; that is,

$$\begin{aligned} \text{For } [H^+] &= 1.0 \times 10^{-1} & \text{pH} &= 1 \\ [H^+] &= 1.0 \times 10^{-3} & \text{pH} &= 3 \\ [H^+] &= 1.0 \times 10^{-7} & \text{pH} &= 7 \\ [H^+] &= 1.0 \times 10^{-10} & \text{pH} &= 10 \end{aligned}$$

A neutral solution ($[H^+] = 1.0 \times 10^{-7}$) has a pH = 7. Values of pH less than 7 correspond to acidic solutions. The lower the pH value the greater the acidity. Values of pH greater than 7 represent basic solutions. The higher the pH value the greater the basicity. The relationships between $[H^+]$, $[OH^-]$, and pH are summarized in Table 14-8. Note from Table 14-8 that a change of one unit in pH corresponds to a tenfold increase or decrease in $[H^+]$. Also note that *lowering* the pH always corresponds to *increasing* the H^+ ion concentration.

Table 14-9 lists the pH values of a number of common substances. Except for gastric

Table 14-8. The pH Scale

pH	$[H^+]$	$[OH^-]$	
0	1	10^{-14}	↑
1	10^{-1}	10^{-13}	
2	10^{-2}	10^{-12}	
3	10^{-3}	10^{-11}	Acidic
4	10^{-4}	10^{-10}	
5	10^{-5}	10^{-9}	
6	10^{-6}	10^{-8}	
7	10^{-7}	10^{-7}	NEUTRAL
8	10^{-8}	10^{-6}	
9	10^{-9}	10^{-5}	
10	10^{-10}	10^{-4}	
11	10^{-11}	10^{-3}	Basic
12	10^{-12}	10^{-2}	
13	10^{-13}	10^{-1}	
14	10^{-14}	1	↓

Table 14-9. Approximate pH Values of Selected Common Substances

	pH	
↑	0	1 *M* HCl (0)
	1	0.1 *M* HCl (1), gastric juice (1.6–1.8), lime juice (1.8–2.0)
	2	soft drinks (2.0–4.0), vinegar (2.4–3.4)
Acidic	3	grapefruit (3.0–3.3), peaches (3.4–3.6)
	4	tomatoes (4.0–4.4), human urine (4.8–8.4)
	5	carrots (4.9–5.3), peas (5.8–6.4)
	6	human saliva (6.2–7.4), cow's milk (6.3–6.6), drinking water (6.5–8.0)
NEUTRAL	7	pure water (7.0), human blood (7.35–7.45), fresh eggs (7.6–8.0)
	8	seawater (8.3), soaps, shampoos (8–9)
	9	detergents (9–10)
	10	milk of magnesia (9.9–10.1)
Basic	11	household ammonia (11.5–12.0)
	12	liquid bleach (12.0)
	13	0.1 *M* NaOH (13.0)
↓	14	1 *M* NaOH (14.0)

juices, most human body fluids have pH values within a couple of units of neutrality. Almost all foods are acidic. Tart taste is associated with food of low pH.

Note that some of the pH values in Table 14-9 are nonintegral. Obtaining such pH values from measured H^+ ion concentrations involves the use of logarithms, a mathematical concept we will not deal with in this text. We can still, however, when given nonintegral pH values, interpret them in a qualitative way without resorting to logarithm use. For example, a pH of 4.75 is between a pH of 4 and a pH of 5. We know that a pH of 4 corresponds to $[H^+] = 1 \times 10^{-4}$ and that a pH of 5 corresponds to $[H^+] = 1 \times 10^{-5}$. Hence, a pH of 4.75 will correspond to a $[H^+]$ in the range bounded by 1×10^{-4} and 1×10^{-5}.

14.12 Acid–Base Titrations

Determining the concentration of acid or base in a solution is a regular activity in many laboratories. The concentration of an acid or base in a solution and the solution's pH are two different entities. The pH of a solution gives information about the concentration of hydrogen (hydronium) ions in solution. Only dissociated molecules influence the pH value. The concentration of an acid or base solution gives information about the *total number* of acid or base molecules present; both dissociated and undissociated molecules are counted.

titration

The procedure most frequently used to determine the concentration of an acidic or basic solution is that of *titration*. **Titration** *is the gradual adding of one solution to another until the solute in the first solution has reacted completely with the solute in the second solution.*

Suppose we want to determine the concentration of an acid solution by titration. We would first measure out a *known volume* of the acid solution into a beaker or flask. A solution of base of *known concentration* is then slowly added to the flask or beaker by means of a buret (see Fig. 14-1). Base addition continues until all the acid has completely reacted with added base. The *volume of base* needed to reach this point is obtained from the buret readings. Knowing the original volume of acid, the concentration of the base, and the volume of added base, the concentration of the acid can be calculated (Sec. 14.13).

indicator

In order to complete a titration successfully, we must be able to detect when the reaction between acid and base is complete. One way to do this is to add an indicator to the solution being titrated. An **indicator** *is a compound that exhibits different colors depending on the pH of its surroundings.* Typically, an indicator is one color in basic solutions and another color in acidic solutions. An indicator is selected that will change color at a pH corresponding as nearly as possible to the pH of the solution when the titration is complete. What this pH will be can be calculated ahead of time knowing the identity of the acid and base involved in the titration.

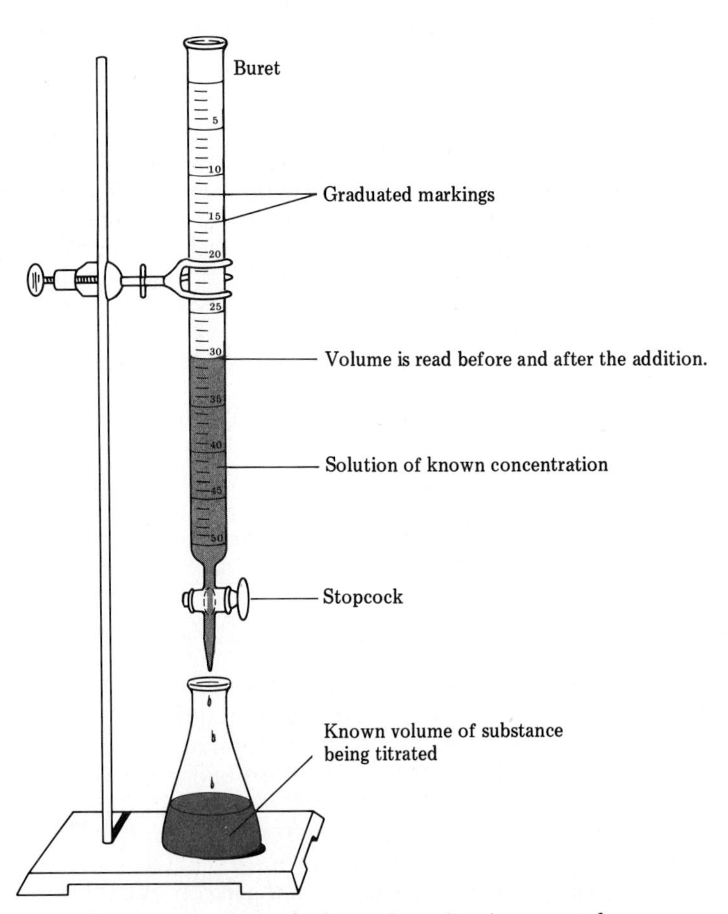

Figure 14-1. Use of a buret in a titration procedure.

14.13 Acid–Base Calculations Using Normality

Normality is the most convenient concentration unit to work with when doing acid–base titration calculations. The reason for this is that it takes into account the fact that all acids and bases do not yield the same number of hydrogen or hydroxide ions per molecule on dissociation. Even though their molar concentrations are equal, 0.10 *M* H_2SO_4 (which can supply two H^+ ions per molecule) will neutralize twice as much base as will the same volume of 0.10 *M* HNO_3 (which can supply only one H^+ ion per molecule). However, equal volumes of 0.10 *N* H_2SO_4 and 0.10 *N* HNO_3 will neutralize exactly the same amount of base.

We have already considered the general definition for normality (Sec. 13.9).

$$\text{Normality} = \frac{\text{equiv of solute}}{\text{L of solution}}$$

In Section 13.9 an equivalent was defined in a way applicable for use when dealing with ionic solutes. Definitions of equivalent for use in acid–base work are as follows. An **equivalent of acid** *is the quantity of acid that will supply one mole of* H^+ *ion upon complete dissociation.* Similarly, an **equivalent of base** *is the quantity of base that will react with one mole of* H^+ *ions.* Examples 14.6 and 14.7 illustrate the use of these two definitions for an equivalent in problem-solving contexts.

equivalent of acid

equivalent of base

EXAMPLE 14.6

Determine the number of equivalents of acid or base in each of the following samples.
(a) 1 mole of H_3PO_4 **(b)** 3 moles of $Ba(OH)_2$ **(c)** 2 moles of HNO_3

SOLUTION

(a) The equation for the complete dissociation of H_3PO_4 is

$$H_3PO_4 \longrightarrow 3\,H^+ + PO_4{}^{3-}$$

Thus, 1 mole of H_3PO_4 gives up 3 moles of H^+ ion on complete dissociation. Three moles of H^+ ion is equal to 3 equiv of H^+ ion by definition. Therefore,

$$1\,\cancel{\text{mole } H_3PO_4} \times \left(\frac{3 \text{ equiv } H_3PO_4}{1\,\cancel{\text{mole } H_3PO_4}}\right) = 3 \text{ equiv } H_3PO_4$$

(b) The equation for the complete dissociation of $Ba(OH)_2$ is

$$Ba(OH)_2 \longrightarrow Ba^{2+} + 2\,OH^-$$

Thus, 1 mole of $Ba(OH)_2$ yields 2 moles of OH^- ion (which can react with 2 moles of H^+). Therefore,

$$3\,\cancel{\text{moles } Ba(OH)_2} \times \left(\frac{2 \text{ equiv } Ba(OH)_2}{1\,\cancel{\text{mole } Ba(OH)_2}}\right) = 6 \text{ equiv } Ba(OH)_2$$

(c) The equation for the complete dissociation of HNO_3 is

$$HNO_3 \longrightarrow H^+ + NO_3{}^-$$

One mole of acid yields only 1 mole (or equiv) of H^+ ion. Therefore,

$$2 \text{ }\cancel{\text{moles } HNO_3} \times \left(\frac{1 \text{ equiv } HNO_3}{1 \text{ }\cancel{\text{mole } HNO_3}}\right) = 2 \text{ equiv } HNO_3$$

EXAMPLE 14.7

Calculate the normality of a solution made by dissolving 25.0 g of NaOH in enough water to give 500.0 mL of solution.

SOLUTION

To calculate normality, we need to know the number of equiv of solute and solution volume (in liters).

For NaOH,

$$1 \text{ equiv} = 1 \text{ mole}$$

since 1 mole of NaOH yields 1 mole of OH^- ion

$$NaOH \longrightarrow Na^+ + OH^-$$

and 1 mole of OH^- reacts with 1 mole of H^+ ion. (Hydrogen and hydroxide ions always react in a one-to-one ratio.)

The number of equivalents of NaOH in 25.0 g of NaOH is obtained by converting the grams of NaOH to moles of NaOH and then using the above mole–equivalent relationship as a conversion factor to obtain equivalents.

$$25.0 \text{ }\cancel{\text{g NaOH}} \times \left(\frac{1 \text{ }\cancel{\text{mole NaOH}}}{40.0 \text{ }\cancel{\text{g NaOH}}}\right) \times \left(\frac{1 \text{ equiv NaOH}}{1 \text{ }\cancel{\text{mole NaOH}}}\right) = 0.625 \text{ equiv NaOH}$$

(calculator and correct answer)

The solution volume of 500.0 mL, changed to liter units, is 0.5000 L.

Both quantities called for in the defining equation for normality, equivalents of solute and liters of solution, are now known. Therefore, substituting into the defining equation, we get

$$N = \frac{0.625 \text{ equiv NaOH}}{0.5000 \text{ L solution}} = 1.25 \frac{\text{equiv NaOH}}{\text{L solution}}$$

$$= 1.25 \text{ N}$$ (calculator and correct answer)

In the laboratory solution concentrations are specified in terms of molarity rather than normality. To convert acid or base molarities to normalities (or vice versa) is a very simple operation. Only the conversion factor relating moles to equivalents is needed.

EXAMPLE 14.8

Express the following molar concentrations as normalities.

(a) 0.030 M H_3PO_4 **(b)** 0.10 M $Ba(OH)_2$ **(c)** 6.0 M HNO_3

SOLUTION

(a) A 0.030 *M* H_3PO_4 solution contains 0.030 mole of H_3PO_4 per liter of solution. Using this fact and the fact (from Ex. 14.6) that

$$1 \text{ mole } H_3PO_4 = 3 \text{ equiv } H_3PO_4$$

we calculate the normality as follows.

$$\underbrace{\left(\frac{0.030 \cancel{\text{mole } H_3PO_4}}{1 \text{ L solution}}\right)}_{\text{Molarity}} \times \left(\frac{3 \text{ equiv } H_3PO_4}{1 \cancel{\text{mole } H_3PO_4}}\right) = \underbrace{\left(\frac{0.09 \text{ equiv } H_3PO_4}{\text{L solution}}\right)}_{\text{Normality}} \quad \text{(calculator answer)}$$

$$= 0.090\ N \quad \text{(correct answer)}$$

Therefore, the solution is 0.090 *N*.

(b) A 0.10 *M* $Ba(OH)_2$ solution contains 0.10 mole of $Ba(OH)_2$ per liter of solution. This information, coupled with the fact (from Ex. 14.6) that

$$1 \text{ mole } Ba(OH)_2 = 2 \text{ equiv } Ba(OH)_2$$

enables us to calculate the normality as follows.

$$\left(\frac{0.10 \cancel{\text{mole } Ba(OH)_2}}{1 \text{ L solution}}\right) \times \left(\frac{2 \text{ equiv } Ba(OH)_2}{1 \cancel{\text{mole } Ba(OH)_2}}\right) = \frac{0.2 \text{ equiv } Ba(OH)_2}{\text{L solution}} \quad \text{(calculator answer)}$$

$$= 0.20\ N\ Ba(OH)_2 \quad \text{(correct answer)}$$

(c) For 6.0 *M* HNO_3, we calculate the normality, using reasoning similar to that in parts (a) and (b), as follows.

$$\left(\frac{6.0 \cancel{\text{moles } HNO_3}}{1 \text{ L solution}}\right) \times \left(\frac{1 \text{ equiv } HNO_3}{1 \cancel{\text{mole } HNO_3}}\right) = \frac{6 \text{ equiv } HNO_3}{\text{L solution}} \quad \text{(calculator answer)}$$

$$= 6.0\ N\ HNO_3 \quad \text{(correct answer)}$$

Molarities may also be converted to normalities, without involvement with dimensional analysis, by the following equations.

$$N_{\text{acid}} = M_{\text{acid}} \times \text{number of acidic hydrogens in acid}$$

$$N_{\text{base}} = M_{\text{base}} \times \text{number of hydroxide ions in base}$$

Both of these equations are based on the Arrhenius system for defining acids and bases. It is not recommended that you use these "shortcut" equations until you thoroughly understand the dimensional-analysis approach to obtaining normalities (Ex. 14.8).

In an acid–base titration, the chemical reaction that occurs is that of neutralization (Sec. 14.7). The H^+ ions from the acid react with the OH^- ions from the base to produce water.

$$H^+ + OH^- \longrightarrow H_2O$$

These two ions always react in a one-to-one ratio; thus, 1 mole of H^+ ions always reacts with 1 mole of OH^- ions. Since 1 mole of each of these ions is an equivalent, it follows that at the endpoint of an acid–base titration an equal number of equivalents of acid and base have been consumed; that is,

$$\text{Equivalents of reacted acid} = \text{equivalents of reacted base}$$

A simple equation, whose derivation is based on the fact that equal numbers of equivalents of acid and base are consumed during an acid–base titration, is employed in titration calculations. It is

$$N_{acid} \times V_{acid} = N_{base} \times V_{base}$$

The derivation of this equation is as follows. The defining equations for the normality of the acid and base are

$$N_{acid} = \frac{E_{acid}}{V_{acid}} \quad \text{and} \quad N_{base} = \frac{E_{base}}{V_{base}}$$

where E is the equivalents of solute and V is the volume of solution (in liters). Each of these equations can be rearranged as follows.

$$E_{acid} = N_{acid} \times V_{acid}$$

$$E_{base} = N_{base} \times V_{base}$$

At the endpoint of the titration, where $E_{acid} = E_{base}$, we, therefore, can write

$$N_{acid} \times V_{acid} = N_{base} \times V_{base}$$

The volumes V_{acid} and V_{base} may be expressed in L or mL, provided the same unit is used for both. Since burets are calibrated in mL, mL is the unit most often used in titration calculations.

The data obtained from an acid–base titration (Sec. 14.12) are (1) the volume of base used, (2) the volume of acid used, and (3) the concentration of the acid or base used to titrate the solution of unknown concentration. These quantities are three of the four variables in the equation

$$N_{acid} \times V_{acid} = N_{base} \times V_{base}$$

Hence, the fourth variable, the concentration of the acid or base that was titrated, can easily be calculated.

EXAMPLE 14.9

In an acid–base titration, 32.7 mL of 0.105 M NaOH is required to neutralize 50.0 mL of HNO_3. What are the normality and the molarity of the HNO_3 solution?

SOLUTION

We will use the equation

$$N_{acid} \times V_{acid} = N_{base} \times V_{base}$$

to solve this problem. Three of the four quantities in this equation are known. The volumes of acid and base are, respectively, 50.0 and 32.7 mL. The normality of the base is not directly given but may easily be calculated from the molarity of the base. Since 1 mole of NaOH contains 1 equiv of OH^- ions, the normality of the base will be the same as the molarity; that is, the base is not only 0.105 M but also 0.105 N.

Substitution of these known values into the equation, rearranged so that N_{acid} is isolated on the left side, gives

$$N_{acid} = \frac{N_{base} \times V_{base}}{V_{acid}}$$

$$= 0.105\, N \times \left(\frac{32.7\, \cancel{mL}}{50.0\, \cancel{mL}}\right) = 0.06867\, N \quad \text{(calculator answer)}$$

$$= \mathbf{0.0687}\, \boldsymbol{N} \quad \text{(correct answer)}$$

The molarity of HNO_3, which was also asked for in this problem, will have the same numerical value as the normality, since HNO_3 is an acid for which equivalents and moles are identical.

EXAMPLE 14.10

What volume of 0.500 M H_3PO_4 would be needed to completely neutralize 275 mL of 0.125 M $Ba(OH)_2$ solution?

SOLUTION

The concentrations of the two solutions must be converted from molarities to normalities. Since H_3PO_4 yields three H^+ ions per acid molecule, 1 mole of H_3PO_4 contains 3 equiv of acid; hence, the normality of the H_3PO_4 solution is three times the molarity.

$$\left(\frac{0.500\, \cancel{\text{mole } H_3PO_4}}{1\text{ L solution}}\right) \times \left(\frac{3\text{ equiv } H_3PO_4}{1\, \cancel{\text{mole } H_3PO_4}}\right) = \frac{1.5\text{ equiv } H_3PO_4}{\text{L solution}} \quad \text{(calculator answer)}$$

$$= \mathbf{1.50}\, \boldsymbol{N}\, \mathbf{H_3PO_4} \quad \text{(correct answer)}$$

For $Ba(OH)_2$, the normality will be double the molarity since $Ba(OH)_2$ yields two OH^- ions per molecule on dissociation.

$$\left(\frac{0.125\, \cancel{\text{mole } Ba(OH)_2}}{1\text{ L solution}}\right) \times \left(\frac{2\text{ equiv } Ba(OH)_2}{1\, \cancel{\text{mole } Ba(OH)_2}}\right) = \frac{0.25\text{ equiv } Ba(OH)_2}{\text{L solution}} \quad \text{(calculator answer)}$$

$$= \mathbf{0.250}\, \boldsymbol{N}\, \mathbf{Ba(OH)_2} \quad \text{(correct answer)}$$

Substituting the known values in the equation

$$V_{acid} = \frac{N_{base} \times V_{base}}{N_{acid}}$$

gives

$$V_{acid} = \frac{(0.250\, \cancel{N})(275\text{ mL})}{(1.50\, \cancel{N})} = 45.833333\text{ mL} \quad \text{(calculator answer)}$$

$$= \mathbf{45.8\ mL} \quad \text{(correct answer)}$$

Learning Objectives

After completing this chapter you should be able to

- State the definitions of acids and bases using (1) the Arrhenius concept and (2) the Brønsted–Lowry concept (Sec. 14.1).
- Differentiate between and give examples of strong and weak Arrhenius acids and bases (Sec. 14.2).
- Differentiate between and give examples of mono-, di-, and triprotic acids, and write equations for the stepwise dissociation of di- and triprotic acids (Sec. 14.3).
- State the meaning of the terms dilute and concentrated when applied to specific common acid and base stock solutions (Sec. 14.4).
- Recognize compounds classified as salts (Sec. 14.5).
- Convert molecular equations to net ionic equations (Sec. 14.6).
- Write equations for the reactions of acids with active metals, bases, and carbonate and bicarbonate salts (Sec. 14.7).
- Write equations for the reactions of salts in solution with selected metals, acids, bases, and other salt solutions (Sec. 14.9).
- Give the value of $[H^+]$ and $[OH^-]$ in pure water at 25°C, and be able to calculate $[OH^-]$ in an aqueous solution given $[H^+]$ and vice versa (Sec. 14.10).
- Tell whether a solution is acidic, basic, or neutral given its $[H^+]$ or pH (Sec. 14.10 and 14.11).
- Define pH and calculate the pH of solutions with selected $[H^+]$ and $[OH^-]$ (Sec. 14.11).
- Understand the steps involved in the titration of an acid or base (Sec. 14.12).
- Express the concentration of an acid or base solution in terms of normality (Sec. 14.13).
- Calculate the volume (or concentration) of an acidic or basic solution, given its concentration (or volume) and the volume and concentration of a second solution with which it is titrated (Sec. 14.13).

Terms and Concepts for Review

The new terms or concepts defined in this chapter are

Arrhenius acid (Sec. 14.1)
Arrhenius base (Sec. 14.1)
Brønsted–Lowry acid (Sec. 14.1)
Brønsted–Lowry base (Sec. 14.1)
strong acid (Sec. 14.2)
weak acid (Sec. 14.2)
monoprotic acid (Sec. 14.3)
diprotic acid (Sec. 14.3)
triprotic acid (Sec. 14.3)
polyprotic acid (Sec. 14.3)
salt (Sec. 14.5)
ionic equation (Sec. 14.6)
net ionic equation (Sec. 14.6)
neutralization (Sec. 14.7)
ion product for water (Sec. 14.10)
acidic solution (Sec. 14.10)
basic solution (Sec. 14.10)
neutral solution (Sec. 14.10)
pH (Sec. 14.11)
titration (Sec. 14.12)
indicator (Sec. 14.12)
equivalent of acid (Sec. 14.13)
equivalent of base (Sec. 14.13)

Questions and Problems

Acid–Base Definitions

14-1 What are the Arrhenius definitions for an acid and a base?

14-2 What are the Brønsted–Lowry definitions for an acid and a base, and what advantages does their use offer over the Arrhenius definitions?

14-3 Write equations for the dissociation of the following Arrhenius acids and bases in water.
- **a.** HBr (hydrobromic acid)
- **b.** $HClO_2$ (chlorous acid)
- **c.** LiOH (lithium hydroxide)
- **d.** CsOH (cesium hydroxide)

14-4 Identify the Brønsted–Lowry acid and base in the following reactions.
- **a.** $HBr + H_2O = H_3O^+ + Br^-$
- **b.** $H_2O + N_3^- = HN_3 + OH^-$
- **c.** $H_2S + H_2O = H_3O^+ + HS^-$
- **d.** $HS^- + H_2O = H_3O^+ + S^{2-}$

14-5 Write equations to illustrate the acid–base reactions that can take place between the following Brønsted–Lowry acids and bases.
- **a.** acid: HOCl; base: NH_3
- **b.** acid: $HClO_4$; base: H_2O
- **c.** acid: H_2O; base: NH_2^-
- **d.** acid: $HC_2O_4^-$; base: H_2O

Strengths of Acids and Bases

14-6 What is the principal distinction between strong and weak acids? Is the distinction between strong and weak bases of a similar nature?

14-7 Identify each of the following compounds as a strong or a weak acid or base.
- **a.** HI **b.** H_3BO_3 **c.** $Sr(OH)_2$
- **d.** H_2SO_4 **e.** NH_3 **f.** RbOH

14-8 What is the distinction between the terms weak acid and dilute acid?

Polyprotic Acids

14-9 Identify the following acids as monoprotic, diprotic, or triprotic.
- **a.** $H_2C_2O_4$ (oxalic acid)
- **b.** $HC_2H_3O_2$ (acetic acid)
- **c.** H_3PO_4 (phosphoric acid)
- **d.** HNO_3 (nitric acid)
- **e.** HClO (hypochlorous acid)
- **f.** $H_3C_6H_5O_7$ (citric acid)

14-10 Write equations showing all steps in the dissociation of the following acids.
a. H_3PO_4 (phosphoric acid)
b. H_2CO_3 (carbonic acid)
c. $HC_4H_7O_2$ (butyric acid)

14-11 The formula for lactic acid is preferably written as $HC_3H_5O_3$ (the symbol for H appears in two places) rather than as $C_3H_6O_3$ (the symbol for H is used only once). Why?

Stock Solutions of Acids and Bases

14-12 What would be the concentration (molarity) of acid or base in containers labeled as follows.
a. dilute H_2SO_4 **b.** dilute HCl
c. dilute aqueous NH_3 **d.** concentrated HNO_3
e. concentrated HCl **f.** NaOH

Salts

14-13 Identify with formula and name the positive and negative ions present in each of the following salts. (You may have to refer back to Secs. 6.5–6.9.)
a. $CaSO_4$ **b.** Li_2CO_3 **c.** NaBr
d. Al_2S_3 **e.** NH_4Cl **f.** $Be(C_2H_3O_2)_2$
g. AgI **h.** KNO_3 **i.** NH_4CN
j. $Mg_3(PO_4)_2$

14-14 Which of the salts listed in Problem 14-13 are soluble in water? For each of the soluble salts indicate the total number of ions produced when one formula unit of the salt dissolves in water.

14-15 Identify each of the following as an acid, a base, or a salt.
a. HCl **b.** NaCl **c.** NH_4Cl
d. $Ca_3(PO_4)_2$ **e.** $Mg(OH)_2$ **f.** $HC_2H_3O_2$
g. KOH **h.** AgCl **i.** $NaNO_3$
j. HNO_3

14-16 The term weak is used in describing certain acids and bases, but is never used in describing salts. Why?

Ionic and Net Ionic Equations

14-17 Classify the following equations for reactions occurring in aqueous solution as *molecular, ionic,* or *net ionic* equations.
a. $MgCO_3 + 2\,HBr \longrightarrow MgBr_2 + H_2O + CO_2$
b. $2\,OH^- + H_2CO_3 \longrightarrow CO_3^{2-} + 2\,H_2O$
c. $K^+ + OH^- + H^+ + I^- \longrightarrow K^+ + I^- + H_2O$
d. $CaCO_3 + 2\,H^+ + 2\,NO_3^- \longrightarrow Ca^{2+} + 2\,Cl^- + H_2O + CO_2$
e. $Ag^+ + Cl^- \longrightarrow AgCl$
f. $Ni + Cu^{2+} \longrightarrow Cu + Ni^{2+}$

14-18 Write a balanced net ionic equation for each of the following reactions, each of which occurs in aqueous solution.
a. $FeS + 2\,HCl \longrightarrow FeCl_2 + H_2S$
b. $HC_2H_3O_2 + NaOH \longrightarrow NaC_2H_3O_2 + H_2O$
c. $Zn + H_2SO_4 \longrightarrow ZnSO_4 + H_2$
d. $HCl + KOH \longrightarrow KCl + H_2O$
e. $Pb(NO_3)_2 + (NH_4)_2S \longrightarrow 2\,NH_4NO_3 + PbS$
f. $2\,Al + 6\,HBr \longrightarrow 2\,AlBr_3 + 3\,H_2$

14-19 Write a balanced net ionic equation for each of the following reactions, each of which occurs in aqueous solution.
a. $2\,K + 2\,H_2O \longrightarrow 2\,KOH + H_2$
b. $Cl_2 + 2\,NaBr \longrightarrow 2\,NaCl + Br_2$
c. $Ba(NO_3)_2 + K_2SO_4 \longrightarrow 2\,KNO_3 + BaSO_4$
d. $Ba(OH)_2 + H_2SO_4 \longrightarrow BaSO_4 + 2\,H_2O$
e. $HNO_3 + NaOH \longrightarrow NaNO_3 + H_2O$
f. $2\,HC_2H_3O_2 + Ca(OH)_2 \longrightarrow Ca(C_2H_3O_2)_2 + 2\,H_2O$

Reactions of Acids and Bases

14-20 On the basis of the activity series (Table 14-6), predict whether each of the following metals will react with the indicated substance to form H_2. If the answer is yes, then write a net ionic equation for the reaction. (Note that Cu, Sn, Zn, and Fe all form +2 ions in solution, Na a +1 ion, and Au a +3 ion.)
a. Cu and H_2SO_4 **b.** Sn and HNO_3
c. Zn and HCl **d.** Fe and H_2O (hot)
e. Na and H_2O **f.** Au and HCl

14-21 Write balanced net ionic equations showing the total neutralization of the following.
a. HCl and NaOH **b.** HNO_3 and NaOH
c. H_2SO_4 and NaOH **d.** $HC_2H_3O_2$ and KOH
e. HBr and $Ba(OH)_2$ **f.** H_3PO_4 and LiOH

14-22 Write a molecular equation for the preparation of the following salts using an acid–base reaction.
a. Li_2SO_4 **b.** NaCl **c.** KNO_3 **d.** $Ba_3(PO_4)_2$

14-23 Write a molecular equation for the action of hydrochloric acid on each of the following. (Zn metal forms a Zn^{2+} ion in solution.)
a. Zn **b.** NaOH **c.** Na_2CO_3 **d.** $NaHCO_3$

14-24 Using the reaction types discussed in Sec. 14.7, write molecular equations to illustrate three different ways that KCl might be prepared from HCl and other appropriate reactants. HCl must be a reactant in every case.

Reactions of Salts

14-25 On the basis of the activity series (Table 14-6), predict whether zinc metal will dissolve in the following aqueous salt solutions. If the answer is yes, then write a net ionic equation for the reaction. (Zinc metal forms a Zn^{2+} ion in solution.)
a. lead(II) nitrate [$Pb(NO_3)_2$] solution
b. sodium chloride (NaCl) solution
c. silver nitrate ($AgNO_3$) solution
d. iron(III) chloride ($FeCl_3$) solution
e. nickel(II) acetate [$Ni(C_2H_3O_2)_2$] solution

14-26 Salts will react with acids, bases, and other salts only if one of three conditions is met. What are these three conditions?

14-27 Indicate the driving force (condition) that causes each of the following acid–salt reactions to occur.
a. $Ba(NO_3)_2 + H_2SO_4 \longrightarrow BaSO_4 + 2\ HNO_3$
b. $3\ CaCl_2 + 2\ H_3PO_4 \longrightarrow Ca_3(PO_4)_2 + 6\ HCl$
c. $AgC_2H_3O_2 + HCl \longrightarrow AgCl + HC_2H_3O_2$
d. $K_2CO_3 + 2\ HNO_3 \longrightarrow 2\ KNO_3 + CO_2 + H_2O$
e. $Na_3PO_4 + 3\ HCl \longrightarrow 3\ NaCl + H_3PO_4$

14-28 Which of the following salt solutions would be expected to react with the base sodium hydroxide (NaOH)? In those cases where a reaction occurs, write the net ionic equation and indicate the driving force for the reaction.
a. Li_2SO_4 solution **b.** $Al_2(SO_4)_3$ solution
c. $LiNO_3$ solution **d.** NH_4Cl solution

14-29 Write a molecular equation for the preparation of each of the following insoluble salts using as reactants aqueous solutions of soluble salts.
a. $PbSO_4$ **b.** $BaCO_3$ **c.** AgCl
d. $Be_3(PO_4)_2$

Hydrogen Ion and Hydroxide Ion Concentrations

14-30 What is the concentration of hydrogen and hydroxide ion in pure water at 25°C?

14-31 Calculate the H^+ ion concentration of a solution if the OH^- ion concentration is
a. $4.0 \times 10^{-3}\ M$ **b.** $6.5 \times 10^{-7}\ M$
c. $8.0 \times 10^{-10}\ M$ **d.** $2.0 \times 10^{-6}\ M$

14-32 Indicate whether the following solutions are acidic, basic, or neutral.
a. $[H^+] = 4.7 \times 10^{-8}$ **b.** $[H^+] = 1.2 \times 10^{-2}$
c. $[H^+] = 3.0 \times 10^{-5}$ **d.** $[OH^-] = 1.0 \times 10^{-7}$
e. $[OH^-] = 2.0 \times 10^{-10}$ **f.** $[OH^-] = 3.0 \times 10^{-3}$

pH Scale

14-33 Calculate the pH of the following solutions.
a. $[H^+] = 1.0 \times 10^{-4}$ **b.** $[H^+] = 1.0 \times 10^{-10}$
c. $[H^+] = 1.0 \times 10^{-7}$ **d.** $[OH^-] = 1.0 \times 10^{-4}$
e. $[OH^-] = 1.0 \times 10^{-8}$ **f.** $[OH^-] = 1.0 \times 10^{-7}$

14-34 Indicate whether the following solutions are very acidic, moderately acidic, slightly acidic, neutral, slightly basic, moderately basic, or very basic.
a. saliva, pH = 6.9
b. sea water, pH = 8.5
c. lime juice, pH = 1.8
d. desert soil, pH = 9.9
e. forest soil, pH = 5.3
f. blood plasma, pH = 7.5
g. household ammonia, pH = 11.7

14-35 What is the H^+ ion concentration in a solution, in moles/L, when the solution has a pH of
a. 1.00 **b.** 7.00 **c.** 3.00 **d.** 12.00

14-36 Solution A has a pH of 4 and solution B has a pH of 6. Which solution is more acidic? How many times more acidic is the one solution than the other?

Normality of Acids and Bases

14-37 How many equivalents of acid or base are present in each of the following samples?
a. 1 mole NaOH **b.** 1 mole H_2SO_4
c. 2 moles $Ba(OH)_2$ **d.** 0.500 mole H_3PO_4
e. 10.0 g H_2SO_4 **f.** 5.00 g $HC_2H_3O_2$
g. 1.00 g HCl **h.** 100 g $Ba(OH)_2$

14-38 Determine the normalities of the following solutions.
a. 250.0 mL of solution containing 5.00 g NaOH
b. 750.0 mL of solution containing 10.0 g H_3PO_4
c. 100.0 mL of solution containing 0.25 mole H_2SO_4
d. 2.00 L of solution containing 0.0400 equiv $H_2C_2O_4$

14-39 Express the following molarities as normalities.
a. 0.0025 *M* H_2SO_4 **b.** 1.70 *M* HNO_3
c. 0.645 *M* NaOH **d.** 6.0 *M* $HC_2H_3O_2$
e. 0.0010 *M* $Ba(OH)_2$ **f.** 3.0 *M* H_3PO_4

14-40 Express the following normalities as molarities.
a. 0.0030 *N* H_2SO_4 **b.** 1.70 *N* HNO_3
c. 0.134 *N* NaOH **d.** 6.0 *N* $HC_2H_3O_2$
e. 0.026 *N* $Ba(OH)_2$ **f.** 3.0 *N* H_3PO_4

Titration Calculations

14-41 How many mL of 0.100 *N* NaOH solution would be needed to neutralize each of the following.
a. 10.00 mL of 0.350 *N* H_2SO_4
b. 50.00 mL of 1.500 *N* HNO_3
c. 5.00 mL of 0.500 *M* H_3PO_4
d. 75.00 mL of 0.00030 *M* HCl

14-42 How many equiv of base would each of the following acid samples neutralize?
a. 100.0 mL of 0.500 *M* H_2SO_4
b. 100.0 mL of 0.400 *M* HCl
c. 100.0 mL of 0.200 *N* H_3PO_4
d. 100.0 mL of 0.300 *N* $HC_2H_3O_2$

14-43 In an acid–base titration, 45.0 mL of 0.200 *M* NaOH is required to neutralize 100.0 mL of H_3PO_4. What are the normality and molarity of the H_3PO_4 solution?

14-44 A 20.00 mL sample of diprotic oxalic acid ($H_2C_2O_4$) is titrated with 0.250 *N* NaOH solution. A total of 27.86 mL of NaOH is required. Calculate
a. the normality of the oxalic acid solution
b. the number of equiv of oxalic acid in the 20.00 mL sample
c. the number of moles of oxalic acid in the 20.00 mL sample
d. the molarity of the oxalic acid solution
e. the number of grams of oxalic acid in the 20.00 mL sample

15 Oxidation and Reduction

15.1 Oxidation–Reduction Terminology

The terms oxidation and reduction, like the terms acid and base (Sec. 14.1), have several definitions. Historically, the word oxidation was first used to describe the reaction of a substance with oxygen. According to this original definition, each of the following reactions involves oxidation.

$$4\,Fe + 3\,O_2 \longrightarrow 2\,Fe_2O_3$$

$$S + O_2 \longrightarrow SO_2$$

$$CH_4 + 2\,O_2 \longrightarrow CO_2 + 2\,H_2O$$

The substance on the far left in each of these equations is said to have been oxidized.

Originally, the term reduction referred to processes where oxygen was removed from a compound. A particularly common type of reduction reaction, according to this original definition, is the removal of oxygen from a metal oxide to produce the free metal.

$$CuO + H_2 \longrightarrow Cu + H_2O$$

$$2\,Fe_2O_3 + 3\,C \longrightarrow 4\,Fe + 3\,CO_2$$

The term reduction comes from the reduction in mass of the metal-containing species; the metal has a mass less than that of the metal oxide.

Today the words oxidation and reduction are used in a much broader sense. Current definitions include the previous examples but also much more. It is now recognized that the changes brought about in a substance from reaction with oxygen can also be caused by reaction with numerous non-oxygen-containing substances. For example, consider the following reactions.

$$2\,Mg + O_2 \longrightarrow 2\,MgO$$
$$Mg + S \longrightarrow MgS$$
$$Mg + F_2 \longrightarrow MgF_2$$
$$3\,Mg + N_2 \longrightarrow Mg_3N_2$$

In each of these reactions magnesium metal is converted to a magnesium compound that contains Mg^{2+} ions. The process is the same — the changing of magnesium atoms, through the loss of two electrons, to magnesium ions; the only difference is the identity of the substance that causes magnesium to undergo the change. All of these reactions are considered to involve oxidation when the modern definition of oxidation is applied. **Oxidation** *is the process whereby a substance in a chemical reaction loses one or more electrons.* The modern definition for reduction involves the use of similar terminology. **Reduction** *is the process whereby a substance in a chemical reaction gains one or more electrons.*

oxidation

reduction

Oxidation and reduction are complimentary processes rather than isolated phenomena. They *always* occur together; you cannot have one without the other. If electrons are lost by one species, they cannot just disappear; they must be gained by another species. Electron transfer, then, is the basis for oxidation and reduction. The collective term **oxidation–reduction reaction** *is used to describe any reaction involving the transfer of electrons between reactants.* This designation is often shortened to simply redox reaction.

oxidation–reduction reaction

There are two different ways of looking at the reactants in a redox reaction. First, the reactants can be viewed as being acted upon. From this viewpoint one reactant is *oxidized* (the one that loses electrons) and one is *reduced* (the one that gains electrons). Second, the reactants can be looked at as bringing about the reaction. In this approach the terms oxidizing agent and reducing agent are used. An **oxidizing agent** *causes oxidation by accepting electrons from the other reactant.* Such acceptance, the gain of electrons, means that the oxidizing agent itself is reduced. Similarly, the **reducing agent** *causes reduction by providing electrons for the other reactant to accept.* As a result of providing electrons, the reducing agent itself becomes oxidized. Note, then, that the reducing agent and substance oxidized are one and the same, as are the oxidizing agent and substance reduced.

oxidizing agent

reducing agent

The terms oxidizing agent and reducing agent sometimes cause confusion because the oxidizing agent is not oxidized (it is reduced) and the reducing agent is not reduced (it is oxidized). By simple analogy, a travel agent is not the one who takes a trip — he or she is the one who causes the trip to be taken.

Table 15-1 summarizes the terms presented in this section.

Table 15-1. Oxidation–Reduction Terminology

Terms Associated with the "Loss of Electrons"	Terms Associated with the "Gain of Electrons"
Process of oxidation	process of reduction
Substance oxidized	substance reduced
Reducing agent	oxidizing agent

15.2 Oxidation Numbers

oxidation number

Oxidation numbers are used to help determine whether oxidation or reduction has occurred in a reaction, and if such is the case, the identity of the oxidizing and reducing agents. Formally defined, an **oxidation number** *is the charge that an atom appears to have when the electrons in each bond it is participating in are assigned to the more electronegative of the two atoms involved in the bond.**

Consider an HCl molecule, a molecule in which there is one bond involving two shared electrons.

```
    ••
H ×• Cl :
    ••
```

According to the definition for oxidation number, the electrons in this bond are assigned to the chlorine atom (the more electronegative atom, Sec. 10.6). This results in the chlorine atom having one more electron than a neutral Cl atom; hence, the oxidation number of chlorine is −1 (one extra electron). At the same time, the H atom in the HCl molecule has one less electron than a neutral H atom; its electron was given to the chlorine. This electron deficiency of one results in an oxidation number of +1 for hydrogen.

As a second example, consider the molecule CF_4.

```
          ××
         ×F×
     ××  •×  ××
    ×F ×C ×F×
     ××  ×•  ××
         ×F×
          ××
```

Fluorine is more electronegative than carbon. Hence, the two shared electrons in each of the four carbon–fluorine bonds are assigned to the fluorine atom. Each F atom, thus, gains an extra electron resulting in F having a −1 oxidation number. The carbon atom loses a total of four electrons, one to each F atom, as a result of the electron "assignments." Hence, its oxidation number is +4, indicating the loss of the four electrons.

As a third example, consider the N_2 molecule, a molecule where like atoms are involved in a triple bond.

```
:N ×××
   ••• N×
```

Since the identical atoms are of equal electronegativity, the shared electrons are "divided" equally between the two atoms; each N receives three of the bonding electrons to count as its own. This results in each N atom having five valence electrons (three from the triple bond and two nonbonding electrons), the same number of valence electrons as in a neutral N atom. Hence, the oxidation number of N in N_2 is zero.

Before going any further in our discussion of oxidation numbers, it should be noted that *calculated* oxidation numbers are *not* actual charges on atoms. This is why the phrase "appears to have" is found in the definition of oxidation number given at the start of this section. In assigning oxidation numbers, we assume when we give the bonding electrons to the more electronegative element that each bond is ionic (complete transfer of electrons). We know that this is not always the case. Sometimes it is a good approximation, sometimes it is not. Why, then, do we do this when we know that

*In some textbooks the term oxidation state is used in place of oxidation number. In other textbooks the two terms are used interchangeably. We will use oxidation number in this textbook.

it does not always correspond to reality? Oxidation numbers, as we shall see shortly, serve as a very convenient device for "keeping track" of electron transfer in redox reactions. Even though they do not always correspond to physical reality, they are very, very useful entities.

In principle, the procedures used to determine oxidation numbers for the atoms in the molecules HCl, CF_4, and N_2 can be used to determine oxidation numbers in all molecules. However, the procedures become very laborious in many cases, especially when complicated electron-dot structures are involved. In practice, an alternate, much simpler procedure that does not require electron-dot structures is used to obtain oxidation numbers. This alternative procedure is based on a set of operational rules that are consistent with and derivable from the general definition for oxidation numbers. The operational rules are

Rule 1. *The oxidation number of an atom in its elemental state is zero.* For example, the oxidation number of Cu is zero, and the oxidation number of Cl in Cl_2 is zero.

Rule 2. *The oxidation number of any monoatomic ion is equal to the charge on the ion.* For example, the Na^+ ion has an oxidation number of +1 and the S^{2-} ion has an oxidation number of −2.

Rule 3. *The oxidation numbers of groups IA and IIA elements are always +1 and +2, respectively.*

Rule 4. *The oxidation number of fluorine is always −1 and that of the other group VIIA elements (Cl, Br, and I) is usually −1.* The exception for these latter elements is when they are bonded to more electronegative elements. In this case they exhibit positive oxidation numbers.

Rule 5. *The usual oxidation number for oxygen is −2.* The exceptions occur when oxygen is bonded to the more electronegative fluorine (O then has a positive oxidation number) or found in compounds containing oxygen-oxygen bonds (peroxides). In peroxides the oxidation number is −1 for oxygen. Peroxides form only between oxygen and hydrogen (H_2O_2), group IA elements (Na_2O_2, etc.), and group IIA elements (BaO_2, etc.).

Rule 6. *The usual oxidation number for hydrogen is +1.* The exception occurs in hydrides, compounds where hydrogen is bonded to a metal of lower electronegativity. In such compounds hydrogen is assigned an oxidation number of −1. Examples of hydrides are NaH, CaH_2, and LiH.

Rule 7. *The algebraic sum of the oxidation numbers of all atoms in a neutral molecule must be zero.*

Rule 8. *The algebraic sum of the oxidation numbers of all atoms in a polyatomic ion is equal to the charge on the ion.*

The use of these rules is illustrated in Example 15.1.

EXAMPLE 15.1

Assign oxidation numbers to each element in each of the following compounds.

(a) NO_2 **(b)** P_2F_4 **(c)** $K_2Cr_2O_7$ **(d)** ClO_2^-

SOLUTION

(a) Oxygen has an oxidation number of −2 (rule 5). The oxidation number of N may be calculated by rule 7. Letting x equal the oxidation number of N, we have

$$\begin{aligned} \text{N:}\quad & 1 \text{ atom} \times (x) = x \\ \text{O:}\quad & 2 \text{ atoms} \times (-2) = \underline{-4} \\ & \text{sum} = 0 \text{ (rule 7)} \end{aligned}$$

Solving for x algebraically, we get

$$x + (-4) = 0$$

$$x = +4$$

Consequently, the oxidation number of N is +4 in the compound NO_2.

(b) Fluorine has an oxidation number of −1 (rule 4). Rule 7 will allow us to calculate the oxidation number of P; the sum of the oxidation numbers must be zero. Letting x equal the oxidation number of P, we have

$$\begin{aligned} \text{F:}\quad & 4 \text{ atoms} \times (-1) = -4 \\ \text{P:}\quad & 2 \text{ atoms} \times (x) = \underline{2x} \\ & \text{sum} = 0 \text{ (rule 7)} \end{aligned}$$

Solving for x algebraically, we get

$$2x + (-4) = 0$$

$$x = +2$$

Thus, the oxidation number of P in P_2F_4 is +2. Note that the oxidation number of P is not +4 (the calculated charge associated with two P atoms). Oxidation number is always specified on a *per atom* basis.

(c) Potassium has an oxidation number of +1 (rule 3) and oxygen has an oxidation number of −2 (rule 5). Letting x equal the oxidation number of chromium and using rule 7, we get

$$\begin{aligned} \text{K:}\quad & 2 \text{ atoms} \times (+1) = +2 \\ \text{Cr:}\quad & 2 \text{ atoms} \times (x) = 2x \\ \text{O:}\quad & 7 \text{ atoms} \times (-2) = \underline{-14} \\ & \text{sum} = 0 \text{ (rule 7)} \end{aligned}$$

Solving for x algebraically, we get

$$(+2) + 2x + (-14) = 0$$

$$x = +6$$

Thus, the oxidation number of chromium in $K_2Cr_2O_7$ is +6.

(d) According to rule 8, the sum of the oxidation numbers must equal −1, the charge on this polyatomic ion. The oxidation number of oxygen is −2 (rule 5). Chlorine will have a positive oxidation number, since it is bonded to a more electronegative element (rule 4). Letting x equal the oxidation number of chlorine, we have

$$\begin{aligned} \text{Cl:}\quad & 1 \text{ atom} \times (x) &= x \\ \text{O:}\quad & 2 \text{ atoms} \times (-2) &= \underline{-4} \\ & \text{sum} &= -1 \text{ (rule 8)} \end{aligned}$$

Solving for x algebraically, we get

$$x + (-4) = -1$$

$$x = +3$$

Thus, chlorine has an oxidation number of +3 in this ion.

Many elements display a range of oxidation numbers in their various compounds. For example, nitrogen exhibits oxidation numbers ranging from −3 to +5 in various compounds. Selected examples are

NH_3	N_2O	NO	N_2O_3	NO_2	HNO_3
−3	+1	+2	+3	+4	+5

As shown in this listing of nitrogen-containing compounds, the oxidation number of an atom is written *underneath* the atom in the formula. This convention is used to avoid confusion with the charge on an ion.

Although not common, nonintegral oxidation numbers are possible. For example, the oxidation number of iron in the compound Fe_3O_4 is +2.67. The oxidation numbers of the oxygens in the compound add up to −8. Therefore, the iron atoms must have an oxidation number sum of +8. Dividing +8 by 3 (the number of iron atoms) gives +2.67.

oxidizing agent

reducing agent

Oxidizing and reducing agents may be defined in terms of changes in oxidation numbers. The **oxidizing agent** *in a redox reaction is the substance that* CONTAINS *the atom that shows a decrease in oxidation number.* Since the oxidizing agent is the substance reduced in a reaction, reduction involves a decrease in oxidation number; the oxidation number is reduced (decreased) in a reduction. The **reducing agent** *in a redox reaction is the substance that* CONTAINS *the atom that shows an increase in oxidation number.* Since the reducing agent is the substance oxidized in a reaction, oxidation involves an increase in oxidation number. Table 15-2 summarizes the rela-

Table 15-2. Oxidation–Reduction Terminology in Terms of Oxidation Number Change

Terms Associated with an "Increase in Oxidation Number"	Terms Associated with a "Decrease in Oxidation Number"
Process of oxidation	process of reduction
Substance oxidized	substance reduced
Reducing agent	oxidizing agent

tionships between oxidation–reduction terms and oxidation-number changes. A comparison of Table 15-2 with Table 15-1 shows that the loss of electrons and oxidation number increases are synonomous as are the gain of electrons and oxidation number decreases. The fact that the oxidation number becomes more positive (increases) as electrons are lost is consistent with our understanding of the proton–electron charge relationships in an atom.

EXAMPLE 15.2

Determine oxidation numbers for each atom in the following reactions, and identify the oxidizing and reducing agents.

(a) $2\,SO_2 + O_2 \longrightarrow 2\,SO_3$ (b) $2\,Fe_2O_3 + 3\,C \longrightarrow 4\,Fe + 3\,CO_2$
(c) $S_2O_8^{2-} + 2\,I^- \longrightarrow I_2 + 2\,SO_4^{2-}$

SOLUTION

The oxidation numbers are calculated by the methods illustrated in Example 15-1.

(a)

$2\,SO_2$	+	O_2	$\longrightarrow$	$2\,SO_3$
+4 −2		0		+6 −2
Rules 5, 7		Rule 1		Rules 5, 7

The oxidation number of S has increased from +4 to +6. Therefore, the substance that contains S, SO_2, has been oxidized and is the reducing agent.

The oxidation number of the O in O_2 has decreased from 0 to −2. Therefore, the O_2 has been reduced and is the oxidizing agent.

(b)

$2\,Fe_2O_3$	+	$3\,C$	$\longrightarrow$	$4\,Fe$	+	$3\,CO_2$
+3 −2		0		0		+4 −2
Rules 5, 7		Rule 1		Rule 1		Rules 5, 7

The oxidation number of C has increased from 0 to +4. An increase in oxidation number is associated with oxidation. Therefore, the element C, since it has been oxidized, is the reducing agent.

The oxidation number of Fe has decreased from +3 to 0. Since a decrease in oxidation number is associated with reduction, the Fe_2O_3, the iron-containing compound, is the oxidizing agent.

(c)

$S_2O_8^{2-}$	+	$2\,I^-$	$\longrightarrow$	I_2	+	$2\,SO_4^{2-}$
+7 −2		−1		0		+6 −2
Rules 5, 8		Rule 2		Rule 1		Rules 5, 7

The oxidation number of I has increased from −1 to 0. Thus, I^-, the iodine-containing reactant, has been oxidized and is the reducing agent.

The oxidation number of S has decreased from +7 to +6. Thus, $S_2O_8^{2-}$, the sulfur-containing reactant, has been reduced and is the oxidizing agent.

15.3 Classes of Chemical Reactions

An almost inconceivable number of chemical reactions is possible. The problems associated with organizing our knowledge about them are diminished considerably by grouping the reactions into classes. Two classification systems are in common use.

1. A system based on *oxidation-number change,* which groups reactions into two categories: oxidation–reduction (redox) and metathetical.
2. A system based on the form of the equation for the reaction, which recognizes four types of reactions: synthesis, decomposition, single replacement, and double replacement.

The two systems are not mutually exclusive and are commonly used together. For example, a particular reaction may be characterized as a single-replacement redox reaction.

metathetical reaction

As we have just learned (Sec. 15.2), reactions in which oxidation numbers change are called oxidation–reduction reactions. A **metathetical reaction** *is one in which there is* NO *oxidation number change.* The word metathetical comes from a Greek word that means "to transpose, or change positions." Many metathetical reactions can be viewed as involving the replacement of an atom or group of atoms with another atom or group of atoms.

synthesis reaction

The first of the four categories of reactions in the second classification system, the one that keys in on the form of the equation, is the synthesis reaction. A **synthesis reaction** *is one in which a single product is produced from two (or more) reactants.* The general equation for a synthesis reaction is

$$X + Y \longrightarrow XY$$

This is a most simple type of reaction, one in which two substances join together to form a more complicated product. The reactants X and Y may be elements or compounds or an element and a compound. The product of the reaction, XY, is always a compound.

Some representative synthesis reactions with elements as the reactants are

$$H_2 + Cl_2 \longrightarrow 2\,HCl$$

$$S + O_2 \longrightarrow SO_2$$

$$Ni + S \longrightarrow NiS$$

Synthesis reactions with only elements as reactants are always oxidation–reduction reactions. Oxidation-number changes must occur because all elements (the reactants) have an oxidation number of zero and all the constituent elements of a compound *cannot* have oxidation numbers of zero.

Some examples of synthesis reactions in which compounds are involved as reactants are

$$SO_3 + H_2O \longrightarrow H_2SO_4$$

$$K_2O + H_2O \longrightarrow 2\,KOH$$

$$2\,NO + O_2 \longrightarrow 2\,NO_2$$

$$2\,NO_2 + H_2O_2 \longrightarrow 2\,HNO_3$$

The first two of these reactions are metathetical, the latter two redox. Most synthesis reactions where both reactants are compounds are metathetical; redox metathetical reactions of this type are extremely rare.

decomposition reaction

A **decomposition reaction** *is one in which a single reactant is converted into two or more simpler substances.* It is, thus, the exact opposite of a synthesis reaction. The general equation for a decomposition reaction is

$$XY \longrightarrow X + Y$$

Both redox and metathetical decomposition reactions are common. At sufficiently high temperatures all compounds may be broken down (decomposed) into their constituent elements. Such reactions, which are always redox reactions, include

$$2\,CuO \longrightarrow 2\,Cu + O_2$$

$$2\,H_2O \longrightarrow 2\,H_2 + O_2$$

Examples of decomposition reactions that result in at least one compound as a product are

$$CaCO_3 \longrightarrow CaO + CO_2$$

$$2\,KClO_3 \longrightarrow 2\,KCl + 3\,O_2$$

The first of these two reactions is metathetical, the second a redox reaction.

single-replacement reaction

A **single-replacement reaction** *is one in which an element replaces another element from its compound.* In this type of reaction there are always two reactants, an element and a compound, and two products, also an element and a compound. The general equation for a single-replacement reaction is

$$X + YZ \longrightarrow Y + XZ$$

Both metals and nonmetals may be replaced from compounds in this manner. Such reactions usually involve aqueous solutions.

Two of the reaction types studied in Chapter 14 are single-replacement reactions. The reaction between an acid and an active metal (Sec. 14.7) and the reaction between a metal and an aqueous salt solution (Sec. 14.9) both qualify. Examples include

$$Zn + H_2SO_4 \longrightarrow H_2 + ZnSO_4$$

$$Ni + 2\,HCl \longrightarrow H_2 + NiCl_2$$

$$Fe + CuSO_4 \longrightarrow Cu + FeSO_4$$

$$Mg + Ni(NO_3)_2 \longrightarrow Ni + Mg(NO_3)_2$$

Reactions of this type are the basis for the formulation of the activity series of metal (Table 14-6).

Replacement series for nonmetals, similar in concept to the activity series for metals, may be established by studying single-replacement reactions where one nonmetal replaces another nonmetal. For example, it is found that the halogens (group VIIA elements) replace each other in the order: fluorine, chlorine, bromine, iodine. Thus, the following reactions do occur.

$$Cl_2 + NiI_2 \longrightarrow I_2 + NiCl_2$$

$$F_2 + 2\,NaCl \longrightarrow Cl_2 + 2\,NaF$$

Fluorine will replace the chloride ion from an aqueous solution of a chloride salt; chlorine will replace bromide ion; and bromine will replace iodide ion.

double-replacement reaction

A **double-replacement reaction** *is one in which two compounds exchange parts of the compounds with each other and form two different compounds.* The general equation for such a reaction is

$$AX + BY \longrightarrow AY + BX$$

Double-replacement reactions generally involve acids, bases, and salts in aqueous solution. Most often the positive ion from one compound exchanges with the positive ion of the other. The process may be thought of as "partner swapping," since each negative ion ends up paired with a new partner (positive ion).

An acid–base neutralization (Sec. 14.7) is a double-replacement reaction; for example,

$$NaOH + HCl \longrightarrow NaCl + HOH$$

The Na and H have exchanged places.

Many of the reactions of dissolved salts (Sec. 14.9) are also double-replacement reactions.

$$AgNO_3 + NaCl \longrightarrow NaNO_3 + AgCl$$

$$NaF + HCl \longrightarrow NaCl + HF$$

$$AgNO_3 + HCl \longrightarrow AgCl + HNO_3$$

In the first of these reactions the formation of insoluble AgCl is the driving force, in the second the formation of the weak acid HF, and in the last the formation again of insoluble AgCl.

EXAMPLE 15.3

Classify each of the following reactions as redox or metathetical. Further classify them as synthesis, decomposition, single replacement, or double replacement.

(a) $2\,C + O_2 \longrightarrow 2\,CO$
(b) $2\,KNO_3 \longrightarrow 2\,KNO_2 + O_2$
(c) $Zn + 2\,AgNO_3 \longrightarrow Zn(NO_3)_2 + 2\,Ag$
(d) $Ni(NO_3)_2 + 2\,NaOH \longrightarrow Ni(OH)_2 + 2\,NaNO_3$

SOLUTION

The oxidation numbers are calculated by the methods illustrated in Example 15.1.

(a) $2\,C + O_2 \longrightarrow 2\,CO$

2 C	O₂	2 CO
0	0	+2 −2
Rule 1	Rule 1	Rules 5, 7

The oxidation numbers of both C and O change, therefore the reaction is a redox reaction. Two substances combine to form a single substance; hence, this reaction is classified as a synthesis reaction. We, thus, have a redox synthesis reaction.

(b) $2\,KNO_3 \longrightarrow 2\,KNO_2 + O_2$

$2\,KNO_3$	$2\,KNO_2$	O_2
+1 +5 −2	+1 +3 −2	0
Rules 3, 5, 7	Rules 3, 5, 7	Rule 1

The oxidation number of N decreases from +5 to +3; the oxidation number for some O atoms increases from −2 to 0. The reaction is a redox reaction. Since two substances are produced from a single substance, it is also a decomposition reaction. We, thus, have a **redox decomposition reaction.**

(c) $Zn + 2\,AgNO_3 \longrightarrow Zn(NO_3)_2 + 2\,Ag$

Zn	$2\,AgNO_3$	$Zn(NO_3)_2$	$2\,Ag$
0	+1 +5 −2	+2 +5 −2	0
Rule 1	Rules 5, 7, 8	Rules 5, 7, 8	Rule 1

This is a redox reaction; zinc is oxidized, silver is reduced. Having an element and a compound as reactants and an element and compound as products is a characteristic of a single-replacement reaction. That is the type of reactant we have here: zinc and silver are exchanging places. We, thus, have a **redox single-replacement reaction.**

(d) $Ni(NO_3)_2 + 2\,NaOH = Ni(OH)_2 + 2\,NaNO_3$

$Ni(NO_3)_2$	$2\,NaOH$	$Ni(OH)_2$	$2\,NaNO_3$
+2 +5 −2	+1 −2 +1	+2 −2 +1	+1 +5 −2
Rules 5, 7, 8	Rules 3, 5, 7	Rules 5, 7, 8	Rules 3, 5, 7

This is a metathetical reaction; there are no oxidation-number changes. The reaction is also a double-replacement reaction; nickel and sodium are changing places, that is, "swapping partners." Thus, we have a **metathetical double-replacement reaction.**

15.4 Balancing Oxidation–Reduction Equations

Balancing an equation is not a new topic to us. In Section 8.3 we learned how to balance equations by the *inspection method.* With this method, you start with the most complicated compound within the equation and balance one of the elements in it. Then the atoms of another element are balanced, then another, and so on until all elements are balanced. This inspection procedure is a useful method for balancing simple equations with small coefficients. However, it breaks down when applied to complicated equations.

Equations for redox reactions are often quite complicated and contain numerous reactants and products and large coefficients. Trying to balance redox equations such as

$$PH_3 + CrO_4^{2-} + H_2O \longrightarrow P_4 + Cr(OH)_4^- + OH^-$$

or

$$As_4O_6 + MnO_4^- + H_2O \longrightarrow AsO_4^{3-} + H^+ + Mn^{2+}$$

by inspection is a tedious, time-consuming, frustrating experience. Balancing such equations is, however, easily accomplished by systematic equation-balancing procedures that use oxidation numbers and focus on the fact that the number of electrons lost and gained in a redox reaction must be equal.

Two distinctly different approaches for systematically balancing redox equations are in common use: the oxidation-number method and the ion–electron method. Each method has advantages and disadvantages to its use. We will consider both methods.

15.5 Using the Oxidation-Number Method to Balance Redox Equations

A useful feature of oxidation numbers is that they provide a rather easy method for balancing redox equations. The steps involved in their use in this balancing process are as follows.

Step 1. *Assign oxidation numbers to all atoms in the equation and determine which atoms are undergoing a change in oxidation number.*

Step 2. *Determine the magnitude of the change in oxidation number* PER ATOM *for the elements undergoing a change in oxidation number.* A large bracket is drawn from the element in the reactant to the element in the product and the increase or decrease in oxidation number is placed at the middle of the bracket. (See the examples that follow.)

Step 3. *When more than one atom of an element that changes oxidation number is present in a formula unit (of either reactant or product), determine the change in oxidation number per* FORMULA UNIT. This change per formula unit is indicated by multiplying the oxidation-number change per atom, already written with the brackets, by an appropriate factor.

Step 4. *Determine multiplying factors that make the total increase in oxidation number equal to the total decrease in oxidation number.* Place them with the bracket also.

Step 5. *Place in front of the oxidizing and reducing agents and their products in the equation coefficients that are consistent with the total number of atoms of the elements undergoing oxidation number change.*

Step 6. *Balance all other atoms in the equation except those of hydrogen and oxygen.* In doing this the coefficients determined in the previous step cannot be altered.

Step 7. *Balance the charge (the sum of all the ionic charges) so that it is the same on both sides of the equation by adding* H^+ *or* OH^- *ions.* (This step is necessary only when dealing with net ionic equations describing aqueous solution reactions.) If the reaction takes place in acidic solution, add H^+ ion to the side deficient in positive charge. If the reaction takes place in basic solution, add OH^- ions to the side deficient in negative charge.

Step 8. *Balance the oxygen atoms.* For net ionic equations, H_2O must usually be added to the appropriate side of the equation to achieve oxygen balance. Water is, of course, present in all aqueous solutions and can be either a reactant or product.

Step 9. *Balance the hydrogen atoms.* If all of the previous steps have been correctly carried out, the hydrogen should automatically have been balanced. If hydrogen does not balance, a mistake has been made in a previous step.

Now let us consider some examples where these rules are applied. The first two examples will involve molecular equations. The third example involves a net ionic equation. In balancing net ionic equations, any H_2O, H^+, or OH^- present is usually left out of the unbalanced equation that we start with and then added as needed during the balancing process.

EXAMPLE 15.4

Balance the following molecular redox equation by the oxidation-number method of balancing.

$$Fe + O_2 + HCl \longrightarrow FeCl_3 + H_2O$$

SOLUTION

Step 1. We identify the elements being oxidized and reduced by assigning oxidation numbers.

$$\underset{0}{Fe} + \underset{0}{O_2} + \underset{+1\,-1}{HCl} \longrightarrow \underset{+3\,-1}{FeCl_3} + \underset{+1-2}{H_2O}$$

Iron (Fe) and oxygen (O) are the elements that undergo an oxidation-number change.

Step 2. The change in oxidation number *per atom* is shown by drawing brackets connecting the oxidizing and reducing agents to their products and indicating the change at the middle of the bracket.

0 (+3) +3

$$Fe + O_2 + HCl \longrightarrow FeCl_3 + H_2O$$

0 (−2) −2

Change in oxidation number per atom

Step 3. For Fe the change in oxidation number per formula unit is the same as the change per atom since both Fe and $FeCl_3$, the two iron-containing species, contain only one atom. For O the change in oxidation number per formula unit will be double that per atom since O_2 contains two atoms. The change per formula unit is indicated by multiplying the per atom change by an appropriate numerical factor, which is 2 in this case.

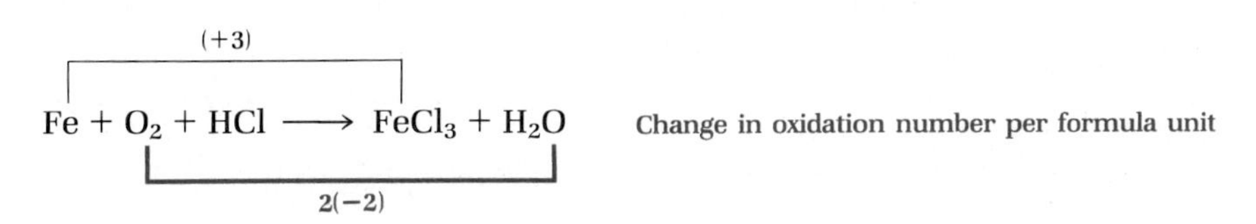

Step 4. For Fe, the total increase in oxidation number per formula unit is +3. For oxygen, the total decrease in oxidation number per formula unit is −4. To make the increase equal to the decrease, we must multiply the oxidation-number change for the element oxidized (Fe) by 4 and the oxidation-number change for the element reduced (O) by 3. This will make the increase and decrease both numerically equal to 12.

4(+3)

$$Fe + O_2 + HCl \longrightarrow FeCl_3 + H_2O$$

3[2(−2)]

Oxidation-number increase equals oxidation-number decrease

Step 5. We are now ready to place coefficients in the equation in front of the oxidizing and reducing agents and their products. The bracket notation indicates that four Fe atoms undergo an oxidation-number change. The coefficient 4 is placed in front of both Fe and $FeCl_3$. The bracket notation indicates that six O atoms (3 × 2) undergo an oxidation number decrease of two units. Thus, we need six oxygen atoms on each side. The coefficient 3 is placed in front of O_2 (6 atoms of O) and the coefficient 6 in front of H_2O (6 atoms of O).

$$4\,Fe + 3\,O_2 + HCl \longrightarrow 4\,FeCl_3 + 6\,H_2O$$

The equation is only partially balanced at this point; only Fe and O atoms are balanced.

Step 6. We next balance the element Cl (by inspection). There are twelve Cl atoms on the right side. Thus, to obtain twelve Cl atoms on the left side we place the coefficient 12 in front of HCl.

$$4\,Fe + 3\,O_2 + 12\,HCl \longrightarrow 4\,FeCl_3 + 6\,H_2O$$

Step 7. This step is not needed when the equation is a molecular equation.

Step 8. In this particular equation the O atoms are already balanced since the O underwent an oxidation number change. There are six oxygen atoms on each side of the equation.

Step 9. If all of the previous procedures (steps) have been carried out correctly, the H atoms should automatically balance. They do. There are twelve H atoms on each side of the equation. The balanced equation is thus

$$4\,Fe + 3\,O_2 + 12\,HCl \longrightarrow 4\,FeCl_3 + 6\,H_2O$$

EXAMPLE 15.5

Balance the following molecular redox equation by the oxidation-number method of balancing.

$$CuO + NH_3 \longrightarrow Cu + N_2 + H_2O$$

SOLUTION

Step 1.

$$\underset{+2\,-2}{CuO} + \underset{-3\,+1}{NH_3} \longrightarrow \underset{0}{Cu} + \underset{0}{N_2} + \underset{+1\,-2}{H_2O}$$

The two elements undergoing oxidation-number changes are Cu and N.

Step 2.

$$\overset{+2}{Cu}O + \overset{-3}{N}H_3 \longrightarrow \overset{0}{Cu} + \overset{0}{N}_2 + H_2O$$

Cu: +2 → 0 (−2); N: −3 → 0 (+3)

Change in oxidation number per atom

Step 3.

$$CuO + NH_3 \longrightarrow Cu + N_2 + H_2O$$

Cu bracket: (−2); N bracket: 2(+3)

Change in oxidation number per formula unit

The nitrogen oxidation-number change per atom had to be multiplied by 2 since there are two nitrogen atoms per molecule in N_2. Thus, a minimum of two N atoms must undergo an oxidation-number increase. This illustrates that *both* reactant and product formulas must be considered when determining the change in oxidation number per formula unit.

Step 4.

$$CuO + NH_3 \longrightarrow Cu + N_2 + H_2O$$

Cu bracket: 3(−2); N bracket: 2(+3)

Oxidation-number increase equals oxidation-number decrease

By multiplying the Cu per formula unit oxidation-number decrease of −2 by 3 we make the oxidation-number increase and decrease equal; both are at six units.

Step 5. The coefficients for CuO, NH_3, Cu, and N_2 in the equation are determined from the bracket information, which indicates that three Cu atoms undergo an oxidation number change for every two N atoms that change.

$$3\,CuO + 2\,NH_3 \longrightarrow 3\,Cu + 1\,N_2 + H_2O$$

Step 6. The only atoms left to balance are H and O.

Step 7. This step is not needed for a molecular equation.

Step 8. We balance the oxygen by placing the coefficient 3 in front of the H_2O on the right side.

$$3\,CuO + 2\,NH_3 \longrightarrow 3\,Cu + 1\,N_2 + 3\,H_2O$$

Step 9. If all of the procedures in previous steps have been carried out correctly, the H atoms should automatically balance. They do. There are six H atoms on each side of the equation.

$$3\,CuO + 2\,NH_3 \longrightarrow 3\,Cu + N_2 + 3\,H_2O$$

EXAMPLE 15.6

Balance the following net ionic redox equation by the oxidation-number method of balancing.

$$Au + Cl^- + NO_3^- \longrightarrow AuCl_4^- + NO_2$$

This reaction occurs in acidic solution.

SOLUTION

Step 1. $\underset{0}{Au} + \underset{-1}{Cl^-} + \underset{+5\,-2}{NO_3^-} \longrightarrow \underset{+3\,-1}{AuCl_4^-} + \underset{+4\,-2}{NO_2}$

The two elements undergoing an oxidation-number change are Au and N.

Step 2.

Au: 0 → +3, change (+3)

$$Au + Cl^- + NO_3^- \longrightarrow AuCl_4^- + NO_2$$ Change in oxidation number per atom

N: +5 → +4, change (−1)

Step 3. For both Au and N the oxidation-number change per formula unit is the same as per atom. Both Au and $AuCl_4^-$ contain only one Au atom; similarly, both NO_3^- and NO_2 contain only one N atom.

Step 4. By multiplying the N oxidation-number decrease by 3 we make the oxidation-number increase and decrease per formula unit the same — three units.

Au → $AuCl_4^-$: +3

$$Au + Cl^- + NO_3^- \longrightarrow AuCl_4^- + NO_2$$ Oxidation-number increase equals oxidation-number decrease

NO_3^- → NO_2: 3(−1)

Step 5. The bracket notation indicates that three N atoms and one Au atom undergo an oxidation-number change. Translating this information into coefficients, we get

$$1\,Au + Cl^- + 3\,NO_3^- \longrightarrow 1\,AuCl_4^- + 3\,NO_2$$

Step 6. We next balance Cl (by inspection). The coefficient 4 is needed on the left side in front of Cl^-. This gives four Cl atoms on each side of the equation.

$$1\,Au + 4\,Cl^- + 3\,NO_3^- \longrightarrow 1\,AuCl_4^- + 3\,NO_2$$

Step 7. Since this is a net ionic equation, the charges must balance; that is, the sum of the ionic charges of all species on each side of the equation must be equal. (They do not have to add up to zero; they just have to be equal.) In acidic solution, which is the case in this example, charge balance is accomplished by adding H^+ ion.

As the equation now stands, we have a charge of −7 on the left side (three nitrate ions each with a −1 charge and four chloride ions each with a −1 charge) and a charge of −1 on the right side (one $AuCl_4^-$ ion). By adding six H^+ to the left side we balance the charge at −1.

$$(-7) + (+6) = (-1)$$

The equation, at this point, becomes

$$1\,Au + 4\,Cl^- + 3\,NO_3^- + 6\,H^+ \longrightarrow 1\,AuCl_4^- + 3\,NO_2$$

Step 8. The oxygen atoms are balanced through the addition of H_2O molecules. There are nine O atoms on the left side (three NO_3^- ions) and only six on the right side (three NO_2 molecules). Addition of three H_2O molecules to the right side will balance the O atoms at nine per side.

$$1\,Au + 4\,Cl^- + 3\,NO_3^- + 6\,H^+ \longrightarrow 1\,AuCl_4^- + 3\,NO_2 + 3\,H_2O$$

Step 9. The H atoms automatically balance at six atoms on each side. This is our double check that previous steps have been correctly carried out. The balanced net ionic equation is thus

$$Au + 4\,Cl^- + 3\,NO_3^- + 6\,H^+ \longrightarrow AuCl_4^- + 3\,NO_2 + 3\,H_2O$$

15.6 Using the Ion–Electron Method to Balance Redox Equations

A second method for balancing redox equations is the ion–electron method. In this method, two separate *partial equations,* called half-reactions, are constructed. One half-reaction describes the oxidation process, the other half-reaction the reduction process. The balanced half-reactions are added together to get the desired balanced redox equation.

The division of the original unbalanced equation into two parts (half-reactions) is done to simplify the process of balancing. It is artificial. The half-reactions do not take place alone; we cannot have oxidation without reduction. Because of its dependence on half-reactions, the ion–electron method is also often referred to as the half-reaction method.

The name "ion–electron" for the method draws attention to the facts that (1) The method is used predominantly for balancing equations of reactions occurring in aqueous solution where ions are reactants and products, and (2) electrons are explicitly shown as reactants or products in the balanced half-reactions. Concerning this latter point, we note that the electrons in the balanced half-reactions cancel out when the half-reactions are combined to give the total reaction. Thus, the equations obtained by the ion–electron method of balancing are identical in appearance to those obtained by the oxidation-number method of balancing.

As we did with the oxidation-number method, we will break the ion–electron balancing process into a series of steps.

Step 1. Write the equation in net ionic form if it is not already in that form and then determine which substances are oxidized and reduced by assigning oxidation numbers to each atom or ion.

Step 2. Write two skeletal partial equations: an *oxidation half-reaction* that includes the formula of the substance containing the element oxidized along with the formulas of the oxidation products, and a *reduction half-reaction* that includes the formula of the substance containing the element reduced along with the formulas of the reduction products. Except for hydrogen and oxygen, the same elements must appear on both sides of the given half-reaction.

Step 3. Balance each half-reaction with respect to the element oxidized or reduced with appropriate coefficients.

Step 4. Balance each half-reaction with respect to all other elements present except for hydrogen and oxygen.

Step 5. Balance each half-reaction with respect to oxygen. How this is accomplished depends on whether the solution in which the reaction is occurring is acidic, basic, or neutral.

In acidic and neutral solutions, oxygen balance is achieved by added H_2O molecules as a source of oxygen to the side of the equation deficient in oxygen.

In basic solution, oxygen balance is achieved by adding, for every deficient oxygen, two OH^- ions to the side deficient in oxygen and one H_2O molecule to the other side of the equation. (The reason for this more complicated procedure is that the two species used to achieve hydrogen–oxygen balance in basic solution, OH^- and H_2O, both contain hydrogen and oxygen.)

Step 6. Balance each half-reaction with respect to hydrogen. Again, how this is done depends on whether the solution is acidic or basic.

In acidic solution, hydrogen balance is achieved by adding H^+ ion to the side deficient in hydrogen.

In basic solution, hydrogen balance is achieved by adding, for every deficient hydrogen, one H_2O molecule to the side deficient in H and then one hydroxide ion to the other side for every H_2O molecule added. (At this point H_2O and OH^- ions may be present on both sides of the half-reaction, since they were also added in step 5; if so, cancel out what you can. For example, if four H_2O are present on the left and two H_2O on the right, then cancel two H_2O from each side to leave two H_2O remaining on the left.)

Step 7. Balance each half-reaction with respect to charge by determining the total charge on each side of the equation and then adding electrons to the side with the total more positive charge such that the charges become equal. This step completes the balancing process for the half-reactions.

Step 8. Multiply each balanced half-reaction by the lowest small whole number that will make the total number of electrons lost equal to the total number of electrons gained. The oxidation half-reaction will contain the electrons lost and the reduction half-reaction the electrons gained. All coefficients in the balanced half-reactions are affected by the multipliers.

Step 9. Add the two half-reactions together to obtain the final balanced equation. The electrons will *always* cancel out at this point since in the previous step electron loss was made equal to electron gain.

Step 10. Cancel out any other species besides electrons common to both sides of the equation. In some, but not all, cases some cancellation of H_2O molecules, H^+ ions, or OH^- ions is possible.

Step 11. Check to see that the equation coefficients are in the lowest possible ratio. Reduce the ratio if needed.

Examples 15.7 through 15.9 illustrate how the above instructions for balancing redox equations are applied.

EXAMPLE 15.7

Balance the following net ionic equation by the ion–electron method of balancing.

$$Cu + NO_3^- \longrightarrow Cu^{2+} + NO_2 \qquad \text{(acidic solution)}$$

SOLUTION

Step 1. Assigning oxidation numbers, we get

$$\underset{0}{Cu} + \underset{+5\ -2}{NO_3^-} \longrightarrow \underset{+2}{Cu^{2+}} + \underset{+4\ -2}{NO_2}$$

Copper is oxidized, increasing in oxidation number from 0 to +2. Nitrogen is reduced, decreasing in oxidation number from +5 to +4.

Step 2. The skeleton half-reactions for oxidation and reduction are

$$\text{Oxidation:} \quad Cu \longrightarrow Cu^{2+}$$

$$\text{Reduction:} \quad NO_3^- \longrightarrow NO_2$$

Step 3. In both half-reactions, the element being oxidized or reduced is already balanced; one atom of Cu on both sides in the first half-reaction and one atom of N on both sides in the second.

Step 4. There are no other elements present other than oxygen.

Step 5. The problem statement indicates the reaction is occurring in acidic solution. Thus, H_2O molecules will be used to achieve any needed O balance.

The oxidation half-reaction does not need to be balanced with oxygen; no oxygen is present.

$$\text{Oxidation:} \quad Cu \longrightarrow Cu^{2+}$$

To achieve oxygen balance in the reduction half-reaction we add a H_2O molecule to the right side.

$$\text{Reduction:} \quad NO_3^- \longrightarrow NO_2 + H_2O$$

Step 6. No H balance is required for the oxidation half-reaction.

$$\text{Oxidation:} \quad Cu \longrightarrow Cu^{2+}$$

Since we are in acidic solution, H balance in the reduction half-reaction is achieved through use of H^+ ions. Two H^+ ions are added to the left side.

$$\text{Reduction:} \quad NO_3^- + 2\,H^+ \longrightarrow NO_2 + H_2O$$

Step 7. The half-reactions are balanced with respect to charge by addition of electrons.

For the oxidation half-reaction, there is no charge on the left side and a charge of +2 (from the copper ion) on the right side. Addition of two electrons to the right side will balance the charge at zero on each side of the half-reaction.

$$\text{Oxidation:}\quad Cu \longrightarrow Cu^{2+} + 2\,e^-$$

For the reduction half-reaction, the two positive charges (the H^+ ions) and one negative charge (the NO_3^- ion) give a net charge of +1 on the left side. There is no charge associated with the species on the right side. Hence, addition of one electron to the left side will balance the charges at zero.

$$\text{Reduction:}\quad NO_3^- + 2\,H^+ + e^- \longrightarrow NO_2 + H_2O$$

Step 8. The oxidation half-reaction involves two electrons and the reduction half-reaction contains one electron. To make the number of electrons in the two reactions equal, the reduction half-reaction is multiplied by the factor 2.

$$\text{Oxidation:}\quad Cu \longrightarrow Cu^{2+} + 2\,e^-$$

$$\text{Reduction:}\quad 2(NO_3^- + 2\,H^+ + e^- \longrightarrow NO_2 + H_2O)$$

Note that when we multiply the reduction half-reaction by 2, all of the coefficients in the equation will be doubled.

Step 9. Adding the two half-reactions together, we get

$$\begin{array}{ll} \text{Oxidation:} & Cu \longrightarrow Cu^{2+} + \cancel{2\,e^-} \\ \text{Reduction:} & 2\,NO_3^- + 4\,H^+ + \cancel{2\,e^-} \longrightarrow 2\,NO_2 + 2\,H_2O \\ \hline & Cu + 4\,H^+ + 2\,NO_3^- \longrightarrow Cu^{2+} + 2\,NO_2 + 2\,H_2O \end{array}$$

The electrons must always cancel at this stage in the balancing process. If they did not cancel that would mean that electron loss was not equal to electron gain. Having these two quantities equal is an absolute requirement in a redox process.

Step 10. There are no additional species to cancel.

Step 11. The coefficients are already in the smallest whole-number ratio possible. Thus, the equation in step 9 is the final balanced equation.

EXAMPLE 15.8

Balance the following net ionic equation by the ion–electron method of balancing.

$$SeO_3^{2-} + Cl_2 \longrightarrow SeO_4^{2-} + Cl^- \qquad \text{(basic solution)}$$

SOLUTION

Step 1.

$$\begin{array}{cccc} SeO_3^{2-} & + \; Cl_2 \longrightarrow & SeO_4^{2-} & + \; Cl^- \\ +4\;-2 & 0 & +6\;-2 & -1 \end{array}$$

Selenium is oxidized, increasing in oxidation number from +4 to +6. Chlorine is reduced, decreasing in oxidation number from 0 to −1.

Step 2. The skeleton half-reactions for oxidation and reduction are

$$\text{Oxidation:} \quad SeO_3^{2-} \longrightarrow SeO_4^{2-}$$

$$\text{Reduction:} \quad Cl_2 \longrightarrow Cl^-$$

Step 3. In the oxidation half-reaction the Se is already balanced.

$$\text{Oxidation:} \quad SeO_3^{2-} \longrightarrow SeO_4^{2-}$$

To balance the Cl in the reduction half-reaction the coefficient 2 must be added on the right side:

$$\text{Reduction:} \quad Cl_2 \longrightarrow 2\,Cl^-$$

Step 4. There are no other elements present besides oxygen.

Step 5. The problem statement indicates that the reaction is occurring in basic solution. In basic solution, oxygen balance is achieved by using both OH^- ions and H_2O molecules. Two OH^- ions are added to the side deficient in O for every needed O, and then one H_2O molecule is added to the other side of the equation for every two OH^- ions that were added.

For the oxidation half-reaction there is an O deficiency of one on the left side. Thus, two OH^- ions are added to the left side and one H_2O molecule to the right side.

$$\text{Oxidation:} \quad SeO_3^{2-} + 2\,OH^- \longrightarrow SeO_4^{2-} + H_2O$$

The reduction half-reaction is oxygen independent; hence, no balance is needed.

$$\text{Reduction:} \quad Cl_2 \longrightarrow 2\,Cl^-$$

Step 6. The procedure for balancing H in basic solution is not needed.

For the oxidation half-reaction, the H is already balanced. It balanced at the same time the O was balanced. Such "automatic" H balancing when the O is balanced is usually not the case.

$$\text{Oxidation:} \quad SeO_3^{2-} + 2\,OH^- \longrightarrow SeO_4^{2-} + H_2O$$

The reduction half-reaction does not contain any hydrogen.

$$\text{Reduction:} \quad Cl_2 \longrightarrow 2\,Cl^-$$

Step 7. Adding electrons to achieve a charge balance gives us

$$\text{Oxidation:} \quad SeO_3^{2-} + 2\,OH^- \longrightarrow SeO_4^{2-} + H_2O + 2\,e^-$$

$$\text{Reduction:} \quad Cl_2 + 2\,e^- \longrightarrow 2\,Cl^-$$

Note that the charge need not balance at zero. In this example the oxidation half-reaction charge balances at −4 and the reduction half-reaction charge balances at −2.

Step 8. Both the oxidation and reduction half-reactions involve two electrons. Thus, we already have electron loss equal to electron gain and do not need to multiply the equations by any small whole numbers to achieve electron balance.

$$\text{Oxidation:} \quad SeO_3^{2-} + 2\,OH^- \longrightarrow SeO_4^{2-} + H_2O + 2\,e^-$$

$$\text{Reduction:} \quad Cl_2 + 2\,e^- \longrightarrow 2\,Cl^-$$

Step 9. Adding the two half-reactions together, we get

$$\text{Oxidation:}\quad SeO_3^{2-} + 2\,OH^- \longrightarrow SeO_4^{2-} + H_2O + \cancel{2e^-}$$

$$\text{Reduction:}\quad Cl_2 + \cancel{2e^-} \longrightarrow 2\,Cl^-$$

$$SeO_3^{2-} + Cl_2 + 2\,OH^- \longrightarrow SeO_4^{2-} + 2\,Cl^- + H_2O$$

The electrons cancelled as they should have.

Step 10. There are no additional species to cancel.

Step 11. The coefficients are already in the smallest whole-number ratio possible. Thus, the equation in step 9 is the final balanced equation.

EXAMPLE 15.9

Balance the following net ionic equation using the ion–electron method of balancing.

$$As_4O_6 + MnO_4^- \longrightarrow H_3AsO_4 + Mn^{2+} \qquad \text{(acidic solution)}$$

SOLUTION

Step 1. $\underset{+3\ -2}{As_4O_6} + \underset{+7\ -2}{MnO_4^-} \longrightarrow \underset{+1\ +5\ -2}{H_3AsO_4} + \underset{+2}{Mn^{2+}}$

Arsenic is oxidized, increasing in oxidation number from +3 to +5. Manganese is reduced, decreasing in oxidation number from +7 to +2.
Step 2. The skeleton half-reactions are

$$\text{Oxidation:}\quad As_4O_6 \longrightarrow H_3AsO_4$$

$$\text{Reduction:}\quad MnO_4^- \longrightarrow Mn^{2+}$$

Step 3. To balance As in the oxidation half-reaction, add the coefficient 4 on the right side.

$$\text{Oxidation:}\quad As_4O_6 \longrightarrow 4\,H_3AsO_4$$

The Mn in the reduction half-reaction is already balanced.

$$\text{Reduction:}\quad MnO_4^- \longrightarrow Mn^{2+}$$

Step 4. There are no other elements present besides H and O.

Step 5. The problem statement indicates that the reaction occurs in acidic solution. Therefore, H_2O molecules are used to achieve the oxygen balance. Both half-reactions need water added to them.

$$\text{Oxidation:}\quad As_4O_6 + 10\,H_2O \longrightarrow 4\,H_3AsO_4$$

$$\text{Reduction:}\quad MnO_4^- \longrightarrow Mn^{2+} + 4\,H_2O$$

Step 6. In acidic solution, H balance is achieved by using H^+ ions. Both half-reactions need H^+ ions added to them.

$$\text{Oxidation:} \quad As_4O_6 + 10\,H_2O \longrightarrow 4\,H_3AsO_4 + 8\,H^+$$

$$\text{Reduction:} \quad MnO_4^- + 8\,H^+ \longrightarrow Mn^{2+} + 4\,H_2O$$

Step 7. Charge balance is achieved using electrons as a source of negative charge. Both half-reactions require electrons to be added.

$$\text{Oxidation:} \quad As_4O_6 + 10\,H_2O \longrightarrow 4\,H_3AsO_4 + 8\,H^+ + 8\,e^-$$

$$\text{Reduction:} \quad MnO_4^- + 8\,H^+ + 5\,e^- \longrightarrow Mn^{2+} + 4\,H_2O$$

Step 8. Eight electrons are lost in the oxidation half-reaction and five electrons are gained in the reduction half-reaction. To make the total number of electrons lost equal to the total number gained, we must multiply the oxidation half-reaction by 5 and the reduction half-reaction by 8. (The lowest common multiple of an 8 and a 5 is 40.)

$$\text{Oxidation:} \quad 5(As_4O_6 + 10\,H_2O \longrightarrow 4\,H_3AsO_4 + 8\,H^+ + 8\,e^-)$$

$$\text{Reduction:} \quad 8(MnO_4^- + 8\,H^+ + 5\,e^- \longrightarrow Mn^{2+} + 4\,H_2O)$$

Remember that the multiplying coefficient affects all of the substances in the half-reaction and not just the electrons.

Step 9. Adding the two half-reactions together, we get

$$\text{Oxidation:} \quad 5\,As_4O_6 + 50\,H_2O \longrightarrow 20\,H_3AsO_4 + 40\,H^+ + \cancel{40\,e^-}$$

$$\text{Reduction:} \quad 8\,MnO_4^- + 64\,H^+ + \cancel{40\,e^-} \longrightarrow 8\,Mn^{2+} + 32\,H_2O$$

$$5\,As_4O_5 + 50\,H_2O + 8\,MnO_4^- + 64\,H^+ \longrightarrow 20\,H_3AsO_4 + 40\,H^+ + 8\,Mn^{2+} + 32\,H_2O$$

The 40 electrons cancel out.

Step 10. Both H_2O and H^+ appear on both sides of the equation. Thirty-two H_2O molecules may be cancelled from each side of the equation, leaving 18 H_2O on the left side and no H_2O on the right side. Forty H^+ ions may be cancelled from each side of the equation leaving 24 H^+ ions on the left side and none on the right side.

$$5\,As_4O_5 + 8\,MnO_4^- + 18\,H_2O + 24\,H^+ \longrightarrow 20\,H_3AsO_4 + 8\,Mn^{2+}$$

Step 11. The coefficients are in the lowest possible ratio. The equation in step 10 is, thus, the final balanced equation.

In each of the three examples we have just considered, the oxidation and reduction half-reactions were simultaneously balanced. This approach was used in the examples to enable us to make comparisons. In practice, particularly when you are thoroughly familiar with the balancing procedure, one half-reaction is usually completely balanced before work begins on balancing the other half-reaction. Usually, it is better to work on just one reaction at a time.

A comparison of the two methods for balancing redox equations is in order. Basic to each method is being able to recognize the elements involved in the actual oxidation–reduction process. The oxidation-number method works on the principle that the increase in oxidation number must equal the decrease in oxidation number. The ion–electron method involves equalizing the number of electrons lost by the substance oxidized with the number of electrons gained by the substance reduced.

The oxidation-number method is usually faster, particularly for simpler equations. This potential speed is considered the major advantage of the oxidation-number method.

A major advantage of the ion–electron method is that all of the coefficients in the equation are generated by systematic procedures. In the oxidation-number method the coefficients of the reducing and oxidizing agents and their products are the only ones generated by systematic procedures; the others are determined by inspection. Another advantage of the ion–electron method is its focus on electron transfer. This feature becomes particularly important in electrochemistry (an area not covered in this text). In this field it is most useful to discuss chemical reactions in terms of half-reactions occurring at different locations (electrodes) in an electrochemical cell.

Learning Objectives

After completing this chapter you should be able to

- Define the terms oxidation, reduction, oxidizing agent, and reducing agent in terms of loss and gain of electrons (Sec. 15.1).
- Determine the oxidation number of an element in a molecule or ion (Sec. 15.2).
- Define the terms oxidation, reduction, oxidizing agent, and reducing agent in terms of increase and decrease in oxidation number (Sec. 15.2).
- Identify, in a given redox equation, the substance oxidized, substance reduced, oxidizing agent, and reducing agent (Sec. 15.2).
- Classify reactions as either redox or metathetical reactions (Sec. 15.3).
- Classify reactions as synthesis, decomposition, single-replacement, or double-replacement reactions (Sec. 15.3).
- Balance a redox equation, in neutral, acidic, or basic solution, by the oxidation-number method of balancing (Sec. 15.5).
- Balance a redox equation, in neutral, acidic, or basic solution, by the ion–electron method of balancing (Sec. 15.6).

Terms and Concepts for Review

The new terms or concepts defined in this chapter are

oxidation (Secs. 15.1 and 15.3)
reduction (Secs. 15.1 and 15.3)
oxidizing agent (Secs. 15.1 and 15.3)
reducing agent (Secs. 15.1 and 15.3)
oxidation–reduction reaction (Sec. 15.1)
oxidation number (Sec. 15.2)
metathetical reaction (Sec. 15.3)
synthesis reaction (Sec. 15.3)
decomposition reaction (Sec. 15.3)
single-replacement reaction (Sec. 15.3)
double-replacement reaction (Sec. 15.3)
oxidation-number method for balancing redox equations (Sec. 15.5)
ion–electron method for balancing redox equations (Sec. 15.6)

Questions and Problems

Oxidation–Reduction Terminology

15-1 Give definitions of *oxidation* and *reduction* in terms of
- **a.** loss and gain of electrons
- **b.** increase and decrease in oxidation number

15-2 Give definitions of *oxidizing agent* and *reducing agent* in terms of
- **a.** loss and gain of electrons
- **b.** increase and decrease in oxidation number
- **c.** substance oxidized and substance reduced

15-3 In each of the following statements, choose the word in parentheses that best completes the statement.
- **a.** An element that has lost electrons in a redox reaction is said to have been (oxidized, reduced).
- **b.** Reduction always results in an (increase, decrease) in the oxidation number.
- **c.** The substance oxidized in a redox reaction is the (oxidizing agent, reducing agent).

d. The reducing agent (gains, loses) electrons during a redox reaction.
e. The reducing agent causes an (increase, decrease) in the oxidation number of the oxidizing agent in a redox reaction.

Calculation of Oxidation Numbers

15-4 What are the oxidation numbers of all elements in the following species.

a. PF_3 b. Cl_2 c. H_2S d. Na_2SO_4
e. $NaOH$ f. N^{3-} g. NH_4^+ h. H_3PO_4

15-5 Determine the oxidation number of Cl in each of the following substances.

a. ClF_4^+ b. $BeCl_2$ c. $Ba(ClO_2)_2$
d. Cl_2O_7 e. $AlCl_4^-$ f. NCl_3
g. ClF h. ClO^-

15-6 Determine the oxidation number of

a. S in S_2O b. N in $(NH_4)_3PO_4$
c. P in P_4O_{10} d. O in OF_2
e. H in NaH f. Ba in Ba^{2+}
g. S in SO_3^{2-} h. Be in Be_3N_2

15-7 Classify the following oxygen-containing compounds into the categories (1) oxygen has a -2 oxidation number, (2) oxygen has a -1 oxidation number (peroxide), or (3) oxygen has a positive oxidation number.

a. Na_2O b. Na_2O_2 c. OF_2 d. BaO
e. CaO_2 f. FeO

15-8 Classify the following hydrogen-containing compounds into the categories (1) hydrogen has a $+1$ oxidation number or (2) hydrogen has a -1 oxidation number (hydrides).

a. CH_4 b. NH_3 c. HCl d. KH
e. CaH_2 f. H_2Se

Oxidation Numbers and Chemical Equations

15-9 In the following unbalanced equations identify the oxidizing agent and reducing agent.

a. $H_2 + Cl_2 \longrightarrow HCl$
b. $SO_2 + O_2 \longrightarrow SO_3$
c. $HBr + Mg \longrightarrow MgBr_2 + H_2$
d. $H_2 + FeCl_3 \longrightarrow HCl + FeCl_2$
e. $H_2S + HNO_3 \longrightarrow S + NO + H_2O$
f. $Zn + Cu^{2+} \longrightarrow Zn^{2+} + Cu$
g. $K_2S + I_2 + KOH \longrightarrow K_2SO_4 + KI + H_2O$
h. $MnO_4^- + H^+ + Hg \longrightarrow Mn^{2+} + H_2O + Hg_2^{2+}$

15-10 For each of the following unbalanced reactions, indicate whether the element in color has been (1) oxidized, (2) reduced, or (3) neither oxidized nor reduced.

a. $MnO_4^- + H_2C_2O_4 \longrightarrow CO_2 + Mn^{2+}$
b. $H_2SO_3 + I_3^- \longrightarrow HSO_4^- + I^-$
c. $Ag^+ + Fe^{2+} \longrightarrow Ag + Fe^{3+}$
d. $Ag^+ + Cl^- \longrightarrow AgCl$
e. $NH_3 + O_2 \longrightarrow N_2O_4 + H_2O$
f. $N_2H_4 + H_2O_2 \longrightarrow N_2 + H_2O$
g. $BaCl_2 + H_2SO_4 \longrightarrow BaSO_4 + HCl$
h. $KClO_3 \longrightarrow KCl + O_2$

Types of Reactions

15-11 Classify each of the following reactions as synthesis, decomposition, single replacement, or double replacement.

a. $SO_3 + H_2O \longrightarrow H_2SO_4$
b. $2\,H_2 + O_2 \longrightarrow 2\,H_2O$
c. $2\,AgNO_3 + K_2SO_4 \longrightarrow Ag_2SO_4 + 2\,KNO_3$
d. $2\,KCl + 3\,O_2 \longrightarrow 2\,KClO_3$
e. $2\,HgO \longrightarrow 2\,Hg + O_2$
f. $Al(OH)_3 + 3\,HCl \longrightarrow AlCl_3 + 3\,H_2O$
g. $Mg + 2\,HCl \longrightarrow MgCl_2 + H_2$
h. $Br_2 + 2\,NaI \longrightarrow 2\,NaBr + I_2$

15-12 Classify each of the reactions of Problem 15-11 as oxidation–reduction or metathetical.

Balancing Redox Equations: Oxidation-Number Method

15-13 Balance the following equations by the oxidation-number method.

a. $F_2 + H_2O \longrightarrow HF + O_3$
b. $BaSO_4 + C \longrightarrow BaS + CO$
c. $I_2O_5 + CO \longrightarrow I_2 + CO_2$
d. $PH_3 + NO_2 \longrightarrow H_3PO_4 + N_2$
e. $Br_2 + H_2O + SO_2 \longrightarrow HBr + H_2SO_4$
f. $Mg + AgNO_3 \longrightarrow Ag + Mg(NO_3)_2$
g. $H_2 + KClO \longrightarrow KCl + H_2O$
h. $Bi(OH)_3 + K_2SnO_2 \longrightarrow Bi + K_2SnO_3 + H_2O$

15-14 Balance the following equations by the oxidation-number method.

a. $PbO_2 + Sb + NaOH \longrightarrow PbO + NaSbO_2 + H_2O$
b. $S + H_2O + Pb(NO_3)_2 \longrightarrow Pb + H_2SO_3 + HNO_3$
c. $SnSO_4 + FeSO_4 \longrightarrow Sn + Fe_2(SO_4)_3$
d. $Na_2TeO_3 + NaI + HCl \longrightarrow NaCl + Te + H_2O + I_2$
e. $MnO_4^- + SO_2 + H_2O \longrightarrow Mn^{2+} + SO_4^{2-} + H^+$
f. $KMnO_4 + AsH_3 + H_2SO_4 \longrightarrow H_3AsO_4 + MnSO_4 + H_2O + K_2SO_4$

Balancing Redox Equations: Ion–Electron Method

15-15 Balance the following half-reactions occurring in an acidic solution.

a. $MnO_4^- \longrightarrow Mn^{2+}$
b. $H_2S \longrightarrow S$
c. $S_2O_3^{2-} \longrightarrow S_2O_6^{2-}$
d. $Pt + Cl^- \longrightarrow PtCl_6^{2-}$
e. $SO_2Cl_2 \longrightarrow SO_3^{2-} + Cl^-$
f. $Cr_2O_7^{2-} \longrightarrow Cr^{3+}$

15-16 Balance the following half-reactions occurring in basic solution.

a. $MnO_4^- \longrightarrow MnO_2$
b. $F_2 \longrightarrow F^-$
c. $PH_3 \longrightarrow P_4$
d. $CN^- \longrightarrow CNO^-$
e. $Zn \longrightarrow Zn(OH)_4^{2-}$
f. $BrO_4^- \longrightarrow Br^-$

15-17 Balance each of the following redox reactions by the ion–electron method. Assume each reaction occurs in acidic solution.

a. $Zn + Ag^+ \longrightarrow Zn^{2+} + Ag$
b. $SO_4^{2-} + Zn \longrightarrow Zn^{2+} + SO_2$
c. $NO_2 + HOCl \longrightarrow NO_3^- + Cl^-$
d. $MnO_4^- + H_2C_2O_4 \longrightarrow Mn^{2+} + CO_2$
e. $As + NO_3^- \longrightarrow H_3AsO_3 + NO$
f. $PH_3 + I_2 \longrightarrow H_3PO_2 + I^-$

15-18 Balance each of the following redox reactions by the ion–electron method. Assume each reaction occurs in basic solution.

a. $MnO_4^- + ClO_2^- \longrightarrow MnO_2 + ClO_4^-$
b. $Al + NO_3^- \longrightarrow NH_3 + AlO_2^-$
c. $BrO_3^- + I^- \longrightarrow Br^- + IO^-$
d. $ClO_3^- + N_2H_4 \longrightarrow NO + Cl^-$
e. $NiO_2 + Mn(OH)_2 \longrightarrow Mn_2O_3 + Ni(OH)_2$
f. $PH_3 + CrO_4^{2-} \longrightarrow Cr(OH)_4^- + P_4$

16 Reaction Rates and Chemical Equilibrium

16.1 Theory of Reaction Rates

In previous chapters we have looked at chemical reactions from two important viewpoints. In Chapter 8 we learned how to write (and balance) chemical equations to represent chemical reactions and then using these equations calculate amounts of products produced and reactants consumed in such reactions. In Chapter 14 we concerned ourselves with the identity of the products resulting from the reaction of specific types of substances (acids, bases and salts). We now concern ourselves with a third important topic relative to chemical reactions: the rate at which they occur.

rate of a chemical reaction

Formally defined, the **rate of a chemical reaction** *is the rate or speed at which reactants are consumed or products are produced.* A number of variables affect the rate of a reaction, some of which we routinely encounter. One variable is temperature. Food is stored in a refrigerator to reduce the rate at which spoiling (a chemical reaction) occurs. State of subdivision of solids is another reaction rate variable. Sawdust and kindling wood burn (a chemical reaction) much faster than do large logs. To understand why these variables and others affect reaction rates we will need to first examine the conditions that are necessary for a reaction to take place.

collision theory

Collision theory *is a set of postulates used to explain* HOW *chemical reactions occur.* Developed from the study of reaction rate information for many different reactions, collision theory contains three fundamental postulates.

1. Reactant particles must collide with each other in order for a reaction to occur.
2. Colliding particles must collide with a certain minimum total amount of energy if the collision is to result in a reaction.
3. In some cases reactants must be oriented in a specific way upon collision if a reaction is to occur.

Let us consider each of these three postulates separately.

Molecular Collisions

When reactions involve two or more reactants, collision theory assumes (postulate 1) that the reactant molecules, ions, or atoms must come in contact (collide) with each other in order for a reaction to occur. The validity of this assumption is fairly obvious. Reactants cannot react with each other if they are miles apart.

Most reactions are carried out in liquid solution or in the gaseous phase. The reason for this is simple. In these situations reacting particles are more free to move about and, thus, it is easier for the reactants to come in contact with each other. Reactions of solids usually take place only on the solid surface and, therefore, include only a small fraction of the total particles present in the solid. As the reaction proceeds, and products dissolve, diffuse, or fall from the surface, fresh solid is exposed. In this way, the reaction eventually can consume all of the solid. The rusting of iron is an example of this type of process.

Activation Energy

activation energy

Not all collisions between reactant particles result in the formation of reaction products. Sometimes reactant particles rebound from a collision unchanged. Postulate 2 of collision theory indicates that for a reaction to occur colliding particles must impact with a certain minimum energy, that is, the sum of the kinetic energies of the colliding particles must add to a certain minimum value. **Activation energy** *is the minimum combined kinetic energy reactant particles must possess in order for their collision to result in a reaction.* Every chemical reaction has a different activation energy.

In a slow reaction, the activation energy is far above the average energy content of the reacting particles. Only a few particles, those with above average energy, will undergo collisions that result in a reaction. Hence, the slowness of the reaction.

It is sometimes possible to start a reaction by providing activation energy and then have it continue on its own. Once the reaction is started, enough energy is released to activate other molecules and keep the reaction going. The striking of a kitchen match is an example of such a situation. Activation energy is initially provided by rubbing the match head against a rough surface; heat is generated by friction. Once the reaction is started, the match continues to burn.

Collision Orientation

Even when activation energy requirements are met, some collisions between reactant particles still do not result in product formation. How can this be? Postulate 3 of collision theory, which deals with the orientation of colliding particles at the moment of collision, relates to this situation. For nonspherical molecules and polyatomic ions, their orientation relative to each other at the moment of collision is a factor in determining whether a collision is effective.

Consider the following hypothetical reaction with the diatomic molecules AB and CD as reactants.

$$\text{A—B} + \text{C—D} \longrightarrow \text{A—C} + \text{B—D}$$

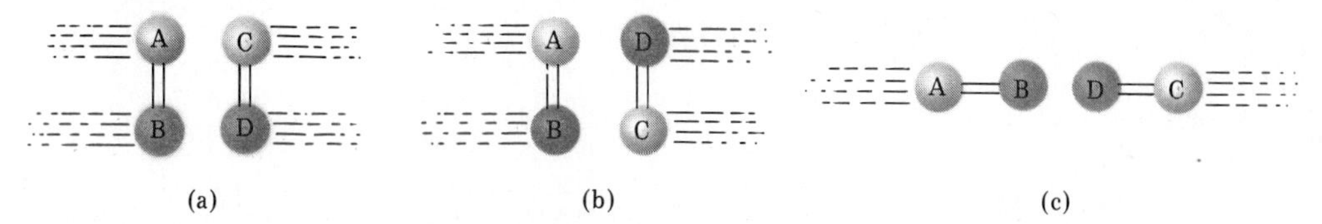

Figure 16-1. Different collision orientations for the reacting molecules AB and CD.

In this reaction B and C exchange places. The most favorable orientation during reactant molecule collisions would be one that simultaneously puts A and C in close proximity to each other (to form the molecule A—C) and B and D near each other (to form the molecule B—D). A possible orientation in which this is the situation is shown in Figure 16-1a. The possibility for a reaction resulting from this orientation is much greater than if the molecules were to collide while oriented as shown in Figure 16-1b or c. In (b), A is not near C nor is B near D. In (c), B is near D, but A and C are far removed from each other. Thus, certain collision orientations are preferred over others. The undesirable collision orientations of Figures 16-1b and c, however, could still result in a reaction if the molecules collided with abnormally high energies.

16.2 Potential Energy Diagrams for Chemical Reactions

potential energy diagram

A **potential energy diagram** *shows graphically the relationship between the activation energy for a chemical reaction and the total potential energy* (*Sec. 11.3*) *of the reactants and products.* Two such diagrams, one representing an exothermic chemical reaction (Sec. 11.8) and the other an endothermic chemical reaction (Sec. 11.8), are shown in Figure 16-2. Note that in each of these diagrams the "hill" or "hump" corresponds to the activation energy. It is the barrier that must be overcome in order for the reaction to proceed. Note also that activation energy is needed independent of whether a given reaction is exothermic or endothermic.

Whether a reaction is exothermic or endothermic is determined by how the total potential energy of the products of a chemical reaction compares with the total potential energy of the reactants. As shown in Figure 16-2a, an exothermic reaction is one in which the products are at a lower potential energy than the reactants; energy has been lost (released) in such a reaction. Conversely, in an endothermic reaction the products are at a higher potential energy than the reactants; thus, an input (absorption) of energy into the system is required.

Further insights about the energy relationships between reactants and products in a chemical reaction can be obtained from a study of Figure 16-3, which gives potential energy diagrams for two specific reactions: the reaction of phosphorus with oxygen and the reaction of sulfur with oxygen.

White phosphorus, a form of elemental phosphorus, spontaneously reacts with oxygen (bursts into flame) at a temperature of 34°C. Sulfur will also spontaneously ignite in the presence of oxygen, but not until it is heated to a temperature of 232°C. The difference in behavior is readily explainable in terms of activation energies. Looking at Figure 16-3 we see that the activation energy for the sulfur–oxygen reaction is a number of times greater than that for the phosphorus–oxygen reaction. It is not until the

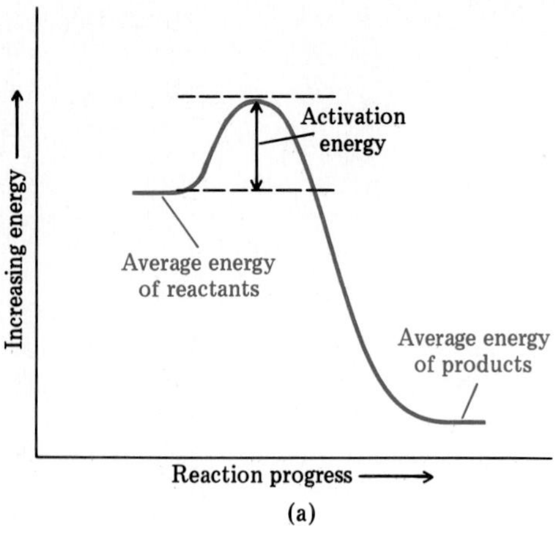

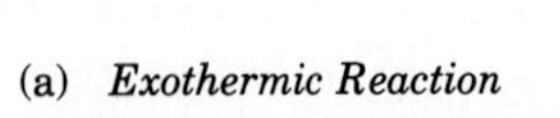

(a) *Exothermic Reaction*

The average energy of the reactants is higher than that of the products indicating energy has been released in the reaction.

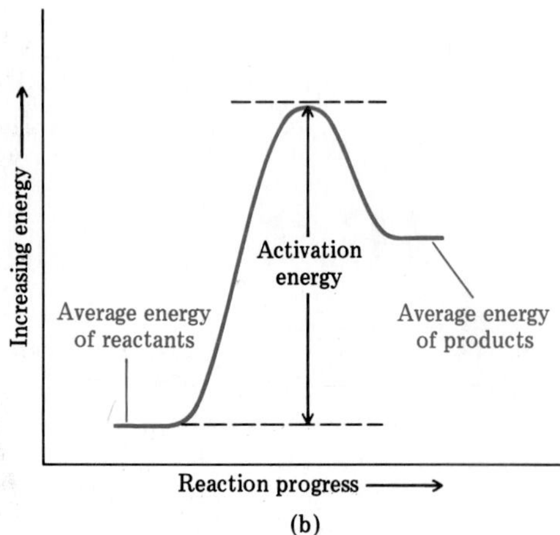

(b) *Endothermic Reaction*

The average energy of the reactants is less than that of the products indicating energy has been absorbed in the reaction.

Figure 16-2. Potential energy diagrams for (a) an exothermic chemical reaction and (b) an endothermic reaction.

temperature reaches 232°C that sulfur and oxygen possess the necessary activation energy.

Figure 16-3 also gives us the information that both the sulfur–oxygen and phosphorus–oxygen reactions are exothermic; the total potential energy of the products, in both cases, is lower than the total potential energy of the reactants. The phosphorus–oxygen reaction is more exothermic than the sulfur–oxygen reaction; that is, much more energy is released in the former than the latter reaction.

16.3 Factors That Influence Reaction Rates

Reaction rates are influenced by a number of different factors. Four that affect the rates of all reactions are (1) the physical nature of the reactants, (2) reactant concentrations, (3) reactant temperature, and (4) the presence of catalysts.

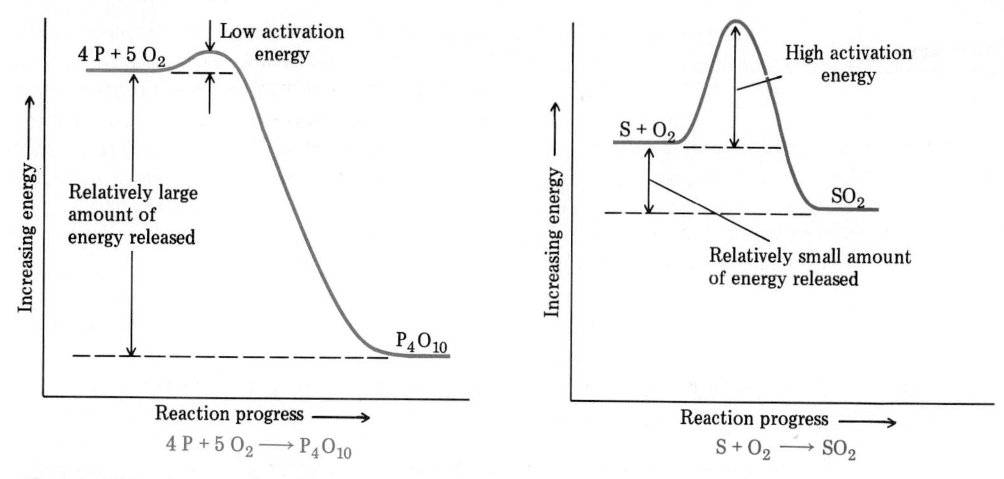

Figure 16-3. Potential energy diagrams for the reactions of the nonmetals phosphorus and sulfur with oxygen.

Physical Nature of Reactants

The physical nature of reactants refers not only to the physical state of each reactant (solid, liquid, or gas) but also to the state of subdivision, that is, particle size. In reactions where the reactants are all in the same physical state, reaction rate is generally faster between liquid state reactants than between solid reactants and faster still between gaseous reactants. Of the three states of matter the gaseous state is the one where there is the most freedom of movement; hence in this state there is a greater frequency of collision (reaction) between reactants.

In reactions where solids are involved, reaction occurs at the boundary surface between the reactions. The greater the amount of boundary surface area, the greater the reaction rate. Subdividing a solid into smaller particles will increase surface area and thus increase reaction rate. A lump of coal is difficult to ignite; coal dust ignites explosively (sometimes spontaneously). The state of subdivision, indeed, makes a difference in reaction rate (see Fig. 16-4).

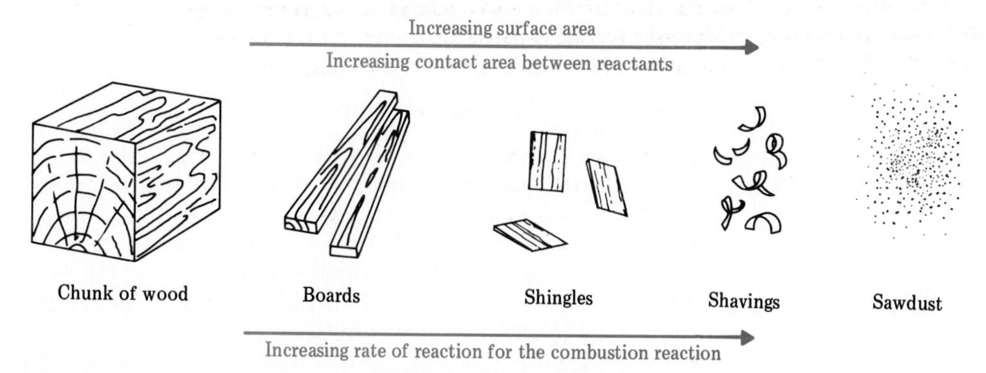

Figure 16-4. Greater surface area increases the reaction rate. [Modified from *Essential Chemistry* by Edward C. Stark. New York: Glencoe Publishing Co., Inc., 1979.]

Reactant Concentration

An increase in the concentration of a reactant causes an increase in the rate of the reaction. Combustible substances burn much more rapidly in pure oxygen than they do in air (20% oxygen). Increasing the concentration of a reactant means that there are more molecules of that reactant present in the reaction mixture and therefore there is a greater possibility for collisions between this reactant and other reactant particles. An analogy to the reaction rate–reactant concentration relationship can be drawn from the game of billiards. The more billiard balls there are on the table, the greater the probability of a moving cue ball striking one of them.

The actual quantitative change in reaction rate as the concentration of reactants is increased varies with the specific reaction. The rate always increases, but not to the same extent in all cases. Simply looking at the balanced equation for a reaction will not enable you to determine how changes in concentration will affect the reaction rate. This must be determined by actual experimentation. Sometimes the rate doubles with a doubling of concentration; however, this is not always the case.

Reaction Temperature

The effect of temperature on reaction rates can also be explained by using the molecular-collision concept. An increase in the temperature of a system results in an increase in the average kinetic energy of the reacting molecules. The increased molecular speed causes more collisions to take place in a given time. Also, since the energy of the colliding molecules is greater, a larger fraction of the collisions will result in reaction from the point of view of activation energy.

As a rough rule of thumb it has been found that the rate of a chemical reaction doubles for every 10°C increase in temperature in the temperature range we normally encounter. The chemical reaction of cooking takes place faster in a pressure cooker because of a higher cooking temperature (Sec. 11.11). On the other hand, foods are cooled or frozen to slow down the chemical reactions that result in the spoiling of food, the souring of milk, and the ripening of fruit.

Presence of Catalysts

catalyst

A **catalyst** *is a substance that, when added to a reaction mixture, increases the rate of the reaction but is itself unchanged after the reaction is completed.* Catalysts may be classified into two categories: (1) homogeneous catalysts and (2) heterogeneous catalysts. **Homogeneous catalysts** *are found in the same phase as the reactants.* They are usually dispersed uniformly throughout the reaction mixture. **Heterogeneous catalysts** *exist as a separate phase from the reactants.* Such catalysts are usually solids.

homogeneous catalysts

heterogeneous catalysts

Catalysts increase reaction rates by providing alternate reaction pathways with lower activation energies than the original uncatalyzed pathway. This lowering of activation energy effect is illustrated in Figure 16-5.

In homogeneous catalysis, it is thought, the alternate pathway involves the formation of an intermediate "complex" that contains the catalyst. This catalyst-containing

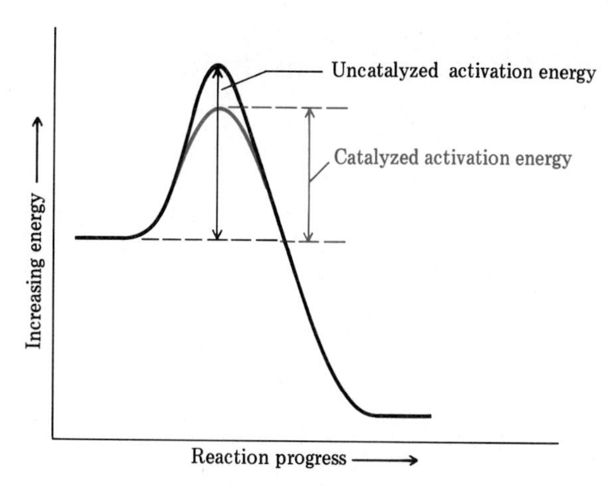

Figure 16-5. The effect of a catalyst on activation energy.

intermediate then breaks up to give the final products and regenerate the catalyst. The following equations illustrate this concept.

Uncatalyzed reaction: A + B ⟶ products

Catalyzed reaction: A + B + catalyst ⟶ A—B (bonded to catalyst) ⟶ products + catalyst

Catalysts that are solids are thought to provide a surface on which impacting reactant molecules are physically attracted and held with a particular orientation. Reactants so held are sufficiently close and favorably oriented toward each other to allow the reaction to take place. The products of the reaction then leave the surface and make it available to catalyze other reactants.

Chemical catalysts are used extensively in the chemical industry. Usually, very specific catalysts are used that accelerate one chemical reaction without influencing other possible competitive reactions. The small amounts of catalysts required, coupled with the fact that they are not used up, make the use of catalysts economically desirable in industrial processes. Often a catalyst is what makes a particular process economically feasible. For example, a catalyst often makes it possible to avoid the high temperatures (costly) that would otherwise be necessary to cause a reaction with high activation energy to proceed.

Catalysts are a key element in the functioning of automobile emission control systems now in use. In such systems heterogeneous catalysts speed up reactions that convert air pollutants present in the exhaust to less harmful products. For example, carbon monoxide is converted to carbon dioxide through reaction with oxygen.

Catalysts are of extreme importance for the proper functioning of the human body and other biological systems. In the human body, catalysts called *enzymes*, which are proteins, cause many reactions to take place rapidly at body temperature and under mild conditions. These same reactions, uncatalyzed, proceed very slowly and then only under harsher conditions.

16.4 Chemical Equilibrium

In our discussions about chemical reactions, up to this point, we have assumed that chemical reactions go to completion, that is, that reactions continue until one or more of the reactants is used up. Strictly speaking, this is not true. Experiments show that in most chemical reactions the complete conversion of reactants to products does not occur regardless of the time allowed for the reactions to take place. The reason for this is that product molecules (provided they are not allowed to escape from the reaction mixture) begin to react with each other to reform reactants. With time, a steady-state situation results where the rate of formation of products and the rate of reformation of reactants are equal. At this point the concentrations of all reactants and all products remain constant; the reaction has reached a state of *chemical equilibrium*. **Chemical equilibrium** *is the process where two opposing chemical reactions occur simultaneously at the same rate.* [We have discussed equilibrium situations in previous chapters, Secs. 11.10 (vapor pressure) and 13.2 (saturated solutions); however, these previous examples involved physical equilibrium rather than chemical equilibrium.]

chemical equilibrium

The conditions that exist in a system in a state of chemical equilibrium can best be seen by considering an actual chemical reaction. Suppose equal molar amounts of gaseous H_2 and I_2 are mixed together in a closed container and allowed to react.

$$H_2 + I_2 \longrightarrow 2\,HI$$

Initially, no HI is present, so the only possible reaction that can occur is that between H_2 and I_2. However, with time, as the HI concentration increases, some HI molecules collide with each other in a way that causes a reverse reaction to occur.

$$2\,HI \longrightarrow H_2 + I_2$$

The initially low concentration of HI makes this reverse reaction slow at first, but as the concentration of HI increases, so does the reaction rate. At the same time the reverse-reaction rate is increasing, the forward-reaction rate (production of HI) is decreasing as reactants are used up. Eventually, the concentrations of H_2, I_2, and HI in the reaction mixture reach a level at which the rates of the forward and reverse reactions become equal. At this point a state of chemical equilibrium has been reached.

Figure 16-6 shows graphically the behavior of reaction rates and reaction concentrations with time for both the forward and reverse reactions in the H_2–I_2–HI system. Figure 16-6a shows that the forward- and reverse-reaction rates become equal as a result of the forward-reaction rate decreasing (as reactants are used up) and the reverse-reaction rate increasing (as product concentration increases). Figure 16-6b shows the important point that reactant and product concentrations are usually not equal at the point at which equilibrium is reached. Rates are equal, but not concentrations. For the H_2–I_2–HI system much more product HI is present than reactants H_2 and I_2 at equilibrium. In Figure 16-6b note that the point at which equilibrium is established is the point where the two curves become straight lines.

The equilibrium involving H_2, I_2, and HI could have been established just as easily by starting with pure HI and allowing it to change into H_2 and I_2 (the reverse reaction). The final position of equilibrium does not depend on the direction from which equilibrium is approached.

Instead of writing separate equations for both the forward and reverse reactions for a system at equilibrium, it is normal procedure to represent the equilibrium by using a

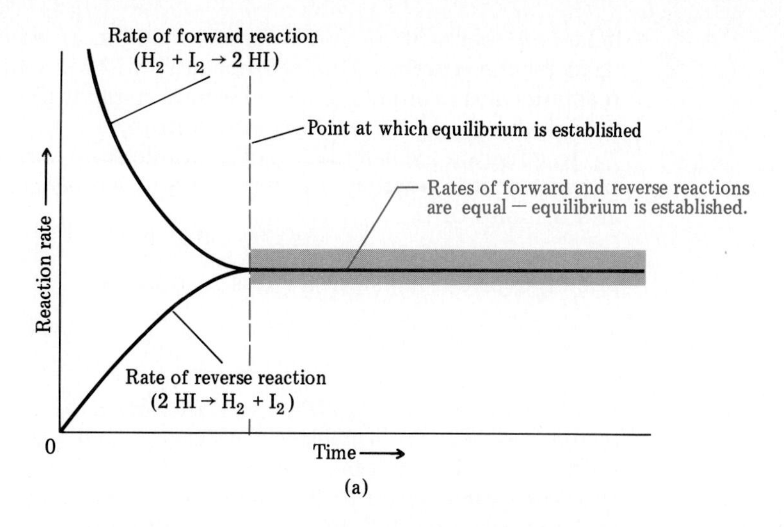

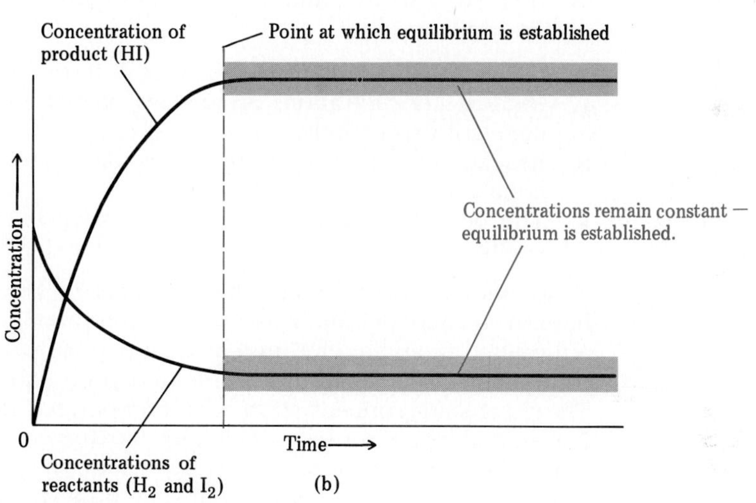

Figure 16-6. Variation of reaction rates and reactant concentrations with time in the chemical system H_2–I_2–HI.

single equation and double arrows. Thus, the reaction between H_2 and I_2, at equilibrium, is written as

$$H_2 + I_2 \rightleftharpoons 2HI$$

16.5 Equilibrium Constants

The concentrations of reactants and products, at the time chemical equilibrium is reached in a chemical reaction, can be experimentally determined through analysis of samples of the equilibrium mixture. By using these equilibrium concentrations and the balanced chemical equations for the reaction an *equilibrium constant* may be calculated for the reaction. This equilibrium constant, which relates the concentrations of reactants and products, gives information about the extent to which the reaction has occurred prior to its reaching equilibrium.

To illustrate the calculation of an equilibrium constant, let us consider a general gas phase reaction in which a moles of A and b moles of B react to produce c moles of C and d moles of D.

$$a\,A(g) + b\,B(g) \rightleftharpoons c\,C(g) + d\,D(g)$$

The equilibrium constant for this reaction is

$$K_{eq} = \frac{[C]^c[D]^d}{[A]^a[B]^b}$$

equilibrium constant

Formally defined, an **equilibrium constant** *is the product of the molar concentrations of the products for a chemical reaction, each raised to the power of its respective coefficient in the equation, divided by the product of the molar concentrations of the reactants, each raised to the power of its respective coefficient in the equation.*

Note that this definition of an equilibrium constant implies the following about equilibrium constants.

1. Concentrations are always expressed in moles/L.
2. Product concentrations are always placed in the numerator of the equilibrium constant expression.
3. Reactant concentrations are always placed in the denominator of the equilibrium constant expression.
4. The powers to which concentrations are raised are always determined by the coefficients in the balanced equation for the reaction.

An added convention in writing equilibrium constants, not apparent from the equilibrium constant definition, is: Only concentrations of gases and substances in solution are written in an equilibrium constant expression. The reason for this convention is that other substances (pure solids and pure liquids) have constant concentrations. These constant concentrations are incorporated into the equilibrium constant itself. For example, pure water in the liquid state has a concentration of 55.5 moles/L. It does not matter whether we have 1.00 mL, 50.0 mL, or 750 mL of liquid water, the concentration will be the same. In the liquid state pure water is pure water and it has only one concentration. Similar reasoning applies to other pure liquids and pure solids. All such substances have constant concentrations.

EXAMPLE 16.1

Write the equilibrium constant expression for each of the following reactions.

(a) $2\,NO_2(g) + 7\,H_2(g) \rightleftharpoons 2\,NH_3(g) + 4\,H_2O(g)$

(b) $C(s) + H_2O(g) \rightleftharpoons CO(g) + H_2(g)$

(c) $NH_4Cl(s) \rightleftharpoons NH_3(g) + HCl(g)$

(d) $Fe_2(SO_4)_3(s) \rightleftharpoons 2\,Fe^{3+}(aq) + 3\,SO_4^{2-}(aq)$

SOLUTION

(a) All of the substances involved in this reaction are gases. Therefore, each reactant and product will appear in the equilibrium constant expression.

The product concentrations, each raised to the power of its coefficient in the balanced equation, are placed in the numerator.

Equation coefficients

$$K_{eq} = \frac{[NH_3]^2[H_2O]^4}{—}$$

The reactant concentrations, each raised to the power of its coefficient in the balanced equation, are placed in the denominator.

$$K_{eq} = \frac{[NH_3]^2[H_2O]^4}{[NO_2]^2[H_2]^7}$$

Note that H_2O as a gas (water vapor or steam) is included in an equilibrium constant expression. The concentration of a gas varies. Water, as a liquid, is never included in equilibrium constant expressions.

(b) The reactant carbon (C) is a solid and thus will not appear in the equilibrium expression. Therefore,

$$K_{eq} = \frac{[CO][H_2]}{[H_2O]}$$

Note that all of the powers in this expression are "one" as a result of all of the coefficients in the balanced equation being equal to unity.

(c) The reactant NH_4Cl is a solid and, thus, will not appear in the equilibrium constant expression. Since NH_4Cl is the only reactant, this means there will be no denominator in the equilibrium constant expression.

$$K_{eq} = [NH_3][HCl]$$

(d) The equilibrium constant expression will contain the concentrations of the ions in aqueous solution but not the concentration of the solid from which they are produced.

$$K_{eq} = [Fe^{3+}]^2[SO_4^{2-}]^3$$

Equilibrium constant values change with temperature changes. A change in temperature changes molecular energies, and molecular energies have a direct effect on the relative amounts of reactants and products present in an equilibrium mixture. At a given temperature, the numerical value of the equilibrium constant for a reaction is obtained by substituting the experimentally determined equilibrium concentrations into the equilibrium constant expression for the reaction.

EXAMPLE 16.2

At a temperature of 350°C, the equilibrium concentrations for the reaction

$$N_2(g) + 3\,H_2(g) \rightleftharpoons 2\,NH_3(g)$$

are

$$[N_2] = [0.885] \qquad [H_2] = [0.665] \qquad \text{and} \qquad [NH_3] = [1.230]$$

Calculate the value of the equilibrium constant for this reaction at 350°C.

SOLUTION

The general expression for the equilibrium constant is

$$K_{eq} = \frac{[NH_3]^2}{[N_2][H_2]^3}$$

Substituting the known equilibrium concentrations into this expression gives

$$K_{eq} = \frac{[1.230]^2}{[0.885][0.665]^3}$$

$$= 5.8130226 \quad \text{(calculator answer)}$$

$$= 5.81 \quad \text{(correct answer)}$$

position of equilibrium

Position of equilibrium *specifies, in a qualitative way, the extent to which a chemical reaction, at equilibrium, has proceeded toward completion.* In equilibrium situations where the concentrations of products are greater than those of reactants, the equilibrium position is said to lie *to the right,* that is, toward the product side of the equation. If, at equilibrium, product concentrations are small relative to reactant concentrations, the equilibrium position lies *to the left.*

The magnitude of the equilibrium constant for a reaction gives information about how far a reaction proceeds toward completion, that is, as to where the equilibrium position lies. A large value of K_{eq} (greater than 10^3) means that the numerical value of the numerator is significantly greater than that of the denominator. In terms of reactants and products, this means that the concentrations of the products are greater than those of the reactants. The position of the equilibrium lies to the right.

Conversely, if the equilibrium constant is small (less than 10^{-3}) we have a situation where reactants will predominate over products in the reaction mixture. The equilibrium position is said to lie to the left in this situation.

For equilibrium conditions where K_{eq} has a value close to unity (between 10^3 and 10^{-3}) appreciable concentrations of both products and reactants are present. The reaction described in Example 16.2 falls into this category.

Table 16-1 summarizes the relationship between equilibrium constant magnitude and the extent to which a reaction proceeds toward completion.

Table 16-1. Comparison of K_{eq} Values and Reactant and Product Concentrations at Equilibrium

Value of K_{eq}	Relative Amounts of Products and Reactants	Description of Equilibrium Position
Very large (10^{30})	essentially all products	far to the right
Large (10^{10})	more products than reactants	to the right
Near unity (between 10^3 and 10^{-3})	significant amounts of both reactants and products	—
Small (10^{-10})	more reactants than products	to the left
Very small (10^{-30})	essentially all reactants	far to the left

EXAMPLE 16.3

Describe qualitatively the position of the equilibrium for each of the following reactions.

(a) $I_2(g) + Cl_2(g) \rightleftharpoons 2\ ICl(g)$ K_{eq}(at 25°C) $= 2 \times 10^5$
(b) $N_2(g) + O_2(g) \rightleftharpoons 2\ NO(g)$ K_{eq}(at 25°C) $= 1 \times 10^{-30}$
(c) $Si(s) + O_2(g) \rightleftharpoons SiO_2(s)$ K_{eq}(at 25°C) $= 2 \times 10^{142}$
(d) $Ag_2CrO_4(s) \rightleftharpoons 2\ Ag^+(aq) + CrO_4^{2-}(aq)$ K_{eq}(at 25°C) $= 9 \times 10^{-12}$

SOLUTION

The guidelines for qualitatively specifying equilibrium position are given in Table 16-1.
(a) The position of equilibrium will lie to the right, that is, more products are present than reactants.
(b) The position of equilibrium lies far to the left. Essentially, the reaction has not occurred. Only a minute amount of product is in the equilibrium mixture.
(c) The position of equilibrium is far to the right. For all intents and purposes the reaction has gone to completion. With such a large K_{eq}, only traces of the reactants would be present in the equilibrium mixture.
(d) The equilibrium position lies to the left. A few, but just a few of the Ag_2CrO_4 formula units will have broken up into ions. The reaction has occurred to only a slight extent.

16.6 Le Châtelier's Principle

A chemical system in a state of equilibrium remains in that state until it is disturbed by some change of condition. Disturbing an equilibrium has one of two results: Either the forward reaction speeds up (to produce additional products) or the reverse reaction speeds up (to produce additional reactants). Then, with time the forward and reverse reactions again become equal and a new equilibrium, not identical to the previous one, is established. If more products have been produced as a result of the disruption, the equilibrium is said to have *shifted to the right.* Similarly, when disruption causes more reactants to form, the equilibrium has *shifted to the left.*

Le Châtelier's principle

Qualitative predictions about the direction in which chemical equilibria shift can be made using a guideline (principle) introduced in 1888 by the French chemist Henry Louis Le Châtelier (1850–1926). **Le Châtelier's principle** *states that if a change of conditions (a stress) is applied to a system in equilibrium, the position of the equilibrium will shift in a direction that best reduces the stress and a new equilibrium position is reached.* We will use this principle in considering how four types of changes affect equilibrium position. The changes are (1) concentration changes, (2) temperature changes, (3) pressure changes, and (4) addition of a catalyst.

Concentration Changes

Adding or removing a reactant or product from a reaction mixture at equilibrium will always upset the equilibrium. Le Châtelier's principle predicts that the reaction will shift in the direction that will minimize the change in concentration caused by the addition or removal. If an additional amount of any reactant or product has been *added* to the system, the stress is relieved by the shifting of the equilibrium in the direction that *consumes* (uses up) some of the added reactant or product. Conversely, if a reactant or product is *removed* from an equilibrium system, the equilibrium will shift in a direction that will *produce* more of the substance that was removed.

Let us consider the effect that selected concentration changes will have on the equilibrium

$$N_2(g) + 3\,H_2(g) \rightleftharpoons 2\,NH_3(g)$$

Suppose some additional H_2 is added to the equilibrium mixture. The equilibrium will shift to the right; that is, the forward reaction rate will increase, in order to use up the additional H_2. Eventually a new equilibrium position will be reached. At this new position the H_2 concentration will still be higher than it was prior to the addition; that is, not all of the added H_2 is consumed. In addition, the N_2 concentration will have decreased (some N_2 had to react with the H_2) and the NH_3 concentration will have increased as it is what forms when H_2 and N_2 react.

Removal of some NH_3 from this newly established equilibrium position would cause a second shift to the right. The concentrations of H_2 and N_2 would decrease as the system attempts to replace the NH_3 that was removed through production of more of it. Again, not all of the removed NH_3 will be replaced. When the new equilibrium position is achieved, the NH_3 concentration will be less than it was prior to the NH_3 removal.

Figure 16-7 shows graphically the effects that the H_2 addition and NH_3 removal just discussed have on the concentrations of all substances present in the N_2–H_2–NH_3 equilibrium mixture.

Temperature Changes

Le Châtelier's principle may be used to predict the influence of temperature changes on an equilibrium provided it is known whether the reaction of concern is endothermic or exothermic.

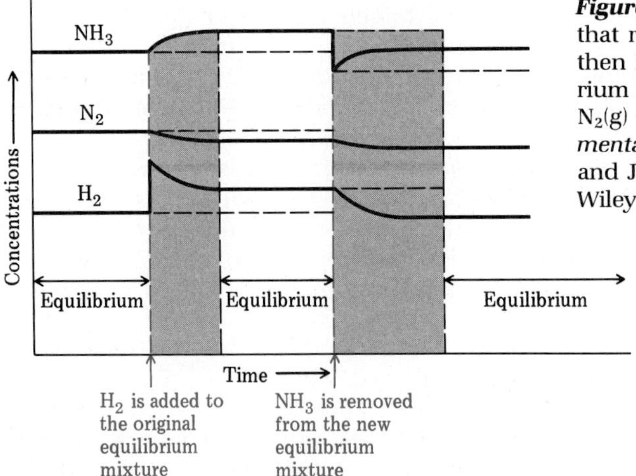

Figure 16-7. Concentration changes that result when H_2 is added to, and then NH_3 is removed from, an equilibrium mixture involving the reaction $N_2(g) + 3\,H_2(g) \rightleftharpoons 2\,NH_3(g)$. [From *Fundamentals of Chemistry* by James E. Brady and John R. Holum. New York: John Wiley & Sons, 1981.]

Consider the following general *exothermic* reaction.

$$A + B \rightleftharpoons C + D + \text{heat}$$

Heat is produced when the reaction proceeds to the right. Thus, if we add heat to an exothermic system at equilibrium (by raising the temperature), the system will shift to the left in an attempt to use up the added heat. When equilibrium is reestablished, the concentrations of A and B will be higher and the concentrations of C and D will have decreased. Lowering the temperature of an exothermic reaction mixture will cause the reaction to shift to the right as the system attempts to replace the lost heat.

The behavior, with temperature change, of an equilibrium reaction mixture involving an endothermic reaction

$$E + F + \text{heat} \rightleftharpoons G + H$$

is just the opposite of that of an exothermic reaction since a shift to the left (rather than the right) produces heat. Consequently, an increase in temperature will cause the equilibrium to shift to the right (to use up the added heat) and a decrease in temperature will produce a shift to the left (to generate more heat).

Pressure Changes

Changes in pressure do not significantly affect the concentrations of solids and liquids (Sec. 11.2), but do alter significantly the concentrations of gases. Thus, pressure changes affect systems at equilibrium only when gases are involved, and then only in cases where the chemical reaction is such that a change in the total number of moles in the gaseous state occurs. This latter point can be illustrated by considering the following two gas phase reactions.

$$2\,H_2(g) + O_2(g) \rightleftharpoons 2\,H_2O(g)$$

$$H_2(g) + Cl_2(g) \rightleftharpoons 2\,HCl(g)$$

In the first reaction the total number of moles of gaseous reactants and products decreases as the reaction proceeds to the right since three moles of reactants combine to give only two moles of products. In the second reaction there is no change in the total number of moles of gaseous substances present as the reaction proceeds since two moles of reactants combine to give two moles of products. Thus, a pressure change will shift the position of the equilibrium in the first reaction but not in the second reaction.

Pressure changes are usually brought about through volume changes. A pressure increase results from a volume decrease and a pressure decrease from a volume increase (Sec. 12.2). The use of Le Châtelier's principle correctly predicts the direction of the equilibrium position shift resulting from a pressure change only when the pressure change is due to volume change. It does not apply to pressure increases caused by the addition of a nonreactive (inert) gas to the reaction mixture. Such an addition has no effect on the equilibrium position. The partial pressures (Sec. 12.12) of each of the gases involved in the reaction remain the same.

According to Le Châtelier's principle, the stress of increased pressure is relieved by decreasing the number of moles of gaseous substances in the system. This is accomplished by the reaction shifting in the direction of the fewer number of moles, that is, to the side of the equation that contains the fewer number of moles of gaseous substances. For the reaction

$$2\ NO_2(g) + 7\ H_2(g) \rightleftharpoons 2\ NH_3(g) + 4\ H_2O(g)$$

an increase in pressure would shift the equilibrium position to the right since there are nine moles of gaseous reactants and only six moles of gaseous products. A stress of decreased pressure will result in an equilibrium system reacting in such a way as to produce more moles of gaseous substances.

Addition of a Catalyst

Catalysts do not change the position of an equilibrium. This fact becomes clear when we remember that a catalyst functions by lowering the activation energy for a reaction (Sec. 16.3). As shown in Figure 16-8, as the activation energy of the forward reaction is lowered so is the energy of the reverse reaction. Hence, a catalyst speeds up both the forward and reverse reactions and has no effect on the position of equilibrium. However, the lowered activation energy allows equilibrium to be established more quickly than if the catalyst were absent.

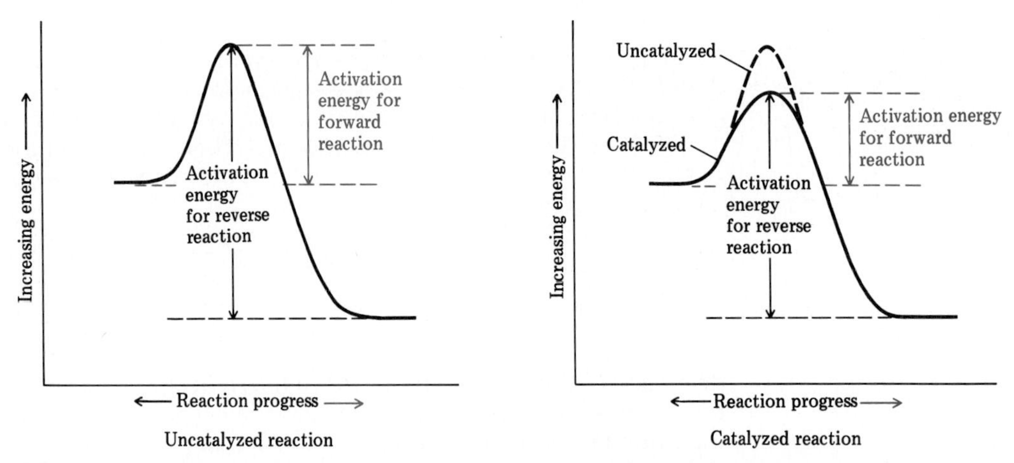

Figure 16-8. Influences of catalysts on the activation energy of forward and reverse reactions.

EXAMPLE 16.4

How will the gaseous phase equilibrium

$$CH_4(g) + 2\ H_2S(g) + \text{heat} \rightleftharpoons CS_2(g) + 4\ H_2(g)$$

be affected by

(a) the removal of $H_2(g)$
(b) the addition of $CS_2(g)$
(c) an increase in the temperature
(d) an increase in the volume of the container (a decrease in pressure)

SOLUTION

(a) The equilibrium will shift to the right, according to Le Châtelier's principle, in an attempt to replenish the H_2 that was removed.
(b) The equilibrium will shift to the left in an attempt to use up the extra $CS_2(g)$ that has been placed in the system.
(c) Raising the temperature means that heat energy has been added. To minimize the effect of this extra heat, the position of the equilibrium will shift to the right, a direction that consumes heat; heat is one of the reactants in an endothermic reaction.
(d) The system will shift to the right in an attempt to produce more moles of gaseous reactants that will increase the pressure. In going to the right the reaction produces five moles of gaseous products for every three moles of gaseous reactants consumed.

16.7 Forcing Reactions to Completion

Reactions that would ordinarily reach a state of equilibrium can be forced to completion by using experimental conditions that place a "continual stress" on the potential equilibrium condition. Let us consider a few ways in which this "forcing" is done.

Continual removal of one or more products of a reaction will force the reaction to completion. To reach equilibrium both reactants and products must be present. The removal of a product continually shifts the reaction to the right, that is, toward completion, according to Le Châtelier's principle. Eventually, one or more of the reactants is depleted, the mark of a completed reaction.

Product removal is very easy to arrange in situations that involve gaseous products. If the reaction is run in an open container, the gaseous products automatically escape to the atmosphere as fast as they are produced. Such a reaction will never reach equilibrium and will continue until the limiting reactant is used up.

Sometimes another chemical reaction is used to remove a product. Consider the situation of a saturated solution of NaCl.

$$NaCl(s) \rightleftharpoons Na^+(aq) + Cl^-(aq)$$

In such a solution, as much NaCl is dissolved as is possible. Adding $AgNO_3$ to the saturated solution will cause more NaCl to dissolve. The Ag^+ ions from the $AgNO_3$ react with the Cl^- ions in the saturated solution to form insoluble AgCl.

$$Ag^+(aq) + Cl^-(aq) \rightleftharpoons AgCl(s)$$

This removes Cl^- ion (one of the products for the original equilibrium) from solution, thus upsetting the equilibrium. More NaCl will dissolve to try and compensate for the loss of the chloride ions (Le Châtelier's principle). Continued addition of $AgNO_3$ will cause all of the NaCl to eventually dissolve.

It is also possible to drive a reaction to completion by insuring that an excess of one of the reactants is always present. The system will continually shift to the right (Le Châtelier's principle) to try and remove the stress caused by the excess reactant. Eventually other reactants will be depleted and the reaction is complete. A procedure such as this is useful in a situation where one reactant is very expensive and others are much cheaper. To insure that none of the expensive reactant goes unreacted, an excess of one of the less expensive reactants is used.

Learning Objectives

After completing this chapter you should be able to

- Explain what is meant by a *reaction rate* (Sec. 16.1).
- Describe how chemical reactions occur using collision theory (Sec. 16.1).
- Draw general potential energy diagrams for exothermic and endothermic reactions (Sec. 16.2).
- List four factors affecting reaction rate and explain how they operate in terms of collision theory (Sec. 16.3).
- Explain what is meant by chemical equilibrium and the conditions necessary for the attainment of such a state (Sec. 16.4).
- Write the equilibrium constant expression for a reaction given a balanced chemical equation (Sec. 16.5).
- Qualitatively interpret values of K_{eq} in terms of position of the equilibrium (Sec. 16.5).
- State Le Châtelier's principle (Sec. 16.6).
- Use Le Châtelier's principle to predict the effect that concentration, temperature, and pressure changes will have on an equilibrium system (Sec. 16.6).
- State several ways in which a reaction may be "forced to completion" (Sec. 16.7).

Terms and Concepts for Review

The new terms or concepts defined in this chapter are

rate of a chemical reaction (Sec. 16.1)
collision theory (Sec. 16.1)
activation energy (Sec. 16.1)
potential energy diagram (Sec. 16.2)
catalyst (Sec. 16.3)
homogeneous catalyst (Sec. 16.3)
heterogeneous catalyst (Sec. 16.3)
chemical equilibrium (Sec. 16.4)
equilibrium constant (Sec. 16.5)
position of equilibrium (Sec. 16.5)
Le Châtelier's principle (Sec. 16.6)

Questions and Problems

Rates of Reactions

16-1 What is collision theory and what are the three basic postulates of this theory?

16-2 List the four factors that influence the rate of a reaction, and describe how each effect can be explained in terms of collision theory.

16-3 Define activation energy.

16-4 Milk will sour in a couple of days when left at room temperature yet will remain unspoiled for two weeks when refrigerated. Explain why this is so.

16-5 Three reactions have, respectively, activation energies of 7 kcal, 10 kcal, and 13 kcal. Which of these reactions will occur most rapidly at 25°C? Explain why you selected the answer that you did.

16-6 What is the difference between a homogeneous and heterogeneous catalyst?

Potential Energy Diagrams

16-7 Sketch energy diagrams to represent exothermic and endothermic reactions. Label the following items in each diagram: (1) average energy of the reactants, (2) average energy of the products, (3) activation energy, and (4) energy liberated or absorbed.

16-8 One reaction occurs at room temperature, and liberates 200 kcal of energy per mole of reactant. Another reaction does not take place until the reaction mixture is heated to 150°C. However, it also liberates 200 kcal of energy per mole of reactant. Draw an energy diagram for each reaction and indicate the similarities and differences between the two diagrams.

16-9 Draw an energy diagram for an exothermic reaction where no catalyst is present. Then draw an energy diagram for the same reaction when a catalyst is present. Indicate the similarities and differences between the two diagrams.

Chemical Equilibrium

16-10 What condition must be met in order for a system to be in a state of chemical equilibrium?

16-11 The concentrations of all species in an equilibrium mixture are necessarily constant with time, yet reactions are still occurring. Explain how this can be so.

16-12 Sketch a graph showing how the concentrations of the reactants and products of a typical reaction vary with time.

Equilibrium Constants

16-13 What is the definition of an equilibrium constant?

16-14 Why are the concentrations of pure solids and pure liquids not included in equilibrium constant expressions?

16-15 Write the expression for the equilibrium constant for each of the following reactions.

a. $4\ HCl(g) + O_2(g) \rightleftharpoons 2\ H_2O(g) + 2\ Cl_2(g)$
b. $2\ SO_2(g) + O_2(g) \rightleftharpoons 2\ SO_3(g)$
c. $C_2H_2(g) + 2\ H_2(g) \rightleftharpoons C_2H_6(g)$
d. $4\ NH_3(g) + 5\ O_2(g) \rightleftharpoons 4\ NO(g) + 6\ H_2O(g)$
e. $CO(g) + H_2O(g) \rightleftharpoons CO_2(g) + H_2(g)$
f. $CH_4(g) + 2\ O_2(g) \rightleftharpoons CO_2(g) + 2\ H_2O(g)$
g. $C_3H_8(g) + 5\ O_2(g) \rightleftharpoons 3\ CO_2(g) + 4\ H_2O(g)$
h. $3\ O_2(g) \rightleftharpoons 2\ O_3(g)$

16-16 Write the expression for the equilibrium constant for each of the following reactions.

a. $2\ Pb(NO_3)_2(s) \rightleftharpoons 2\ PbO(s) + 4\ NO_2(g) + O_2(g)$
b. $2\ KClO_3(s) \rightleftharpoons 2\ KCl(s) + 3\ O_2(g)$
c. $2\ Ag(s) + Cl_2(g) \rightleftharpoons 2\ AgCl(s)$
d. $PCl_5(s) \rightleftharpoons PCl_3(l) + Cl_2(g)$
e. $Bi_2S_3(s) \rightleftharpoons 2\ Bi^{3+}(aq) + 3\ S^{2-}(aq)$
f. $H_2SO_4(l) \rightleftharpoons SO_3(g) + H_2O(l)$
g. $2\ FeBr_3(s) \rightleftharpoons 2\ FeBr_2(s) + Br_2(g)$
h. $2\ Na_2O(s) \rightleftharpoons 4\ Na(l) + O_2(g)$

16-17 What is the value of the equilibrium constant for the reaction

$$N_2(g) + 2\ O_2(g) \rightleftharpoons 2\ NO_2(g)$$

if the concentrations of each species at equilibrium are as follows.

$[N_2] = 0.0013$ mole/L $\quad [O_2] = 0.0024$ mole/L
and $\quad [NO_2] = 0.00065$ mole/L

16-18 What is the value of the equilibrium constant for the reaction

$$2\ HI(g) \rightleftharpoons H_2(g) + I_2(g)$$

if the concentrations of each species at equilibrium are as follows.

$[HI] = 2.20$ moles/L $\quad [H_2] = 0.355$ mole/L
and $\quad [I_2] = 0.355$ mole/L

16-19 Explain the significance of

a. a very large value of K_{eq}
b. a very small value of K_{eq}
c. a value of K_{eq} of about 1.0

16-20 What are the relative sizes of the equilibrium constants for a reaction if

a. Very small amounts of products are present at equilibrium.
b. Nearly all of the reactants have been used up at equilibrium.
c. Product and reactant concentrations are about the same at equilibrium.

16-21 Describe qualitatively the position of the equilibrium for each of the following reactions at the given temperatures.

a. $AgCl(s) \rightleftharpoons Ag^+(aq) + Cl^-(aq)$ $\quad K_{eq} = 10^{-10}$ at 25°C
b. $NO_2(g) + CO(g) \rightleftharpoons NO(g) + CO_2(g)$ $\quad K_{eq} = 10^{28}$ at 127°C
c. $N_2(g) + O_2(g) \rightleftharpoons 2\ NO(g)$ $\quad K_{eq} = 10^{-31}$ at 25°C
d. $CO(g) + H_2O(g) \rightleftharpoons CO_2(g) + H_2(g)$ $\quad K_{eq} = 10^2$ at 327°C

Le Châtelier's Principle

16-22 List the three changes in conditions that can affect the position of equilibrium, according to Le Châtelier's principle.

16-23 For the reaction

$$2\ Cl_2(g) + 2\ H_2O(g) \rightleftharpoons 4\ HCl(g) + O_2(g)$$

determine the direction that the equilibrium will be shifted by the following changes.

a. increasing the concentration of Cl_2
b. increasing the concentration of HCl
c. increasing the concentration of O_2
d. decreasing the concentration of Cl_2
e. decreasing the concentration of H_2O
f. decreasing the concentration of HCl

16-24 For the reaction

$$C_6H_6(g) + 3\ H_2(g) \rightleftharpoons C_6H_{12}(g) + \text{heat}$$

determine the direction that the equilibrium will be shifted by the following changes.

a. increasing the concentration of H_2
b. decreasing the concentration of C_6H_{12}
c. increasing the temperature
d. decreasing the temperature
e. increasing the pressure by decreasing the volume of the container
f. decreasing the pressure by increasing the volume of the container

16-25 List five ways in which the following gas phase equilibrium can be shifted to the right.

$$2\ N_2(g) + 6\ H_2O(g) \rightleftharpoons 4\ NH_3(g) + 3\ O_2(g)$$

16-26 Consider the following chemical system at equilibrium.

$$CO(g) + H_2O(g) + \text{heat} \rightleftharpoons CO_2(g) + H_2(g)$$

For each of the following adjustments of conditions indicate the effect on the position of the equilibrium — shifts left, shifts right, no effect.

a. heating the equilibrium mixture
b. adding CO to the equilibrium mixture
c. adding H_2 to the equilibrium mixture
d. increasing the pressure on the equilibrium mixture by adding an inert gas
e. increasing the pressure on the equilibrium mixture by decreasing the size of the reaction container
f. removing CO_2 from the equilibrium mixture
g. adding a catalyst to the equilibrium mixture
h. increasing the size of the reaction container

17 Nuclear Chemistry

17.1 Atomic Nuclei

Early in the text (Sec. 5.4) we considered the structure of the atom. An atom is composed of two regions: (1) a nuclear region where all protons and neutrons are found and (2) an extranuclear region where the electrons are found. Since this initial discussion of atomic structure, we have focused primarily on the behavior of electrons in atoms, ions, and molecules, since the arrangement of electrons determines the physical and chemical properties of substances. Little has been said about the nucleus, since it remains unchanged in ordinary chemical reactions and is important only insofar as it influences the electrons.

nuclear reaction

In this chapter we examine a group of processes known as nuclear reactions. A **nuclear reaction** *is one that involves changes in the nucleus of an atom.* Nuclear reactions are not considered to be ordinary chemical reactions. It is in the field of nuclear reactions that we encounter the terms "radioactivity," "radioisotope," "nuclear power plant," "A-bomb", and "H-bomb." All of these terms are now part of our everyday vocabulary.

A brief review of what we already know about atomic nuclei as well as some new material concerning them will serve as the starting point for our discussion of nuclear reactions.

Atomic nuclei are the very dense positively charged centers of atoms about which the electrons move. All nuclei of atoms of a given element contain the same number of protons. It is this characteristic number of protons that determines the identity of the element. The *atomic number* for an atom gives the number of protons in the nucleus. The number of neutrons associated with the nuclei of a given element may vary within a small range. Atoms of a given element that differ in the number of nuclear neutrons they contain are called *isotopes.* The *mass number* of an atom is equal to the total number of protons and neutrons present in the nucleus. Isotopes of an element have different mass numbers but the same atomic number.

In order to identify uniquely a nucleus, or atom for that matter, both the atomic number and mass number must be specified. Two different notation systems exist for doing this. Consider an isotope of nitrogen with seven protons and eight neutrons. This isotope may be denoted as $^{15}_{7}N$ or nitrogen-15. In the first notation the superscript is the mass number and the subscript is the atomic number. In the second notation the mass number is placed immediately after the name of the element. An advantage of the first notation is that the atomic number is shown; a disadvantage is the need for super- and subscripts. Both types of notation will be used in this chapter.

Some naturally occurring isotopes as well as all synthetically produced isotopes (Sec. 17.6) possess nuclei that are *unstable*. To achieve stability such unstable nuclei emit energy (radiation). **Radioactive** *is the term used to describe isotopes that possess unstable nuclei that spontaneously emit energy (radiation)*. Approximately one-fifth of the elements have at least one naturally radioactive isotope. In addition, many synthetic radioactive isotopes have been produced in the laboratory (Sec. 17.6). Radioactive isotope is often shortened to simply radioisotope.

radioactive

17.2 The Discovery of Radioactivity

The fact that certain naturally occurring isotopes are radioactive was unexpectedly (accidentally) discovered by the French physicist and engineer Antoine-Henri Becquerel (1852–1908) while he was studying certain minerals called phosphors, which glow in the dark (phosphoresce) after exposure to radiation such as sunlight or ultraviolet light. The rays emitted by these phosphorescing minerals, like visible light, darken a photographic plate. One day, in 1896, while working with a uranium ore sample that phosphoresced in the normal manner, he was interrupted. As he left, he inadvertently placed the uranium ore sample on top of an unexposed photographic plate packaged to protect it from light. Later it was determined that this photographic plate had been exposed by the uranium ore despite its being protectively wrapped. Becquerel correctly concluded that this plate exposure was due to radiation emitted by the uranium ore without external stimulus. As a result of this incident Becquerel is credited with having discovered the phenomenon we now call radioactivity. Further studies by Becquerel showed that this "radioactivity" was not unique to the one uranium ore he had used, but was a characteristic of all uranium-containing substances independent of whether they phosphoresced or not. (It seems strange now that this phenomenon was not detected earlier, since the element uranium had been isolated more than 100 years before Becquerel's discovery.)

Becquerel shared his discovery with two of his colleagues Marie Sklodowska Curie (1862–1934) and her husband Pierre (1859–1906). The Curies conducted a systematic search of the then known elements to see how widespread this phenomenon was. The only other radioactive element they found was thorium. In further investigations, the Curies discovered a uranium ore sample that exhibited four times the radioactivity of a similar quantity of pure uranium or thorium, thus indicating the presence of a new substance more radioactive than either of these elements. From this ore the Curies were able to isolate, in 1898, two new radioactive elements: polonium, 400 times more radioactive than uranium, and radium, over a million times more radioactive than uranium. It was the Curies who coined the word radioactive to describe elements that spontaneously emit radiation.

17.3 The Nature of Radioactive Emissions

The first information concerning the nature of the radiation emanating from naturally radioactive materials was obtained by the British scientist Ernest Rutherford (1871–1937) in the years 1898–1899. Using an apparatus similar to that shown in Figure 17-1, he found that if a beam of radiation is passed between electric plates it is split into three components, indicating the presence of three different types of emissions from radioactive materials. A closer analysis of Rutherford's experiment reveals that one radiation component is positively charged (because it is attracted to the negative plate), a second component is negatively charged (because it is attracted to the positive plate), and the third component carries no charge (because it is unaffected by either charged plate). Rutherford chose to call the three radiation components alpha rays (α rays) (the positive component), beta rays (β rays) (the negative component), and gamma rays (γ rays) (the uncharged component). (Alpha, beta, and gamma are the first three letters of the Greek alphabet.) We mention Rutherford's nomenclature system because it "stuck"; we still use these names today for these radiation types even though we know much, much more about their identity.

Additional research has substantiated Rutherford's conclusion that three distinct types of radiation are present in the emissions from naturally radioactive substances. This research has also supplied the necessary information for their complete characterization. Complete identification required many years. Early researchers in the field were hampered by the fact that many of the details concerning atomic structure were not yet known. (Recall, Sec. 5.4, that the neutron was not identified until 1932.)

alpha rays

Alpha rays *consist of a stream of positively charged particles (alpha particles), each of which is made up of two protons and two neutrons.* The notation used to represent an alpha particle is ${}^{4}_{2}\alpha$. The numerical subscript indicates that the charge on the particle is +2 (from the two protons). The numerical superscript indicates a mass of 4 amu. On the atomic mass scale (Sec. 5.8) protons and neutrons both have masses almost exactly equal to 1.0 amu. Thus, the total mass of an alpha particle (two protons and two neutrons) is 4 amu. Alpha particles are identical with the nuclei of helium-4 (${}^{4}_{2}He$) atoms.

beta rays

Beta rays *consist of a stream of negatively charged particles (beta particles) whose*

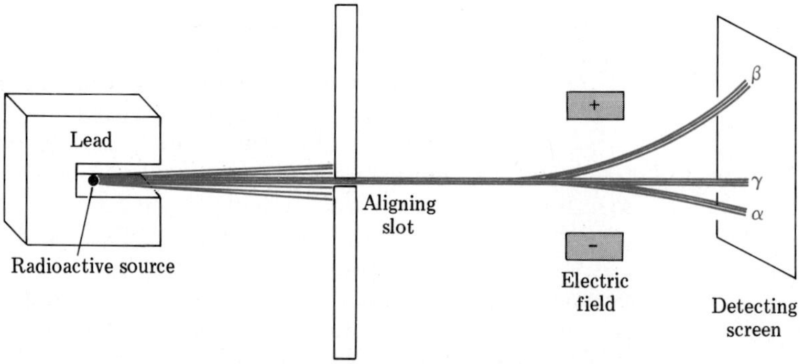

Figure 17-1. Effect of an electric field on radiation emanating from a naturally radioactive substance. Gamma rays are unaffected. The lighter beta rays are deflected considerably more than the heavier alpha rays.

charge and mass are identical to that of an electron. However, beta particles are not extranuclear electrons; they are electrons that have been produced inside the nucleus and then ejected. More concerning this process will be given in Section 17.4. The symbol used to represent a beta particle is ${}_{-1}^{0}\beta$. The numerical subscript indicates that the charge on the beta particle is -1, that of an electron. The use of the superscript zero for the mass of a beta particle is not to be interpreted as meaning that a beta particle has no mass, but rather that the mass is very close to zero amu. The actual mass of a beta particle on the atomic mass scale is 0.00055 amu.

gamma rays

Gamma rays *are not considered to be particles but rather pure energy without mass or charge.* They are very high energy radiation somewhat like x rays. The symbol for gamma rays is ${}_{0}^{0}\gamma$.

17.4 Equations for Nuclear Reactions

Alpha, beta, and gamma emissions come from the nucleus of an atom. These spontaneous emissions alter nuclei; obviously, if a nucleus loses an alpha particle (two protons and two neutrons), it will not be the same as it was before the departure of the particle. In the case of alpha and beta emissions the nuclear alteration causes the identity of the atom to change; that is, a new element is formed. Nuclear reactions thus differ dramatically from ordinary chemical reactions. In the latter the identity of the elements is always maintained. This is not the case in nuclear reactions.

The term decay or disintegration is used in describing a nuclear process where an element disappears, that is, changes into another element as a result of radiation emission. **Radioactive decay** *is the process whereby a radioisotope is transformed into an isotope of another element as a result of the emission of radiation.*

radioactive decay

Radioactive decay occurs only in certain ways, called *modes of decay.* For naturally occurring radioactive substances the modes of decay are (1) alpha-particle decay and (2) beta-particle decay. Separate consideration of each mode of decay provides further insights into the nature of nuclear reactions and also illustrates how equations for nuclear reactions are written.

Alpha-Particle Decay

Alpha-particle decay, the emission of an alpha particle from a nucleus, *always* results in the formation of an isotope of a different element. The product nucleus of such decay has an atomic number that is two less than that of the original nucleus and a mass number that is four less. We can represent alpha particle decay in general terms by the equation

$${}_{Z}^{A}X \longrightarrow {}_{2}^{4}\alpha + {}_{Z-2}^{A-4}Y$$

where X is the symbol for the nucleus of the original element undergoing decay and Y is the symbol of the element formed as a result of the decay.

To introduce us to actual nuclear equations let us write such equations for two alpha-particle-decay processes. Both ${}_{83}^{211}Bi$ and ${}_{92}^{238}U$ are alpha emitters; that is, they are radioisotopes that undergo alpha-particle decay. The nuclear equations for these two decay processes are

$$^{211}_{83}\text{Bi} \longrightarrow \,^{4}_{2}\alpha + \,^{207}_{81}\text{Tl}$$

$$^{238}_{92}\text{U} \longrightarrow \,^{4}_{2}\alpha + \,^{234}_{90}\text{Th}$$

How do these two equations differ from ordinary chemical equations? First of all, nuclear equations convey a different type of information from that found in ordinary chemical equations. The symbols in nuclear equations stand for nuclei rather than atoms. (We do not worry about electrons when writing nuclear equations.) Second, mass numbers and atomic numbers (nuclear charge) are always used in conjunction with elemental symbols in nuclear equations. Third, the elemental symbols on both sides of the equation need not be the same in nuclear equations.

balanced nuclear equation

The procedures for balancing nuclear equations are different from those used for ordinary chemical equations. A **balanced nuclear equation** *has the sum of the subscripts (atomic numbers or particle charge) on each side of the equation equal and the sum of the superscripts (mass number) on each side of the equation equal.* Both of our examples are balanced. In the α decay of $^{211}_{83}\text{Bi}$, the subscripts on both sides total 83 and the superscripts total 211. For the decay of $^{238}_{92}\text{U}$, the subscripts total 92 on both sides and the superscripts total 238 on both sides.

parent isotope

daughter isotope

The terms parent isotope and daughter isotope are often used in describing radioactive decay processes. The **parent isotope** *is the isotope undergoing decay.* The **daughter isotope** *is the product isotope resulting from the decay.* In our two previous equations thallium-207 and thorium-234 are the daughter isotopes.

Beta-Particle Decay

Beta-particle decay also always results in the formation of an isotope of a different element. The mass number of the new isotope is the same as that of the original atom. The atomic number, however, has increased by one unit. The general equation for beta decay is

$$^{A}_{Z}\text{X} \longrightarrow \,^{0}_{-1}\beta + \,^{A}_{Z+1}\text{Y}$$

Specific examples of beta-particle decay are

$$^{10}_{4}\text{Be} \longrightarrow \,^{0}_{-1}\beta + \,^{10}_{5}\text{B}$$

$$^{234}_{90}\text{Th} \longrightarrow \,^{0}_{-1}\beta + \,^{234}_{91}\text{Pa}$$

Both of these nuclear equations are balanced; superscripts and subscripts add to the same sums on each side of the equation.

At this point in the discussion you may be wondering how a nucleus, composed only of neutrons and protons, ejects a negative particle (beta particle) when no such particle is present in the nucleus. The accepted explanation is that through a complex series of steps a neutron in the nucleus is transformed into a proton and a beta particle; that is,

$$^{1}_{0}\text{n} = \,^{1}_{1}\text{p} + \,^{0}_{-1}\beta$$

Once formed within the nucleus, the beta particle is ejected with a high velocity. The net result of beta-particle formation is an increase by one in the number of protons present in the nucleus and a decrease by one in the number of neutrons present in the nucleus. Note in our two examples of beta emission that the daughter isotope has one more proton than the parent as evidenced by the atomic number of the daughter being greater than that of the parent by one unit. Subtraction of the atomic number of the daughter isotope from its mass number in each case — to get the number of neutrons — will reveal that the daughter isotope contains one less neutron than the parent. A consideration of why and when beta-particle formation occurs is delayed until Section 17.9.

Gamma-Ray Emission

Gamma-ray emission always occurs in conjunction with an alpha- or beta-decay process; it never occurs independently. Such gamma rays are most often not included in the nuclear equation, since they do not affect the balancing of the equation nor the identity of the decay product.

Because gamma rays are usually left out of nuclear equations it should not be assumed that gamma rays are not important. On the contrary, gamma rays are more important than alpha and beta particles when the effects of external radiation exposure on living organisms are considered (Sec. 17.5).

EXAMPLE 17.1

Write a balanced nuclear equation for the decay of each of the following radioactive isotopes. The mode of decay is indicated in parentheses.

(a) ${}^{70}_{31}Ga$ (beta emission) **(b)** ${}^{144}_{60}Nd$ (alpha emission)
(c) ${}^{248}_{100}Fm$ (alpha emission) **(d)** ${}^{113}_{47}Ag$ (beta emission)

SOLUTION

In each case the atomic and mass numbers of the daughter nucleus are obtained by first writing the symbols of the parent nucleus and the particle emitted by the nucleus (alpha or beta particle) and then balancing the equation.

(a) Let X represent the product of the radioactive decay, that is, the daughter isotope. Then

$$ {}^{70}_{31}\text{Ga} \longrightarrow {}^{0}_{-1}\beta + \text{X} $$

Since the sum of the superscripts on each side of the equation must be equal, the superscript for X must be 70. In order for the sum of the subscripts on each side of the equation to be equal, the subscript for X must be 32. Then $31 = (-1) + (32)$. As soon as the subscript of X is determined, the identity of X may be determined from a periodic table. The element with an atomic number of 32 is germanium (Ge). Therefore,

$$ {}^{70}_{31}\text{Ga} \longrightarrow {}^{0}_{-1}\beta + {}^{70}_{32}\text{Ge} $$

(b) Similarly, letting X represent the product of the radioactive decay, we have for the alpha decay of ${}^{144}_{60}Nd$

$$ {}^{144}_{60}\text{Nd} \longrightarrow {}^{4}_{2}\alpha + \text{X} $$

Balancing the equation, making the superscripts on each side of the equation total 144 and the subscripts total 60, we get

$$ {}^{144}_{60}\text{Nd} \longrightarrow {}^{4}_{2}\alpha + {}^{140}_{58}\text{Ce} $$

(c) Similarly, we write

$$^{248}_{100}\mathrm{Fm} \longrightarrow {}^{4}_{2}\alpha + \mathrm{X}$$

Balancing superscripts and subscripts, we get

$$^{248}_{100}\mathrm{Fm} \longrightarrow {}^{4}_{2}\alpha + {}^{244}_{98}\mathrm{Cf}$$

(d) Finally, we write

$$^{113}_{47}\mathrm{Ag} \longrightarrow {}^{0}_{-1}\beta + \mathrm{X}$$

In beta emission the atomic number of the daughter isotope always increases by one and the mass number does not change from that of the parent. The balancing procedure gives us this result.

$$^{113}_{47}\mathrm{Ag} \longrightarrow {}^{0}_{-1}\beta + {}^{113}_{48}\mathrm{Cd}$$

17.5 Effects of Radiation on Living Organisms

Alpha, beta, and gamma radiations produced from radioactive decay travel outward from their nuclear sources into the material surrounding the radioactive substance where they lose their energy through collisions with the atoms of the material. Because of the extremely high energy the radiations possess, the collisions they undergo cause the atoms involved to lose electrons; that is, ionization occurs. This ionization process is not the voluntary transfer of electrons that occurs during ionic compound formation, but rather a nonchemical, involuntary removal of electrons from atoms to form ions. Radiation damage to tissue happens when the structures of important biological molecules are changed as a result of ionization, disrupting normal cellular functions. Our bodies can repair small amounts of radiation-damaged tissue, but cellular functions are overwhelmed when large radiation dosages are experienced.

Alpha particles are the most massive and also the slowest of the particles involved in natural radioactive decay, being emitted from nuclei at a velocity about one-tenth of the speed of light. They have low penetrating power and cannot penetrate the body's outer layers of skin. The major danger from alpha radiation is internal exposure resulting when alpha-emitting radioisotopes are ingested, for example, in contaminated food. Within the body there are no protective layers of skin.

Beta particles are emitted from nuclei at speeds of about 0.9 that of light. With their greater velocity, they can penetrate much deeper than alpha particles and can cause severe skin burns if their source remains in contact with the skin for an appreciable time. They do not ionize molecules as readily as do alpha particles because of their much smaller size. An alpha particle is approximately 8000 times heavier than a beta particle. Internal beta exposure is serious as was the case with internal alpha exposure.

Gamma radiation is released at a velocity equal to that of the velocity of light. Gamma rays readily penetrate deeply into organs, bone, and tissue.

Figure 17-2 contrasts the abilities of alpha, beta, and gamma radiations to penetrate paper, aluminum, and a lead–concrete mixture.

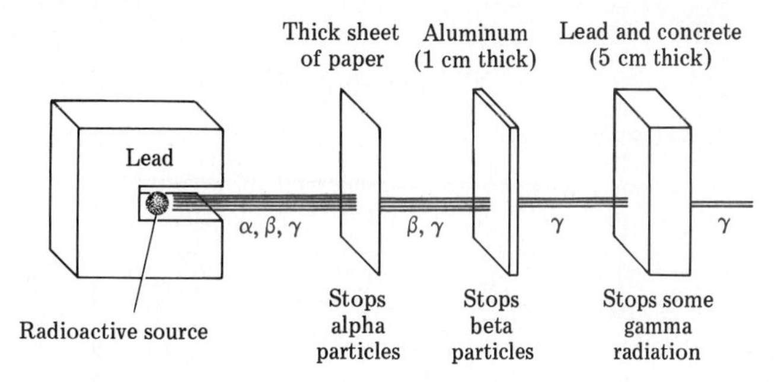

Figure 17-2. Penetrating abilities of alpha, beta, and gamma radiation.

17.6 Bombardment Reactions and Artificial Radioactivity

transmutation reaction

bombardment reaction

A **transmutation reaction** *is a nuclear reaction in which one isotope is changed into an isotope of another element.* Radioactive decay, discussed in the last section, is an example of a natural transmutation process. An artificial process that causes transmutation also exists. This process involves the use of bombardment reactions. A **bombardment reaction** *is a nuclear reaction in which small particles traveling at very high speeds are collided with stable nuclei causing them to undergo nuclear change.*

The first successful bombardment reaction was carried out in 1919, twenty-five years after the discovery of radioactive decay, by Ernest Rutherford, the same Rutherford who earlier had investigated the nature of alpha, beta, and gamma rays (Sec. 17.3). Rutherford's initial successful bombardment experiment consisted of letting alpha particles from a natural source (radium) bombard nitrogen gas. In this process he found that two new stable nuclides were formed: oxygen-17 and hydrogen-1. The nuclear equation for this transmutation is

$$^{14}_{7}\text{N} + ^{4}_{2}\alpha \longrightarrow ^{17}_{8}\text{O} + ^{1}_{1}\text{H}$$

artificial radioactivity

Since this initial successful reaction, further research carried out by many investigators has shown that numerous nuclei experience change under the stress of small particle bombardment. In most cases the new isotope produced as a result of the transmutation is radioactive (unstable) rather than stable as was the case with oxygen-17. **Artificial radioactivity** (*or induced radioactivity*) *is the name given to the radioactivity associated with radioisotopes produced from nonradioactive isotopes through bombardment reactions.*

Over 1600 synthetically produced radioisotopes that do not occur naturally are now known. Included in this total is at least one radioisotope of every naturally occurring element. In addition, isotopes of eighteen elements that do not occur in nature have been produced in small quantities as a result of bombardment reactions. These synthetic elements will be discussed in Section 17.10.

The number of synthetically produced isotopes is over five times greater than the number of naturally occurring isotopes. Figure 17-3 shows the contrast between the number of synthetic and naturally occurring isotopes for each of the known elements.

Significant uses exist for some synthetic radioisotopes. Many of these uses lie in the

Element	Group	Naturally occurring stable isotopes[a]	Naturally occurring radioactive isotopes[a]	Synthetic radioactive isotopes
H	IA	2	0	1
He	VIIIA	2	0	2
Li	IA	2	0	3
Be	IIA	1	0	4
B	IIIA	2	0	4
C	IVA	2	0	6
N	VA	2	0	5
O	VIA	3	0	5
F	VIIA	1	0	6
Ne	VIIIA	3	0	6
Na	IA	1	0	13
Mg	IIA	3	0	7
Al	IIIA	1	0	8
Si	IVA	3	0	7
P	VA	1	0	7
S	VIA	4	0	6
Cl	VIIA	2	0	8
Ar	VIIIA	3	0	9
K	IA	2	1	12
Ca	IIA	6	0	8
Sc	IIIB	1	0	11
Ti	IVB	5	0	8
V	VB	1	1	8
Cr	VIB	4	0	7
Mn	VIIB	1	0	8
Fe	VIIIB	4	0	8
Co	VIIIB	1	0	11
Ni	VIIIB	5	0	8
Cu	IB	2	0	11
Zn	IIB	5	0	15
Ga	IIIA	2	0	20
Ge	IVA	5	0	16
As	VA	1	0	19
Se	VIA	5	1	17
Br	VIIA	2	0	21
Kr	VIIIA	6	0	18
Rb	IA	1	1	24
Sr	IIA	4	0	19
Y	IIIB	1	0	19
Zr	IVB	5	0	17
Nb	VB	1	0	20
Mo	VIB	7	0	13
Tc	VIIB	0	0	21
Ru	VIIIB	7	0	14
Rh	VIIIB	1	0	20
Pd	VIIIB	6	0	16
Ag	IB	2	0	21
Cd	IIB	7	1	16
In	IIIA	1	1	27
Sn	IVA	10	0	19
Sb	VA	2	0	27
Te	VIA	5	3	21
I	VIIA	1	0	26
Xe	VIIIA	9	0	23
Cs	IA	1	0	30
Ba	IIA	7	0	23
La	IIIB	1	1	21
Hf	IVB	5	1	18
Ta	VB	1	1	19
W	VIB	5	0	22
Re	VIIB	1	1	19
Os	VIIIB	6	1	21
Ir	VIIIB	2	0	26
Pt	VIIIB	5	1	23
Au	IB	1	0	28
Hg	IIB	7	1[b]	22
Tl	IIIA	2	4[b]	21
Pb	IVA	4	4[b]	22
Bi	VA	1	4[b]	22
Po	VIA	0	7[b]	19
At	VIIA	0	3[b]	21
Rn	VIIIA	0	3[b]	24
Fr	IA	0	1[b]	26
Ra	IIA	0	4[b]	21
Ac	IIIB	0	2[b]	22
Rf	IVB	0	0	9
Ha	VB	0	0	2
—	VIB	0	0	2
Ce		4	0	20
Pr		1	0	22
Nd		6	1	17
Pm		0	0	23
Sm		5	2	17
Eu		2	0	21
Gd		6	1	13
Tb		1	0	18
Dy		6	1	14
Ho		1	0	20
Er		6	0	17
Tm		1	0	23
Yb		7	0	17
Lu		1	1	18
Th		0	6[b]	16
Pa		0	2[b]	17
U		0	3[b]	12
Np		0	0	13
Pu		0	0	15
Am		0	0	15
Cm		0	0	14
Bk		0	0	11
Cf		0	0	17
Es		0	0	14
Fm		0	0	17
Md		0	0	11
No		0	0	10
Lr		0	0	6

Metals ↙ ↘ Nonmetals

[a] Isotopes that occur in nature in only trace amounts (less than 0.01% isotopic abundance) are not listed as naturally occurring.

[b] Members of the uranium-238, thorium-232, or uranium-235 decay series (see Sec. 17-9).

Figure 17-3. Synthetic and naturally occurring isotopes of the elements. [Data from *Nuclear and Radiochemistry*, 3rd ed., by G. Friedlander, J. W. Kennedy, E. S. Macias, and J. M. Miller, John Wiley & Sons, New York, 1981.]

field of medicine. The synthetic radioisotopes cobalt-60, yttrium-90, iodine-131, and gold-198 all find use in radiotherapy treatments for cancer.

Many of the early bombardment reactions were carried out with alpha particles ejected from naturally radioactive materials. Today, many other types of particles, generated in the laboratory by particle accelerators, are available to bombard nuclei. They include protons (hydrogen-1 nuclei), neutrons, and deuterons (hydrogen-2 nuclei). Gamma rays have also been used successfully to produce artificially radioactive isotopes through bombardment. Examples of such bombardment reactions now carried out in laboratories include

$$^{44}_{20}\text{Ca} + ^{1}_{1}\text{H} \longrightarrow ^{44}_{21}\text{Sc} + ^{1}_{0}\text{n}$$

$$^{23}_{11}\text{Na} + ^{2}_{1}\text{H} \longrightarrow ^{21}_{10}\text{Ne} + ^{4}_{2}\alpha$$

Gold can be produced from platinum by the bombardment technique. However, the process is astronomically expensive compared to the worth of the gold so produced. Platinum-196 is bombarded with deuterons ($^{2}_{1}\text{H}$) to produce platinum-197.

$$^{196}_{78}\text{Pt} + ^{2}_{1}\text{H} \longrightarrow ^{197}_{78}\text{Pt} + ^{1}_{1}\text{H}$$

The platinum-197 decays through beta-particle emission to produce gold-197.

$$^{197}_{78}\text{Pt} \longrightarrow ^{0}_{-1}\beta + ^{197}_{79}\text{Au}$$

Synthetically produced radioisotopes undergo radioactive decay just as do naturally occurring radioisotopes. In many cases the mode of decay involves the previously discussed (Sec. 17.4) alpha- and beta-particle modes of decay. Frequently, however, two modes of decay not found among the naturally occurring radioisotopes are encountered: positron emission and electron capture. *A* **positron**, *designated by the symbol* ${}^{0}_{1}\beta$, *is identical to an electron or a beta particle except that it has a positive charge.* Its production in the nucleus is due to the conversion within the nucleus of a proton to a neutron.

positron

$${}^{1}_{1}p = {}^{1}_{0}n + {}^{0}_{1}\beta$$

This process is just the opposite of that occurring during beta-particle emission (Sec. 17.4). The net effect of positron emission is thus to decrease the atomic number (number of protons), while the mass number remains constant. An example of a radioactive decay process involving positron emission is

$${}^{30}_{15}P = {}^{0}_{1}\beta + {}^{30}_{14}Si$$

electron capture

In **electron capture** *an electron in a low energy orbital, such as the 1s orbital, is pulled into the nucleus, converting a proton to a neutron.*

$${}^{0}_{-1}e + {}^{1}_{1}p \longrightarrow {}^{1}_{0}n$$

An example of such a process is the reaction

$${}^{87}_{37}Rb + {}^{0}_{-1}e \longrightarrow {}^{87}_{36}Kr$$

More information concerning electron capture and also positron emission will be given in Section 17.9.

A summary of the notation used and the characteristics of the various types of radiation and small particles involved in transmutation reactions is given in Table 17-1.

Table 17-1. Notation for and Characteristics of Nuclear Radiation

Type of Radiation	Symbol	Mass Number	Charge	Composition
Alpha	${}^{4}_{2}\alpha$ or ${}^{4}_{2}He$	4	+2	helium-4 nucleus (two protons and two neutrons)
Beta	${}^{0}_{-1}\beta$	0	−1	electron ejected from an unstable nucleus
Deuteron	${}^{2}_{1}H$ or ${}^{2}_{1}D$	2	+1	hydrogen-2 nucleus (one neutron and one proton)
Electron (extranuclear)	${}^{0}_{-1}e$	0	−1	electron
Gamma	${}^{0}_{0}\gamma$	0	0	radiation of high energy
Neutron	${}^{1}_{0}n$	1	0	neutron
Positron	${}^{0}_{1}\beta$	0	+1	positive electron
Proton	${}^{1}_{1}H$ or ${}^{1}_{1}p$	1	+1	hydrogen-1 nucleus (a proton)
Triton	${}^{3}_{1}H$ or ${}^{3}_{1}T$	3	+1	hydrogen-3 nucleus (two neutrons and one proton)

EXAMPLE 17.2

Write a balanced nuclear equation for the decay of each of the following radioactive isotopes. The mode of decay is indicated in parentheses.

(a) $^{23}_{12}Mg$ (positron emission) **(b)** $^{73}_{33}As$ (electron capture)
(c) $^{195}_{79}Au$ (electron capture) **(d)** $^{51}_{25}Mn$ (positron emission)

SOLUTION

In each case the atomic number and mass number of the daughter nucleus are obtained by first writing the symbols of the parent nucleus and the particle emitted (positron) or absorbed (electron) and then balancing the equation.
(a) Let X represent the product of the radioactive decay, that is, the daughter isotope. Then

$$^{23}_{12}Mg \longrightarrow {}^{0}_{1}\beta + X$$

Note that the Greek letter beta is used to denote not only a beta particle but also a positron. The difference between the two particles is that the former is negatively charged ($_{-1}^{0}\beta$) and the latter is positively charged ($^{0}_{1}\beta$).

Since the sum of the superscripts on each side of the equation must be equal, the superscript for X must be 23. In order for the sum of the subscripts on each side of the equation to be equal, at 12, the subscript for X must be 11. As soon as the subscript of X is determined, the identity of X is known. Looking at a periodic table we determine that the element with an atomic number of 11 is sodium (Na). Therefore,

$$^{23}_{12}Mg \longrightarrow {}^{0}_{1}\beta + {}^{23}_{11}Na$$

(b) Similarly, letting X represent the product daughter isotope of the radioactive decay, we have for $^{73}_{33}As$ decaying by the electron capture mechanism

$$^{73}_{33}As + {}_{-1}^{0}e \longrightarrow X$$

Note that in electron capture the electron appears on the reactant side of the equation. This makes equations for electron capture different from those for alpha, beta, and positron emissions where in each case the small particle involved is placed on the product side of the equation.

Balancing the above equation, making the superscripts on each side of the equation total 73 and the subscripts total 32, we get

$$^{73}_{33}As + {}_{-1}^{0}e \longrightarrow {}^{73}_{32}Ge$$

(c) Similarly, for this electron capture we write

$$^{195}_{79}Au + {}_{-1}^{0}e \longrightarrow X$$

Balancing superscripts and subscripts, we get

$$^{195}_{79}Au + {}_{-1}^{0}e \longrightarrow {}^{195}_{78}Pt$$

(d) Finally, we write

$$^{51}_{25}Mn \longrightarrow {}^{0}_{1}\beta + X$$

In both positron emission and electron capture the atomic number of the daughter isotope decreases by one and the mass number does not change from that of the parent. The balancing process gives results consistent with this generalization in this part as well as in the previous three parts of this problem.

$$^{51}_{25}Mn \longrightarrow {}^{0}_{1}\beta + {}^{51}_{24}Cr$$

17.7 Rate of Radioactive Decay

All radioactive isotopes do not decay at the same rate. Some decay very rapidly; others undergo disintegration at extremely slow rates. This indicates that all radioisotopes are not equally unstable. The faster the decay rate, the lower the stability.

half-life

The concept of *half-life* is used to quantitatively express nuclear stability. The **half-life** *is the time required for one-half of any given quantity of a radioactive substance to undergo decay.* For example, if a radioisotope's half-life is 12 days and you have a 4.00 g sample of it, then after 12 days (one half-life) only 2.00 g of the sample (one-half the original amount) will remain undecayed; the other half will have decayed into some other substance.

Half-lives as long as billions of years and as short as a fraction of a second have been determined. Table 17-2 contains examples of the wide range of half-life values that occur.

Table 17-2. Range of Half-lives Found for Naturally Occurring and Synthetic Radioisotopes

Element	Half-life
Naturally Occurring Radioisotopes	
Vanadium-50	6×10^{15} yr
Platinum-190	6.9×10^{11} yr
Uranium-238	4.5×10^{9} yr
Uranium-235	7.1×10^{8} yr
Thorium-230[a]	7.5×10^{4} yr
Lead-210[a]	22 yr
Bismuth-214[a]	19.7 min
Polonium-212[a]	3.0×10^{-7} sec
Synthetic Radioisotopes	
Iodine-129	1.7×10^{7} yr
Nickel-63	92 yr
Gold-195	200 days
Lead-200	21 hr
Silver-106	24 min
Oxygen-19	29.4 sec
Fermium-246	1.2 sec
Beryllium-8	3×10^{-16} sec

[a]A product in a naturally occurring decay series (Sec. 17.9).

Most naturally occurring radioisotopes have long half-lives. Some radioisotopes with *short* half-lives, however, are also found in nature. Such short-lived species, since they decay rapidly, must be continually produced in order to be present. Processes that result in their production are (1) the decay of naturally occurring long-lived isotopes, (2) the decay of short-lived isotopes (daughter isotopes) that have been produced in the previous manner, and (3) bombardment reactions involving cosmic rays, which take place naturally in the upper atmosphere. Examples of the second method of producing short-lived isotopes are presented in Section 17.9.

The decay rate (half-life) of a radioisotope is constant. It is independent of outward conditions such as temperature, pressure, and state of chemical combination. It is dependent only on the identity of the radioisotope. For example, radioactive sodium-24, whether incorporated into NaCl, NaBr, Na_2SO_4, or $NaC_2H_3O_2$, decays at the same rate. Once something is radioactive, nothing will stop it from decaying and nothing will increase or decrease its decay rate.

Figure 17-4 shows graphically the meaning of half-life. After one half-life has passed, one-half of the original atoms have decayed, so half remain. During the next half-life, one-half of the remaining half will decay, and one-fourth of the original atoms remain undecayed. After three half-lives, $\frac{1}{2} \times \frac{1}{2} \times \frac{1}{2} = \frac{1}{8}$ of the original atoms remain undecayed, and so on. Note from Figure 17-3 that only a very small amount of original material (less than 1%) remains after seven half-lives have elapsed.

Calculations involving amounts of radioactive material decayed, amounts remaining undecayed, and time elapsed can be carried out by using the following equation.

$$\left(\begin{array}{c}\text{Amount of radioisotope undecayed}\\ \text{after } n \text{ half-lives}\end{array}\right) = \left(\begin{array}{c}\text{original amount}\\ \text{of radioisotope}\end{array}\right) \times \left(\frac{1}{2^n}\right)$$

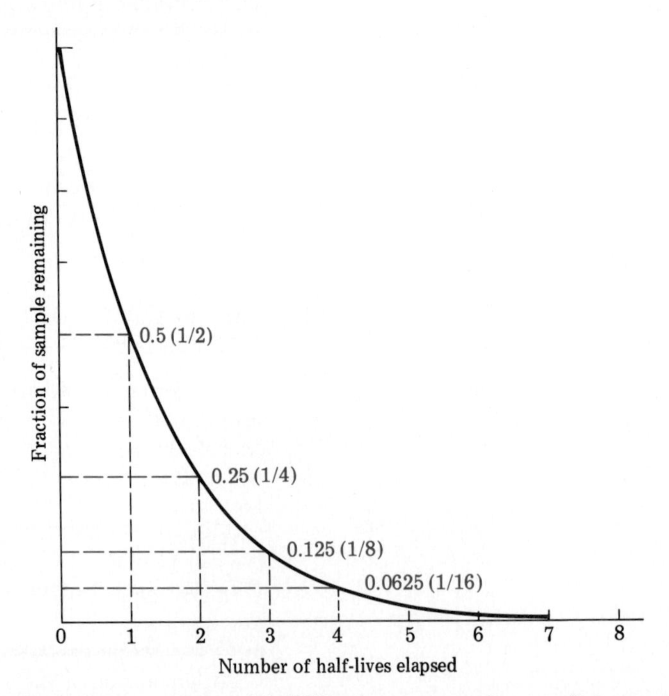

Figure 17-4. A general half-life decay curve for a radioisotope.

EXAMPLE 17.3

The half-life of iodine-131 is 8.0 days. If 12 g of iodine-131 is allowed to decay for a period of 32 days, how many grams of iodine-131 remain?

SOLUTION

First, we must determine the number of half-lives that have elapsed.

$$32\ \cancel{\text{days}} \times \frac{1\ \text{half-life}}{8.0\ \cancel{\text{days}}} = 4\ \text{half-lives}$$

Knowing the number of elapsed half-lives and the original amount of radioactive iodine present, we can use the equation

$$\begin{pmatrix}\text{Amount of radioisotope undecayed}\\ \text{after } n \text{ half-lives}\end{pmatrix} = \begin{pmatrix}\text{original amount}\\ \text{of radioisotope}\end{pmatrix} \times \left(\frac{1}{2^n}\right)$$

to get our answer.

$$\begin{pmatrix}\text{Amount of radioisotope undecayed}\\ \text{after } n \text{ half-lives}\end{pmatrix} = 12\ \text{g} \times \frac{1}{2^4} \longleftarrow \text{Four half-lives}$$

$$= 12\ \text{g} \times \frac{1}{16}$$

$$= 0.75\ \text{g} \quad \text{(calculator and correct answer)}$$

EXAMPLE 17.4

Strontium-90 is an isotope found in radioactive fallout from nuclear weapon explosions. Its half-life is 28.0 yr. How long will it take for 93.75% (15/16) of the strontium-90 atoms in the sample to undergo decay?

SOLUTION

If 15/16 of the sample has decayed, then 1/16 of the sample remains undecayed. In terms of $1/2^n$, 1/16 is equal to $1/2^4$; that is,

$$\frac{1}{2} \times \frac{1}{2} \times \frac{1}{2} \times \frac{1}{2} = \frac{1}{2^4} = \frac{1}{16}$$

Thus four half-lives have elapsed in reducing the amount of strontium-90 to 1/16 of its original amount.

Since the half-life of strontium-90 is 28 yr, the total time elapsed will be

$$4\ \cancel{\text{half-lives}} \times \frac{28.0\ \text{yr}}{1\ \cancel{\text{half-life}}} = 112\ \text{yr} \quad \text{(calculator and correct answer)}$$

In both Examples 17.3 and 17.4 the time elapsed was equivalent to a whole number of half-lives. In order to work problems involving a fractional number of half-lives, more complicated equations with logarithms must be used. Such equations will not be presented in this text; hence you will be expected to be able to work only problems that involve a whole number of half-lives.

The half-life concept can be used to determine the age of objects that contain a radioisotope. The most well known of radiochemical dating techniques is that of radiocarbon dating. Radiocarbon dating involves measuring the amount of carbon-14 present in an object. Many archeological artifacts can be dated by carbon-14 techniques, since they were made from or contain once-living carbon-containing materials.

The isotope carbon-14, the only naturally occurring radioisotope of carbon, has a half-life of 5730 yr. It is continually produced in the upper atmosphere as a result of cosmic ray bombardment.

$$^{14}_{7}N + ^{1}_{0}n \longrightarrow ^{14}_{6}C + ^{1}_{1}H$$

The steady-state (equilibrium) concentration of carbon-14, which reflects both its rate of formation and its rate of decay, is 1 carbon-14 atom/10^{12} nonradioactive carbon atoms. This trace amount of carbon-14 in the atmosphere reacts with oxygen to give carbon dioxide in the same manner that nonradioactive carbon does. Thus, approximately 1 out of every 10^{12} carbon dioxide molecules is radioactive. This radioactive carbon is incorporated into the structure of plants through photosynthesis and into animals and human beings through the food chain. A steady-state concentration of carbon-14, equal to that found in the atmosphere, is thus found in all living organisms. Upon the death of an organism, the intake of carbon-14 ceases and the natural level of radioactive carbon present within the structure begins to decrease as the result of carbon-14 decay.

$$^{14}_{6}C \longrightarrow ^{0}_{-1}\beta + ^{14}_{7}N$$

In carbon-14 dating the ratio of carbon-14 to total carbon in an object that contains once-living material (parchment, cloth, charcoal, etc.) is compared to that for living matter. A wooden object with a ratio of carbon-14 to total carbon ratio one-fourth that of a living tree would be approximately 11,400 years (two half-lives) old. An important assumption in the carbon-14 dating method is that the flow of carbon-14 into the biosphere is constant with time. There is some evidence, such as the carbon-14 content of the growth rings in older trees, to indicate that this is approximately true.

17.8 Factors Affecting Nuclear Stability

Some nuclei are stable; others are not. This is true even for isotopes of the same element. For example, sodium-23 is stable, while sodium-24 is radioactive. What determines whether a given isotope is stable or unstable? Is it possible to predict which isotopes will be stable?

As a step in the direction of understanding the more basic requirements for nuclear stability we can turn to some empirical observations of naturally occurring stable nuclei. These observations are presented in graphic form in Figure 17-5. Two generalizations concerning nuclear stability can be made after close study of this graph.

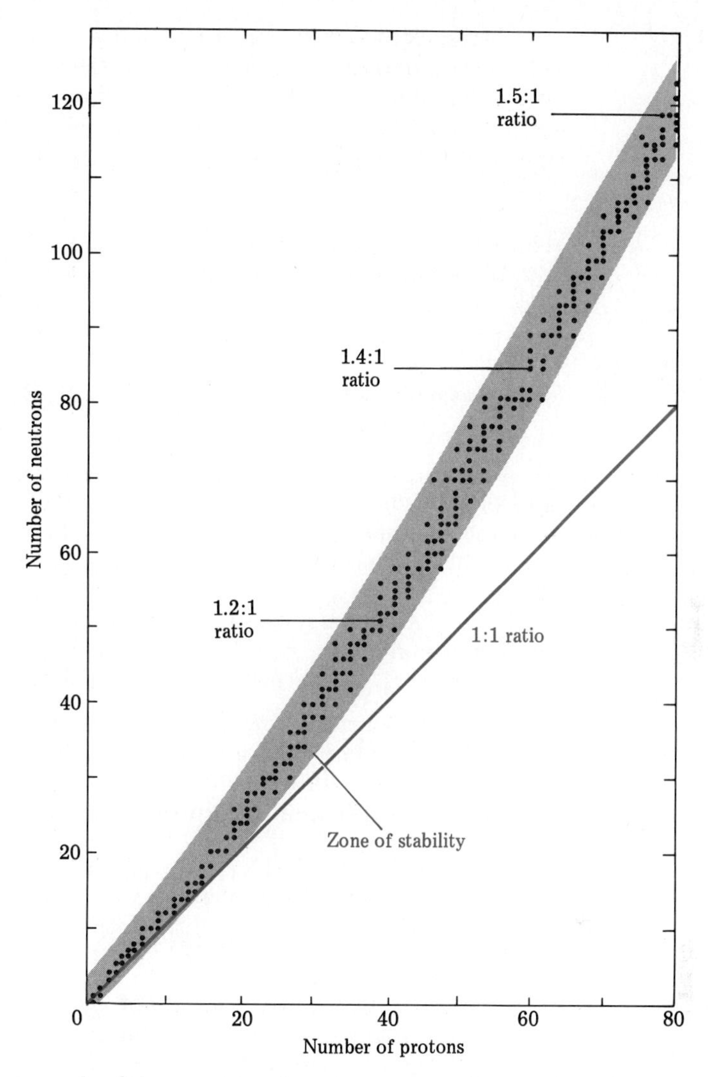

Figure 17-5. A graph of the number of neutrons versus the number of protons for stable nuclei.

1. *There is a correlation between nuclear stability and the neutron-to-proton ratio.* For elements of low atomic number neutron-to-proton ratios of stable isotopes are very close to one. For heavier elements, stable isotopes have higher neutron-to-proton ratios, with the ratio reaching approximately 1.5 for the heaviest stable elements.
2. *There is a correlation between nuclear stability and the total number of nucleons present.* Every known isotope of every element with an atomic number greater than 83 is unstable (radioactive). It thus appears that there is a limit to the number of nucleons that can be packed into a stable nucleus. That limit is 209, reached with $^{209}_{83}Bi$, the largest known stable nucleus.

The fact that stable nuclei fall into a rather narrow zone of stability defined by neutron/proton ratios strongly suggests that neutrons are at least partially responsible for the stability of the nucleus. It should be remembered that like charges repel each other and that most nuclei contain many protons (with identical positive charges) squeezed together into a very small volume. As the number of protons increases, the electrostatic forces of repulsion between protons increase sharply. Therefore a greater number of neutrons is required to counteract the increased repulsive forces. Finally, at element 84, the repulsive forces become great enough so that the nuclei are not stable regardless of the number of neutrons present.

17.9 Neutron/Proton Ratio and Mode of Decay

The mode by which a radioisotope decays depends upon how the neutron-to-proton ratio for the radioisotope compares with that of stable nuclei containing approximately the same number of nucleons. The driving force for radioactive decay is the tendency of unstable nuclei to adjust neutron-to-proton ratios in such a way that stability is achieved.

zone of stability

Unstable nuclei may be divided into three categories on the basis of their position on the graph in Figure 17-4 relative to the zone of stability. The **zone of stability** *is the area in Figure 17-4 where the stable nuclei are located.* These categories are

1. *Unstable nuclei in which the neutron-to-proton ratio is too high.* Such nuclei, which may be considered proton-poor, lie to the left of the zone of stability in Figure 17-4.
2. *Unstable nuclei in which the neutron-to-proton ratio is too low.* Such nuclides, which may be considered proton-rich, lie to the right of the zone of stability in Figure 17-4.
3. *Unstable nuclei in which the total number of nucleons exceeds 209 — the limit for a stable nucleus.* Such nuclei lie beyond the zone of stability in Figure 17-4.

Nuclei in each of these categories have a particular mode of decay, which predominates over others.

Beta emission is the predominant decay mode for nuclides having neutron-to-proton ratios that are *too high* for stability (category 1). In almost all cases, nuclei in this category have mass numbers greater than the atomic weight of the element. In the following example of beta emission, note that the mass number of the radioactive manganese isotope (56) is greater than the atomic weight of manganese (54.9 amu).

$$^{56}_{25}\text{Mn} \longrightarrow {}^{56}_{26}\text{Fe} + {}^{0}_{-1}\beta$$

As discussed previously (Sec. 17.4), beta emission involves the transformation of a neutron into a proton. This increases the number of protons, decreases the number of neutrons, and causes a decrease in the neutron/proton ratio — the desired result. Note that in the above reaction for the beta decay of manganese-56, the neutron/proton ratio of the parent isotope is 1.24 and that of the daughter iron-56 is 1.15.

Radioisotopes lying to the right of the zone of stability, that is, those with *too low* a neutron-to-proton ratio, decay by converting a proton into a neutron — just the opposite of the process that occurs in beta-particle emission. Radionuclei in this category generally have mass numbers that are lower than the atomic weight of the element. The conversion of a proton into a neutron may be accomplished in two ways: (1) by

Table 17-3. The Four Known Radioactive Decay Series

Parent Member	Number of Decay Steps	Final Product of Series
Uranium-238	14	lead-206
Thorium-232	10	lead-208
Uranium-235	11	lead-207
Plutonium-241	13	bismuth-209

positron emission and (2) by electron capture. Both of these decay modes were previously described in Section 17.5. The process of electron capture seems to be preferred over positron emission for isotopes of high atomic number. For lighter isotopes numerous examples of both types of decay processes are known.

decay series

In elements that lie beyond the zone of stability — radionuclei containing more than 209 nucleons — usually more than one decay step is required to reach stability. This results in the formation of a decay series. A **decay series** *is a series of elements produced from the successive emission of alpha and beta particles.* A decay series always starts with a long-lived radioisotope and ends with a stable isotope. Three naturally occurring decay series are known. A fourth series was discovered after the synthesis of certain elements not found in nature (Sec. 17.10). Because the parent of this fourth series, plutonium-241, does not occur in nature to a measurable extent, the series is not classified as a naturally occurring series. General characteristics of the four decay series are given in Table 17-3. Figure 17-6 shows all of the members of the uranium-238 decay series. It is representative of the other three series.

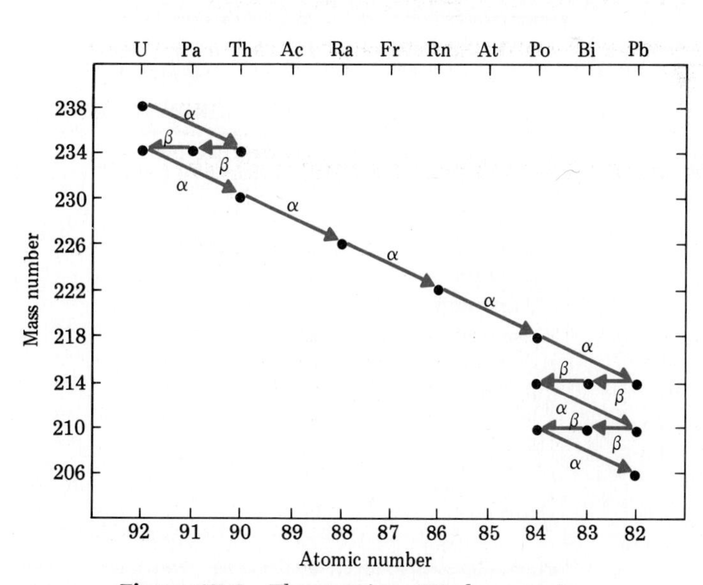

Figure 17-6. The uranium-238 decay series.

17.10 Synthetic Elements

One of the most interesting facets of bombardment reaction research (Sec. 17.6) is the production of elements that do not occur in nature, that is, *synthetic elements.* Four synthetic elements, produced between 1937 and 1941, filled gaps in the periodic table for which no naturally occurring element had been found. These four are technicium (Tc, element 43), promethium (Pm, element 61), astatine (At, element 85), and francium (Fr, element 87). The reactions for their production are

$$^{96}_{42}\text{Mo} + {}^{2}_{1}\text{H} \longrightarrow {}^{97}_{43}\text{Tc} + {}^{1}_{0}\text{n} \qquad (\text{half-life} = 2.6 \times 10^{6}\ \text{yr})$$

$$^{142}_{60}\text{Nd} + {}^{1}_{0}\text{n} \longrightarrow {}^{143}_{61}\text{Pm} + {}^{0}_{-1}\beta \qquad (\text{half-life} = 265\ \text{days})$$

$$^{209}_{83}\text{Bi} + {}^{4}_{2}\alpha \longrightarrow {}^{210}_{85}\text{At} + 3\,{}^{1}_{0}\text{n} \qquad (\text{half-life} = 8.3\ \text{hr})$$

$$^{230}_{90}\text{Th} + {}^{1}_{1}\text{p} \longrightarrow {}^{223}_{87}\text{Fr} + 2\,{}^{4}_{2}\alpha \qquad (\text{half-life} = 22\ \text{min})$$

Bombardment reactions have also been the source for elements 93 to 106. These elements are called the *transuranium elements,* since they occur immediately following uranium in the periodic table. (Uranium is the highest atomic numbered, naturally occurring element.) All isotopes of all of the transuranium elements are radioactive.

Collision between the nucleus of a light atom and a very large nucleus is one of the methods used to prepare transuranium elements. For example, element 104 can be prepared by the reaction

$$^{249}_{98}\text{Cf} + {}^{12}_{6}\text{C} \longrightarrow {}^{257}_{104}\text{Rf} + 4\,{}^{1}_{0}\text{n} \qquad (\text{half-life} = 4.5\ \text{sec})$$

Information concerning the stability of the transuranium elements is given in Table 17-4. Note that in many cases the synthetic isotopes are very short-lived, having half-

Table 17-4. Stability of Transuranium Element Isotopes

Name	Symbol	Atomic Number	Atomic Weight of Most Stable Isotope	Half-life of Most Stable Isotope	Date of Discovery
Neptunium	Np	93	237	2.14×10^{6} yr	1940
Plutonium	Pu	94	244	7.6×10^{7} yr	1940
Americium	Am	95	243	8.0×10^{3} yr	1944
Curium	Cm	96	247	1.6×10^{7} yr	1944
Berkelium	Bk	97	247	1400 yr	1950
Californium	Cf	98	251	900 yr	1950
Einsteinium	Es	99	252	472 days	1952
Fermium	Fm	100	257	100 days	1953
Mendelevium	Md	101	258	56 days	1955
Nobelium	No	102	259	1 hr	1958
Lawrencium	Lr	103	260	3 min	1961
[a] Rutherfordium	Rf	104	261	70 sec	1969
[a] Hahnium	Ha	105	262	40 sec	1970
—	—	106	263	0.9 sec	1974

[a] Name and symbol not yet officially accepted.

lives in the seconds range. Only small amounts of these short-lived isotopes have been produced — only hundreds of atoms in some cases. Since these elements, one produced, quickly disappear (decay), they cannot be detected and identified on the basis of chemical properties. Identification of these elements is made with instruments that analyze the characteristic radiation emitted by each new element.

The transuranium element of greatest practical importance is plutonium (element 93), which is produced in fission reactors (Sec. 17.11) in sizable quantities. Plutonium is also under study for use as a nuclear reactor fuel (Sec. 17.11).

17.11 Nuclear Fission

Shortly before World War II several research groups studied the products obtained by bombarding uranium-235 with neutrons. They hoped to discover new synthetic elements (Sec. 17.10) with atomic numbers greater than 92. To their surprise, they isolated from the reaction products an isotope of the element that has an atomic number of 56 (barium). A lighter element rather than a heavier element had been produced. Further study led to the characterization of a new type of nuclear reaction — nuclear fission.

nuclear fission

Nuclear fission *is the process in which a heavy element nucleus splits into two or more medium-sized nuclei as the result of bombardment.*

A number of heavy element nuclei are now known to undergo fission when struck with a variety of particles, including neutrons, alpha particles, protons, and gamma rays. However, at present the fission reaction is considered to be of practical importance only in the case of three isotopes: naturally occurring uranium-235, synthetically produced plutonium-239, and synthetically produced uranium-233. All three of these isotopes will undergo fission when bombarded by low energy neutrons, and all have long half-lives. Many of the other fissionable isotopes require higher energy bombarding particles, are not available in large quantities, or have half-lives that are too short to be of much significance.

atomic energy

When fission occurs, *very large* amounts of energy are released, many times greater than that from ordinary radioactive decay. This large amount of released energy is the most important aspect of the fission reaction. **Atomic energy** *is the name given to energy released during a nuclear fission reaction.* Two important results of the use of atomic energy are the atomic bomb and nuclear power plants.

Atomic Bombs

The first fissionable isotope to be discovered, and still the most important one, was uranium-235. Intensive study of the fission characteristics of this radioisotope, during the period 1940–1945, led to the production of the atomic bombs used to end World War II. Uranium-235 undergoes fission when bombarded with slow-moving neutrons (thermal neutrons). There is no unique way in which the uranium-235 nucleus splits, and more than 200 nuclides of 35 different elements have been identified as products of the reaction. Typically, one fragment has an atomic number between 48 and 58 and the other between 35 and 45. The following are examples of the numerous ways in which this particular uranium nucleus can split.

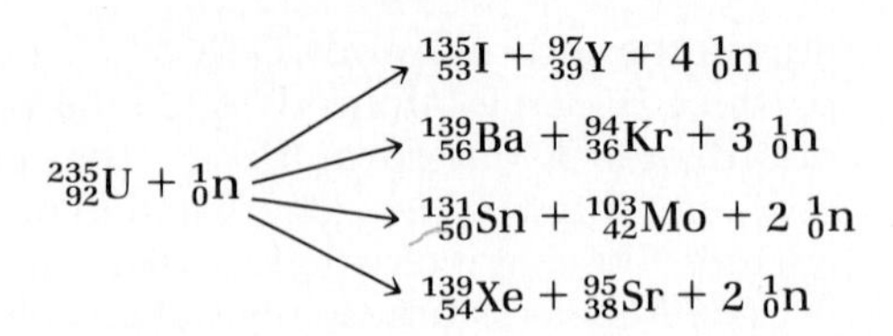

$$
{}^{235}_{92}U + {}^{1}_{0}n \rightarrow
\begin{cases}
{}^{135}_{53}I + {}^{97}_{39}Y + 4\ {}^{1}_{0}n \\
{}^{139}_{56}Ba + {}^{94}_{36}Kr + 3\ {}^{1}_{0}n \\
{}^{131}_{50}Sn + {}^{103}_{42}Mo + 2\ {}^{1}_{0}n \\
{}^{139}_{54}Xe + {}^{95}_{38}Sr + 2\ {}^{1}_{0}n
\end{cases}
$$

Note from the equations that a neutron is required to induce the fission of a uranium-235 atom, and that neutrons are also produced as a result of fission. On the average 2.4 neutrons are produced per uranium-235 fission. When small quantities of uranium-235 undergo fission, the newly produced neutrons escape from the sample into the surroundings. When a sufficiently large mass of uranium-235 is present, the neutrons collide with other uranium-235 atoms and cause more fissions. If only one of the

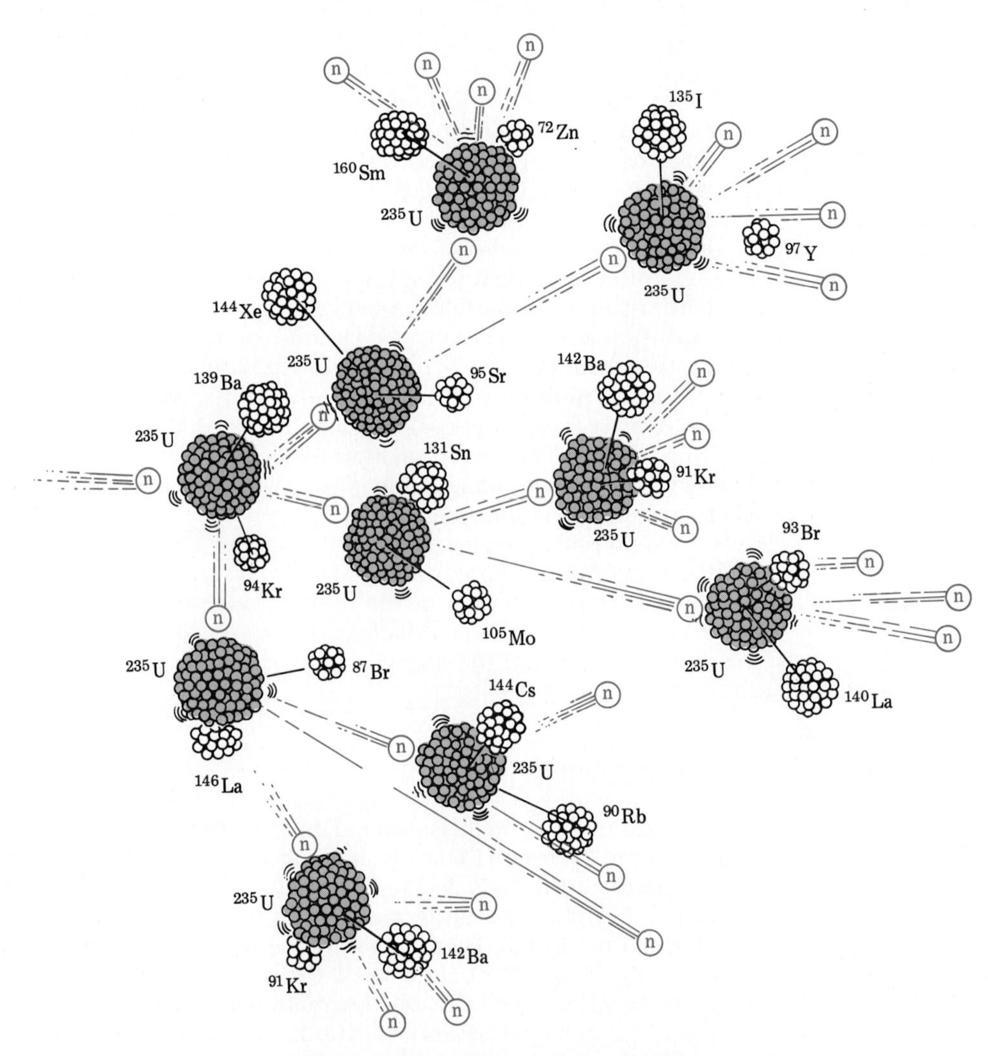

Figure 17-7. A fission chain reaction.

neutrons produced each time reacts with another uranium-235 atom, the process will continue at a constant rate. When this condition prevails, the process is said to be self-propagating or *critical.* If more than one of the neutrons generated per disintegration produces another disintegration, the reaction becomes an expanding or branching chain reaction. Such a situation, which is known as a *supercritical* condition, leads to an explosion. In an atomic bomb, small pieces of uranium, each incapable of sustaining a chain reaction, are brought together (at the appropriate time) to form one piece capable of sustaining a chain reaction. Figure 17-7 illustrates a fission chain reaction.

Nuclear Power Plants

Even before the first atomic bomb was exploded, scientists began to speculate about the use of nuclear fission as a source of energy for peaceful purposes. This is now a reality. Each year a small but increasing amount of electricity used in the United States is generated by nuclear power plants whose operation involves nuclear fission.

The basic concepts in the production of electricity from a nuclear power plant are the same as those for a traditional coal-fired generating plant. In both cases heat is used to generate steam, which in turn causes a steam turbine generator to function. The difference between the two types of plants is that nuclear fuel replaces fossil fuel as a source of heat. The essential components of a fission-powered generating plant are given in Figure 17-8. The uranium present in the reactor core is in the form of pellets of the oxide U_3O_8 enclosed in long steel tubes. The uranium is a mixture of all isotopes of uranium, not just uranium-235, because of the cost of isotope separation. The energy from fission appears as heat, which is drawn out of the core by circulating primary coolant. A heat exchanger is then used to cause water to turn to steam, which then turns the electric generator. The spent steam is condensed back to water as it passes through a condenser. It then recirculates back to the heat exchanger.

The use of nuclear power plants has created a large amount of controversy in recent years. The major focus of attention is on problems associated with environmental damage and nuclear waste disposal. Large amounts of waste heat are a necessary by-product of nuclear reactor operations. The way in which this waste heat is released into the environment has been a matter of concern. This problem, often referred to as

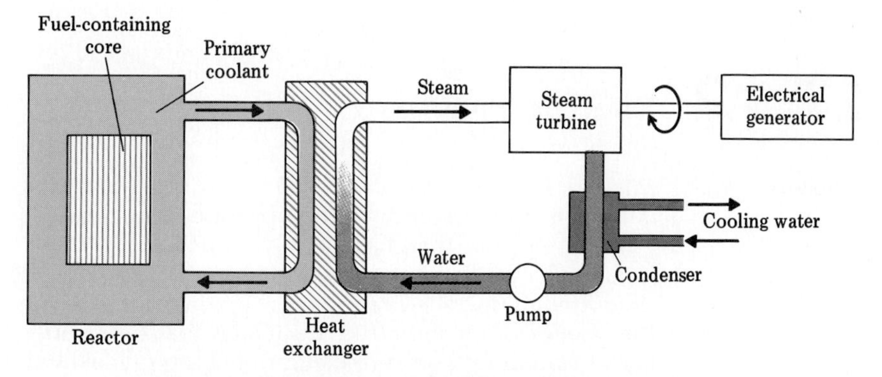

Figure 17-8. Components of a nuclear power plant.

thermal pollution, is solvable through the use of devices such as cooling towers. In addition to heat, radioactive wastes are also produced during reactor operation. They must be removed periodically, and they create disposal problems. They cannot and are not released into the environment, but must be stored indefinitely. Where and how to safely store these wastes has been and still is a matter of much debate.

Breeder Reactors

Uranium-235, the only known naturally occurring fissionable isotope, represents only about 0.71% of naturally occurring uranium. Nonfissionable uranium-238 is the abundant isotope of uranium (99.28%). At the current and projected rate of use, known sources of uranium-235 will be exhausted near the end of the century. Breeder reactors represent a solution to this problem. Such reactors produce or "breed" fissionable isotopes by nuclear transformations. Most research in breeder technology has been directed toward the use of uranium-238 and thorium-232 as sources for the fissionable fuel. Both are naturally occurring abundant isotopes that have long half-lives. Thorium is three times more abundant than uranium. The breeder reactor consumes uranium-235 as fuel, but produces fissionable plutonium-239 or uranium-233 during the operation of the reactor, as shown in the following equations.

$$\underset{}{{}^{238}_{92}\text{U}} + \underset{\text{(from fission)}}{{}^{1}_{0}\text{n}} \longrightarrow {}^{239}_{92}\text{U} \xrightarrow{\beta} {}^{239}_{93}\text{Np} \xrightarrow{\beta} \underset{\text{(fissionable)}}{{}^{239}_{94}\text{Pu}}$$

$$\underset{}{{}^{232}_{90}\text{Th}} + \underset{\text{(from fission)}}{{}^{1}_{0}\text{n}} \longrightarrow {}^{233}_{90}\text{Th} \xrightarrow{\beta} {}^{233}_{91}\text{Pa} \xrightarrow{\beta} \underset{\text{(fissionable)}}{{}^{233}_{92}\text{U}}$$

Experimental breeder systems are now under study in many parts of the world, although none are yet commercially producing energy. The technical problems associated with building breeder reactors have been considerable. Much research and testing still need to be carried out. There is no question about the ability to produce these two fissionable isotopes. It has been and is being done. The problem occurs in the design of an economical process that generates more fuel than it consumes.

17.12 Nuclear Fusion

nuclear fusion

Nuclear fusion *is the process in which small atoms are put together to make larger ones.* It is, thus, essentially the opposite of nuclear fission. Fusion is a higher energy yielding reaction than fission.

In order for fusion to occur, a very high temperature, on the order of several hundred million degrees, is required. One place that is hot enough for fusion to occur is the interior of the sun. It is postulated that the energy of the sun is derived from the conversion of hydrogen nuclei into helium nuclei by nuclear fusion. The overall reaction is thought to occur in steps.

1. Two hydrogen-1 nuclei fuse to produce a hydrogen-2 nucleus and a positron.

$$^{1}_{1}H + ^{1}_{1}H \longrightarrow ^{2}_{1}H + ^{0}_{1}\beta$$

2. The hydrogen-2 nucleus fuses with another hydrogen-1 nucleus to give helium-3.

$$^{2}_{1}H + ^{1}_{1}H \longrightarrow ^{3}_{2}He$$

3. Pairs of helium-3 nuclei fuse to form helium-4 plus two hydrogen-1 nuclei.

$$^{3}_{2}He + ^{3}_{2}He \longrightarrow ^{4}_{2}He + 2\,^{1}_{1}H$$

The net overall reaction for the production of energy on the sun (the sum of reactions in steps 1, 2, and 3) is represented by the equation

$$4\,^{1}_{1}H \longrightarrow ^{4}_{2}He + 2\,^{0}_{1}\beta$$

The use of nuclear fusion on earth might seem impossible because of the high temperatures required. It has, however, been accomplished in a hydrogen bomb. In such a bomb, a *fission* device (an atomic bomb) is used to achieve the high temperatures needed to start the following fusion reaction.

$$^{3}_{1}H + ^{2}_{1}H \longrightarrow ^{4}_{2}He + ^{1}_{0}n$$

The hydrogen-3 needed in the reaction does not occur naturally and is produced by bombarding lithium-6 with neutrons.

$$^{6}_{3}Li + ^{1}_{0}n \longrightarrow ^{3}_{1}H + ^{4}_{2}He$$

The necessary hydrogen-2 may be obtained from sea water. Hydrogen-1 and hydrogen-2 are the only naturally occurring isotopes of hydrogen. One out of approximately 5000 hydrogen atoms is hydrogen-2.

At the high temperature of fusion reactions, electrons completely separate from nuclei. Neutral atoms cannot exist. This high temperature gas-like mixture of nuclei and electrons is called a *plasma* and is considered by some scientists to represent a fourth state of matter (in addition to solids, liquids, and gases).

In principle, fusion reactions can be used as controlled energy sources just as fission reactions are. Fusion reactors would have a major advantage over fission reactors. No significant amount of radioactive wastes would be produced during operation. Unfortunately, fusion reactors have not yet been proven feasible.

In order to start a fusion reaction a temperature of about 100 million degrees must be achieved and the resulting plasma must be contained. Several approaches to solving the heating and containment problems are under active investigation. The use of laser beams to achieve the high temperature needed to initiate a fusion reaction is an approach under study.

Approaches to the containment problem deal with a nonmaterial containment system called a "magnetic bottle." Charged particles (the plasma) have difficulty crossing magnetic lines of force, so a "magnetic bottle" can be created by surrounding the plasma with a suitable magnetic field. Magnetic fields of various shapes can be generated by passing an electric current through an appropriately arranged pattern of wires or other conductors, as shown in Figure 17-9. The use of a nonmaterial container is sometimes erroneously thought to be necessary because the hot plasma would destroy any material it contacted. However, the plasma density is so low that the total energy content would be insufficient to damage container walls. On the contrary, a nonmaterial container is necessary to keep the plasma from the walls to avoid conductive heat losses that would result in plasma temperatures too low to support fusion. In

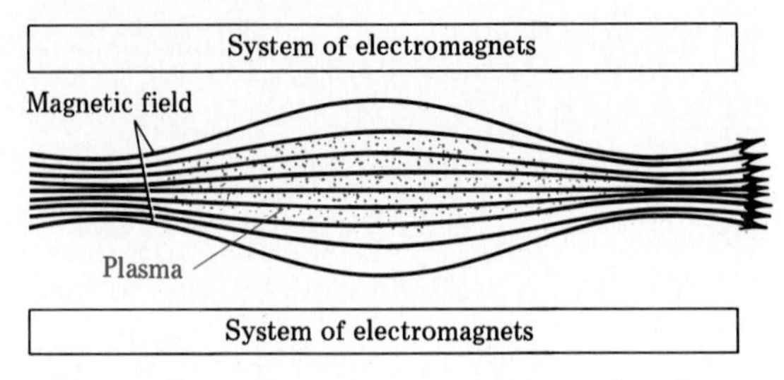

Figure 17-9. Magnetic containment of a plasma.

addition, contact with container walls introduces into the plasma heavy contaminants, which would cause additional problems.

Although major technological advances might be made suddenly, it is now estimated that fusion power will not become available until at least the year 2000, if it is proved feasible at all.

17.13 A Comparison of Nuclear and Chemical Reactions

As can be seen from the discussions of the previous sections of this chapter, nuclear chemistry is a field quite different from ordinary chemistry. Many of the laws of chemistry must be modified when we consider nuclear reactions. The major differences between nuclear reactions and ordinary chemical reactions are listed in Table 17-5. This table serves as a summary of the concepts presented in this chapter.

Table 17-5. Differences Between Nuclear and Chemical Reactions

Chemical Reaction	Nuclear Reaction
1. Different isotopes of an element have practically identical chemical properties.	1. Different isotopes of an element have different properties in nuclear processes.
2. The chemical reactivity of an element depends on the element's state of combination (free element, compound, etc.).	2. The nuclear reactivity of an element is independent of the state of chemical combination.
3. Elements retain their identity in chemical reactions.	3. Elements may be changed into other elements during nuclear reactions.
4. Energy changes that accompany chemical reactions are relatively small.	4. Nuclear reactions involve energy changes a number of orders of magnitude larger than those in chemical reactions.

Learning Objectives

After completing this chapter you should be able to

- Define the term radioactive (Sec. 17.1).
- Name and write symbols that indicate the composition of the three types of radiation given off by naturally occurring radioactive materials (Sec. 17.3).
- Write balanced nuclear equations for various alpha and beta decay processes (Sec. 17.4).
- Indicate the changes that occur in the atomic number and mass number of a radioisotope as a result of alpha or beta decay (Sec. 17.4).
- Contrast the relative penetrating abilities of alpha, beta, and gamma radiations (Sec. 17.5).
- Define the terms transmutation reaction and bombardment reaction (Sec. 17.6).
- Balance nuclear equations representing bombardment reactions (Sec. 17.6).
- Write balanced nuclear equations for various positron and electron capture decay processes (Sec. 17.6).
- Indicate the changes that occur in the atomic number and mass number of a radioisotope as a result of positron emission or electron capture (Sec. 17.6).
- Define the term half-life and be able to calculate the fraction of a radioisotope left after a given whole number of half-lives have elapsed or vice versa (Sec. 17.7).
- Relate nuclear stability to the total number of nucleons present and to neutron/proton ratio (Sec. 17.8).
- Relate mode of decay for a radioisotope to neutron/proton ratio (Sec. 17.9).
- State general methods of production for and stability characteristics of transuranium elements (Sec. 17.10).
- Describe the general characteristics of a nuclear fission reaction and indicate how the fission process finds use in nuclear weapons and nuclear power plants (Sec. 17.11).
- Describe the general characteristics of a nuclear fusion reaction and indicate the types of problems that must be overcome if this process is to be used in a practical manner (Sec. 17.12).
- Contrast the major differences between nuclear reactions and "ordinary" chemical reactions (Sec. 17.13).

Terms and Concepts for Review

The new terms or concepts defined in this chapter are

nuclear reaction (Sec. 17.1)
radioactive (Sec. 17.1)
alpha rays (Sec. 17.3)
beta rays (Sec. 17.3)
gamma rays (Sec. 17.3)
radioactive decay (Sec. 17.4)
balanced nuclear equation (Sec. 17.4)
parent isotope (Sec. 17.4)
daughter isotope (Sec. 17.4)
transmutation reaction (Sec. 17.6)
bombardment reaction (Sec. 17.6)
artificial or induced radioactivity (Sec. 17.6)
positron (Sec. 17.6)
electron capture (Sec. 17.6)
half-life (Sec. 17.7)
zone of stability (Sec. 17.9)
decay series (Sec. 17.9)
nuclear fission (Sec. 17.11)
atomic energy (Sec. 17.11)
nuclear fusion (Sec. 17.12)

Questions and Problems

Notation and Terminology in Nuclear Reactions

17-1 Use two different notations to denote each of the following radioactive isotopes.
a. contains 4 protons, 4 electrons, 6 neutrons
b. contains 20 protons, 20 electrons, 18 neutrons
c. contains 41 protons, 41 electrons, 55 neutrons
d. contains 99 protons, 99 electrons, 157 neutrons

17-2 Supply a complete symbol, including superscript and subscript, for each of the following nuclear particles.
a. alpha particle **b.** beta particle
c. gamma ray **d.** positron
e. neutron **f.** proton

17-3 Supply a correct name for each of the following nuclear chemistry symbols.
a. $^{4}_{2}X$ **b.** $^{0}_{-1}X$ **c.** $^{1}_{0}X$ **d.** $^{0}_{1}X$ **e.** $^{1}_{1}X$
f. $^{0}_{0}X$

17-4 Group the commonly encountered nuclear particles of Table 17-1 into the following categories.
a. those with a mass number of zero
b. those with a positive charge
c. those with a negative charge
d. those with zero charge
e. those with a mass number greater than one

Equations for Radioactive Decay Processes

17-5 Write balanced nuclear equations for the alpha decay of the following.
a. $^{200}_{84}Po$ **b.** $^{244}_{96}Cm$ **c.** thorium-229
d. fermium-252

17-6 Write balanced nuclear equations for the beta decay of the following.
a. $^{25}_{11}Na$ **b.** $^{104}_{45}Rh$
c. germanium-77 **d.** uranium-237

17-7 Write balanced nuclear equations for the positron decay of the following.
a. $^{8}_{5}B$ **b.** $^{103}_{47}Ag$ **c.** rubidium-79
d. barium-127

17-8 Write balanced nuclear equations for the electron capture decay of the following.
a. $^{59}_{28}Ni$ **b.** $^{161}_{67}Ho$ **c.** palladium-100
d. gold-189

17-9 What is the effect upon the mass number and atomic number when the following nuclear transmutations occur?

- **a.** An alpha particle is emitted.
- **b.** A beta particle is emitted.
- **c.** An electron is captured.
- **d.** A positron is emitted.

17-10 Identify X in each of the following radioactive decay equations.

- **a.** $^{10}_{4}Be \longrightarrow {}^{10}_{5}B + X$
- **b.** $^{210}_{83}Bi \longrightarrow {}^{4}_{2}\alpha + X$
- **c.** $^{41}_{20}Ca + X \longrightarrow {}^{41}_{19}K$
- **d.** $^{15}_{8}O \longrightarrow {}^{15}_{7}N + X$
- **e.** $^{44}_{22}Ti + {}^{0}_{-1}e \longrightarrow X$
- **f.** $X \longrightarrow {}^{4}_{2}\alpha + {}^{222}_{86}Rn$

17-11 Write the nuclear equation for each of the following radioactive decay processes.

- **a.** Tantulum-181 is formed by electron capture.
- **b.** Cesium-139 is formed by beta emission.
- **c.** Beta emission from bromine-84
- **d.** Positron emission from copper-59
- **e.** Mercury-200 is formed by alpha emission.
- **f.** Sulfur-33 is formed by positron emission.

17-12 Identify the mode of decay for each of the following isotopes given the identity of the daughter isotope.

- **a.** $^{190}_{78}Pt$ (daughter = $^{186}_{76}Os$)
- **b.** $^{19}_{8}O$ (daughter = $^{19}_{9}F$)
- **c.** $^{22}_{11}Na$ (daughter = $^{22}_{10}Ne$) (Give two possible modes of decay for this process.)

Equations for Bombardment Reactions

17-13 Identify X in each of the following bombardment reactions.

- **a.** $^{24}_{12}Mg + X \longrightarrow {}^{27}_{14}Si + {}^{1}_{0}n$
- **b.** $^{27}_{13}Al + {}^{2}_{1}H \longrightarrow X + {}^{4}_{2}\alpha$
- **c.** $X + {}^{1}_{1}H \longrightarrow 2\,{}^{4}_{2}He$
- **d.** $^{14}_{7}N + {}^{4}_{2}\alpha \longrightarrow X + {}^{1}_{1}H$
- **e.** $X + {}^{11}_{5}B \longrightarrow {}^{257}_{103}Lr + 4\,{}^{1}_{0}n$
- **f.** $^{9}_{4}Be + X \longrightarrow {}^{12}_{6}C + {}^{1}_{0}n$

17-14 Write equations for the following nuclear bombardment processes.

- **a.** Beryllium-9 captures an alpha particle and emits a neutron.
- **b.** Nickel-58 is bombarded with a proton and an alpha particle is emitted.
- **c.** Bombardment of a radioisotope with an alpha particle results in the production of curium-242 and one neutron.
- **d.** Bombardment of curium-246 with a small particle results in the production of $^{254}_{102}No$ and 4 neutrons.

Rate of Radioactive Decay

17-15 What is meant by the term half-life when it is applied to a radioisotope?

17-16 Explain the fallacy in the conclusion that the whole-life of a radioisotope is equal to twice the half-life.

17-17 An isotope of lead, $^{194}_{82}Pb$, has a half-life of 11 min. What fraction of lead-194 atoms in a sample will remain after

- **a.** 44 min
- **b.** 66 min
- **c.** 1 hr 50 min
- **d.** three half-lives
- **e.** five half-lives
- **f.** eight half-lives

17-18 If a sample contains 4.00 g of iodine-131 (half-life = 8.0 days), how many grams of iodine-131 will have decayed after four half-lives?

17-19 After 126 days have elapsed 1/64 of a radioisotope sample remains undecayed. What is the half-life of the radioisotope?

17-20 Some objections to nuclear power plants are based on the need to store the radioactive wastes that result. One radioisotope found in fission reactor wastes is strontium-90, with a half-life of 28 yr. How long would this isotope have to be stored in order to reduce its amount to about 1/1000 that in the waste originally?

Neutron-to-Proton Ratio and Mode of Decay

17-21 What is the predominant mode of decay for a radionuclide in which

- **a.** the neutron-to-proton ratio is too high for stability?
- **b.** the neutron-to-proton ratio is too low for stability?

17-22 What mode or modes of decay can be thought of as resulting from

- **a.** the conversion of a neutron into a proton?
- **b.** the conversion of a proton into a neutron?

17-23 Calculate the neutron-to-proton ratio before and after each of the following processes take place.

- **a.** positron decay of $^{43}_{22}Ti$
- **b.** beta decay of $^{24}_{10}Ne$
- **c.** electron capture by $^{71}_{32}Ge$
- **d.** alpha decay of $^{218}_{85}At$

17-24 When elements lie beyond the region of stability, decay occurs in a series of steps. The steps in the decay of thorium-232 result in the emission of the following particles in the order given.

alpha
beta
beta
alpha
alpha
alpha
alpha
beta
beta
alpha

Using this information, draw a diagram similar to Figure 17-6 for the decay of thorium-232. The stable end product of the decay series is lead-208.

17-25 One member of each of the following pairs of radioisotopes decays by beta particle emission and the other by positron emission. Which is which? Explain your reasoning.

a. $^{79}_{37}Rb$, $^{90}_{37}Rb$ **b.** $^{44}_{19}K$, $^{37}_{19}K$ **c.** $^{87}_{34}Se$, $^{59}_{29}Cu$

17-26 Which of the following would most likely decay by (1) alpha emission, (2) beta emission, (3) electron capture or positron emission.

a. $^{100}_{41}Nb$ **b.** $^{56}_{28}Ni$ **c.** $^{120}_{48}Cd$ **d.** $^{33}_{17}Cl$
e. $^{250}_{100}Fm$

Synthetic and Naturally Occurring Elements

17-27 Using Figure 17-3 as your source of information, determine the following.

a. elements for which no stable isotopes exist
b. elements for which only one stable isotope exists
c. elements for which more stable isotopes exist than unstable ones
d. element for which the greatest number of stable isotopes exists
e. element for which the greatest number of unstable isotopes exists
f. element for which the greatest total number of isotopes exist (both stable and unstable)

Nuclear Fission and Nuclear Fusion

17-28 Identify which of the following characteristics apply to the fission process, the fusion process, or both processes.

a. A high temperature is required to start the process.
b. An example of the process occurs on the sun.
c. Transmutation of elements occurs.
d. Radiation is emitted as a result of the process occurring.
e. Neutrons are involved.
f. The process is now used to generate some electrical power in the United States.

17-29 Identify the following as fission or fusion reactions or as neither.

a. $^{3}_{2}He + ^{3}_{2}He \longrightarrow ^{4}_{2}He + 2\,^{1}_{1}H$
b. $^{239}_{92}U \longrightarrow ^{239}_{93}Np + ^{0}_{-1}\beta$
c. $^{235}_{92}U + ^{1}_{0}n \longrightarrow ^{144}_{55}Cs + ^{90}_{37}Rb + 2\,^{1}_{0}n$
d. $^{230}_{90}Th + ^{1}_{1}H \longrightarrow ^{223}_{87}Fr + 2\,^{4}_{2}\alpha$
e. $^{3}_{1}H + ^{2}_{1}H \longrightarrow ^{4}_{2}He + ^{1}_{0}n$

17-30 Why is ^{235}U, which is much less abundant than ^{238}U, considered more important than ^{238}U in nuclear fission processes?

17-31 What role do each of the following isotopes play in breeder reactor technology?

a. uranium-233 **b.** uranium-235
c. uranium-238 **d.** thorium-232
e. plutonium-239

17-32 What role do each of the following substances play in fusion technology and from where are they obtained?

a. hydrogen-2 **b.** hydrogen-3

17-33 A "fourth state of matter" called a plasma is encountered in studying fusion reactions. What are the characteristics of matter in this state?

18 Introduction to Organic Chemistry

18.1 Organic Chemistry — A Historical Perspective

During the latter part of the 18th century and early part of the 19th century, chemists began to distinguish between two types of compounds: organic compounds and inorganic compounds. The distinction between the two types of compounds involved origins. Did a compound come from a "living" or "nonliving" source? The term organic was used for those compounds that came from plants and animals (organisms) and the term inorganic for those compounds obtained from the mineral constituents of the earth.

Closely connected with this organic–inorganic terminology was the thinking of that time period that organic compounds contained a special "vital force" that only a living organism could supply. Thus, the only source of such compounds was thought to be nature itself. The fact that no scientists of the time were able successfully to synthesize a known organic compound from inorganic starting materials gave credence to this "vital force" theory.

The "vital force" theory is now known to be incorrect. Routinely today, in many laboratories throughout the world, compounds found in living organisms are synthesized from compounds not found in living organisms. Despite the fall of the "vital force" theory, with its emphasis on living and nonliving origins of substances, the terminology associated with the theory (organic and inorganic) is still in use. The original definitions for these terms, however, have been changed.

organic chemistry

Today, **organic chemistry** *is defined as the study of hydrocarbons (binary compounds of hydrogen and carbon) and their derivatives.* Interestingly, almost all compounds found in living organisms still fall in the field of organic chemistry when this modern definition is applied. In addition, many compounds synthesized in the laboratory, which have never been found in nature or in living organisms, are considered to be organic compounds.

In a less rigorous manner, organic chemistry is often defined as the study of carbon-containing compounds. It is true that almost all carbon-containing compounds qualify as organic compounds. There are, however, some exceptions. The oxides of carbon, carbonates, cyanides, and metallic carbides are all considered to be inorganic compounds rather than organic compounds. The field of *inorganic chemistry* encompasses the study of all noncarbon-containing compounds (the other 105 elements) plus the few carbon-containing compounds just mentioned.

In essence, organic chemistry is the study of one element (carbon) and inorganic chemistry the study of 105 elements. Why, relative to their study, is there such an unequal partitioning of the elements? The answer is simple. The chemistry of carbon is so much more extensive than that of the other elements that there is justification in making its study a field by itself. Approximately 4 million organic compounds are known. Fewer than 250,000 inorganic compounds exist. This is an approximate 15-to-1 ratio between organic and inorganic compounds.

Why does carbon form fifteen times as many compounds as all of the other elements combined? The reason is that carbon possesses the unique ability to bond to itself in long chains, rings, and complex combinations of both. Chains and rings of all lengths are possible. All such chains and rings may contain carbon-atom side chains as well.

Literally, the number of possible arrangements for carbon atoms bonded to each other is limitless. It has been calculated that there are 366,319 different ways of arranging twenty carbon atoms based on a chain of atoms and allowing for side chains. Figure 18-1 illustrates some of the possible ways of arranging carbon atoms to form organic molecules. Each of the carbon atoms in the structures of Figure 18-1 will also be involved in additional bonds to those shown. Most often the additional bonds will be to hydrogen atoms.

Organic compounds are the chemical basis for life itself, as well as the basis for our current high standard of living. Not only are proteins, carbohydrates, enzymes, and hormones organic molecules, but so are natural gas, petroleum, coal, gasoline, and many synthetic materials such as plastics, dyes, and fibers such as rayon, nylon, and dacron.

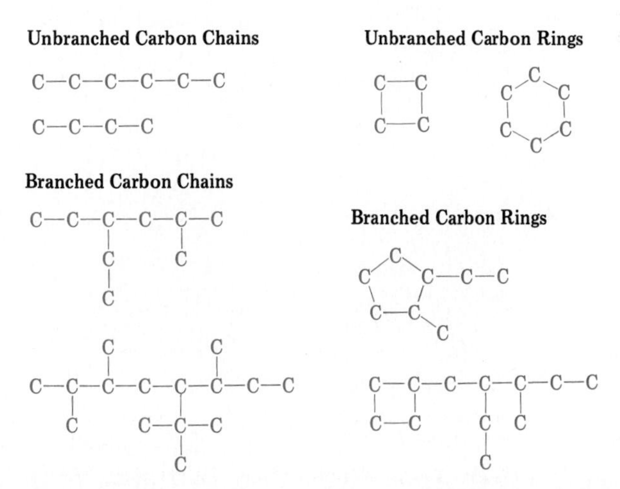

Figure 18-1. Simple and complex chains and rings of carbon atoms.

18.2 Hydrocarbons

The formal definition of organic chemistry, given in Section 18.1, suggests a logical way for organizing our study of the many organic compounds that are known. First, we will consider *hydrocarbons* and then *derivatives of hydrocarbons*. This section, and those that immediately follow it, deal with hydrocarbons, the simplest type of organic compound. Beginning in Section 18.10, hydrocarbon derivatives will occupy our attention.

hydrocarbons

As the name implies, **hydrocarbons** *are compounds that contain only the two elements hydrogen and carbon.* Several different series of hydrocarbons are known. These include alkanes, alkenes, alkynes, and aromatic hydrocarbons, each of which are discussed in this chapter.

18.3 Alkanes

alkanes

saturated compound

Alkanes *are hydrocarbons in which all chemical bonds are single bonds.* Straight-chain, branched-chain, and cyclic structures are possible for such compounds. Since an alkane molecule contains only single bonds, each carbon atom in such a molecule is bonded to the maximum number of atoms possible — four. Alkanes are also referred to as *saturated* hydrocarbons. The term **saturated compound** *describes any organic compound that contains only single bonds between carbon atoms.*

All noncyclic alkanes have molecular formulas that fit the general formula C_nH_{2n+2}, where n is the number of carbon atoms present. Cyclic alkanes always contain two less hydrogen atoms than their noncyclic counterparts and, hence, have molecular formulas that fit the general formula C_nH_{2n}.

The first member of the noncyclic alkane series is *methane,* which contains one carbon atom ($n = 1$) and therefore has the formula CH_4. The methane molecule has a tetrahedral structure. The carbon atom is found at the center of the tetrahedron and the four hydrogen atoms bonded to the carbon are at the corners of the tetrahedron. Different ways of showing this tetrahedral structure for methane are shown in Figure 18-2.

The next member of the noncyclic alkane series, with $n = 2$ and a molecular for-

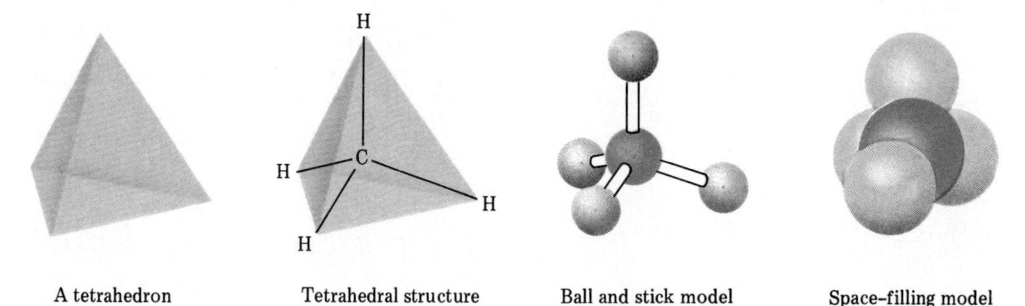

Figure 18-2. Different ways of showing the tetrahedral arrangement of atoms in the methane (CH_4) molecule.

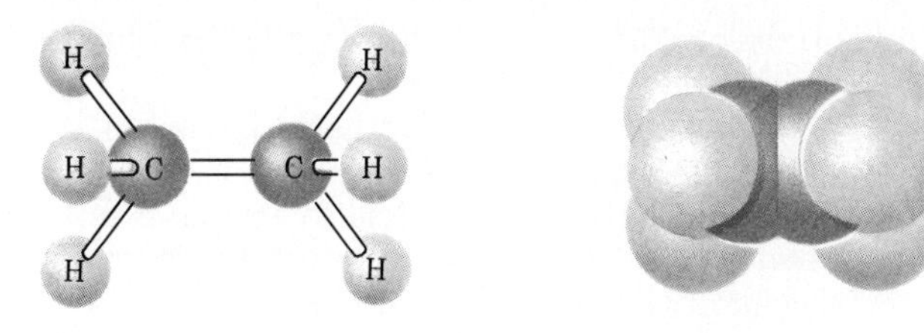

Ball and stick model   Space-filling model

Figure 18-3. Ball and stick and space-filling models of the hydrocarbon ethane (C_2H_6).

mula of C_2H_6, is *ethane*. This molecule may be thought of as a methane molecule with one hydrogen atom removed and a —CH_3 group put in its place. Perspective drawings of the ethane molecule, given in Figure 18-3, illustrate that the carbon bonds still have a tetrahedral geometry. Carbon atoms in alkanes, and also in all other kinds of organic molecules, always form four bonds. Since carbon is in group IVA of the periodic table and thus has four valence electrons, it must form a total of four bonds in order to achieve an octet of outer shell electrons (Sec. 10.5).

Propane, the third alkane, has the molecular formula C_3H_8 ($n = 3$). Once again, we can produce this formula by removing a hydrogen atom from the preceding compound (ethane) and substituting a —CH_3 group in its place. All six hydrogen atoms of ethane are equivalent, so it makes no difference which one we choose to replace. Both ball-and-stick and space-filling models of the propane molecule are given in Figure 18-4.

homologous series

The noncyclic alkanes methane, ethane, and propane are the first three members of a homologous series of compounds. In a **homologous series** *of compounds, each compound in the series differs from the previous one in the series by a constant amount.* For the noncyclic alkanes this factor by which consecutive series members differ is a —CH_2 group.

18.4 Structural Isomerism

It should be apparent that the procedures outlined in the last section in establishing the structures of the ethane and propane molecules (replacement of a hydrogen with a —CH_3 group) can be used to generate other members of the homologous noncyclic alkane series. However, a complication arises when four or more carbon atoms are

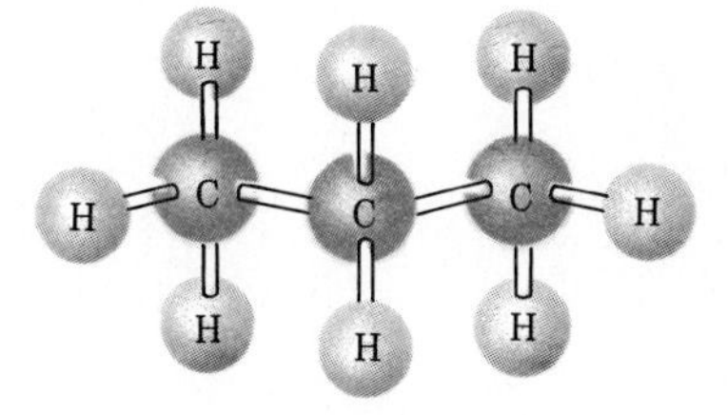

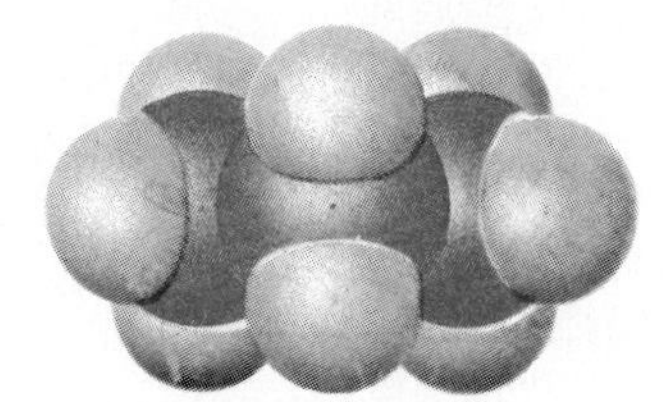

Ball and stick model   Space-filling model

Figure 18-4. Ball and stick and space-filling models of the hydrocarbon propane (C_3H_8).

present: Different structures may be obtained depending upon which hydrogen is replaced.

Two different structural arrangements of four carbons can be produced by removing a hydrogen atom from propane and replacing it with a —CH_3 group. This is the result of all the hydrogens in propane not being geometrically equivalent. The two hydrogens attached to the central carbon atom in propane (Fig. 18-4) are equivalent to each other but distinct from the six associated with the end carbons, which in turn are all equivalent to each other. Replacement of a hydrogen on an end carbon gives *butane*, the compound shown in Figure 18-5a. The compound in Figure 18-5b, *isobutane*, is the result of a —CH_3 group replacing a hydrogen on the central carbon atom.

Butane and isobutane, although they both have the same molecular formula of C_4H_{10}, are two different compounds with different properties. The melting point of butane is −138.3°C and that of isobutane is −160°C. The boiling points of the two compounds are, respectively, −0.5°C and −12°C. Their densities at 20°C also differ: 0.579 g/mL for butane and 0.557 g/mL for isobutane.

structural isomers

Compounds such as butane and isobutane are called structural isomers. **Structural isomers** *are compounds that have the same molecular formula but different structural formulas, that is, different arrangements of atoms within the molecule.* Structural isomers, as we shall see, are not rare in organic chemistry; in fact, they are the rule rather than the exception. The phenomenon of structural isomerism is one of the major reasons why there are so many organic compounds.

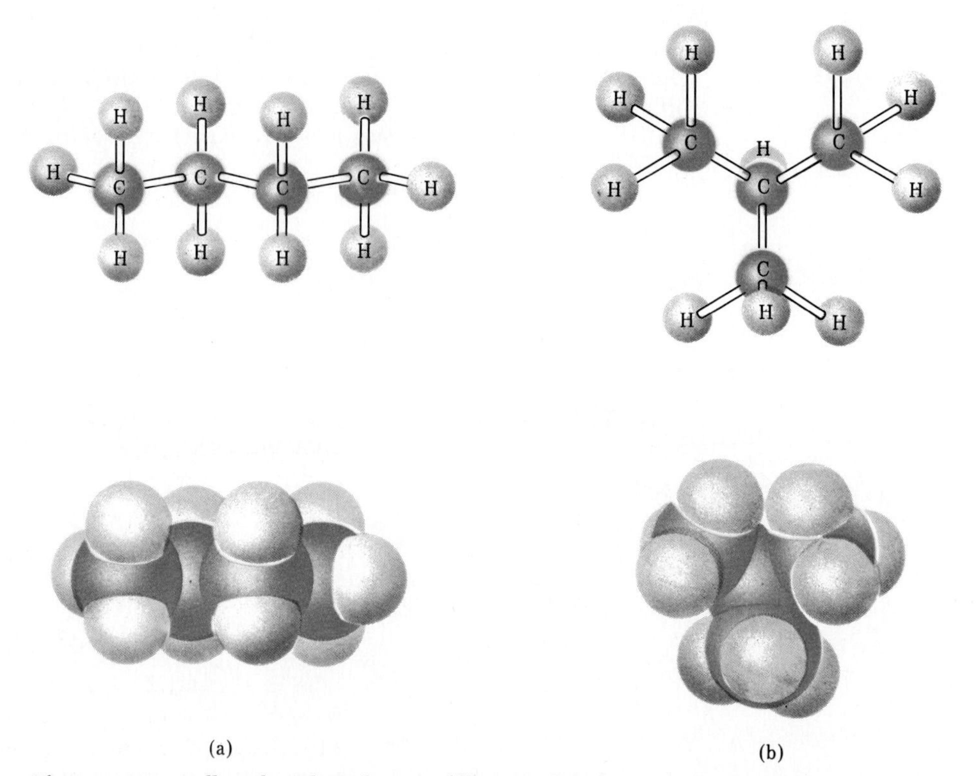

Figure 18-5. Ball and stick and space-filling models for (a) butane and (b) isobutane.

In Figure 18-5, three-dimensional representations were used to represent the structures of butane and isobutane. Such representations give both the arrangement and spatial orientation of the atoms in a molecule. However, it is often difficult to draw such structures, especially if artistic talent is lacking, and it is also time consuming. Because of these drawbacks, an easier system for indicating structure has been developed. This alternate system involves structural formulas. **Structural formulas** *are two-dimensional (planar) representations of the arrangement of the atoms in molecules.* Such formulas give complete information about the arrangement of the atoms in a molecule, but not the spatial orientation of the atoms. The structural formulas for butane and isobutane are

structural formulas

$$CH_3{-}CH_2{-}CH_2{-}CH_3 \qquad \begin{array}{c} CH_3{-}CH{-}CH_3 \\ | \\ CH_3 \end{array}$$

Butane | Isobutane

Note that in writing structural formulas each carbon atom is followed by the attached hydrogens. The number of attached hydrogens is determined by the number of other carbon atoms to which a given carbon atom is bonded. Since each carbon atom must have four bonds (Sec. 18.3), carbon atoms attached to only one other carbon will need three hydrogen atoms (CH_3). Carbons bonded to two other carbons will need two hydrogens (CH_2), those bonded to three other carbons need only one hydrogen (CH), and those bonded to four other carbons need no hydrogens.

When we consider noncyclic alkanes with five carbon atoms, we find that three structural isomers exist.

$$CH_3{-}CH_2{-}CH_2{-}CH_2{-}CH_3 \qquad \begin{array}{c} CH_3{-}CH{-}CH_2{-}CH_3 \\ | \\ CH_3 \end{array} \qquad \begin{array}{c} CH_3 \\ | \\ CH_3{-}C{-}CH_3 \\ | \\ CH_3 \end{array}$$

Pentane | Isopentane | Neopentane

These three C_5H_{12} isomers, like the two C_4H_{10} isomers, are distinctly different compounds with different properties.

The number of possible structural isomers increases rapidly with the number of carbon atoms in an alkane. There are five noncyclic alkane isomers with six carbons (all C_6H_{14}) and nine isomers with seven carbons (all C_7H_{16}). A listing of the number of possible noncyclic alkane isomers, as a function of the number of carbon atoms present, is given in Table 18-1. Obviously, no one has prepared all the isomers for compounds that contain a large number of carbon atoms. However, methods are available for their synthesis if the need arises.

The three isomers of C_5H_{12} were called pentane, isopentane, and neopentane. What prefixes do we use to name the five isomers of C_6H_{14}? What about names for the 9 C_7H_{16} isomers and the 75 $C_{10}H_{22}$ isomers? Obviously, using prefixes to distinguish between various isomers is not a workable system when the number of isomers becomes large. We would rapidly run out of prefixes to use. Even if we had a sufficient number of prefixes, how could we remember the exact meaning of each prefix? It would be an almost impossible task. There must be a better way and there is. By use of a systematic nomenclature system called the IUPAC system (for the International Union of Pure and Applied Chemistry) organic compounds are assigned names that relate directly to their structures. This system keeps prefix use to a minimum. The logic of the system

Table 18-1. Possible Number of Isomers for Selected Noncyclic Alkanes

Molecular Formula	Possible Number of Isomers
CH_4	1
C_2H_6	1
C_3H_8	1
C_4H_{10}	2
C_5H_{12}	3
C_6H_{14}	5
C_7H_{16}	9
C_8H_{18}	18
C_9H_{20}	35
$C_{10}H_{22}$	75
$C_{15}H_{32}$	4,347
$C_{20}H_{42}$	336,319
$C_{30}H_{62}$	4,111,846,763

also minimizes the amount of rote memorization required. The IUPAC rules for naming noncyclic alkanes (both branched and unbranched) are given in Section 18.5.

18.5 IUPAC Nomenclature for Noncyclic Alkanes

When only a relatively few organic compounds were known, chemists named them by what are today called common names. Such names were just arbitrarily selected. Isopentane and neopentane (Sec. 18.4) are examples of such names. As more and more compounds became known, it became obvious that a systematic method for naming organic compounds was needed. The IUPAC nomenclature system meets that need.

normal chain

branched chain

For nomenclatural purposes we will classify noncyclic alkanes into two categories: normal chain and branched chain. In **normal chain** *noncylic hydrocarbons all carbon atoms are connected in a continuous nonbranching chain. In* **branched chain** *noncyclic hydrocarbons one or more side chains of carbon atoms are attached at some point to a continuous chain of carbon atoms.* The two four-carbon noncyclic alkane isomers (Sec. 18.4) illustrate this classification system.

$$CH_3—CH_2—CH_2—CH_3 \qquad CH_3—\underset{\displaystyle CH_3}{\underset{|}{CH}}—CH_3$$

Normal chain — Branched chain

In the branched chain compound the longest continuous chain of carbon atoms is three to which is attached (on the middle carbon of the three) a $—CH_3$ group (the branch).

Let us first consider IUPAC names for normal chain noncyclic alkanes, compounds

Table 18-2. ***IUPAC Names for Normal Chain Noncyclic Alkanes Through Decane***

CH_4	methane
C_2H_6	ethane
C_3H_8	propane
C_4H_{10}	butane
C_5H_{12}	pentane
C_6H_{14}	hexane
C_7H_{16}	heptane
C_8H_{18}	octane
C_9H_{20}	nonane
$C_{10}H_{22}$	decane

that are structurally the simplest of all alkanes. IUPAC names for such alkanes with one through ten carbons are given in Table 18-2. Note that all of the names end in -ane, the characteristic ending for all alkanes. The first four names do not indicate the number of carbon atoms in the molecule, but beginning with five carbon atoms Greek numerical prefixes are used that give directly the number of carbon atoms present. It is important to memorize the names for these normal chain alkanes because they are the basis for the entire IUPAC nomenclature system. (The IUPAC system also includes names for normal chain alkanes with more than ten carbon atoms, but we will not consider them. The names given in Table 18-2 will be sufficient for our purposes.)

The prefix *n*- (for normal) is often attached to the name of a continuous chain alkane, when four or more carbon atoms are present. Thus, the names hexane and *n*-hexane are both correct names for the compound

$$CH_3—CH_2—CH_2—CH_2—CH_2—CH_3$$

Use of the prefix *n*- starts when four carbon atoms are present because it is at this point that isomers become possible.

alkyl group

Branched chain noncyclic alkanes always contain alkyl groups. The alkyl groups are the branches. Formally defined, an **alkyl group** *is the fragment produced by the removal of one hydrogen atom from an alkane.* The general formula for an alkyl group is C_nH_{2n+1}. Alkyl groups do not lead a stable independent existence, but rather are always found attached to another group of carbon atoms.

The two most commonly encountered alkyl groups are the two simplest ones, the methyl group and the ethyl group.

$$——CH_3 \qquad ——CH_2—CH_3$$

Methyl group   Ethyl group

(The long bond in the alkyl group structures denotes the point of attachment to the carbon chain.) Note how alkyl groups are named: the stem of the name of the parent alkane plus the ending *-yl*.

Two different three-carbon alkyl groups exist. They differ from each other in the point of attachment to the main carbon chain.

$$——CH_2—CH_2—CH_3 \qquad ——\underset{\displaystyle CH_3}{\underset{|}{CH}}—CH_3$$

Propyl or *n*-propyl group   Isopropyl group

A propyl group is attached to a carbon chain through an end carbon atom, whereas an isopropyl group is attached via the middle carbon of the three carbons of the propyl group. The isopropyl group is an example of a branched chain alkyl group, that is, a branched branch. Other branched alkyl groups, containing more carbon atoms, also exist. Rules for naming these more complex groups will not be considered in this text.

To name branched chain noncyclic alkanes, that is, alkanes where the carbon atoms are not arranged in a continuous chain, the following rules are used.

Rule 1. *Select the longest continuous carbon-atom chain in the molecule as the base for the name.* This longest carbon-atom chain is named as in Table 18-2.

Example: $CH_3—CH_2—\underset{\displaystyle CH_3}{\underset{|}{CH}}—CH_2—CH_2—CH_2—CH_3$

In this example the longest continuous carbon atom chain (shown in color) is seven carbon atoms. Therefore, the base name (but not the complete name) for this compound is heptane.

Example:

```
CH3—CH2—CH—CH2—CH—CH3
        |       |
        CH3     CH2
                |
                CH2
                |
                CH3
```

In this example the longest continuous carbon chain (shown in color) possesses eight carbon atoms. Note that the carbon atoms in the longest continuous chain do not necessarily have to lie in a straight line. The base name (but not the complete name) for the alkane will be octane.

Rule 2. *The carbon atoms in the longest continuous chain of carbon atoms are numbered consecutively from the end that will give the lowest number(s) to any carbon(s) to which an alkyl group is attached.* If only one alkyl group is present, the end to choose to start numbering at is the one closest to the alkyl group. When there are two or more alkyl groups, number the carbons so that the carbons with the alkyl groups are given numbers with the lowest possible sum.

Example:

```
                    CH3 <———————— Alkyl group
                    |
CH3—CH2—CH2—CH—CH2—CH3
 1    2    3   [4]   5    6    (left to right)
 6    5    4   [3]   2    1    (right to left)
                ↑
                └——— Carbon atom to which the
                     alkyl group is attached
```

Since a chain always has two ends there are always two ways to number the chain: either from left to right or right to left. In this example, the left–right numbering system assigns the number 4 to the carbon atom bearing the alkyl group. If the chain is numbered the other way, the alkyl group is on carbon number 3. Since 3 is lower than 4, the right to left numbering system is the one used.

Example:

```
             CH3
             |
CH3—CH2—C——CH—CH—CH2—CH3
             |    |    |
             CH3  CH3  CH3
 1    2    3    4    5    6    7   (left to right)
 7    6    5    4    3    2    1   (right to left)
```

In this example there are four alkyl groups (side chains) attached to the main chain. If the chain is numbered from left to right, the alkyl groups are found on carbons 3, 3, 4, and 5, the sum of which is 15. (Note that when a carbon atom carries two alkyl groups that carbon's number must be counted twice in the sum.) Numbering from right to left, we generate the numbers 3, 4, 5, and 5, the sum of which is 17. The left to right numbering system is the one used for this compound, since 15 is smaller than 17.

Rule 3. *The complete name for a noncyclic alkane contains the location by number and the name of each alkyl group and the name of the longest carbon chain.* The alkyl group names with their locations always precede the name of the base chain of carbon atoms.

Example: $\overset{1}{CH_3}—\underset{\displaystyle CH_3}{\overset{2}{CH}}—\overset{3}{CH_2}—\overset{4}{CH_2}—\overset{5}{CH_3}$ is 2-methylpentane

Example: $\overset{1}{CH_3}—\overset{2}{CH_2}—\underset{\displaystyle CH_3}{\overset{3}{CH}}—\overset{4}{CH_2}—\overset{5}{CH_3}$ is 3-methylpentane

Note that the names are written as one word with a hyphen used between the number (location) and the name of the alkyl group.

Rule 4. *If two or more of the same kind of alkyl group (two methyl groups or two ethyl groups, and so forth) are present in a molecule, the number of them is indicated by the prefixes* DI-, TRI-, TETRA-, PENTA-, *and so on, and the location of each is again indicated by a number.* These position numbers, separated by commas, are put just before the numerical prefix, with hyphens before and after the numbers when necessary.

Example: $\overset{1}{CH_3}—\underset{\displaystyle CH_3}{\overset{2}{CH}}—\overset{3}{CH_2}—\underset{\displaystyle CH_3}{\overset{4}{CH}}—\overset{5}{CH_3}$ is 2,4-dimethylpentane

Example: $\overset{1}{CH_3}—\overset{2}{CH_2}—\underset{\displaystyle CH_3}{\overset{\displaystyle CH_3}{\overset{3}{C}}}—\overset{4}{CH_2}—\overset{5}{CH_3}$ is 3,3-dimethylpentane

Note that the prefix "di" must always be accompanied by two numbers, "tri" by three, and so on, even if the same number must be written twice, as in 3,3-dimethylpentane.

Rule 5. *When two different kinds of alkyl groups are present on the same carbon chain, each group is separately numbered with the names of the alkyl groups being listed in alphabetical order.*

Example: $\overset{5}{CH_3}—\overset{4}{CH_2}—\underset{\displaystyle \underset{\displaystyle CH_3}{CH_2}}{\overset{3}{CH}}—\underset{\displaystyle CH_3}{\overset{2}{CH}}—\overset{1}{CH_3}$ is 3-ethyl-2-methylpentane

Note that ethyl is named first in accordance with the alphabetical rule determining the order in which alkyl groups are listed.

Example:

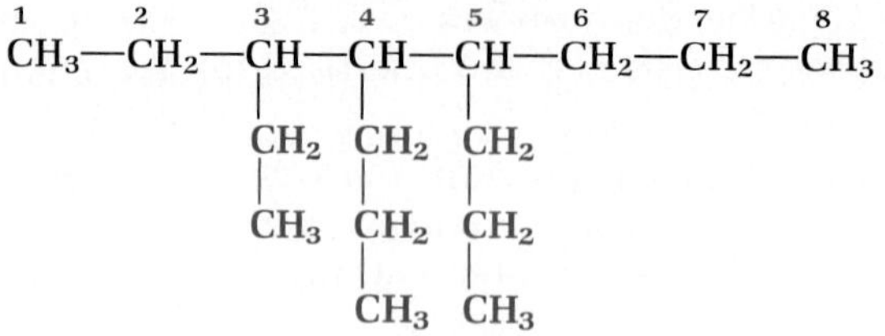

```
       1      2      3    4    5    6      7      8
      CH3—CH2—CH—CH—CH—CH2—CH2—CH3   is  3-ethyl-4,5-dipropyloctane
                    |    |    |
                   CH2  CH2  CH2
                    |    |    |
                   CH3  CH2  CH2
                         |    |
                        CH3  CH3
```

Note that the prefix di- does not affect the alphabetical order for alkyl groups; "e" from ethyl is compared to "p" from propyl.

EXAMPLE 18.1

In Section 18.4 common names were used in naming the four-carbon noncyclic alkane isomers (butane and isobutane) and the five-carbon noncyclic alkane isomers (pentane, isopentane, and neopentane). What is the IUPAC name for each of these compounds?

SOLUTION

(a) Butane, $CH_3—CH_2—CH_2—CH_3$

The IUPAC name is **butane** or ***n*-butane**. Thus, the common and IUPAC names correspond.

(b) Isobutane,

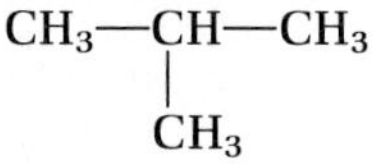

```
CH3—CH—CH3
     |
    CH3
```

The longest chain contains three carbon atoms, and the attached methyl group is at carbon 2, so the name is **2-methylpropane**.

(c) Pentane, $CH_3—CH_2—CH_2—CH_2—CH_3$

The IUPAC name is **pentane** or ***n*-pentane**. Thus, the common and IUPAC names correspond.

(d) Isopentane,

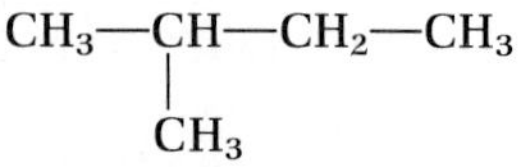

```
CH3—CH—CH2—CH3
     |
    CH3
```

The longest chain contains four carbon atoms, and a methyl group is attached to the chain at carbon 2. Thus, the IUPAC name is **2-methylbutane**.

(e) Neopentane,

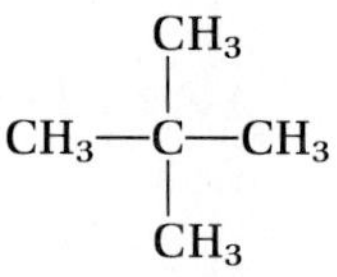

```
    CH3
     |
CH3—C—CH3
     |
    CH3
```

The longest chain contains only three carbons and two alkyl groups are attached to it. Both alkyl groups are methyl groups and both are attached to carbon 2. Thus, the IUPAC name is **2,2-dimethylpropane**.

Structural formulas can easily be obtained from correct IUPAC names since all the information necessary to draw a structure is contained within the IUPAC name. Example 18.2 illustrates the process of going from an IUPAC name to a structural formula.

EXAMPLE 18.2

Draw the structural formula of the noncyclic alkane whose IUPAC name is 4,4-diethyl-2,5-dimethyloctane.

SOLUTION

Step 1. The IUPAC name indicates that the base chain contains eight carbon atoms (octane). Draw an octane skeleton (no hydrogens) and number it.

```
1   2   3   4   5   6   7   8
C—C—C—C—C—C—C—C
```

Step 2. Place two ethyl groups on carbon number 4.

```
             CH3
             |
             CH2
             |
C—C—C—C—C—C—C—C
             |
             CH2
             |
             CH3
```

Step 3. Place methyl groups on carbons 2 and 5.

```
             CH3
             |
             CH2
             |
C—C—C—C——C—C—C—C
   |         |     |
   CH3       CH2   CH3
             |
             CH3
```

Step 4. Add necessary hydrogen atoms to the carbon base chain so that each carbon atom has four bonds.

```
                       CH3
                       |
                       CH2
                       |
CH3—CH—CH2—C——CH—CH2—CH2—CH3
     |                 |     |
     CH3               CH2   CH3
                       |
                       CH3
```

This structure has 14 carbon atoms. It should therefore have 30 hydrogen atoms (C_nH_{2n+2}). Counting the number of hydrogen atoms enables you to check the correctness of the structure.

18.6 Structure and Nomenclature of Cycloalkanes

cycloalkanes

In addition to normal chain and branched chain noncyclic alkanes, a third type exists, the cycloalkanes. **Cycloalkanes** *contain rings of carbon atoms and have the general formula* C_nH_{2n}. Cycloalkanes thus contain two fewer hydrogen atoms than the other types of alkanes. The reason for this deficiency can be visualized by considering that cycloalkanes arise from the removal of one hydrogen atom from each of the terminal carbons in a linear alkane. The two end carbons, which now need to form one more bond, then join together to give a cyclic structure.

The simplest cycloalkane possible contains three carbon atoms. Figure 18-6 shows ball-and-stick models of cycloalkanes containing three, four, and five carbons.

For convenience, geometric figures are often used to represent cycloalkanes: a triangle for a three-carbon ring, a square for a four-carbon ring, and so forth. When such figures are used, it is assumed that each corner of the figure represents a carbon atom together with the number of hydrogens needed to give the carbon four bonds. Examples of this geometric figure type notation along with IUPAC names for the example compounds are

CH_3 CH_3 CH_3 CH_2 CH_3

1,2-Dimethylcyclopropane 1-Ethyl-4-methylcyclohexane Cyclobutane

As can be seen from the above examples, IUPAC nomenclature for cycloalkanes is very similar to that for noncyclic alkanes. The only modifications to the noncyclic alkane rules are

1. The prefix *cyclo-* is placed before the name that corresponds to the noncyclic chain that has the same number of carbon atoms as the ring.

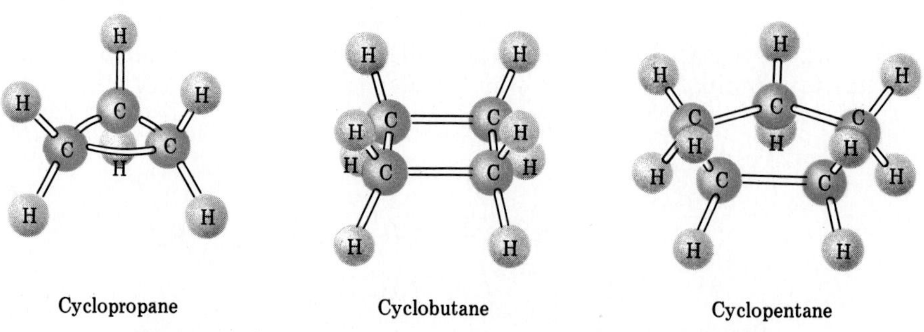

Figure 18-6. Ball and stick models of simple cycloalkanes.

2. Alkyl groups, when present, are located by numbering the carbons in the ring according to a system that yields the lowest numbers for the carbons at which the alkyl groups are attached.

18.7 Structure and Nomenclature of Alkenes and Alkynes

alkenes

alkynes

Alkenes *are hydrocarbons in which there is one carbon–carbon* DOUBLE *bond per molecule.* Numerous noncyclic and cyclic alkenes are known. **Alkynes** *are hydrocarbons in which there is one carbon–carbon* TRIPLE *bond per molecule.* Both noncyclic and cyclic alkynes are known although cyclic alkynes are not common.

unsaturated compound

Alkenes and alkynes are classified as unsaturated hydrocarbons, as contrasted to alkanes, which are saturated hydrocarbons. An **unsaturated compound** *is an organic compound that contains fewer hydrogen atoms than the maximum possible owing to the presence of one or more multiple (double or triple) carbon–carbon bonds.*

The simplest alkene is ethene (common name, ethylene), which has the formula C_2H_4. Obviously a one-carbon alkene cannot exist because two carbon atoms are required to have a carbon–carbon double bond. The simplest alkyne is also a two-carbon species with the formula C_2H_2. The name of this compound is ethyne (common name of acetylene). Ball-and-stick models as well as space-filling models of ethene and ethyne are given in Figure 18-7.

The general formula for a noncyclic alkene is C_nH_{2n}, and for a noncyclic alkyne C_nH_{2n-2}. The general formula for a noncyclic alkene is, thus, identical to that for a cycloalkane (Sec. 18.6). This observation shows that compounds belonging to different hydrocarbon classes may be isomeric with each other. Both alkenes and cycloalkanes contain two fewer hydrogen atoms than the maximum number possible for a hydro-

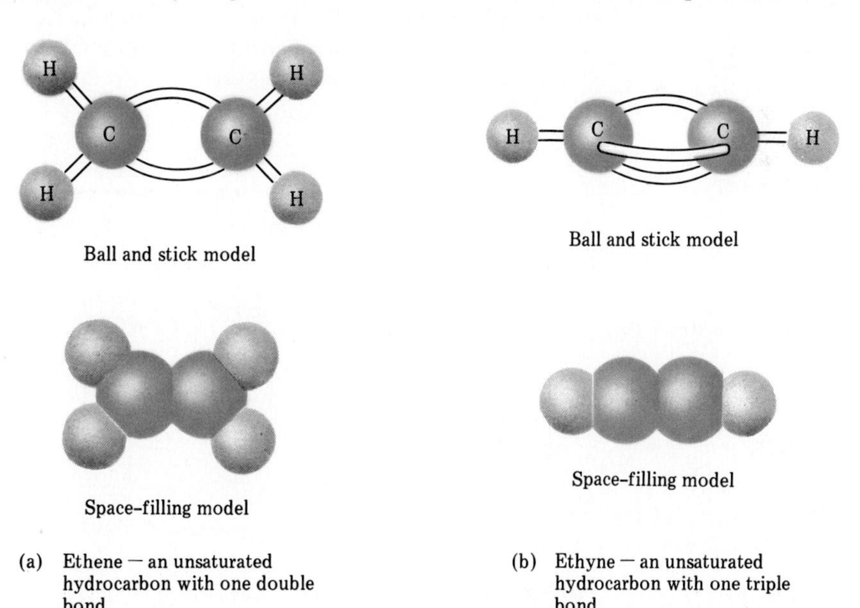

Figure 18-7. Ball and stick and space-filling models for (a) ethene and (b) ethyne.

carbon; noncyclic alkanes contain the maximum number. For alkenes, the lower number of hydrogen atoms is caused by the presence of the double bond. For cycloalkanes, the cyclization process brings about the loss of two hydrogens.

The rules we have previously developed for naming alkanes, with slight modifications, can be used to name alkenes and alkynes. The modifications are

1. The *-ane* ending characteristic of alkanes is changed to *-ene* for alkenes and to *-yne* for alkynes.
2. For noncyclic molecules containing more than three carbon atoms, the position of the multiple bond must be specified since there is more than one position it may occupy. The position is given by a single number (the number of the lower numbered carbon involved in the multiple bond), which is placed immediately in front of the base chain name.

Examples: $\overset{1}{C}H_2{=}\overset{2}{C}H{-}\overset{3}{C}H_2{-}\overset{4}{C}H_3$ is **1-butene**

and

$\overset{1}{C}H_3{-}\overset{2}{C}H{=}\overset{3}{C}H{-}\overset{4}{C}H_3$ is **2-butene**

The compounds 1-butene and 2-butene are isomers. Note that in the case of 1-butene the double bond involves carbons 1 and 2, but we only write the lower of the two numbers in the name.

Example: $\overset{4}{C}H_3{-}\underset{\underset{CH_3}{|}}{\overset{3}{C}H}{-}\overset{2}{C}H{=}\overset{1}{C}H_3$ is **3-methyl-1-butene**

The chain is always numbered in such a way as to give the multiple bond the lowest number possible, even if this means that alkyl groups must get higher numbers; that is, multiple bonds always take precedence over alkyl groups when deciding upon a numbering system.

EXAMPLE 18.3

Name the following unsaturated hydrocarbons.

(a) $CH{\equiv}C{-}\underset{\underset{CH_3}{|}}{CH}{-}CH_3$ (b) $CH_3{-}\underset{\underset{CH_3}{|}}{C}{=}CH_2$ (c) [cyclobutane square] (d) [cyclohexene ring with CH_3]

SOLUTION

(a) $\overset{1}{C}H{\equiv}\overset{2}{C}{-}\underset{\underset{CH_3}{|}}{\overset{3}{C}H}{-}\overset{4}{C}H_3$

The longest continuous chain containing the triple bond has four carbons; therefore, the base name for the compound is butyne. We number from the left (the chain end closest to the triple bond) to the right. Thus, the triple bond position is 1, and the location of the methyl group 3. The IUPAC name is **3-methyl-1-butyne.**

(b)
$$\overset{3}{CH_3}-\overset{2}{\underset{\displaystyle CH_3}{\underset{|}{C}}}=\overset{1}{CH_2}$$

The base name is propene since the longest carbon chain containing the double bond has three carbons in it. Numbering from right to left, we get the number 1 for the position of the double bond and the number 2 for the location of the methyl group. The IUPAC name is thus **2-methyl-1-propene.**

(c) This compound is simply **cyclobutene.** In cyclic structures containing only one multiple bond no number is needed to specify bond position. It is understood that the multiple bond involves carbons 1 and 2.

(d) 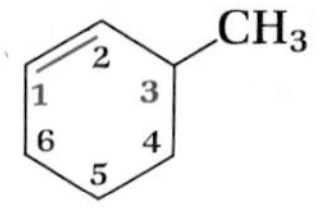 [cyclohexene ring numbered 1–6 with CH_3 on carbon 3]

This compound is a cyclohexene. The double-bonded carbons become carbons 1 and 2. We number in the clockwise direction in order for the alkyl group to be on the lowest numbered carbon possible. (Numbering in the counterclockwise direction would put the methyl group on carbon 6.) The IUPAC name of the compound is **3-methylcyclohexene.**

Numerous unsaturated hydrocarbons contain more than one multiple bond, for example, two double bonds, or two triple bonds. Compounds containing two double bonds are properly referred to as alkadienes and those with two triple bonds as alkadiynes. If three double bonds are present, we would have an alkatriene.

The general family names just mentioned give us the key to naming compounds containing more than one multiple bond. A prefix is added to the base chain ending indicating the number of double bonds — *di*ene for two double bonds, *tri*ene for three double bonds, *di*yne for two triple bonds, and so on. In addition, a number is used to locate each multiple bond. If there are three multiple bonds, three numbers are needed.

EXAMPLE 18.4

Name the following compounds, each of which contains more than one multiple bond.

(a) $CH_2=CH-CH=CH_2$

(b)
$$CH_2=\underset{\displaystyle CH_3}{\underset{|}{C}}-\overset{\displaystyle CH_3}{\overset{|}{CH}}-CH=CH_2$$

SOLUTION

(a) The base name for this hydrocarbon is butadiene, since there are two double bonds present in a chain of four carbons. It does not matter which way we number the chain because of the symmetrical nature of the molecule. Either way, the double bonds will be listed as being on carbons 1 and 3. Therefore, the name of the compound is 1,3-butadiene.

(b)
$$\overset{1}{CH_2}=\underset{\displaystyle CH_3}{\overset{2}{C}}-\overset{\displaystyle CH_3}{\overset{3}{CH}}-\overset{4}{CH}=\overset{5}{CH_2}$$

This molecule is a pentadiene with two methyl groups on it. It does not matter which end we number from relative to the double bonds because of their symmetrical positioning within the molecule. It does, however, matter where we start relative to the alkyl groups. Numbering from the left gives a lower set of numbers for the methyl groups. According to the left–right numbering system, the IUPAC name for this compound becomes 2,3-dimethyl-1,4-pentadiene.

Unsaturated *cyclic* hydrocarbons may also contain more than one multiple bond. Introducing a second double bond into the molecule cyclohexene

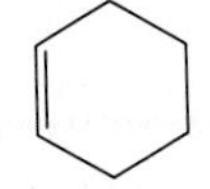

would produce either

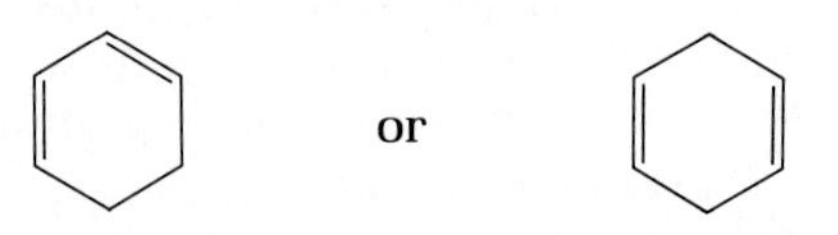

1,3-Cyclohexadiene 1,4-Cyclohexadiene

depending on where the additional double bond is positioned relative to the first one. These two compounds are isomers, differing only in the positions of the double bonds (1,3- versus 1,4-).

Introduction of an additional double bond into a cyclohexadiene can also be done in a number of ways. One of the compounds that could be produced is

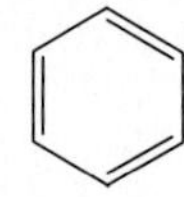

Using the nomenclature rules of this section, we would expect to call this compound 1,3,5-cyclohexatriene. However, the compound is not called that. Instead, its name is *benzene.* The reason for this "rule violation" is that benzene does not possess the chemical properties normally associated with cyclic hydrocarbons containing multiple bonds. Its properties are different enough from those of these other compounds that it is considered to be the first member of a new series of hydrocarbons called aromatic hydrocarbons. The reason for benzene's different behavior is the subject of Section 18.8. Benzene is more than just a cycloalkatriene.

18.8 Aromatic Hydrocarbons

aromatic hydrocarbons

The cyclic hydrocarbon benzene, C_6H_6, is the parent molecule for a large family of hydrocarbons called aromatic hydrocarbons. The structural feature in aromatic hydrocarbons that is responsible for their distinctive properties is the *benzene ring*. **Aromatic hydrocarbons** *are hydrocarbons that contain the characteristic benzene ring.* An understanding of the uniqueness of the benzene ring is the key to understanding aromatic hydrocarbon chemistry.

At the end of Section 18.7, the structure of benzene was given as

meaning

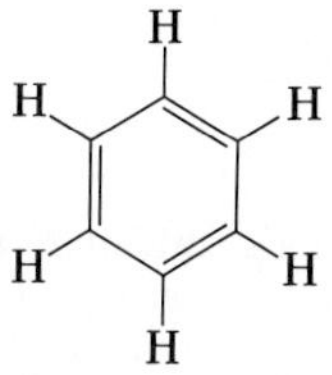

It was noted at that time that this compound was not named as a cyclohexatriene. What is the reason for this?

The cyclohexatriene interpretation of the benzene structure implies that alternating single and double carbon–carbon bonds are present about the carbon ring. With this interpretation, we would predict that not all carbon–carbon bonds in the ring would be of the same length, since carbon–carbon double bonds are known to be shorter in length than carbon–carbon single bonds. This prediction is contrary to the experimental information available on bond lengths in the benzene molecule. Experimental studies indicate that all carbon–carbon bonds in the benzene molecule are equal in length. Thus, the bonding present in the carbon ring must be something different from alternating single and double bonds.

If all of the carbon–carbon bonds in benzene were of the same kind, that is, all single bonds or all double bonds, we would have a molecule whose carbon–carbon bond lengths were consistent with experimental bond length data. However, neither of these structures is acceptable; in both structures the carbon atoms violate the octet rule (Sec. 10.3). The all single bond suggestion leaves each carbon atom with only three bonds instead of the needed four bonds. In an all double bond molecule, each carbon atom would have too many bonds — five per carbon atom.

delocalized bonding

To obtain a structure for benzene that is consistent with both the octet rule and experimental bond length information, a new type of bonding concept must be invoked, that of delocalized bonding. In **delocalized bonding** *three or more atoms are involved in the sharing of the same valence electrons.* (Up to now all bonding discussions in the text have involved localized bonds, that is, bonds in which electrons are shared between *two* atoms.) Let us now consider how the delocalized bond concept helps explain the characteristics of benzene and other aromatic compounds.

Carbon atoms have four valence electrons for use in bonding. Three of the four electrons on each carbon atom in benzene will be used in forming the "structural framework" of a benzene molecule, as shown in the following diagram.

In this structure each carbon atom has formed three bonds. Thus, each carbon atom still has one more electron available for bonding, which is shown as a dot in the following structure.

This additional electron, in each case, is in a *p* orbital (Sec. 9.4) whose orientation is such that it may interact simultaneously with two other *p* orbitals — the *p* orbitals on each side of it in the carbon ring. This results in a continuous interaction around the ring and a delocalized bond that "runs" completely around the ring, as shown in Figure 18-8. In this delocalized bond, the six electrons (one from each carbon atom) are considered to be shared equally by all six carbon atoms.

A circle drawn inside the benzene ring is used to denote the delocalized bond present in this molecule.

Note that in this structure each carbon atom has four bonds: one bond to a H atom, two localized bonds to other C atoms, and one bond participating in the delocalized bond.

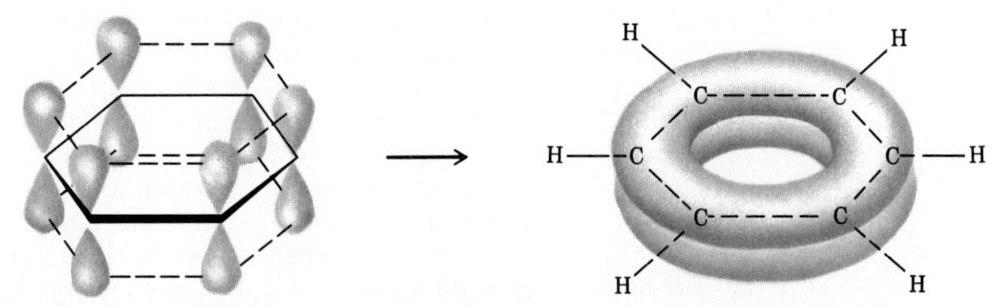

Figure 18-8. The six *p* orbitals of benzene interact to form a delocalized bond "running" completely around the ring.

Delocalized bonding is possible only for specific orientations of atomic orbitals. Aromatic hydrocarbons are the only hydrocarbon class where proper orbital orientation is present.

The presence of a delocalized bond in a molecule gives it extra stability. It is this extra stability that causes benzene and the other aromatic hydrocarbons to have chemical properties that are different from those of other hydrocarbons.

The hydrogen atoms on benzene may be replaced by alkyl groups without destroying the delocalized bonding present in the carbon ring. Replacements of this type give rise to numerous other members of the aromatic hydrocarbon family. The simplest aromatic hydrocarbon (next to benzene) is methylbenzene (common name of toluene), a compound in which one hydrogen on the ring has been replaced by a methyl group.

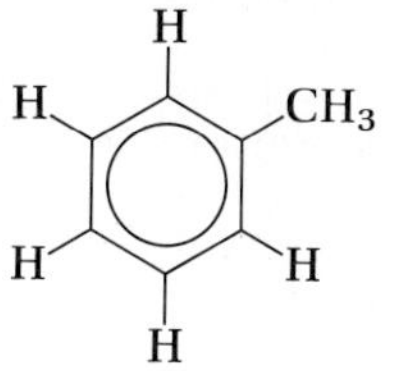

H H CH_3 H H H or CH_3

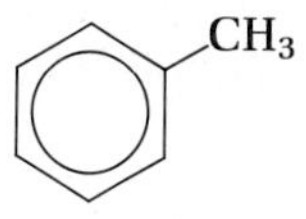

Methylbenzene
or
Toluene

Three isomeric compounds are possible when two methyl groups are placed on a benzene ring. IUPAC names for these compounds are 1,2-, 1,3- and 1,4-dimethylbenzene. Their structures are, respectively,

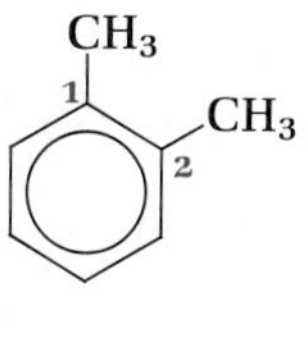

CH_3 1 CH_3 2 CH_3 1 2 3 CH_3 CH_3 1 2 3 4 CH_3

1,2-Dimethylbenzene 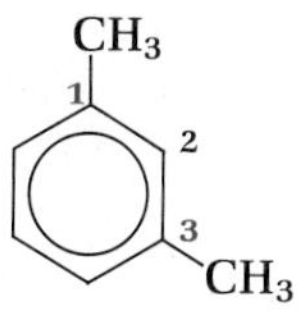 1,3-Dimethylbenzene 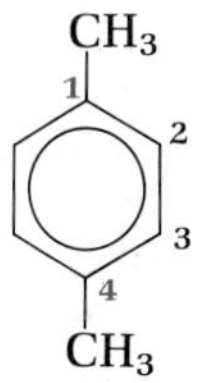 1,4-Dimethylbenzene

A common nomenclature system for indicating positioning on the carbon ring in disubstituted benzenes uses the prefixes *ortho-* (substituents that are adjacent to each other), *meta-* (substituents that are one carbon removed from each other), and *para-* (substituents that are two carbons removed from each other). The prefixes are often abbreviated as *o*-, *m*-, and *p*-. Thus, we have for the three dimethylbenzenes (common name of xylene) the following additional acceptable names.

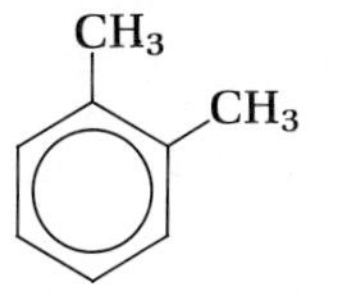

CH_3 CH_3 CH_3 CH_3 CH_3 CH_3

ortho-Dimethylbenzene
ortho-Xylene
o-Xylene

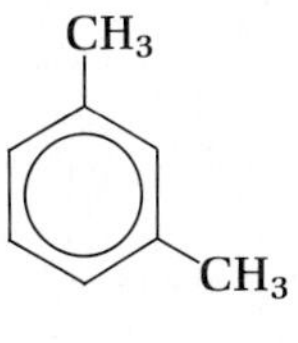

meta-Dimethylbenzene
meta-Xylene
m-Xylene

para-Dimethylbenzene
para-Xylene
p-Xylene

When three or more groups are attached to a benzene ring, the carbon atoms in the ring are given numbers, which are used to indicate the position of the groups. The number system that gives the lowest possible sum is the system selected.

$CH_2—CH_3$ (4), H_3C (1), CH_3 (2), 3

4-Ethyl-1,2-dimethylbenzene

Note that the attachments to the benzene ring need not be identical as illustrated in the above example.

Another type of aromatic hydrocarbon contains *fused* benzene rings. The simplest example of such a compound is naphthalene, which is commonly sold as moth repellent.

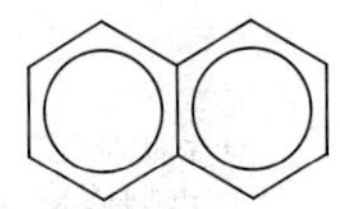

Naphthalene

Another fused ring aromatic hydrocarbon is benzpyrene, which has been identified as one of the cancer-causing agents in tobacco smoke.

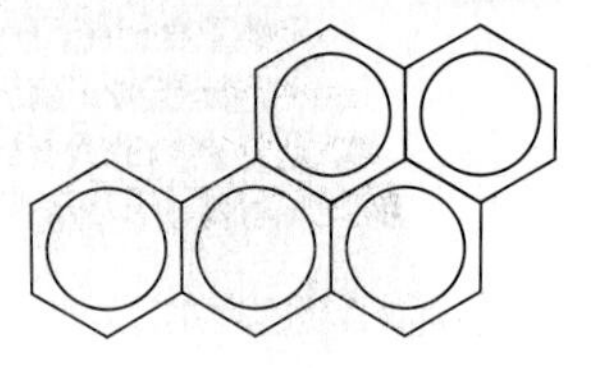

Benzpyrene

18.9 Sources and Uses of Hydrocarbons

There are three major natural sources of hydrocarbons: natural gas, crude oil (petroleum), and coal.

Unprocessed natural gas contains methane (50–90%) and ethane (1–10%) with smaller amounts of propane and butanes. Processed natural gas, the "natural gas" used in homes for cooking and as a heating fuel, is mainly methane.

Crude petroleum is an extremely complex mixture of hydrocarbons. Some crudes consist chiefly of noncyclic alkanes; others contain as much as 40% of other types of hydrocarbons — cycloalkanes and aromatics.

In its natural state, crude petroleum has very few uses, but this complex hydrocarbon mixture can be separated into various useful fractions through refining. The resulting fractions are still hydrocarbon mixtures, but each one is simpler (fewer compounds are present). During refining, the physical separation of the crude into component fractions is accomplished by *fractional distillation*, a process that takes advantage of boiling-point differences between the various components of the crude petroleum. Each fraction contains hydrocarbons within a specific boiling point range. Table 18-3 characterizes some of these specific fractions in terms of size of hydrocarbon molecule, boiling point range, and uses. Note the many important and familiar products obtained from petroleum. All of these products are mixtures of hydrocarbons.

Although petroleum is by far the leading natural source of hydrocarbons, coal is another significant source. When coal is heated in the absence of air, a process called *destructive distillation,* the coal breaks down into three components: (1) coal gas, (2) coke, and (3) coal tar (a liquid). Coal gas is mainly methane and hydrogen and finds use as a fuel. Coke, the main destructive distillation product, is essentially pure carbon and is a vital raw material for the production of iron and steel. Coal tar is a mixture of numerous hydrocarbons, many of which are aromatic. Coal tar is a major source of aromatic hydrocarbons.

There are no large natural sources of unsaturated hydrocarbons (alkenes and alkynes), although small amounts of alkenes are present in some crude petroleums. Most alkenes and alkynes are produced industrially from alkanes during the refining of crude petroleum.

When hydrocarbons are burned as fuel, their major use, the products of combustion (reaction with oxygen) are carbon dioxide, water, and heat energy. Representative combustion reactions for methane (natural gas) and *n*-octane (a component of gasoline) are

$$CH_4 + 2\,O_2 \longrightarrow CO_2 + 2\,H_2O + \text{energy}$$

$$2\,C_8H_{18} + 25\,O_2 \longrightarrow 16\,CO_2 + 18\,H_2O + \text{energy}$$

A small, but significant, percentage of hydrocarbons find use as chemical feedstock for the petrochemical industry. In this industry many of the consumer products that are considered to be necessities today are manufactured. Examples of materials produced from hydrocarbon starting materials are plastics, synthetic fibers, adhesives, dyes, and many pharmaceuticals.

Table 18-3. ***Hydrocarbon Fractions Obtained from Crude Oil by Fractional Distillation***

Fraction Name	Size of Hydrocarbon Molecule	Boiling Point Range (°C)	Uses
Natural gas	1 to 4 carbon atoms	below room temperature	home heating, cooking fuel
Petroleum ether	5 to 7 carbon atoms	20–60	solvents, dry cleaning
Gasoline	6 to 12 carbon atoms	50–200	automobile fuel
Kerosene	12 to 16 carbon atoms	175–275	jet fuel, diesel fuel
Fuel oil	15 to 18 carbon atoms	250–400	furnace oil
Lubricating oils	16 to 20 carbon atoms	350 and above	lubricants
Greases	more than 18 carbon atoms	semisolid	lubricants
Asphalt	high number of carbon atoms	solid	paving, roofing

18.10 Derivatives of Hydrocarbons

hydrocarbon derivative

functional group

Hydrocarbons, as numerous as they are, make up only a small fraction of known organic compounds. The majority of organic compounds are *hydrocarbon derivatives.* A **hydrocarbon derivative** *is a hydrocarbon molecule in which one or more of the hydrogen atoms has been replaced with a new atom or group of atoms.* The newly attached part of the hydrocarbon derivative serves as a basis for characterizing the molecule and is called the *functional group* for the molecule. A **functional group** *is that part of a hydrocarbon derivative that contains the elements other than carbon and hydrogen.* The name functional group is very appropriate because it is at the functional group site that most chemical reactions occur. The functional group literally determines how the hydrocarbon derivative functions in chemical reactions.

The use of the functional group concept greatly simplifies the study of hydrocarbon derivatives. All compounds containing the same functional group are studied as a class because of the similarities in their chemistry. Table 18-4 lists the classes of hydrocarbon derivatives most commonly encountered in organic chemistry as well as the functional group that characterizes each class.

Note that the element oxygen is a particularly common component of functional

Table 18-4. Classes of Hydrocarbon Derivatives

Class Name	Functional Group	General Formula for Class[a]	Example
Halide	—X (X = F, Cl, Br, or I)	R—X	CH_3—Br
Alcohol	—OH	R—OH	CH_3—OH
Ether	—O—	R—O—R′	CH_3—O—CH_3
Aldehyde	—C(=O)—H	R—C(=O)—H	CH_3—C(=O)—H
Ketone	—C(=O)—	R—C(=O)—R′	CH_3—C(=O)—CH_3
Carboxylic acid	—C(=O)—OH	R—C(=O)—OH	CH_3—C(=O)—OH
Ester	—C(=O)—O—	R—C(=O)—O—R′	CH_3—C(=O)—O—CH_3
Amine	—NH_2	R—NH_2	CH_3—NH_2
Amide	—C(=O)—NH_2	R—C(=O)—NH_2	CH_3—C(=O)—NH_2
Thiol	—SH	R—SH	CH_3—SH
Nitrile	—CN	R—CN	CH_3—CN
Nitro compound	—NO_2	R—NO_2	CH_3—NO_2

[a]The symbol R′ represents a hydrocarbon group that may or may not be the same as R.

groups. The element nitrogen is the next most frequently encountered functional group constituent. The symbol R, used in the general formulas for a class of hydrocarbon derivatives (third column of Table 18-4), is a designation for the hydrocarbon part of any hydrocarbon derivative. For example, the general formula for a carboxylic acid

$$\mathrm{R{-}\overset{\overset{\displaystyle O}{\|}}{C}{-}OH}$$

collectively designates the following compounds,

$$\mathrm{CH_3{-}\overset{\overset{\displaystyle O}{\|}}{C}{-}OH} \qquad (\mathrm{R = CH_3{-}})$$

$$\mathrm{CH_3{-}CH_2{-}\overset{\overset{\displaystyle O}{\|}}{C}{-}OH} \qquad (\mathrm{R = CH_3{-}CH_2{-}})$$

$$\mathrm{CH_3{-}\underset{\underset{\displaystyle CH_3}{|}}{CH}{-}\overset{\overset{\displaystyle O}{\|}}{C}{-}OH} \qquad (\mathrm{R = CH_3{-}\underset{\underset{\displaystyle CH_3}{|}}{CH}{-}})$$

$$\mathrm{C_6H_5{-}\overset{\overset{\displaystyle O}{\|}}{C}{-}OH} \qquad (\mathrm{R = C_6H_5{-}})$$

as well as hundreds of other similar compounds that differ from each other only in the identity of the hydrocarbon part of the molecule.

We shall not try to discuss all of the functional groups listed in Table 18-4. As representative of the field of hydrocarbon derivatives, we will consider briefly five classes of such derivatives: halides, alcohols, ethers, carboxylic acids, and esters.

18.11 Halogenated Hydrocarbons (Organic Halides)

halogenated hydrocarbons

Halogenated hydrocarbons *(or organic halides) are hydrocarbon derivatives in which one or more halogen atoms have replaced hydrogen atoms in the parent hydrocarbon.* (The group name for the elements in group VIIA of the periodic table — fluorine, chlorine, bromine, and iodine — is *halogen.*)

In the IUPAC system for naming organic halides, the prefixes fluoro-, chloro-, bromo-, and iodo- are used to designate the various halogen atoms. These prefixes (and location numbers for the halogen atoms if isomers are possible) are attached to the name of the longest continuous hydrocarbon chain. Thus,

$$\mathrm{H{-}\underset{\underset{\displaystyle H}{|}}{\overset{\overset{\displaystyle H}{|}}{C}}{-}\underset{\underset{\displaystyle Cl}{|}}{\overset{\overset{\displaystyle H}{|}}{C}}{-}\underset{\underset{\displaystyle H}{|}}{\overset{\overset{\displaystyle H}{|}}{C}}{-}\underset{\underset{\displaystyle H}{|}}{\overset{\overset{\displaystyle H}{|}}{C}}{-}H} \quad \text{or} \quad \mathrm{CH_3{-}\underset{\underset{\displaystyle Cl}{|}}{CH}{-}CH_2{-}CH_3}$$

would be named 2-chlorobutane.

Many halogenated hydrocarbons are commercially important. The structures for some of these compounds are given in Table 18-5.

Freons — chlorofluoromethanes — are used as refrigerants (air conditioning, refrigeration, and so on) and also as propellants in aerosol cans. Considerable controversy has developed recently concerning this latter use for Freons. There is a possibility that these compounds are negatively affecting the ozone layer in the earth's upper atmosphere. (The function of the ozone layer is to filter out most of the ultraviolet radiation given off from the sun.) The use of Freons to pressurize aerosol cans has decreased markedly because of the concern relative to the ozone layer.

Table 18-5. Some Commercially Important Halogenated Hydrocarbons

Structural Formula	Common Name	IUPAC Name	Uses
F F—C—Cl Cl	freon-12	dichlorodifluoromethane	refrigerant, aerosol can propellant
Cl Cl C═C Cl H	trichloroethylene	trichloroethene	drycleaning solvent
Cl Cl—C—Cl Cl	carbon tetrachloride	tetrachloromethane	solvent
Cl Cl	*p*-dichlorobenzene	1,4-dichlorobenzene or *p*-dichlorobenzene	moth repellent
Cl Cl H—C—C—Cl Cl Cl	DDT (dichlorodiphenyl trichloroethane)	1,1,1-trichloro-2,2-bis(*p*-chlorophenyl) ethane	insecticide

Chlorinated ethenes (trichloroethylene and tetrachloroethylene) are very important solvents in the drycleaning industry. Carbon tetrachloride finds use as a stain remover because of its ability to dissolve greases.

Some halogenated hydrocarbons, because of their toxic nature, find use as insecticides. Para-dichlorobenzene, a compound we call mothballs, is such a compound. The halogenated hydrocarbon that has been used the most as an insecticide is DDT (dichlorodiphenyltrichloroethane). For many years it was the dominant chemical insecticide throughout the world. Its use has, however, been discontinued because it does not readily decompose once it is deposited in the environment. Consequently, high concentrations can build up in certain environs. Its use was discontinued when it was discovered that DDT concentrations were such that fish and birds were being adversely affected. Despite having now fallen into disfavor, DDT is responsible for having saved hundreds of thousands of human lives as it was used in the control of insect-disease carriers, particularly in underdeveloped countries in the world.

The plastics polyvinyl chloride (PVC) and Teflon are made from the halogenated hydrocarbon starting materials chloroethene and tetrafluoroethene, respectively,

```
H  H        F  F
|  |        |  |
C==C        C==C
|  |        |  |
H  Cl       F  F
```

Chloroethene (vinyl chloride)   Tetrafluoroethene

polymer

Both PVC and Teflon are polymeric materials. *A* **polymer** *is a giant molecule that contains thousands upon thousands of small molecules (monomers) linked together in long chains.* Chloroethene and tetrafluoroethene are the monomers for the two plastics we have mentioned. In the process of linking the monomers together to generate the polymers, the double bonds revert to single bonds. Thus, the structures of PVC and Teflon are

```
        H  H  H  H  H  H  H  H  H  H  H  H  H  H
        |  |  |  |     |  |  |  |  |  |  |  |  |
<---...C--C--C--C--C--C--C--C--C--C--C--C--C--C...--->
        |  |  |  |     |  |  |  |  |  |  |  |  |
        H  Cl H  Cl H  Cl H  Cl H  Cl H  Cl H  Cl
```

Polyvinyl chloride (PVC)

```
        F  F  F  F  F  F  F  F  F  F  F  F  F  F  F  F
        |  |  |  |  |  |  |  |  |  |  |  |  |  |  |  |
<---...C--C--C--C--C--C--C--C--C--C--C--C--C--C--C--C...--->
        |  |  |  |  |  |  |  |  |  |  |  |  |  |  |  |
        F  F  F  F  F  F  F  F  F  F  F  F  F  F  F  F
```

Teflon

18.12 Alcohols

alcohols

Alcohols *are hydrocarbon derivatives in which one or more hydroxyl groups (—OH groups) have replaced hydrogen atoms in the parent hydrocarbon.* Besides being hydrocarbon derivatives, alcohols may also be thought of as being derivatives of water. If one of the hydrogen atoms in water is replaced with an R group, an alcohol results.

H—O—H	R—O—H
Water	An alcohol

The IUPAC name of an alcohol containing only one hydroxyl group is obtained from the name of the parent alkane by replacing the *e* ending with *ol*. When necessary the position of the —OH functional group is specified by a number. The simpler alcohols are also often known by common names, where the alkyl group name is followed by the word alcohol.

$CH_3—OH$	$CH_2(OH)—CH_2—CH_2—CH_3$
Methanol or Methyl alcohol	1-Butanol or *n*-Butyl alcohol

Alcohols that contain more than one hydroxyl group are called *polyhydric alcohols*. Dihydric alcohols (two —OH groups) are called glycols or diols. Trihydric alcohols (three —OH groups) are called triols.

Five of the most common alcohols are methanol, ethanol, 2-propanol, 1,2-ethanediol, and 1,2,3-propanetriol. Structural formulas and common names (which are used more often than the IUPAC names) for these alcohols are given in Table 18-6.

Methanol, which is the simplest of all alcohols, is a colorless liquid with a characteristic odor. Its common name, wood alcohol, comes from the fact that its principal source for many years was the destructive distillation of wood. Today, it is synthetically produced in large quantities from hydrogen and carbon monoxide gases.

Methanol is used as a chemical raw material in the production of some plastics, as a solvent for shellacs and varnishes, and as a fuel for motor vehicles. It is a highly toxic substance and should never be used for medicinal purposes. If taken internally, even

Table 18-6. Some Important Common Alcohols

IUPAC Name	Structural Formula	Common Names
Methanol	$CH_3—OH$	methyl alcohol wood alcohol
Ethanol	$CH_3—CH_2—OH$	ethyl alcohol grain alcohol drinking alcohol
2-Propanol	$CH_3—CH(OH)—CH_3$	isopropyl alcohol rubbing alcohol
1,2-Ethanediol	$CH_2(OH)—CH_2(OH)$	ethylene glycol
1,2,3-Propanetriol	$CH_2(OH)—CH(OH)—CH_2(OH)$	glycerol glycerin

small amounts can cause permanent blindness and paralysis, and large amounts can be fatal.

Ethanol is the compound commonly referred to as simply "alcohol" by the layperson. Besides being the physiologically active ingredient in alcoholic beverages, it is also used in the pharmaceutical industry as a solvent (tinctures are ethanol solutions), as a medicinal ingredient (cough syrups often contain as much as 20% ethanol), and as an industrial solvent. A 70% ethanol solution is an excellent antiseptic. Of all the monohydroxy alcohols, only ethyl alcohol has toxic effects mild enough to render it "safe" for human consumption. Long-term, excessive use, however, is known to cause undesirable effects such as cirrhosis of the liver.

Gasohol, a mixture of 10% ethanol and gasoline, is a fuel for automobiles that has been developed in response to the energy crisis. The ethanol is prepared by fermenting cereal grains.

Ethanol destined for human consumption is carefully controlled and heavily taxed. Ethanol for use in industry is not taxed, and substances are usually added to it to make it unfit to drink—thus removing any temptation. Such alcohol is called *denatured alcohol.* Common denaturants are wood alcohol and formaldehyde. The small amounts added do not interfere with the industrial uses made of ethanol, but definitely make it unfit to drink.

Isopropyl alcohol is the alcohol that is commonly called "rubbing alcohol." It is also used as an astringent, a medication used externally to contract tissue and decrease the size of blood vessels.

Ethylene glycol is the most important and also the simplest of the dihydric alcohols. It is used extensively as the basic ingredient in permanent antifreeze for automobile radiators. It is nonvolatile (will not evaporate), completely miscible with water, noncorrosive, and relatively inexpensive to produce; hence its use as antifreeze. Ethylene glycol is also used to make polymers, the most common of which is the synthetic fiber Dacron.

The most important trihydroxy alcohol is glycerin or glycerol. It has an affinity for water and finds use as a moistening agent in tobacco and many food products such as candy and shredded coconut. Florists use glycerin on cut flowers to retain water and maintain freshness. Glycerin is also used for its soothing qualities in shaving and toilet soaps and in many cosmetics.

Alcohols in general are very important biologically, since the hydroxyl group occurs in a variety of compounds associated with living systems. For example, sugars contain several hydroxy groups, and starch and cellulose contain thousands of hydroxyl groups; both starch and cellulose are polymeric materials.

18.13 Ethers

ethers

Ethers *are compounds with the general formula of R—O—R′, where the R groups may or may not be the same.* When both R groups are the same, the compound is called a simple ether. When R and R′ are different, a mixed ether results.

$$\underset{\text{A simple ether}}{CH_3—O—CH_3} \qquad \underset{\text{A mixed ether}}{CH_3—CH_2—O—CH_3}$$

Common names for ethers are obtained by first naming the two hydrocarbon groups attached to the oxygen and then adding the word *ether*. If the two hydrocarbon groups are identical, the prefix di- is often omitted.

$$\underset{\substack{\text{Dimethyl ether}\\ \text{(methyl ether)}}}{CH_3-O-CH_3} \qquad \underset{\text{Ethyl methyl ether}}{CH_3-CH_2-O-CH_3}$$

IUPAC names are obtained by calling an —O—R group an alkoxy group. Thus, $-O-CH_3$ is called a methoxy group. The alkoxy group is then treated as a substituent on a parent hydrocarbon molecule. In mixed ethers, the smallest alkyl group becomes the alkoxy group.

$$\underset{\text{Methoxymethane}}{CH_3-O-CH_3} \qquad \underset{\text{2-Methoxybutane}}{CH_3-CH_2-\underset{\substack{|\\ O-CH_3}}{CH}-CH_3}$$

Ethers can be considered to be organic derivatives of water in which both hydrogens have been replaced by R groups. In addition, they may be treated as derivatives of alcohols in which the hydrogen of the hydroxyl group has been replaced by an R group.

$$\underset{\text{Water}}{H-O-H} \xrightarrow[\text{hydrogen}]{\text{replace one}} \underset{\text{An alcohol}}{R-O-H} \xrightarrow[\text{hydrogen}]{\text{replace hydroxyl}} \underset{\text{An ether}}{R-O-R'}$$

Many ethers have been found to have general anesthetic properties. Diethyl ether, which is often called simply "ether," has in the past been used extensively as an anesthetic. It was easy to administer and caused excellent relaxation of the muscles. Its

Table 18-7. ***Some Common Inhalation Anesthetics That Are Ethers***

Structure	Chemical Name	Trade Name
$CH_3-O-CH_2-CH_2-CH_3$	methyl propyl ether	Neothyl
$CH_2=CH-O-CH=CH_2$	divinyl ether	Vinethene
$H-\underset{F}{\overset{Cl}{C}}-\underset{F}{\overset{F}{C}}-O-\underset{F}{\overset{F}{C}}-H$	2-chloro-1,1,2-trifluoroethyl difluoromethyl ether	Enflurane or Ethrane
$H-\underset{Cl}{\overset{Cl}{C}}-\underset{F}{\overset{F}{C}}-O-\underset{H}{\overset{H}{C}}-H$	2,2-dichloro-1,1-difluoroethyl methyl ether	Methoxyflurane or Penthrane
$F-\underset{F}{\overset{F}{C}}-\underset{H}{\overset{H}{C}}-O-\overset{H}{C}=\underset{H}{\overset{H}{C}}$	2,2,2-trifluoroethyl vinyl ether	Fluoxene or Fluoromar

drawbacks were that it caused a slight respiratory-passage irritation and some postanesthetic nausea. In addition, it is a very flammable compound; hence, extreme care had to be taken in administering it. Other anesthetics that do not have these disadvantages have now taken its place. Many of the newer ethers are halogenated ethers. It has been found that introduction of halogen atoms into an ether in many cases cuts down on the flammability of the compounds. Table 18-7 gives the structures and names of some of the ethers now used in producing anesthesia.

18.14 Carboxylic Acids

carboxylic acids

Carboxylic acids *are compounds that contain the carboxyl functional group, that is, the* $-\overset{\overset{\displaystyle O}{\|}}{C}-OH$ *group.* They were some of the first organic compounds studied in detail because of their wide distribution and abundance in natural products. Many of the tart or sour tastes we encounter in foods are due to the presence of carboxylic acids. Some of the unpleasant odors associated with spoiled or spoiling foods are also due to carboxylic acids.

IUPAC names for carboxylic acids are derived by replacing the *-e* of the name of the longest carbon chain containing the functional group with *-oic* and then adding the word *acid.* The functional group carbon atom is counted as part of the longest chain and is assigned the number 1 when a numbering system is needed. The IUPAC names and structural formulas for the two simplest carboxylic acids are

$$H-\overset{\overset{\displaystyle O}{\|}}{C}-OH \qquad CH_3-\overset{\overset{\displaystyle O}{\|}}{C}-OH$$

Methanoic acid   Ethanoic acid

Since many carboxylic acids were isolated from natural sources long before systematic nomenclature was established, we still often refer to them by their common names, which usually indicate their early sources. The common names for methanoic and ethanoic acid are, respectively, formic acid and acetic acid. Formic acid is an active irritant in both ant and bee stings. (The Latin word for ant is *formica.*) Vinegar is a dilute solution of acetic acid, with the pungent flavor of this substance being due to the acetic acid present. (The Latin word for vinegar is *acetum.*) Butyric acid, a four-carbon acid, which occurs in rancid butter, gets its common name from the Latin *butyrum* for butter.

Some carboxylic acids contain more than one carboxyl group per molecule. Such acids are known almost exclusively by their common names. The tart (sour) taste associated with citrus fruits is due to the presence of citric acid, a carboxylic acid containing three carboxyl groups. Oxalic acid, a poisonous material found in the leaves of the rhubarb plant, is a dicarboxylic acid.

The structures of all of the previously mentioned acids, as well as those of selected other commonly encountered acids, are given in Table 18-8. The table also includes both common and IUPAC names for each acid. Note how dicarboxylic acids are named by the IUPAC system.

*Table 18-8. **Selected Examples of Carboxylic Acids***

Common Name	IUPAC Name	Structural Formula	Characteristics and Typical Uses
Formic acid	methanoic acid	$H-C(=O)-OH$	stinging agents of red ants, bees, and nettles
Acetic acid	ethanoic acid	$CH_3-C(=O)-OH$	active ingredient in vinegar
Propionic acid	propanoic acid	$CH_3-CH_2-C(=O)-OH$	salts of this acid are used as mold inhibitors in breads and cereals
Butyric acid	butanoic acid	$CH_3-(CH_2)_2-C(=O)-OH$	odor-causing agent in rancid butter; present in human perspiration
Caproic acid	hexanoic acid	$CH_3-(CH_2)_4-C(=O)-OH$	characteristic odor of limburger cheese
Oxalic acid	ethanedioic acid	$HO-C(=O)-C(=O)-OH$	poisonous material in leaves of some plants such as rhubarb; used as cleaning agent for rust stains on fabric and porcelain
Citric acid	3-hydroxy-3-carboxy-pentanedioic acid	$HO-C(=O)-CH_2-C(OH)(C(=O)-OH)-CH_2-C(=O)-OH$	present in citrus fruits, used as a flavoring agent in foods
Lactic acid	2-hydroxypropanoic acid	$CH_3-CH(OH)-C(=O)-OH$	found in sour milk and sauerkraut; formed in muscles during exercise

18.15 Esters

esters

Esters *are compounds that contain the* $-\overset{\overset{\displaystyle O}{\|}}{C}-O-R$ *functional group.* Thus, esters are structurally very closely related to carboxylic acids. The ester functional group may be visualized to be a carboxyl group in which an R group has been substituted for the hydrogen atom in the carboxyl group.

Names of esters bear a direct relation to the names of the acids from which they may be considered to be derived. The name of the ester is formed by changing the *-ic* ending of the acid name (either common or IUPAC name) to *-ate* and preceding this name with the name of the R group attached to the oxygen atom as a separate word.

Table 18-9. Fruit Odor or Flavor Associated with Selected Ester Flavoring Agents

Name	Structural Formula	Characteristic Flavor and Odor
Isobutyl methanoate (isobutyl formate)	$H-\overset{\overset{\displaystyle O}{\|}}{C}-O-CH_2-\overset{\overset{\displaystyle CH_3}{\mid}}{CH}-CH_3$	raspberry
n-Pentyl ethanoate (*n*-pentyl acetate)	$CH_3-\overset{\overset{\displaystyle O}{\|}}{C}-O-(CH_2)_4-CH_3$	banana
n-Octyl ethanoate (*n*-octyl acetate)	$CH_3-\overset{\overset{\displaystyle O}{\|}}{C}-O-(CH_2)_7-CH_3$	orange
n-Pentyl propanoate (*n*-pentyl propionate)	$CH_3-CH_2-\overset{\overset{\displaystyle O}{\|}}{C}-O-(CH_2)_4-CH_3$	apricot
Methyl butanoate (methyl butyrate)	$CH_3-(CH_2)_2-\overset{\overset{\displaystyle O}{\|}}{C}-O-CH_3$	apple
Ethyl butanoate (ethyl butyrate)	$CH_3-(CH_2)_2-\overset{\overset{\displaystyle O}{\|}}{C}-O-CH_2-CH_3$	pineapple
Methyl 2-aminobenzoate (methyl anthranilate)	benzene ring bearing $-\overset{\overset{\displaystyle O}{\|}}{C}-O-CH_3$ and, on the adjacent carbon, NH_2	grape

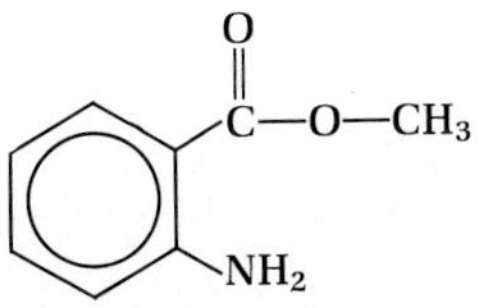

The relatively simple ester

$$CH_3-\overset{\overset{\displaystyle O}{\|}}{C}-O-CH_2-CH_3$$

has the IUPAC name ethyl acetate (from acetic acid). Many nail polish removers contain the ester ethyl acetate. It is an excellent solvent. Ethyl acetate is also the compound responsible for the characteristic odor of nail polish remover. Most simple esters are pleasant-smelling substances.

Esters are found widely distributed in nature. Many of the fragrances of flowers and fruits as well as the flavors of many fruits are due to the presence of esters. Esters are also the chemical components of many of the artificial fruit flavors that are used in cakes, candies, ice cream, and soft drinks. Table 18-9 lists some common ester flavoring agents and the odors or tastes we associate with them. Note from the table entries how a relatively small change in the size of the R group of the ester functional group significantly alters our perception of flavor. A five-carbon R group (*n*-pentyl acetate) is perceived by us to be banana flavor, whereas an eight-carbon R group (*n*-octyl acetate) registers as orange flavor. The difference between apple and pineapple flavor is a methyl group versus an ethyl group.

Learning Objectives

After completing this chapter you should be able to

- Distinguish between organic and inorganic chemistry in terms of modern-day definitions, elements studied, and number of known compounds (Sec. 18.1)
- Define the terms alkane, unsaturated hydrocarbon, and homologous series (Sec. 18.3).
- Define the term structural isomerism and give examples of such isomerism in the alkane hydrocarbon series (Sec. 18.4).
- Given the structural formula for an alkane or cycloalkane, name it using IUPAC rules, and vice versa (Secs. 18.5 and 18.6).
- Define the terms unsaturated hydrocarbon, alkene, and alkyne (Sec. 18.7).
- Given the structural formula for an unsaturated hydrocarbon, name it using IUPAC rules, and vice versa (Sec. 18.7).
- Describe how the bonding in aromatic hydrocarbons differs from that in nonaromatic hydrocarbons (Sec. 18.8).
- Name, using IUPAC or common names, the simpler aromatic hydrocarbons (Sec. 18.8).
- List the major natural sources and uses of hydrocarbons (Sec. 18.9).
- Define the terms hydrocarbon derivative and functional group and know the meaning associated with the symbol R (Sec. 18.10).
- Know the general structural features, including functional group, associated with the following classes of hydrocarbon derivatives: halogenated hydrocarbons, alcohols, ethers, carboxylic acids, and esters (Secs. 18.11 through 18.15).
- Be able to give both IUPAC and common names for the simpler halogenated hydrocarbons, alcohols, ethers, carboxylic acids, and esters (Secs. 18.11 through 18.15).
- List uses for selected halogenated hydrocarbons, alcohols, ethers, carboxylic acids, and esters (Secs. 18.11 through 18.15).

Terms and Concepts for Review

The new terms or concepts defined in this chapter are

organic chemistry (Sec. 18.1)
hydrocarbon (Sec. 18.2)
alkane (Sec. 18.3)
saturated compound (Sec. 18.3)
homologous series (Sec. 18.3)
structural isomerism (Sec. 18.4)
structural formula (Sec. 18.4)
normal chain (Sec. 18.5)
branched chain (Sec. 18.5)
alkyl group (Sec. 18.5)
cycloalkane (Sec. 18.6)
alkene (Sec. 18.7)
alkyne (Sec. 18.7)
unsaturated compound (Sec. 18.7)
aromatic hydrocarbon (Sec. 18.8)
delocalized bonding (Sec. 18.8)
hydrocarbon derivative (Sec. 18.10)
functional group (Sec. 18.10)

halogenated hydrocarbon (Sec. 18.11)
polymer (Sec. 18.11)
alcohol (Sec. 18.12)
ether (Sec. 18.13)
carboxylic acid (Sec. 18.14)
ester (Sec. 18.15)

Questions and Problems

18-1 Using the general formula for a noncyclic alkane, calculate the number of hydrogen atoms that would be present in a noncyclic alkane molecule containing

a. 9 carbon atoms **b.** 13 carbon atoms
c. 18 carbon atoms **d.** 32 carbon atoms

18-2 The following structural formulas for noncyclic alkanes are incomplete in that hydrogen atoms attached to each carbon are not shown. Complete each of these formulas by writing in the correct number of hydrogen atoms attached to each carbon atom.

a.
```
C—C—C—C
  |
  C
```

b.
```
C—C—C—C—C—C
  | | |
  C C C
```

c.
```
C—C—C—C—C—C
```

d.
```
  C
  |
C—C—C—C
  |
  C
```

18-3 What are structural isomers?

18-4 Write structural formulas (showing only carbon atoms) for all the isomers for each of the following molecular formulas.

a. C_6H_{14} (five isomers are possible)
b. C_7H_{16} (nine isomers are possible)

18-5 The first step in naming an alkane is to identify the longest continuous carbon-atom chain. Give the number of carbon atoms in the longest continuous chain in each of the following carbon-atom arrangements.

a.
```
        C
        |
C—C—C—C—C—C—C
    |
    C
    |
    C
```

b.
```
C—C—C—C—C—C
  |
  C
  |
  C
```

c.
```
      C—C
      |
  C—C—C—C
    |
C—C—C
```

d.
```
C—C—C
    |
  C—C—C—C
    |
    C
    |
    C—C
```

e.
```
C—C—C—C—C—C—C
    |   |
    C   C
    |   |
    C   C
    |   |
    C   C
```

18-6 Using the IUPAC system, name the following alkanes.

a.
```
CH3—CH2—CH—CH2—CH3
        |
        CH3
```

b.
```
    CH3
    |
CH3—C—CH2—CH3
    |
    CH3
```

c.
```
       CH3
       |
CH3—CH—CH—CH3
    |
    CH2
    |
    CH3
```

d.
```
CH3—CH—CH2—CH2—CH3
    |
    CH2
    |
    CH3
```

e.
```
            CH3
            |
CH3—CH2—CH2—CH—CH2—CH—CH3
                   |
                   CH3
```

f.
```
    CH3       CH3
    |         |
CH3—C—CH2—C—CH3
    |         |
    CH3       CH3
```

18-7 Draw the structural formula for each of the following compounds.

a. 3-methylpentane
b. 2,2-dimethylpropane

c. 3-ethyl-4,4-dimethyl-6-propyldecane
d. 2,4-dimethyloctane
e. 3-isopropylheptane

18-8 The following names, although incorrect according to IUPAC rules, contain sufficient information to enable you to draw structural formulas. Draw the structural formulas and then tell why each name is incorrect. Write the correct IUPAC name for each compound.
a. 3,3-dimethylbutane
b. 4-ethyl-5-methylhexane
c. 2,5-dimethyl-2-propylhexane
d. 1,2-dimethylpropane

18-9 Indicate the total number of alkyl groups present in each of the molecules in Prob. 18-6.

Cycloalkanes

18-10 All of the bonds in both *n*-butane (C_4H_{10}) and cyclobutane (C_4H_8) are single bonds, yet the latter compound contains two less hydrogens than the former. Why?

18-11 How many hydrogen atoms are present in each of the following molecules?

a. 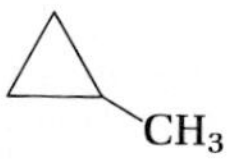 (cyclopropane ring) CH_3

b. 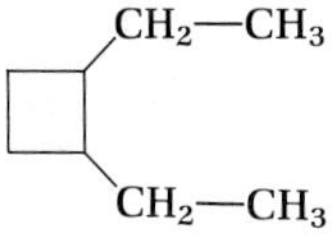 (cyclobutane ring) $CH_2—CH_3$, $CH_2—CH_3$

c. 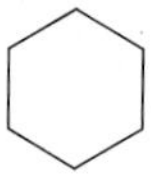 (cyclohexane ring)

d. 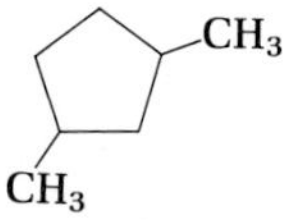 (cyclopentane ring) CH_3, CH_3

18-12 Assign IUPAC names to each of the compounds in Prob. 18-11.

18-13 Write structural formulas for each of the following cycloalkanes using geometrical figures.
a. 1,2,4-trimethylcyclohexane
b. 3-ethyl-1,1-dimethylcyclopentane
c. *n*-propylcyclobutane
d. isopropylcyclobutane

18-14 Draw and name the three possible dimethylcyclobutane isomers.

Alkenes and Alkynes

18-15 How does the general formula for a noncyclic alkene differ from that for a noncyclic alkane?

18-16 Why are the general formulas for noncyclic alkenes and cycloalkanes identical?

18-17 Classify each of the following hydrocarbons as unsaturated or saturated. In addition, classify each unsaturated compound as an alkene, alkyne, cycloalkene, or cycloalkyne.

a. $CH_3—CH{=}CH_2$

b. 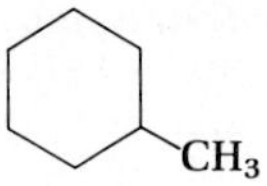 (cyclohexane ring) CH_3

c. $CH_3—CH(CH_3)—CH_2—C{\equiv}CH$

d. 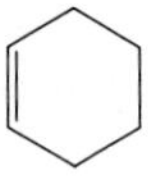 (cyclohexene ring)

e. 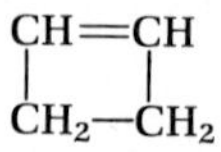 $CH{=}CH$ / $CH_2—CH_2$ (cyclobutene ring)

f. $CH_3—CH_2—CH_2—CH_2—CH_3$

18-18 Using the IUPAC system, name each of the following unsaturated hydrocarbons.

a. $CH_3—CH_2—CH{=}CH—CH_3$

b. $CH_3—C{\equiv}C—CH(CH_3)—CH_3$

c. $CH_3—CH(CH_3)—CH{=}CH_2$

d. $CH_3—CH(CH_2—CH_3)—C{\equiv}C—CH_3$

e. 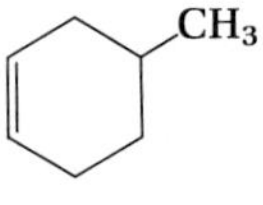 (cyclohexene ring) CH_3

f. 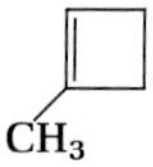 (cyclobutane ring) CH_3

g. $CH_3—CH{=}CH—CH{=}CH_2$

h. 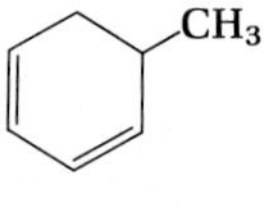 (cyclohexadiene ring) CH_3

18-19 Draw structural formulas (show carbon atoms only) for the following unsaturated hydrocarbons.
a. 2-methyl-3-hexyne
b. 3-ethyl-1,4-pentadiene
c. 3-methylcyclopentene
d. 4,4,5-trimethyl-2-heptyne
e. 1,3-butadiene
f. 3,3-dimethyl-1,4-pentadiyne

18-20 Cycloalkanes and noncyclic alkenes with the same number of carbon atoms are isomeric. Draw all isomers for both types of compounds which have the formula C_4H_8. (There is a combined total of five isomers.)

Aromatic Hydrocarbons

18-21 Classify each of the following cyclic hydrocarbons into the categories aromatic, cyclic unsaturated, or cyclic saturated.

a. 
b. 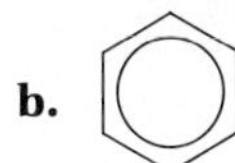
c.
d.
e.
f.

18-22 How many hydrogen atoms are present in each of the molecules in Prob. 18-21?

18-23 Give the IUPAC name for each of the following aromatic hydrocarbons using the word benzene in the name.

a.
b. CH_3, CH_3
c. CH_3, CH_3
d. CH_3, CH_3, CH_3
e. 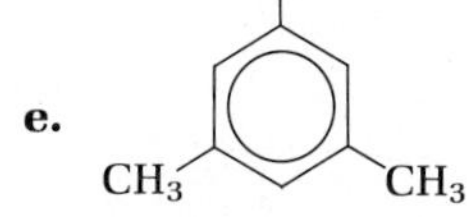
CH_3, CH_3, CH_3
f. 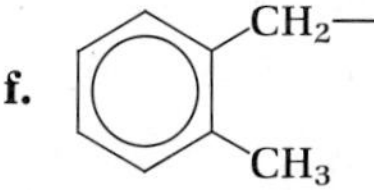
$CH_2—CH_3$, CH_3

18-24 When and how are the prefixes ortho-, meta-, and para- used in naming aromatic compounds?

18-25 Write structural formulas for the following aromatic compounds.
a. 1,3-diethylbenzene
b. *o*-xylene
c. toluene
d. *p*-ethylmethylbenzene
e. 1,4-diethyl-2,5-dichlorobenzene
f. naphthalene

18-26 Draw structural formulas for all of the isomeric tetramethylbenzenes.

18-27 Describe how the bonding in benzene differs from that in cyclohexane. Both compounds have a ring of six carbon atoms.

Sources of Hydrocarbons

18-28 List the major types of hydrocarbons found in
a. natural gas b. crude oil (petroleum)
c. coal

Derivatives of Hydrocarbons

18-29 In each of the following hydrocarbon derivatives circle the functional group present.

a. $CH_3—CH(Br)—CH_2—CH_3$

b. $CH_3—CH_2—C(=O)—OH$

c. $CH_3—CH_2—O—CH_2—CH_3$

d. $CH_3—CH_2—CH_2(OH)$

e. $CH_3—CH_2—C(=O)—O—CH_2—CH_3$

f. $CH_3—CH_2—CH_2—C(=O)—O—CH_3$

18-30 Name the functional group present in each of the compounds in Prob. 18-29.

18-31 Name the following halogenated hydrocarbons according to IUPAC rules.

a. $H—C(H)(H)—C(H)(Br)—Br$

b. $H—C(H)(Br)—C(H)(H)—Br$

c. $CH_3—CH(CH_3)—CH(I)—CH_3$

d. $CH_3—I$

e. $H—C(H)(H)—C(Cl)(Br)—C(Cl)(H)—H$

f. $H—C(H)(H)—C(H)(H)—F$

18-32 Name the following alcohols according to IUPAC rules.

a. $CH_3—CH_2—CH(CH_3)—OH$

b. $CH_3—CH_2—CH_2—C(CH_3)(OH)—CH_3$

c. $CH_2(CH_3)—CH(OH)—CH_2(CH_3)$

d. $CH_3—CH(CH_3)—CH_2—C(CH_3)(CH_3)—OH$

18-33 Write structural formulas for the following alcohols.
a. 4-methyl-2-pentanol
b. 2,4,4-trimethyl-2-heptanol
c. ethyl alcohol
d. isopropyl alcohol
e. cyclohexanol
f. 2-chloro-4-propylheptanol

18-34 Assign IUPAC names to the following polyhydric alcohols.

a.
$$\begin{array}{ccccc} CH_2 & — & CH_2 & — & CH_2 \\ | & & & & | \\ OH & & & & OH \end{array}$$

b.
$$\begin{array}{ccccccc} CH_2 & — & CH_2 & — & CH & — & CH_2 \\ | & & & & | & & | \\ OH & & & & OH & & OH \end{array}$$

c.
$$\begin{array}{ccccccc} CH_3 & — & CH & — & CH & — & OH \\ & & | & & | & & \\ & & OH & & CH_3 & & \end{array}$$

18-35 Assign a common name and an IUPAC name to each of the following ethers.

a. $CH_3—O—CH_2—CH_2—CH_3$

b.
$$\begin{array}{ccccccccccc} CH_3 & — & CH_2 & — & CH_2 & — & O & — & CH & — & CH_3 \\ & & & & & & & & | & & \\ & & & & & & & & CH_3 & & \end{array}$$

c. $CH_3—CH_2—O—CH_2—CH_3$

18-36 Using IUPAC rules, name the following carboxylic acids.

a.
$$\begin{array}{ccccccccccc} & & & & & & & & O & & \\ & & & & & & & & \| & & \\ CH_3 & — & CH_2 & — & CH_2 & — & CH_2 & — & C & — & OH \end{array}$$

b.
$$\begin{array}{ccccccccccc} & & & & & & & & O & & \\ & & & & & & & & \| & & \\ CH_3 & — & CH_2 & — & CH & — & CH_2 & — & C & — & OH \\ & & & & | & & & & & & \\ & & & & CH_3 & & & & & & \end{array}$$

c.
$$\begin{array}{ccccccccccccc} & & & & & & & & & & O & & \\ & & & & & & & & & & \| & & \\ CH_3 & — & CH & — & CH_2 & — & CH & — & CH_2 & — & C & — & OH \\ & & | & & & & | & & & & & & \\ & & CH_3 & & & & CH_2 & & & & & & \\ & & & & & & | & & & & & & \\ & & & & & & CH_3 & & & & & & \end{array}$$

d.
$$\begin{array}{ccccc} & & O & & \\ & & \| & & \\ CH_3 & — & C & — & OH \end{array}$$

18-37 Draw structural formulas to represent the following carboxylic acids.
a. 2,3-dimethylbutanoic acid
b. acetic acid (common name)
c. 5-chloro-2-pentenoic acid
d. ethanoic acid
e. oxalic acid (common name)
f. citric acid (common name)

18-38 Draw structural formulas for each of the following esters.
a. ethyl propanoate **b.** *n*-propyl acetate
c. ethyl methanoate **d.** methyl ethanoate

18-39 Using IUPAC rules, name the following esters.

a.
$$\begin{array}{ccccccc} & & O & & & & \\ & & \| & & & & \\ CH_3 & — & C & — & O & — & CH_3 \end{array}$$

b.
$$\begin{array}{ccccccccccccc} & & & & O & & & & & & & & \\ & & & & \| & & & & & & & & \\ CH_3 & — & CH_2 & — & C & — & O & — & CH_2 & — & CH_2 & — & CH_3 \end{array}$$

c.
$$\begin{array}{ccccccccccccccc} & & & & & & O & & & & & & & & \\ & & & & & & \| & & & & & & & & \\ CH_3 & — & CH_2 & — & CH_2 & — & C & — & O & — & CH_2 & — & CH_2 & — & CH_3 \end{array}$$

18-40 Using IUPAC rules, name the following hydrocarbon derivatives.

a.
$$\begin{array}{ccccccccc} & & & & & & O & & \\ & & & & & & \| & & \\ CH_3 & — & CH_2 & — & CH_2 & — & C & — & OH \end{array}$$

b. $CH_3—CH_2—O—CH_3$

c. $CH_3—CH_2—OH$

d. $CH_3—CH_2—Br$

e.
$$\begin{array}{ccccccccccc} & & & & O & & & & & & \\ & & & & \| & & & & & & \\ CH_3 & — & CH_2 & — & C & — & O & — & CH_2 & — & CH_3 \end{array}$$

f. $CH_3—CH_2—CH_2—O—CH_2—CH_3$

g.
$$\begin{array}{ccccccccccc} CH_3 & — & CH_2 & — & CH_2 & — & CH & — & CH_2 & — & CH_3 \\ & & & & & & | & & & & \\ & & & & & & OH & & & & \end{array}$$

h.
$$\begin{array}{ccccccccc} CH_3 & — & CH_2 & — & CH_2 & — & CH & — & CH_3 \\ & & & & & & | & & \\ & & & & & & O & & \\ & & & & & & | & & \\ & & & & & & CH_3 & & \end{array}$$

i.
$$\begin{array}{ccccccccccc} & & & & & & O & & & & \\ & & & & & & \| & & & & \\ CH_3 & — & CH_2 & — & CH_2 & — & C & — & O & — & CH_3 \end{array}$$

Selected Answers

Chapter 1

1-5 **a.** True **c.** True **e.** False
1-6 **a.** Fact **c.** Law

Chapter 2

2-2 **a.** 4 **c.** 5 **e.** 4 **g.** 3
2-5 **a.** ±1000 **c.** ±100 **e.** ±0.001
2-6 **a.** Infinite — a defined number **c.** 3
e. Infinite — a counted number
g. Infinite — a counted number
2-7 **a.** 3880 **c.** 1,030,000 **e.** 4.40
g. 12,300
2-8 **a.** 0.1 **c.** 0.00013 **e.** 0.100 **g.** 1.22
2-9 **a.** 14 **c.** 17.954 **e.** 20 **g.** 8,675,000
2-10 **a.** 825 **c.** 3 **e.** 900.00 **g.** 1.6
2-11 **a.** 15 **c.** 0.203
2-12 **a.** 34 **c.** 0.1 **e.** 1000.1 **g.** 0.2285
2-14 **a.** 10^8 **c.** 10^{-2}
2-15 **a.** 2.34×10^2 **c.** 3.03002×10^1
e. 2.30×10^{-1} **g.** 3.235×10^3
2-16 **a.** 2,340,000 **c.** $1{,}2\bar{0}0{,}000$ **e.** 0.03
g. $3\bar{0}$
2-17 **a.** 1×10^6 **c.** 1.0000×10^6
2-19 **a.** 2.34×10^5 **c.** 1.00×10^3 **e.** 1.12×10^6
g. 3.33×10^{-4}
2-20 **a.** 6×10^6 **c.** 9×10^{-4} **e.** 4.69×10^4
g. 7.2×10^1
2-21 **a.** 2.0×10^2 **c.** 2.0×10^{-6} **e.** 3.32×10^1
g. 2×10^{10}
2-22 **a.** 6.32×10^5 **c.** 3.2×10^1 **e.** 6.32×10^{-3}
g. 3.2×10^{-3}
2-23 **a.** 2×10^6 **c.** 2.4×10^{-12} **e.** 2×10^2
g. 2×10^1
2-24 **a.** 5.794×10^3 **c.** 4.0×10^6 **e.** 1.980×10^6
g. 3.1×10^5

Chapter 3

3-5 **a.** 2.5×10^{-2} g **c.** 1×10^2 μg
e. 7.53×10^1 cm **g.** 3×10^{12} nL
3-6 1.425 g
3-8 4,184,000 cm
3-10 7×10^{-6} lb
3-12 **a.** 7.47×10^3 cm^3 **b.** 7.89 qt
3-15 0.0625 gal
3-17 3.56×10^3 dollars
3-19 **a.** 18 g **c.** 5.31 g
3-20 **a.** 2.0 mL **c.** 46.3 mL
3-22 331 g Cu
3-24 0.534
3-26 5.2 g
3-28 24
3-29 **a.** 773 K **c.** 144 K **e.** 261°F
3-31 6098°F
3-33 3253 F°, 1807 K°
3-35 −60°F

Chapter 4

4-2 **a.** Solid, solid, liquid **d.** Solid, liquid, gas
4-4 **a.** Physical **c.** Chemical **e.** Physical
g. Chemical **i.** Chemical
4-5 **a.** Physical **c.** Physical **e.** Physical
g. Physical

4-6 **a.** Freezing **c.** Sublimation **e.** Sublimation **g.** Evaporation
4-7 **a.** Heterogeneous mixture, homogeneous mixture **c.** Pure substance
4-8 **a.** Heterogeneous mixture **c.** Pure substance
4-9 **a.** Pure substance, one phase **c.** Heterogeneous mixture, two phases **e.** Heterogeneous mixture, three phases
4-10 **a.** Compound **d.** No classification possible **g.** No classification possible
4-11 **a.** True **c.** False **f.** True

Chapter 5

5-2 **a.** Compound **d.** Compound **g.** Element
5-3 **b.** $C_8H_{10}N_4O_2$
5-4 **a.** 1 **d.** 3 **g.** 6
5-6 **a.** Proton **e.** Neutron, proton **j.** Neutron
5-8 **a.** $^{121}_{51}Sb$ **c.** $^{64}_{30}Zn$
5-9 **a.** $^{15}_{7}N$ **c.** $^{103}_{45}Rh$
5-12 **a.** $^{50}_{25}X$, $^{49}_{25}X$ **c.** $^{48}_{26}X$, $^{48}_{24}X$
5-15 90.905 amu
5-17 1.743
5-19 **a.** 35.46 amu **c.** 12.0 amu
5-21 3.7×10^2 atoms $^{17}_{8}O$, 2.0×10^3 atoms $^{18}_{8}O$
5-22 **a.** $_{19}K$ **c.** $_{13}Al$
5-24 **a.** Zero **c.** +1
5-26 **a.** 20 p, 23 n, 18 e **c.** 7 p, 8 n, 10 e
5-27 **a.** $^{56}_{26}Fe^{3+}$ **d.** $^{14}_{7}N^{3-}$

Chapter 6

6-5 **a.** Nonmetal **d.** Metal **g.** Nonmetal
6-6 **a.** Ionic, formula unit **d.** Ionic, formula unit **g.** Molecular, molecule
6-7 **a.** Binary **d.** Binary **g.** Ternary
6-8 **a.** 3 lost **d.** 1 lost **g.** 1 gained
6-9 **a.** Variable-charge **d.** Variable-charge **g.** Fixed-charge
6-10 **a.** Na_2O **d.** Zn_3N_2 **g.** Cu_3P
6-11 **a.** LiBr **d.** AlP **g.** CaS
6-12 **a.** Na^+, S^{2-} **d.** Fe^{2+}, N^{3-} **g.** Al^{3+}, X^{2-}
6-14 **a.** Beryllium chloride **d.** Sodium chloride **g.** Tin(IV) chloride
6-15 **a.** KI **d.** Ba_3P_2 **g.** $CdCl_2$
6-16 **a.** +1 **d.** +3 **g.** +2
6-17 **a.** Nitrate **d.** Cyanide **g.** Hydrogen carbonate
6-18 **a.** ClO_3^- **d.** SO_3^{2-} **g.** ClO_4^-
6-19 **a.** None **d.** Nitrate **g.** Dichromate
6-20 **a.** 11 **c.** 17
6-21 **a.** $Ba(NO_3)_2$ **d.** $NaClO_4$ **g.** $Co(H_2PO_4)_2$
6-22 **a.** $KClO_4$ **d.** $Cu(OH)_2$ **g.** $AuC_2H_3O_2$
6-23 **a.** XO_3^- **c.** XO_2^-
6-24 **a.** Copper(II) sulfate **d.** Chromium(II) hydrogenphosphate **g.** Sodium phosphite
6-25 **a.** $Co_2(SO_4)_3$ **d.** $NaHCO_3$ **g.** $Au(NO_3)_3$
6-26 **a.** Sulfur tetrafluoride **d.** Carbon tetrabromide **g.** Ammonia
6-27 **a.** ICl **d.** OF_2 **g.** SO_2
6-28 **a.** Nonoxy **d.** Oxy **g.** Nonoxy
6-29 **a.** Hydrobromic acid **d.** Sulfurous acid **g.** Hydrochloric acid
6-30 **a.** $HC_2H_3O_2$ **d.** $HClO_4$ **g.** $H_2C_2O_4$
6-31 **a.** Binary ionic **d.** Oxyacid **g.** Ternary ionic
6-32 **a.** PF_5 **d.** HF **g.** HNO_2
6-33 **a.** Hydrogen iodide **d.** Lead(IV) carbonate **g.** Aluminum oxide

Chapter 7

7-1 **a.** 84.01 amu **d.** 176.14 amu
7-3 **a.** 24.42% Ca, 17.07% N, 58.50% O **d.** 47.433% C, 2.56% H, 50.003% Cl
7-5 32.37% Na, 22.58% S, 45.05% O
7-8 $(NH_2)_2CNH$
7-9 **a.** 6.02×10^{23} molecules **d.** 6.02×10^{23} molecules
7-11 **a.** 44.01 g **d.** 342.14 g
7-12 **a.** 32.0 g **d.** 271 g
7-15 **a.** 2.00 moles C, 8.00 moles H **d.** 1.50 moles Ba, 3.00 moles Cl
7-16 **a.** 4 moles **d.** 11.3 moles
7-17 **a.** 3.0×10^{23} S atoms **d.** 9.03×10^{23} S atoms
7-18 **a.** 4.00 moles CO **d.** 1.33 moles C_3H_6
7-19 **a.** 6.022×10^{23} Be atoms **d.** 8.27×10^{22} Fe atoms
7-20 **a.** 2.15×10^{23} O atoms **d.** 2.12×10^{23} O atoms
7-21 **a.** 63.5 g Cu **d.** 3.19×10^{-21} g NH_3
7-22 **a.** 2.50×10^{-7} mole He **d.** 2.41×10^{-22} mole He
7-23 **a.** 3.27 g S **d.** 5.33×10^{-23} g S
7-24 **a.** 0.1418 mole **d.** 6.833×10^{23} atoms C
7-26 4.25 g Al
7-27 **a.** 1.66×10^{-18} mole PH_3 **d.** 5.15×10^{-17} g P
7-28 **a.** 33.16 moles $C_{12}H_{22}O_{11}$ **c.** 4.001×10^{-23} cent
7-30 197 amu, gold
7-32 **a.** $PNCl_2$ **d.** $(NH_4)_2SO_4$
7-33 **a.** C_2H_4 **d.** N_2O_4
7-35 **a.** 3 to 5 ratio **d.** 4 to 5 to 10 ratio
7-36 **a.** H_2O **d.** Cr_3S_4
7-38 $NaC_5H_8NO_4$
7-40 **a.** $FeCrO_4$ (d) SO_2

Chapter 8

8-7 **a.** $N_2 + O_2 \longrightarrow 2\,NO$ **c.** $2\,H_2O \longrightarrow 2\,H_2 + O_2$ **e.** $4\,Al + 3\,O_2 \longrightarrow 2\,Al_2O_3$
8-8 **a.** $CH_4 + 2\,O_2 \longrightarrow CO_2 + 2\,H_2O$ **d.** $2\,C_4H_{10} + 13\,O_2 \longrightarrow 8\,CO_2 + 10\,H_2O$

8-9 **a.** $P_4O_{10} + 6\,H_2O \longrightarrow 4\,H_3PO_4$
c. $Na_2NH + 2\,H_2O \longrightarrow NH_3 + 2\,NaOH$
e. $Cl_2 + H_2O \longrightarrow HCl + HClO$

8-10 **a.** $2\,NaHCO_3 + H_2SO_4 \longrightarrow Na_2SO_4 + 2\,H_2O + 2\,CO_2$
c. $2\,KClO_3 + 4\,HCl \longrightarrow 2\,KCl + 2\,ClO_2 + Cl_2 + 2\,H_2O$
e. $2\,C_8H_{18}O_4 + 21\,O_2 \longrightarrow 16\,CO_2 + 18\,H_2O$

8-11 **a.** $Ca(OH)_2 + 2\,HNO_3 \longrightarrow Ca(NO_3)_2 + 2\,H_2O$
d. $2\,Al + 3\,Sn(NO_3)_2 \longrightarrow 2\,Al(NO_3)_3 + 3\,Sn$

8-14 **a.** 1.50 moles H_2O **c.** 0.750 mole $MgCl_2$
8-15 **a.** 1.68 moles O_2 **c.** 2.58 moles NH_3
8-16 **a.** 102 g NH_3 **c.** 96.0 g N_2H_4
8-17 **a.** 6.46 g CS_2 **c.** 19.6 g Fe_3O_4
8-18 **a.** 246 g $KClO_3$ **c.** 3.986×10^{-19} g $KClO_3$
e. 1.03 g $KClO_3$
8-19 **a.** 7.25 g N_2H_4 **c.** 22.8 g H_2O
8-20 **a.** 1005 g LiOH
8-21 **a.** 0.4634 mole Ag_2CO_3 **c.** 3.764×10^{24} formula units Ag_2CO_3
8-22 **a.** N_2 **c.** Mg **e.** Mg
8-23 **a.** 3.6 g NH_3 **c.** 44.9 g NH_3
8-24 **a.** 131 g SO_2
8-26 **a.** 208 g Al_2S_3 **b.** 60.1%
8-28 42.9 g ZnS
8-29 37.5 g CO_2
8-31 39.1 g SO_2
8-33 349.9 g Fe

Chapter 9

9-2 **a.** Orbital **d.** Subshell **g.** Subshell
9-3 **a.** Dependent on **d.** Dependent on
9-4 **a.** True **d.** False **g.** False **j.** False
9-6 **a.** 6 **d.** 8
9-8 **a.** $1s^22s^22p^4$
d. $1s^22s^22p^63s^23p^64s^23d^{10}4p^65s^24d^{10}5p^66s^24f^{14}5d^9$
g. $1s^22s^22p^63s^23p^3$
9-9 **a.** ${}_7N$ **c.** ${}_{34}Se$
9-11 **a.** 2, 8, 8, 1 **c.** 2, 8, 2
9-13 **a.** 1, 2, 3 **c.** −1, 0, +1
9-14 **a.** 2, 0 **d.** 3, 2 **g.** 4, 2
9-15 **a.** 2*s* **c.** 3*p*
9-16 $(3, 1, 1, +\frac{1}{2}), (3, 1, -1, -\frac{1}{2}), (3, 1, -1, +\frac{1}{2})$
9-19 **a.** $(1, 0, 0, -\frac{1}{2}), (1, 0, 0, +\frac{1}{2}), (2, 0, 0, -\frac{1}{2}), (2, 0, 0, +\frac{1}{2})$
9-20 **a.** 2 **d.** 9
9-22 **a.** **d.** **g.**

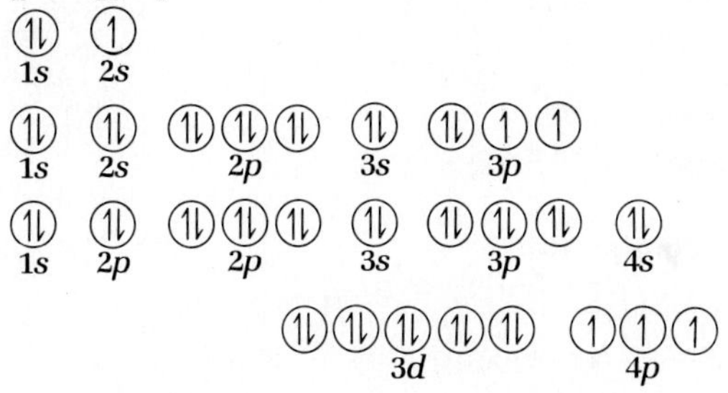

1s
2s
2p
3s
3p
4s
3d
4p

9-23 **a.** Zero **d.** 3
9-24 **a.** Diamagnetic **d.** Paramagnetic
9-25 **a.** 1*s* **c.** 4*d*
9-27 **a.** ${}_{11}Na$ **d.** ${}_{54}Xe$
9-28 **a.** $1s^22s^22p^63s^23p^64s^2$
d. $1s^22s^22p^63s^23p^64s^23d^{10}4p^65s^24d^3$
9-29 **a.** 4 **d.** 10
9-30 **a.** 2*s* **d.** 5*p*
9-31 **a.** ${}_{56}Ba$ **d.** ${}_{103}Lr$
9-32 **a.** ${}_{82}Pb$ **d.** ${}_{33}As$
9-34 **a.** Rare gas **d.** Transition element
g. Representative element

Chapter 10

10-1 **a.** 3 **d.** 2
10-2 **a.** 2 **d.** 5
10-4 **a.** ·Cl: **d.** Rb·
10-8 **a.** Gain electrons **c.** Lose electrons
10-9 **a.** −3 **d.** −1
10-11 **a.** $1s^22s^22p^6$
d. $1s^22s^22p^63s^23p^64s^23d^{10}4p^65s^24d^{10}5p^6$
10-12 **b.** As^{3-}, Se^{2-}, Br^-, Rb^+, Sr^{2+}, In^{3+}
10-13 **c.** S^{2-} and P^{3-}, Rb^+ and Sr^{2+}
10-14 **a.**

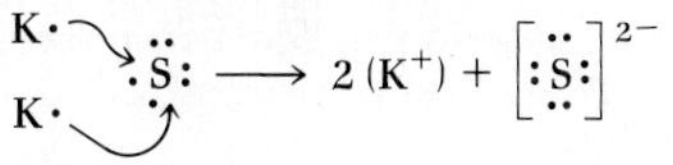

c.

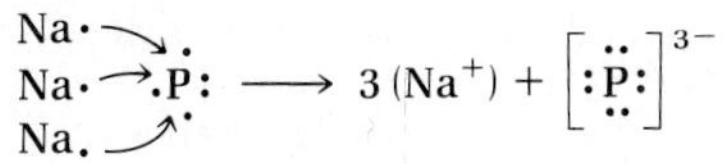

10-16 **a.** **d.**

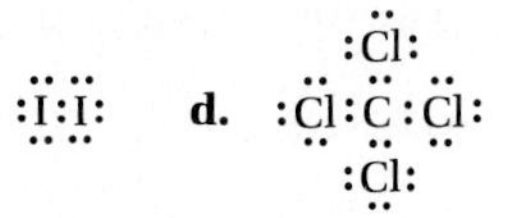

10-17 **a.** **c.**

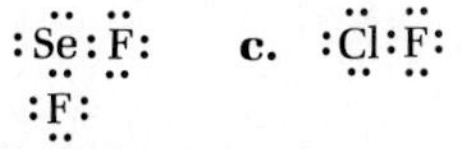

10-18 **a.** Na, Mg, Al, P **d.** Rb, Sr, I, F
10-19 **a.** H—Br, H—Cl, H—O
d. C—N, Al—Cl, H—F
10-21 **a.** Polar covalent **d.** Ionic
10-23 **a.** **d.**

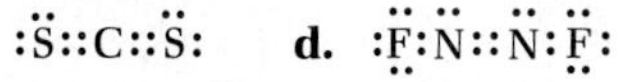

10-24 **b.**

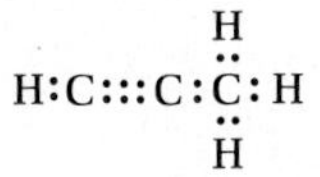

10-26 **b.**

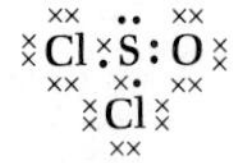

10-27 **a.**

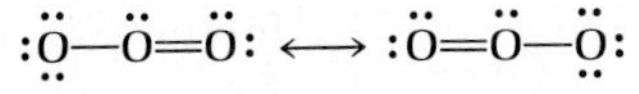

c.

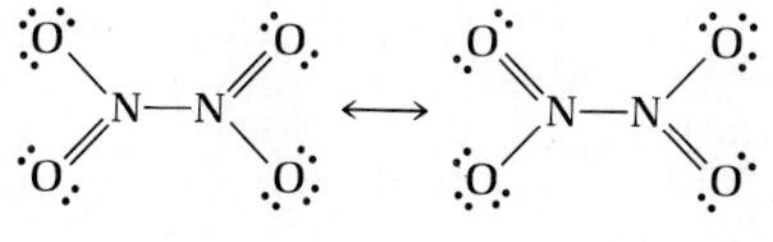

10-30 **a.** 24 **d.** 32

10-31 **a.**
```
 [ ..  ..  .. ]⁻
 [:O::N:O:    ]
 [  ..  ..    ]
 [   :O:      ]
 [    ..      ]
```
d.
```
 [ ..   ]⁻
 [:O:H  ]
 [ ..   ]
```

10-33 **a.**
```
      ..
     :O:
   ..  ..  ..
 H:O:P:O:H
   ..  ..  ..
     :O:
      ..
      H
```
d.
```
 [     ..      ]⁻
 [    :O:      ]
 [ ..  ..  ..  ]
 [:O:Cl:O:     ]
 [ ..  ..  ..  ]
 [    :O:      ]
 [     ..      ]
```
g. :N:::C:C:::N:

10-35 **a.** Polar bonds, nonpolar molecule **d.** Polar bonds, polar molecule

10-36 **a.** Polar **d.** Polar

10-37 **a.** Trigonal pyramidal **d.** Angular **g.** Linear

10-38 **a.** Trigonal planar **d.** Trigonal planar

Chapter 11

11-1 **a.** Gaseous **d.** Gaseous

11-2 Through collisions with each other

11-8 **a.** Endothermic **d.** Exothermic

11-9 **a.** Boiling point **d.** Boiling point

11-12 Temperature, surface area

11-14 Intermolecular — forces between molecules; intramolecular — forces within molecules (chemical bonds)

11-16 **b.** CO_2 — electrons are more susceptible to polarization

11-18 Crystalline — regular arrangement of particles; amorphous — random, nonrepetitive arrangement of particles

11-21 **a.** 350 cal **c.** 1500 cal

11-23 85,300 cal

11-25 1900 g Cu

Chapter 12

12-1 **a.** 735 torr **c.** 549 torr

12-2 **a.** 1.02 atm **c.** 1.02 atm

12-5 **a.** 0.733 L **c.** 1.95 L

12-6 **a.** 1580 torr **c.** 4260 torr

12-8 **a.** 4.55 L **c.** 13.4 L

12-9 **a.** 55°C **c.** 2280°C

12-11 **a.** 3.46 atm **c.** 11.4 atm

12-12 **a.** 228°C **c.** 65°C

12-13 177°C

12-15 **a.** 29.6 L **c.** 85.5 L

12-17 20.5 L

12-18 **a.** 10.3 L **c.** 34.8 L

12-19 **a.** 0.0873 atm **c.** 0.0216 atm

12-21 2.50 L

12-23 **a.** 14.4 L N_2 **c.** 9.16 L CO_2

12-24 **a.** 1.17 moles H_2 **c.** 0.136 mole O_2

12-25 **a.** 0.0902 g/L **c.** 1.96 g/L

12-26 **a.** 2.24 L **c.** 0.317 L

12-28 397°C

12-30 5.18 atm

12-31 **a.** 24.0 g CH_4 **c.** 5.71 g O_2

12-33 **a.** 1.59 g/L **c.** 15.9 g/L

12-34 4.56 g NO

12-36 99.1 amu, $C_2H_4Cl_2$

12-37 **a.** 2.80 L NH_3 **c.** 0.62 L NH_3

12-39 $16\bar{0}0$ g KNO_3

12-42 $14\overline{00}$ torr

12-44 1245 torr

12-45 **a.** 726 torr **c.** $72\bar{0}$ torr

12-46 36.6 mL O_2

Chapter 13

13-2 **a.** Soluble **c.** Immiscible or insoluble

13-3 **a.** Saturated **c.** Unsaturated

13-4 **a.** Soluble **c.** Very soluble

13-5 **a.** Dilute **c.** Dilute

13-6 **a.** Slightly soluble **d.** Very soluble

13-7 **a.** Soluble **d.** Soluble **g.** Insoluble

13-8 **a.** 7.20 wt % **c.** 7.30 wt %

13-9 **a.** 15.0 g NaI **c.** 11 g KNO_3

13-11 **a.** 26.2 wt % **c.** 55.0 wt %

13-12 **a.** 6.00 vol % **c.** 16.9 vol %

13-14 112 mL CH_3OH

13-16 **a.** 2.0 wt-vol % **c.** 95.3 wt-vol %

13-18 **a.** 4.0 *M* **c.** 0.124 *M*

13-19 **a.** 18 moles HNO_3 **c.** 0.656 mole $BaCl_2$

13-20 **a.** 171 mL **c.** 1,200,000 mL

13-22 0.090 *M*

13-24 11.6 *M*

13-26 a. 1.97 *m* **c.** 0.0545 *m*

13-27 a. 3.45 g KNO_3 **c.** 0.0561 g KI

13-28 835 g H_2O

13-30 4.46 *m*

13-31 2.16 *m*

13-32 **a.** 55.5 g/equiv **d.** 80.0 g/equiv **g.** 59.5 g/equiv

13-33 **a.** 1.85 *N* **c.** 0.0388 *N*

13-34 **a.** 0.333 *N* **c.** 4.68 *N*

13-36 352 mL

13-37 **a.** 96 mL **c.** $1{,}2\bar{0}0{,}000$ mL

13-38 **a.** $15\bar{0}0$ mL **c.** 85,600 mL

13-39 **a.** 3.3 *M* **c.** 8.86 *M*

13-41 3.00 L H_2S

13-42 1.10 L HBr

13-44 51.5 g AgCl

13-46 6.7 L H_2

13-48 288 g $BaCrO_4$

Chapter 14

14-3 **a.** $HBr(g) \xrightarrow{H_2O} H^+(aq) + Br^-(aq)$

c. $LiOH(s) \xrightarrow{H_2O} Li^+(aq) + OH^-(aq)$

14-4 **a.** HBr (acid), H_2O (base) **c.** H_2S (acid), H_2O (base)
14-5 **a.** $HOCl + NH_3 \longrightarrow NH_4^+ + OCl^-$
c. $H_2O + NH_2^- \longrightarrow NH_3 + OH^-$
14-7 **a.** Strong acid **d.** Strong acid
14-9 **a.** Diprotic **d.** Monoprotic
14-10 **b.** $H_2CO_3 \longrightarrow H^+ + HCO_3^-$; $HCO_3^- \longrightarrow H^+ + CO_3^{2-}$
14-12 **a.** 3 *M* **d.** 16 *M*
14-13 **a.** Calcium ion (Ca^{2+}), sulfate ion (SO_4^{2-})
d. Aluminum ion (Al^{3+}), sulfide ion (S^{2-})
g. Silver ion (Ag^+), iodide ion (I^-)
j. Magnesium ion (Mg^{2+}), phosphate ion (PO_4^{3-})
14-15 **a.** Acid **d.** Salt **g.** Base **j.** Acid
14-17 **a.** Molecular **d.** Net ionic
14-18 **a.** $FeS + 2\,H^+ \longrightarrow Fe^{2+} + H_2S$
d. $H^+ + OH^- \longrightarrow H_2O$
14-19 **a.** $2\,K + 2\,H_2O \longrightarrow 2\,K^+ + 2\,OH^- + H_2$
d. $Ba^{2+} + 2\,OH^- + 2\,H^+ + SO_4^{2-} \longrightarrow BaSO_4 + 2\,H_2O$
14-20 **a.** No reaction **d.** No reaction
14-21 **a.** $H^+ + OH^- \longrightarrow H_2O$
d. $HC_2H_3O_2 + OH^- \longrightarrow C_2H_3O_2^- + H_2O$
14-22 **a.** $2\,LiOH + H_2SO_4 \longrightarrow Li_2SO_4 + 2\,H_2O$
c. $KOH + HNO_3 \longrightarrow KNO_3 + H_2O$
14-23 **a.** $Zn + 2\,HCl \longrightarrow ZnCl_2 + H_2$
c. $Na_2CO_3 + 2\,HCl \longrightarrow 2\,NaCl + CO_2 + H_2O$
14-25 **a.** $Zn + Pb^{2+} \longrightarrow Zn^{2+} + Pb$
d. $3\,Zn + 2\,Fe^{3+} \longrightarrow 3\,Zn^{2+} + 2\,Fe$
14-27 **a.** $BaSO_4$ (insoluble product) **d.** CO_2 (gaseous product)
14-31 **a.** $2.5 \times 10^{-12}\,M$ **c.** $1.2 \times 10^{-5}\,M$
14-32 **a.** Basic **d.** Neutral
14-33 **a.** 4 **d.** 10
14-34 **a.** Strongly acidic **d.** Moderately basic
g. Very basic
14-35 **a.** $1.00 \times 10^{-1}\,M$ **c.** $1.00 \times 10^{-3}\,M$
14-37 **a.** 1 equiv NaOH **d.** 1.50 equiv H_3PO_4
g. 0.274 equiv HCl
14-38 **a.** 0.500 *N* **c.** 5.0 *N*
14-39 **a.** 0.0050 *N* **d.** 6.0 *N*
14-40 **a.** 0.0015 *M* **d.** 6.0 *M*
14-41 **a.** 35.0 mL NaOH **c.** 75.0 mL H_3PO_4
14-42 **a.** 0.100 equiv **c.** 0.0200 equiv
14-44 **a.** 0.348 *N* **d.** 0.174 *M*

Chapter 15

15-3 **a.** Oxidized **d.** Loses
15-4 **a.** +3 (P), −1 (F) **d.** +1 (Na), +6 (S), −2 (O)
g. −3 (N), +1 (H)
15-5 **a.** +5 **d.** +7 **g.** +1
15-6 **a.** +1 **d.** +2 **g.** +4
15-7 **a.** −2 **d.** −2
15-8 **a.** +1 **d.** −1
15-9 **a.** Cl (ox. agent), H (red. agent)
d. Fe (ox. agent), H (red. agent)
g. I (ox. agent), S (red. agent)
15-10 **a.** Reduced **d.** Neither oxidized or reduced **g.** Neither oxidized or reduced
15-11 **a.** Synthesis **d.** Synthesis
g. Single replacement
15-13 **a.** $3\,F_2 + 3\,H_2O \longrightarrow 6\,HF + O_3$
d. $PH_3 + 2\,NO_2 \longrightarrow H_3PO_4 + N_2$
g. $H_2 + KClO \longrightarrow KCl + H_2O$
15-14 **a.** $3\,PbO_2 + 2\,Sb + 2\,NaOH \longrightarrow 3\,PbO + 2\,NaSbO_2 + H_2O$
d. $Na_2TeO_3 + 4\,NaI + 6\,HCl \longrightarrow 6\,NaCl + Te + 3\,H_2O + 2\,I_2$
15-15 **a.** $8\,H^+ + MnO_4^- + 5\,e^- \longrightarrow Mn^{2+} + 4\,H_2O$
d. $Pt + 6\,Cl^- \longrightarrow PtCl_6^{2-} + 4\,e^-$
15-16 **a.** $MnO_4^- + 2\,H_2O + 3\,e^- \longrightarrow MnO_2 + 4\,OH^-$
d. $CN^- + 2\,OH^- \longrightarrow CNO^- + H_2O + 2\,e^-$
15-17 **a.** $Zn + 2\,Ag^+ \longrightarrow Zn^{2+} + 2\,Ag$
d. $5\,H_2C_2O_4 + 2\,MnO_4^- + 6\,H^+ \longrightarrow 10\,CO_2 + 2\,Mn^{2+} + 8\,H_2O$
15-18 **a.** $3\,ClO_2^- + 2\,H_2O + 4\,MnO_4^- \longrightarrow 3\,ClO_4^- + 4\,MnO_2 + 4\,OH^-$
d. $3\,N_2H_4 + 4\,ClO_3^- \longrightarrow 6\,NO + 4\,Cl^- + 6\,H_2O$

Chapter 16

16-2 Physical nature of reactants, reactant concentrations, reactant temperature, presence of catalysts
16-6 Homogeneous catalyst—same phase as reactants; heterogeneous catalyst—separate phase from reactants
16-10 Rate of forward reaction must equal rate of reverse reaction
16-11 Rate of formation of products equals rate of reformation of reactants
16-15 **a.** $\dfrac{[Cl_2]^2[H_2O]^2}{[O_2][HCl]^4}$ **d.** $\dfrac{[H_2O]^6[NO]^4}{[O_2]^5[NH_3]^4}$
g. $\dfrac{[H_2O]^4[CO_2]^3}{[O_2]^5[C_3H_8]}$
16-16 **a.** $[O_2][NO_2]^4$ **d.** $[Cl_2]$ **g.** $[Br_2]$
16-18 0.0260
16-21 **a.** To the left **c.** Far to the left
16-23 **a.** Right **c.** Left **e.** Left
16-24 **a.** Right **c.** Left **e.** Right
16-26 **a.** Right **d.** No effect **g.** No effect

Chapter 17

17-3 **a.** Alpha particle **d.** Positron
17-5 **a.** ${}^{200}_{84}Po \longrightarrow {}^{4}_{2}\alpha + {}^{196}_{82}Pb$
c. ${}^{229}_{90}Th \longrightarrow {}^{4}_{2}\alpha + {}^{225}_{88}Ra$
17-6 **a.** ${}^{25}_{11}Na \longrightarrow {}^{0}_{-1}\beta + {}^{25}_{12}Mg$
c. ${}^{77}_{32}Ge \longrightarrow {}^{0}_{-1}\beta + {}^{77}_{33}As$
17-7 **a.** ${}^{8}_{5}B \longrightarrow {}^{0}_{1}\beta + {}^{8}_{4}Be$ **c.** ${}^{79}_{37}Rb \longrightarrow {}^{0}_{1}\beta + {}^{79}_{36}Kr$
17-8 **a.** ${}^{59}_{28}Ni + {}^{0}_{-1}e \longrightarrow {}^{59}_{27}Co$
c. ${}^{100}_{46}Pd + {}^{0}_{-1}e \longrightarrow {}^{100}_{45}Rh$

17-9 **a.** Mass number decreases by 4, atomic number decreases by 2 **c.** Mass number does not change, atomic number decreases by 1

17-10 **a.** ${}_{-1}^{0}\beta$ **d.** ${}_{1}^{0}\beta$

17-11 **a.** ${}_{74}^{181}W + {}_{-1}^{0}e \longrightarrow {}_{73}^{181}Ta$
d. ${}_{29}^{59}Cu \longrightarrow {}_{1}^{0}\beta + {}_{28}^{59}Ni$

17-12 **a.** Alpha decay

17-13 **a.** ${}_{2}^{4}\alpha$ **d.** ${}_{8}^{17}O$

17-14 **a.** ${}_{4}^{9}Be + {}_{2}^{4}\alpha \longrightarrow {}_{0}^{1}n + {}_{6}^{12}C$
c. ${}_{94}^{243}Pu + {}_{2}^{4}\alpha \longrightarrow {}_{96}^{246}Cm + {}_{0}^{1}n$

17-17 **a.** 1/16 **d.** 1/8

17-18 0.250 g

17-20 280 years

17-23 **a.** 0.95 (before), 1.05 (after)
c. 1.22 (before), 1.29 (after)

17-25 **a.** ${}_{37}^{79}Rb$ (positron emission since mass number is less than atomic weight), ${}_{37}^{90}Rb$ (beta emission since mass number is greater than atomic weight)

17-26 **a.** Beta **d.** Electron capture or positron emission

17-28 **a.** Both processes **d.** Fission

17-29 **a.** Fusion **d.** Neither

Chapter 18

18-1 **a.** 20 **c.** 38

18-2 **a.**
```
CH3—CH—CH2—CH3
     |
    CH3
```
c. $CH_3—CH_2—CH_2—CH_2—CH_2—CH_3$

18-4 **a.**
```
C—C—C—C—C—C, C—C—C—C—C,
                 |
                 C

                                         C
                                         |
C—C—C—C—C, C—C—C—C, C—C—C—C
     |          |  |         |
     C          C  C         C
```

18-5 **a.** 7 **d.** 7

18-6 **a.** 3-Methylpentane **d.** 3-Methylhexane

18-7 **a.**
```
C—C—C—C—C
     |
     C
```
d.
```
C—C—C—C—C—C—C—C
  |     |
  C     C
```

18-11 **a.** 8 **c.** 12

18-13 **a.** (cyclohexane ring with CH_3 groups at positions 1, 2 and 4) **c.** (cyclobutane ring with $CH_2—CH_2—CH_3$ substituent)

18-17 **a.** Unsaturated, alkene
d. Unsaturated, cycloalkene

18-18 **a.** 2-Pentene **d.** 4-Methyl-2-hexyne
g. 1,3-Pentadiene

18-19 **a.**
```
C—C—C≡C—C—C
  |
  C
```
d.
```
            C
            |
C—C≡C—C—C—C—C
            |  |
            C  C
```

18-21 **a.** Cyclic saturated **d.** Cyclic unsaturated

18-23 **a.** Benzene **d.** 1,2,4-Trimethylbenzene

18-25 **a.** (benzene ring with C_2H_5 groups in 1,3 positions) **d.** (benzene ring with CH_3 and C_2H_5 groups in 1,4 positions)

18-29 **a.**
```
CH3—CH—CH2—CH3
     |
    (Br)
```
d.
```
CH3—CH2—CH2
         |
        (OH)
```

18-31 **a.** 1,1-Dibromoethane **d.** Iodomethane

18-32 **a.** 2-Butanol **c.** 3-Pentanol

18-33 **a.**
```
C—C—C—C—C
  |     |
  OH    C
```
d.
```
C—C—C
     |
     OH
```

18-34 **a.** 1,3-Propanediol

18-35 **a.** Methyl propyl ether, 1-methoxypropane

18-36 **a.** Pentanoic acid **c.** 3-Ethyl-5-methylhexanoic acid

18-37 **a.**
```
           O
           ‖
C—C—C—C—OH
  |  |
  C  C
```
d.
```
   O
   ‖
C—C—OH
```

18-38 **a.**
```
      O
      ‖
C—C—C—O—C—C
```
d.
```
   O
   ‖
H—C—O—C—C
```

18-39 **a.** Methyl ethanoate

18-40 **a.** Butanoic acid **d.** Bromoethane
g. 3-Hexanol

Boldface entries and page numbers indicate defined words and their location.